Development and Environment

AF212873

Warren Burggren
Benjamin Dubansky
Editors

Development and Environment

Springer

Editors
Warren Burggren
University of North Texas
Department of Biological Sciences
Denton, Texas
USA

Benjamin Dubansky
University of North Texas
Department of Biological Sciences
Denton, Texas
USA

ISBN 978-3-030-09357-0 ISBN 978-3-319-75935-7 (eBook)
https://doi.org/10.1007/978-3-319-75935-7

© Springer International Publishing AG, part of Springer Nature 2018
Softcover re-print of the Hardcover 1st edition 2018
This work is subject to copyright. All rights are reserved by the Publisher, whether the whole or part of the material is concerned, specifically the rights of translation, reprinting, reuse of illustrations, recitation, broadcasting, reproduction on microfilms or in any other physical way, and transmission or information storage and retrieval, electronic adaptation, computer software, or by similar or dissimilar methodology now known or hereafter developed.
The use of general descriptive names, registered names, trademarks, service marks, etc. in this publication does not imply, even in the absence of a specific statement, that such names are exempt from the relevant protective laws and regulations and therefore free for general use.
The publisher, the authors, and the editors are safe to assume that the advice and information in this book are believed to be true and accurate at the date of publication. Neither the publisher nor the authors or the editors give a warranty, express or implied, with respect to the material contained herein or for any errors or omissions that may have been made. The publisher remains neutral with regard to jurisdictional claims in published maps and institutional affiliations.

This Springer imprint is published by Springer Nature, under the registered company Springer International Publishing AG
The registered company address is: Gewerbestrasse 11, 6330 Cham, Switzerland

Preface

The genesis of this book lay in our discovery that, despite our quite different backgrounds in the diverse field of developmental biology, there was nonetheless substantial overlap in approaches, techniques, interpretations, and assessment of importance of our experimental findings. Such a connection appeared to be widespread across a diversity of fields that utilize the study of developing organisms to understand effects of environmental influences on ecosystem health. Yet, we also realized that there was no one written source for understanding how researchers from our respective fields and other connected fields approached important topics in development and how the environment influences this process over individual life cycles or through evolutionary time—hence this book.

We thank the many professionals at Springer who indulged our requests and forgave our tardiness while putting together this book. We also wish to thank those who wrote chapters in this book for their contribution and expertise. A special thanks goes to Maria Rojas-Antich of the University of North Texas, who cheerfully provided flawless and energetic administrative support. Finally, this book project was made possible in part by a grant from the Gulf of Mexico Research Initiative.

Denton, TX, USA — Warren Burggren
Denton, TX, USA — Benjamin Dubansky

Contents

Abbreviations

2,3,7,8-TCDD	2,3,7,8-dibenzo-*p*-dioxin
3β-HSD	3β-hydroxysteroid dehydrogenase
ΣPAH	Total PAH
AAS	Absolute aerobic scope
ABC	ATP-binding cassette transporters
ABP	Androgen binding-protein
AC	Adenylate (adenylyl) cyclase
ACTH	Adrenocorticotropic hormone
ADF	Anatomical diffusion factor
AFLP	Amplified fragment length polymorphisms
AH	Aryl hydrocarbon
AhR	Aryl hydrocarbon receptor
AHRR	Aryl hydrocarbon receptor repressor
AIP	AHR interacting protein
AKT	Protein kinase B
ALN	Asynchronous Learning Networks
ALPM	Anterior later plate mesoderm
amhc	Atrial myosin heavy chain isoform genes
ANP	Atrial natriuretic peptide
ANPa	Atrial natriuretic peptide type A
AnRE	Androgen-response elements
AP-1	Activation protein 1
ap1s1	Adaptor protein gene
ARNT	Aryl hydrocarbon receptor nuclear translocator
Ars	Androgen receptors
AVC	Atrioventricular canal
AVP	Vasopressin
AW	Atlantic Woods
BaP	Benzo[*a*]pyrene
βAR	β-adrenoreceptors
Bdnf	Brain-derived neurotrophic factor
BkF	Benzo[*k*]fluoranthene
BMP	Bone morphogenic protein
bmp4	Bone morphogenic protein 4

BOEM	Bureau of Energy Management
BP	Bipolar
BPA	Bisphenol A
BPDE	BaP diolepoxide
BtB	Bench to Bassinet
BTEX	Benzene, toluene, ethylbenzene, and xylenes
BWM	Brian W. Matherne
cacna1c	Calcium channel 1C subunit gene
CaM	Calmodulin
cAMP	Cyclic 3′5′-adenosine monophosphate
CaN	Calcineurin
CAP	Comparative animal physiology
casq1	Ca^{2+}-buffering proteins calsequestrin 1
casq2	Ca^{2+}-buffering proteins calsequestrin 2
CaTF	Calcium-sensitive transcription factors
ccnd1	Cyclin D1 gene
Cdk1	B-cyclin-dependent kinase 1 complex
CF	Cystic fibrosis
cfos	Proto-oncogene
Clo	*Cloche* gene
cmlc2	Cardiac myosin light chain 2 gene
CNS	Central nervous system
CO	Carbon monoxide
CO_2	Carbon dioxide
CO_3^{2-}	Carbonate ions
CoA	Co-activators
CpG	Cytosine-phosphate-guanine
CRH	Corticotropin-releasing hormone
ctr1	Copper transporter
CYP17	Cytochrome p450 17
CYP1A	Cytochrome P4501A
CYP1A1	Cytochrome P450 1A1
DA	Dopamine
DAG	Diacylglycerol
DART	Direct analysis in real time
DBD	DNA-binding domain
DDT	Dichlorodiphenyl-trichloroethane
DESI	Desorption electrospray ionization
DHEA	Dehydroepiandrosterone
DHOS	Deepwater Horizon oil spill
DLCs	Dioxin-like compounds
DMR	Differentially methylated regions
dmt1	Divalent metal transporter
DNMTs	DNA methyltransferases
E2	17ß-estradiol

E-C	Excitation-contraction
ED	Endocrine disrupter
EDCs	Endocrine disrupting chemicals (compounds)
EDSP	Endocrine Disruptor Screening Program
EE	Environmental enrichment
EES	Extended Evolutionary Synthesis
EET	Extended Evolutionary Theory
eNOS	Endothelial nitric oxide synthase
EPA	Environmental Protection Agency
ER	Estrogen receptor
EREs	Estrogen response elements
erg	Ether-á-go-go related gene
ERK	Extracellular signal-regulated kinase
Erα	Estrogen receptor α
ESI	Electrospray ionization
E-T	Excitation-transcription
FAS	Fetal alcohol syndrome
fhl2	Four-and-a-half LIM domain family protein 2 gene
FSH	Follicle-stimulating hormone
GC	Gas chromatography
GCMS	Gas chromatography mass spectrometry
GH	Growth hormone
GHRH	Growth hormone-releasing hormone
GnIH	Gonadotropin-inhibitory hormone
GnRH	Gonadotropin releasing hormone
GoMRI	Gulf of Mexico Research Initiative
GPER	G-protein coupled estrogen receptor
GR or Nrc31	Glucocorticoid receptor
GtHs	Gonadotropins
GW	George Washburn
H_2CO_3	Carbonic acid
H_2S	Hydrogen sulfide
HAHs	Halogenated aromatic hydrocarbons
hand2	Helix–loop–helix transcription factor gene
Hb Bart's	Alpha-Thalassemia major
Hb S/C	Hemoglobin S/C
Hb S/β	Hemoglobin S/β
HCO_3^-	Bicarbonate
HELLP	Hemolysis, Elevated Liver enzymes, Low Platelet count Syndrome
HEWAF	High-energy WAF
HIF	Hypoxia inducible factor
HPA	Hypothalamic-pituitary-adrenal
hpf	Hours post-fertilization
HPG	Hypothalamic–pituitary– gonadal
HSC	Houston Ship Channel

HSP	Heat shock protein
hsp90	Heat shock protein 90
ICR	Imprinting center regions
IDA	Iron deficiency anemia
IP3	Inositol 1,4,5-triphosphate
IUCRCs	Industry/University Cooperative Research Centers
IUGR	Intrauterine growth restriction
IVF	In vitro fertilization
LBD	Ligand-binding domain
LC	Liquid chromatography
LC_{50}	Lethal concentration, where 50% of the population is killed by contaminant
LDLR	Low-density lipoprotein receptor
LH	Luteinizing hormone
LOE	Loss-of-equilibrium
LTCC	L-type Ca^{2+} channels
MC252	Mississippi Canyon Block 252
MDD	Major depressive disorder
MDR	Multidrug resistance ABC transporters
meHg	Methylmercury
mil	Miles apart gene mutation
MIMS	Membrane inlet mass spectrometry
miRNA	MicroRNAs
MMR	Maximal aerobic metabolic rate
MOA	Mechanisms of action
MR	Mineralocorticoid receptor
MS	Mass spectrometry
MSH	Melanophore stimulating hormone
MSUS	Maternal separation and stress
NBH	New Bedford Harbor
NCX1	Sodium/calcium exchanger 1
NIST	National Institute of Standards and Technology
NKA	Na^+/K^+-ATPase
nkx2.5	NK2 homeobox 5 gene
NO	Nitric oxide
NOAA	National Oceanic and Atmospheric Administration
NOS	Nitric oxide synthase
NP	Neurophysins
NRDA	Natural Resource Damage Assessment
OAT	Organic anion transporters
oatp	Organic anion transporting polypeptide
OCLTT	Oxygen- and capacity-limited thermal tolerance
OCT	Organic cation transporters
ORNs	Ontogenetic reaction norms
OSA	Ontogenic sequence analysis

OT	Oxytocin
PAHs	Polycyclic aromatic hydrocarbons
PCBs	Polychlorinated biphenyls
PCDD	Polychlorinated dibenzodioxin
PCDD/Fs	Polychlorinated dibenzo-*p*-dioxins and furans
P_{crit}	Critical oxygen partial pressure
PFASs	Perfluoroalkylated substances
PFNA, C9	Perfluorononanoic acid
PFOA	Perfluorooctanoic acid
PFOS	Perfluorooctane sulfonate
PFOS, C8	Perfluorooctanoic sulfonate
PI3K	Phosphoinositide 3-kinase
PIP_2	Phosphatidylinositol 4,5-bisphosphate
PIs	Principle investigators
PKA	Protein kinase A
PLB	Phospholamban
PLC	Phospholipase C
PNNL	Pacific Northwest Nation Laboratory
POPs	Persistent organic pollutants
ppb	Parts per billion
PPI	Paired pulse inhibition
ppm	Parts per million
ppt	Parts per trillion
Prkcc	Protein kinase C γ
PSM	Presomatic (unsegemented) mesodermal tissue
PTMs	Post-translational modifications
PTSD	Posttraumatic stress disorder
PXR	Pregnane X receptor
QTL	Quantitative trait loci analysis
ROI	Return on investment
ROS	Reactive oxygen species
RR	Rogue researcher
RW	Rogue writer
RXRs	Retinoid X receptors
RyR	Ryanodine receptor
SA:V	Surface area to volume ratio
SERCA	Sarcoplasmic reticulum Ca^{2+} ATPase
SERCA2	Sarcoplasmic-endoplasmic reticulum calcium ATPase 2
SIDS	Sudden infant death syndrome
sih	Silent heart gene
slco2b1	Oatp transporter gene
Sloan-C	Sloan Consortium
smolt	Ocean migrant stage
SMR	Standard metabolic rate
SNPs	Single nucleotide polymorphisms

SOPs	Standard operating procedures
SPE	Solid phase extraction
SR	Sarcoplasmic reticulum
SS	Somatostatin
StAR	Steroidogenic acute regulatory protein
SVR	Systemic vascular resistance
T3	Triiodothyronine
T4	Thyroxin
tasiRNA	Trans-acting small interfering RNA
tbx5	T-box 5 protein
TCDD	2,3,7,8-Tetrachlorodibenzo-*p*-dioxin
tcf3a	Transcription factor gene
Tg	Thyroglobulin
TGF-β	Transforming growth factor β
TGF β1	Transforming growth factor β1
TNF	Tumor necrosis factor
toh	Two of hearts gene
TRA	Technology Research Associations
TRAIL	TNF related apoptosis ligand
TRE	Thyroid response element
TREs	Thyroid regulatory elements
TRH	Thyrotropin-releasing hormone
TRs	Thyroid hormone receptors
TSH	Thyroid-stimulating hormone
TSS	Transcription start sites
UCP	Uncoupling protein
U_{crit}	Critical swim speed
URM	Under-represented minorities
UV	Ultraviolet
Ve/VO_2	Ventilation to O_2 consumption
VGM	Variation-generating mechanism
vhmc	Ventricular myosin heavy chain isoform genes
WAFs	Water-accommodated fractions
WET	Whole effluent toxicity
XRE	Xenobiotic response elements
YOY	Young of the year
YSL	Yolk syncytial layer
ZEDTA	Zebrafish development toxicity assay

The Nexus of Development and Environment

1

Warren Burggren and Benjamin Dubansky

Abstract

The relationship between environment and animal development has been recognized since the time of Aristotle, but the urgency of creating a thorough understanding of this relationship is emerging as environments in which animals develop are changing as a result of pollution, climate change and other anthropogenic activities. This book, an overview of which is provided in this first chapter, is organized along three key themes. Each theme predominates in its own section – Part I: Plasticity in developmental and evolutionary time and space, Part II: Contemporary experimental approaches, and Part III. Environmental effects and experimental outcomes. Each of its chapters provides a comprehensive, up-to-date assessment of how development and environment are inextricably woven together, and points to future directions for research at their nexus.

That environment can influence organismal development is one of the basic tenets of the discipline of developmental biology. This relationship between a developing organism and its surroundings has roots in Aristotle's *On the Parts of Animals* in 350 BC (Leroi 2014). Indeed Aristotle recognized that "things are best studied as they come into being," and this viewpoint pervades this book. Since Aristotle's time, literally centuries of research have revealed that both short- or long-term shifts in

W. Burggren (✉) · B. Dubansky
Developmental Integrative Biology Group, Department of Biological Sciences, University of North Texas, Denton, TX, USA
e-mail: Burggren@unt.edu; Benjamin.Dubansky@unt.edu

© Springer International Publishing AG, part of Springer Nature 2018

W. Burggren, B. Dubansky (eds.), *Development and Environment*,
https://doi.org/10.1007/978-3-319-75935-7_1

the physiochemical environment can impact organismal development and growth, with potentially severe consequences at the individual and population level—for an entry into this vast and venerable literature, including human research (Weis and Weis 1987; Burggren and Warburton 2005; Warburton et al. 2006; Miller and Marty 2010; Yeung et al. 2011; Dubansky et al. 2013; Wright and Brunst 2013; Dubansky et al. 2014; Liao et al. 2014; Veldhoen et al. 2014; Heindel et al. 2015; Lee 2015; Vieira 2015; Grandjean 2016; Lovato et al. 2016; Luo et al. 2016; Patrick 2017; Robert 2017, as well as, of course, multiple chapters in this book).

Against this backdrop of an extensive field of study of correlative and causational relationships between organismal development and environmental stressors, *Development and Environment* pushes beyond the existing conceptual base to explore both general and specific aspects of the nexus between organismal development (including in humans) and environmental stressors, using contemporary approaches and tools. In doing so, this book is divided into four major parts, each with distinct themes.

Part I: Plasticity in Developmental and Evolutionary Time and Space

This initial part takes the reader from very short-term time considerations—early organismal development—through to an evolutionary time frame. It begins with Chap. 2 and Benjamin Dubansky's discussion of the often complex relationship between chronological and developmental time and how environmental change can confound this relationship. Casey Mueller in Chap. 3 then considers critical windows in development. While critical windows (also known as sensitive periods) have been long recognized, Mueller shows how more nuanced views of the relationship between environment, phenotype, and time can reveal important components of development and its processes. Building on this, Bernd Pelster and Warren Burggren explore phenotypic plasticity in developing animals in Chap. 4. Such plasticity has benefits, but it also has costs to the organisms, and there can frequently be imbalance depending upon the severity of the environmental challenge. John Spicer, Oliver Tills, Manuela Truebano, and Simon Rundle then continue the discussion of developmental plasticity and evoke the concept of heterokairy in Chap. 5. Contrary to the view that developmental sequence is genetically dictated, this chapter shows that environmental cues can potentially alter the developmental sequence. This part of the book then moves to consideration of individual and population level fitness in the context of early developmental exposure to stressors in Chap. 6 by Kathleen Annunziato and Keith Cooper. Using examples of aquatic organisms, these authors show that xenobiotics can dramatically impact populations through their effects on reproductive success. Part I concludes with Chap. 7 by Thomas Dickins and Benjamin Dickins' exploration of the modern synthesis, putting development into a contemporary evolutionary framework that challenges models of causation within the modern synthesis. Taken together, the chapters in Part I provide a rigorous time framework and set of time-correlated definitions for viewing the interactions between development and environment.

Part II: Contemporary Experimental Approaches
Many new and emerging approaches, as well as those time honored, are being employed to study the nexus of development and environment. This second part is replete with approaches and tools of use to researchers involved in environmental effects on development (and beyond!). This part begins with Chap. 8 by Fernando Galvez, who recognizes that developing organisms often face multiple simultaneous environmental stressors, rather than a single stressor which has traditionally been employed in experimental research on development and environment. Galvez indicates how physiological and genomic mechanisms are employed to deal with multiple stressors in the environment. Andrew Esbaugh, Alexis Khursigara, and Jacob Johansen in Chap. 9 then continue the theme of the complex developmental effects of multiple stressors by considering the "cocktail effect." They reveal important synergies and antagonisms among multiple simultaneously experienced environmental stressors. A case study of development and environment based on the Deepwater Horizon oil spill is then presented by John Incardona and Nathaniel Scholz in Chap. 10. Focusing on cardiovascular development, this chapter documents the often devastating effects of oil exposure on key developmental processes in larval fishes, exemplifying how developmental model organisms can be effectively used to understand ecological effects of novel stressors. Chapter 11 by Guido Verbeck then considers some of the latest analytical methods and trends in environmental forensics. Optical- and radiation-based analysis versus mass analysis both have applications in unique scenarios and can powerfully be used in tandem to provide forensic links between environmental stressors and biological effects. This part concludes with Chap. 12 and a change of direction to consider the increasing extent of collaborative studies, often under the umbrella of consortia. In this chapter Linda Hooper-Bùi, Brian Matherne, and George Washburn consider the practical, managerial dynamics of large-scale experimental collaborations that are often important in achieving meaningful goals when combining developmental and environmental sciences.

Part III: Environmental Effects and Experimental Outcomes
This third part of *Development and Environment* then focuses on experimental outcomes, putting forward both general and specific examples of the nexus of development and environment.

This part begins with Graciel Diamante and Daniel Schlenk's Chap. 13 that considers endocrine disruptors as environmental stressors. Using cardiac development as an exemplar, these authors describe how developing organisms deal with the challenges of endocrine disruption. Elias Oziolor and Cole Matson in Chap. 14 reveal how population-level adaptation can be driven by highly polluted waters using the model organism the Gulf killifish. Populations of these fishes have high levels of pollution resistance and comprise useful models for understanding toxicant resistance and its evolution. Concluding this part is a case study by Michael Sackville and Colin Brauner in Chap. 15 discussing developmental plasticity in the face of environmental stressors in larval fishes. Here, the developing gills, the interface between environment and the interior milieu, provide a front line against environmental challenge that is illustrative of developmental plasticity.

Part IV: The Human Component
This concluding part moves to specifically examine the impact of environment on human development. Damian Hutter in Chap. 16 begins by considering the development of the human embryo and how in utero conditions can impact physiological and other outcomes. Oxygen has been long recognized as a powerful determinant of development, and Hutter discusses multiple implications of in utero hypoxia. In Chap. 17 Mallory Bowers and Rachel Yahuda review the genesis of human psychopathologies. They reveal how biological responses, including those to trauma, can be inherited through epigenetic mechanisms. This part finishes with Chap. 18 and a case study by Alexandra Joachmans-Lemoine and Vincent Joseph of how high-altitude exposure during postnatal development can alter human physiology. Cold exposure in conjunction with hypoxia is also shown to have detrimental effects, continuing the theme of how multiple environmental challenges can act synergistically on developing organisms.

Moving Forward
Collectively, these four parts of *Development and Environment* and the different emphases of their 18 chapters within reveal the often powerful, often subtle, but always important influence of environment the organismal development. As you read the various contributions to this book, bear in mind that many, many aspects of development comprise a highly conserved suite of processes (e.g., Fritzsch et al. 2000; Orgeig et al. 2003; Hartenstein 2006; Nakamura and Hamada 2012; Shklover et al. 2015; Jones and Wilkinson 2017; Tseng et al. 2017; Xie and Dorsky 2017). Consequently, while several authors explore animal models or focus on specific organisms, the lessons learned from each chapter are likely to be broadly applicable in our study of the nexus of development and environment.

References

Burggren W, Warburton S (2005) Comparative developmental physiology: an interdisciplinary convergence. Annu Rev Physiol 67:203–223

Dubansky B, Whitehead A, Miller J, Rice CD, Galvez F (2013) Multi-tissue molecular, genomic, and developmental effects of the Deepwater Horizon oil spill on resident Gulf killifish (*Fundulus grandis*). Environ Sci Technol 47(10):5074–5082

Dubansky B, Whitehead A, Rice CD, Galvez F (2014) Multi-tissue molecular, genomic, and developmental effects of the Deepwater Horizon oil spill on resident Gulf killifish (*Fundulus grandis*). Environ Sci Technol 47(10):5074–5082

Fritzsch B, Beisel KW, Bermingham NA (2000) Developmental evolutionary biology of the vertebrate ear: conserving mechanoelectric transduction and developmental pathways in diverging morphologies. Neuroreport 11(17):R35–R44

Grandjean P (2016) Paracelsus revisited: the dose concept in a complex world. Basic Clin Pharmacol Toxicol 119(2):126–132

Hartenstein V (2006) Blood cells and blood cell development in the animal kingdom. Annu Rev Cell Dev Biol 22:677–712

Heindel JJ, Balbus J, Birnbaum L, Brune-Drisse MN, Grandjean P, Gray K, Landrigan PJ, Sly PD, Suk W, Cory Slechta D, Thompson C, Hanson M (2015) Developmental origins of health and disease: integrating environmental influences. Endocrinology 156(10):3416–3421

Jones SH, Wilkinson M (2017) RNA decay, evolution, and the testis. RNA Biol 14(2):146–155

Lee HS (2015) Impact of maternal diet on the epigenome during in utero life and the developmental programming of diseases in childhood and adulthood. Forum Nutr 7(11):9492–9507

Leroi AM (2014) The lagoon. Viking, New York

Liao PH, Chu SH, Tu TY, Wang XH, Lin AY, Chen PJ (2014) Persistent endocrine disruption effects in medaka fish with early life-stage exposure to a triazole-containing aromatase inhibitor (letrozole). J Hazard Mater 277:141–149

Lovato AK, Creton R, Colwill RM (2016) Effects of embryonic exposure to polychlorinated biphenyls (PCBs) on larval zebrafish behavior. Neurotoxicol Teratol 53:1–10

Luo S, Wu B, Xiong X, Wang J (2016) Short-term toxicity of ammonia, nitrite, and nitrate to early life stages of the rare minnow (*Gobiocypris rarus*). Environ Toxicol Chem 35(6):1422–1427

Miller MD, Marty MA (2010) Impact of environmental chemicals on lung development. Environ Health Perspect 118(8):1155–1164

Nakamura T, Hamada H (2012) Left-right patterning: conserved and divergent mechanisms. Development 139(18):3257–3262

Orgeig S, Daniels CB, Johnston SD, Sullivan LC (2003) The pattern of surfactant cholesterol during vertebrate evolution and development: does ontogeny recapitulate phylogeny? Reprod Fertil Dev 15(1–2):55–73

Patrick B (2017) Robustness and plasticity in development. WIREs Cogn Sci 8:e1386. https://doi.org/10.1002/wcs.1386

Robert L (2017) Developmental evolution. WIREs Cogn Sci 8:e1422. https://doi.org/10.1002/wcs.1422

Shklover J, Levy-Adam F, Kurant E (2015) Apoptotic cell clearance in development. Curr Top Dev Biol 114:297–334

Tseng WC, Munisha M, Gutierrez JB, Dougan ST (2017) Establishment of the vertebrate germ layers. Adv Exp Med Biol 953:307–381

Veldhoen N, Skirrow RC, Brown LL, van Aggelen G, Helbing CC (2014) Effects of acute exposure to the non-steroidal anti-inflammatory drug ibuprofen on the developing North American bullfrog (*Rana catesbeiana*) tadpole. Environ Sci Technol 48(17):10439–10447

Vieira SE (2015) The health burden of pollution: the impact of prenatal exposure to air pollutants. Int J Chron Obstruct Pulmon Dis 10:1111–1121

Warburton SJ, Burggren WW, Pelster B, Reiber CL, Spicer JI (2006) Comparative developmental physiology: contributions, tools and trends. Oxford University Press, Oxford

Weis JS, Weis P (1987) Pollutants as developmental toxicants in aquatic organisms. Environ Health Perspect 71:77–85

Wright RJ, Brunst KJ (2013) Programming of respiratory health in childhood: influence of outdoor air pollution. Curr Opin Pediatr 25(2):232–239

Xie Y, Dorsky RI (2017) Development of the hypothalamus: conservation, modification and innovation. Development 144(9):1588–1599

Yeung BH, Wan HT, Law AY, Wong CK (2011) Endocrine disrupting chemicals: multiple effects on testicular signaling and spermatogenesis. Spermatogenesis 1(3):231–239

Part I

Plasticity in Developmental and Evolutionary Time and Space

2 The Interaction of Environment and Chronological and Developmental Time

Benjamin Dubansky

Abstract

As organisms develop from embryos to adults, adaptive phenotypes must emerge to meet the demands of the habitat. This physiological and morphological transformation occurs along a continuum, where the emerging traits are often illustrated as landmarks charted along a predictable chronological timeline. Chronological time is measured according to astronomical phenomena, in hours, days, weeks, months, and years. However, developmental timing is largely driven by innate molecular oscillators that are independent of chronological time and species-specific. Environmental stressors can alter the timing of emergence of developmental phenotypes, creating further discord between developmental time and chronological time. Further, variation in the timing of emergence of developmental phenotypes is the norm, rather than the exception both within and between species. Such examples of environmentally driven variation of developmental timing abound, including alteration in development speed within chronological time and alteration of the sequence of emergence of landmark traits. It follows that if treatment groups (or species) are at different developmental stages, experimental comparisons become increasingly complicated. Within the context of the growing use of developmental model organisms in environmental sciences, the effect of alteration of developmental timing is particularly relevant. This is well illustrated in several examples utilized here to describe how alteration of developmental time can be a maladaptive consequence, or an adaptive compensatory response to environmental stressors, which can be selected for during evolution. Thus, such alterations of developmental timing can be measured and accounted for when designing developmental studies and in the interpretation of resultant data.

B. Dubansky
Department of Biological Sciences, University of North Texas, Denton, TX, USA
e-mail: Benjamin.Dubansky@unt.edu

© Springer International Publishing AG, part of Springer Nature 2018

W. Burggren, B. Dubansky (eds.), *Development and Environment*,
https://doi.org/10.1007/978-3-319-75935-7_2

2.1 Overview

Environmental stressors can exert greater effects on early life stages when the physiological capacities to deal with stressors may not yet be present or fully developed. As such, developing organisms have gained popularity for use in assessing the effects of environmental stressors on organismal health since stressors that alter developmental success might ultimately affect population success and ecosystem health (Dubansky et al. 2013, 2014; Grandjean 2016; Heindel et al. 2015). However, environmental stressors have a range of influence during different developmental stages that affect the etiology and susceptibility to diseases of environmental origin (Carlos-Wallace et al. 2016; Mahler and Butcher 2011). Since physiological capacities develop over time to meet the demands of a habitat, exposure to environmental stressors may also affect organisms differently depending on when, during development, they are exposed (Burggren and Reyna 2011; Heindel et al. 2015). With this in mind, the rate of ontogeny of a species can also become altered in different rearing conditions (Burggren 2018; Dubansky and Dubansky 2017; Fuiman et al. 1998; Goldberg et al. 2016; Gomula and Koziel 2015; Lewis et al. 2008; Wankowska and Polkowska 2010), shifting the timing of emergence of physiological capacities. Thus, the timing of experimental exposure to environmental stressors requires consideration of the developmental timeline with respect to when during ontogeny the physiological capacity to respond to a stressor arises. Knowing the ontogeny of physiological capacity can define critical windows of sensitivity during development that determines when an organism may be most sensitive to a stressor (see Chap. 3 and references therein for a comprehensive evaluation of critical windows of development) (Burggren and Reyna 2011). However, such information is not always available for model species, despite the importance of such knowledge.

2.1.1 Plasticity and Development

A key compensatory mechanism to respond to stressors is phenotypic plasticity, where alterations in morphological or physiological phenotype occur in response to environmental pressure to enable survival. Phenotypic plasticity may result in a permanent change, or it may also be transitory. In contrast to homeostasis, phenotypic plasticity must involve either a remodeling of architecture to enable a response to a stressor or an alteration in a physiological mechanism such as modification of biochemical phenotype to increase homeostatic abilities (Burggren and Reyna 2011). Developmental plasticity is a form of phenotypic plasticity, which occurs during development, when the typical developmental plan is altered, in a way that may or may not be permanent, and likewise may or may not affect progression to adulthood, or impact the health of the adult organism (see Chap. 5 for an in-depth look at developmental phenotypic plasticity) (Burggren 2018; Burggren and Reyna 2011).

During development, both homeostatic mechanisms and plasticity are permissible only within the context of the developmental plan of an organism. That is, plasticity isn't possible if the organism has not gained the ability to conduct such compensatory responses. Indeed, the constant change in morphology and physiological capacity that occurs as organisms develop creates a situation where effects of stressors depend on when, during development, the stressor is encountered. With all of this in mind, it is clear that in comparison to studying effects of stressors on adult organisms, there is added complexity in the study of developing organisms, where developmental trajectories may become altered, thus altering the downstream phenotype (Burggren 2018; Burggren and Reyna 2011). Despite the wide use of animal models in biological sciences, few model systems have a foundation of data describing the development of physiological capacity.

2.1.2 Developmental Model Systems

The understanding of convergent mechanisms of alteration of development, and conceptual understanding of development in higher vertebrates, has been accelerated by using established animal models in biomedical sciences (Albertson et al. 2009; Felter et al. 2015; Garcia et al. 2016; Schartl 2014; Warner 2014). Since the mechanisms that guide development in the embryo are mostly conserved, many unique model systems have been developed to study basic science and human health (Albertson et al. 2009), and several important species have emerged as popular subjects. The zebrafish (*Danio rerio*) is a premier developmental model with a considerable foundation of knowledge that facilitated discovery of many human pathologies and advanced our understanding of basic scientific principles (Phillips and Westerfield 2014; Vornanen and Hassinen 2016). For nearly a century, amphibian developmental models such as the frog (i.e., *Xenopus* sp.) have also been heavily studied (Burggren and Warburton 2007; Warkman and Krieg 2007). Mammalian models, avian models, and a number of other metazoan models have also been foundational in understanding early development (e.g., mouse, rat, chicken, fruit fly, nematodes, etc.) (Burggren et al. 2016; Felter et al. 2015; Nimmo and Slack 2009; Webb and Miller 2012; Wittig and Münsterberg 2016). The emergence and widespread use of established model organisms has indeed led to a better understanding of developmental physiology. Such established model systems have enabled the development of molecular and experimental tools and standard methodologies, and produced data foundations for others working within related fields of study.

2.1.3 Novel Organisms: Working Beyond Traditional Models

Of great use in environmental biology are developmental model organisms that retain properties that are relevant to a region/location or stressor of interest. Certainly, if a certain ecosystem is of interest, it may be more relevant to study an organism from that habitat, than an established model organism. For example, following the Deepwater Horizon oil spill (DHOS), several studies have investigated the effects of DHOS oil on developing endemic species that were likely affected by the DHOS, including several fish and crustacean species, among others (Dubansky et al. 2013; Duffy et al. 2016; Lotufo et al. 2016; Pasparakis et al. 2016; Pie et al. 2015; Ransom et al. 2016; Stefansson et al. 2016). Likewise, when studying the response to a stressor of interest, organisms that show particular resilience to that stressor may be of use in comparative studies (Love and Rees 2002; McDonald et al. 2010; Ramakrishnan 2007). Such species can be most useful as *novel developmental models*, if they are amenable to experimentation and are easy to collect and culture. With this in mind, new model organisms are emerging for use in environmental science, especially since anthropogenic alteration of habitats can be unique. As such, site-specific model species may prove useful for understanding the site-specific alterations to the environment (Dubansky et al. 2013, 2017; Oziolor et al. 2016b). In addition to understanding the effects on the environment, novel model organisms present opportunities to better understand how animals are able to respond to the environment through conserved mechanisms of phenotypic and developmental plasticity. Since environmental stressors such as pollution or agents of climate change can alter habitats, thus redefining species' niches, studying the effects of habitat alteration is best facilitated by using local species, whenever possible.

2.1.4 Setting the Stage

As new models are brought into use to study the effects of novel environmental stressors, establishing a stable experimental platform for understanding the relevance of sublethal effects on developmental physiology requires understanding of the species' developmental timeline. So, as new model species emerge for use in environmental biology, a chronological timeline that corresponds to the experimental time is often superimposed on the developmental timeline to characterize effects of experimental conditions on development. In doing so, we assign an age to experimental organisms in terms of time after fertilization, time after hatch, etc. However, a reasonable timeline for experiments may not be in harmony with the developmental program, since developmental trajectories can become altered by stressors and manifest as an alteration of the timing of emergence of physiological capacities. For example, two organisms that may be 3 weeks old may not be at the same developmental state, if held in different conditions. Since a developmental timeline may not follow a practical chronological timeline for experimentation, thoughtful consideration of the developmental program should be well integrated into methodologies to provide physiological context of observations and to provide comparative context between experiments, treatment groups, and species.

It is important, for the time being, to not focus too heavily on the difference between developmental and chronological time (though this difference will emerge later in the chapter). Instead, this chapter starts by illustrating the importance of developmental timing in environmental developmental biology, after providing a conceptual foundation. Then several forms of alteration of developmental time will be discussed. Though multiple examples are utilized here to illustrate concepts, repeated references to the author's work with teleost models are utilized to illustrate concepts that are not species-specific. Implications of alteration of developmental timing will be interspersed throughout to illustrate how interpretation of data can be variable, depending on the experimental plan and the interpretation of the produced data, within the context of developmental time. By the conclusion, examples intend to illustrate, using the author's past and emerging data, how a conceptual understanding of timing mechanisms and the alteration of developmental time can be (and could have been) utilized to better plan experimental timelines, to account for developmental timing.

2.2 Developmental Timing

Developmental events are highly coordinated, beginning the moment fertilization occurs. At the earliest stages of embryonic development, cells divide through the stages of mitosis, dividing at regular intervals (Webb and Oates 2016). Across phyla, at the first cell cycle, when male and female pronuclei fuse, developmental time proceeds relatively slowly toward the first cell division. During the next 12 divisions, cells divide more quickly and with rhythm, in near synchronicity with less than 5% variation in timing (Tsai et al. 2014). Such biological rhythms require an *oscillator* to keep time. An oscillator can be any signal that produces a periodic indicator of change, and it can be either extrinsic or intrinsic. For example, the oscillator for the circadian clock is the light/dark cycle caused by the daily rotation of the earth in relation to the sun (Christ et al. 2012; Jolma et al. 2010; Monsalve and Frand 2012) and the seasonal cues caused by change in photoperiod as the earth makes its the yearly revolution around the sun (Amaral et al. 2014). The circadian oscillator is the most well-known example of an extrinsic biological timer (Monsalve and Frand 2012). However, intrinsic developmental oscillators are coded into the genome, which also act as regulators of biological rhythm. For example, there is an intrinsic molecular oscillator that times early cell division in the embryo. In this well-conserved mechanism, the cyclin B-cyclin-dependent kinase 1 (Cdk1) complex operates through both negative and positive feedback loops to attenuate the frequency and amplitude of molecular signals that regulate the timing of cell division (Tsai et al. 2008, 2014; Webb and Oates 2016). The segmentation clock is another intrinsic timer, which is driven by rhythmic expression of transcription factors, that oscillate to drive the temporal and spatial distribution of cells and tissues during somitogenesis (see Sect 2.2.1) (Monsalve and Frand 2012; Pourquie 2011). Though the mechanism of oscillation of these molecular signals is mostly known, the pacemaker that sets the rhythm of this signal is not yet understood.

Molecular, cellular, and tissue-level timing mechanisms are in place to measure time in the embryo, but in addition to the timing of *rhythms*, timing also occurs in the form of *ordering* and *interval timing* (Fig. 2.1) (Webb and Oates 2016). Within

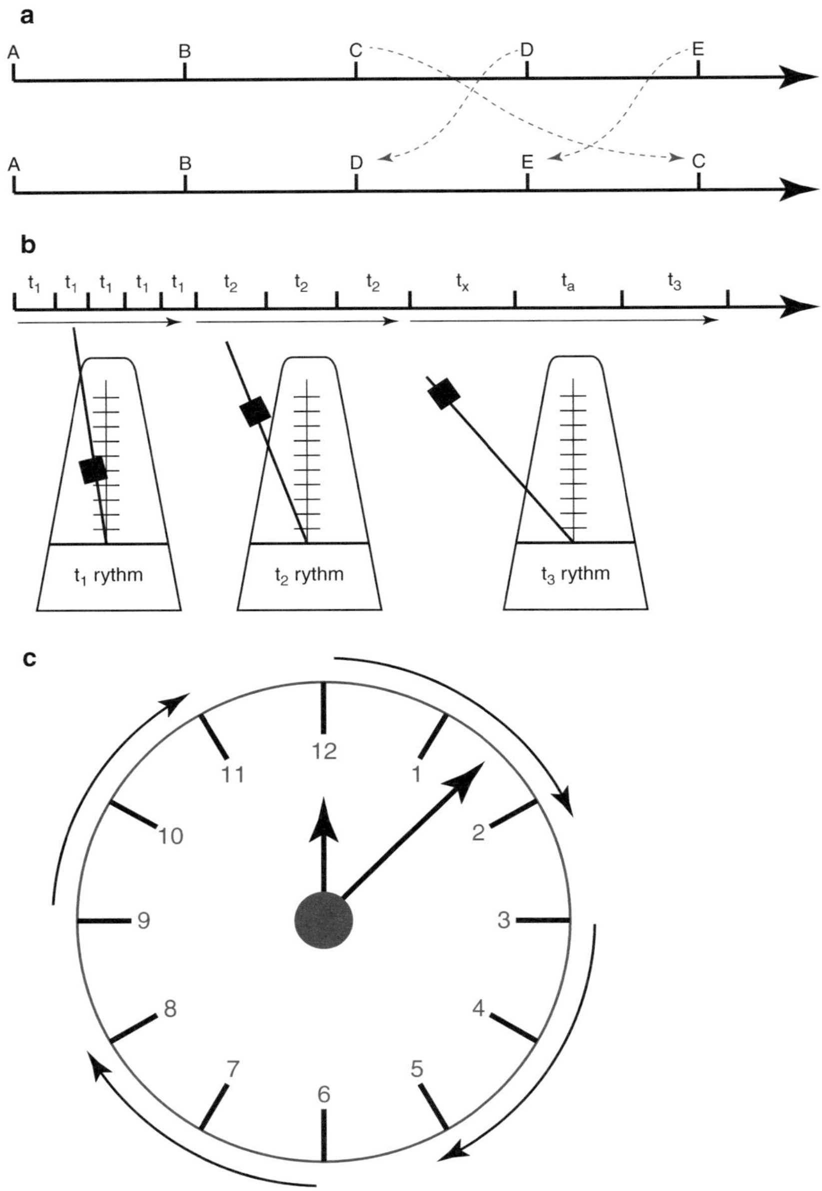

Fig. 2.1 Timing mechanisms. (**a**) *Ordering* during development is driven by the scheduling of when phenotypes arise. Though many events rely on the completion of a previous event, stressors can alter sequence of phenotype, resulting in a rearranged ordering of the expected developmental schedule. (**b**) *Interval timing* controls the time interval between events, such as the duration between molts and the formation of somites in the embryo. (**c**) *Intervals* can occur with rhythm, such as in early embryonic cell division and in instances where phenotype is driven by an extrinsic clock (e.g., circadian rhythm), but intervals can also be nonrhythmic, such as progression through the stages of mitosis

the context of development, ordering occurs as successive phenotypes arise during development, in a defined sequence. However, ordering might be best thought of as a scheduling mechanism, rather than a timed activity, since many events are conditional on the completion of earlier events (Keyte and Smith 2014). Still, ordering occurs along the timeline of development and will be discussed here, as such. *Interval timing*, on the other hand, is when a discrete time interval between events is set, such as the duration between molts in *Drosophila* and *C. ecdysis* larvae (see below and (Webb and Oates 2016) for a succinct review of timing and rhythms). As described above, intervals can (but don't have to) occur with rhythm (discussed below in Sect. 2.2.2).

The development of organisms has evolved to be well timed and ordered, but the programmed timeline of development of a species is susceptible to environmental influence. So, when the developmental environment is altered by a novel stressor, mechanisms of developmental plasticity can alter the developmental trajectory, such that the organism may ultimately end up with a modified phenotype. Alternately, if the stressor occurs while self-repair mechanisms are in place, the organism may take an alternate developmental trajectory but still end up with an otherwise normal adult phenotype (Fig. 2.2) (Burggren and Reyna 2011). In either scenario, stressors that affect the developmental trajectory can present as effects that may *or may not* result in any downstream consequence.

2.2.1 Heterochrony

Alterations in developmental timing can result in severe birth defects but can also have adaptive value (Moss 2007; Wilson et al. 1988). Heterochrony is a concept originally coined by Ernst Haeckel, who observed shifts in the sequence of appearance of morphological traits *during development* in some species, compared to the sequence of appearance of traits that appear *during the phylogeny* of the species' ancestors (Gould 1977; Keyte and Smith 2014). Interestingly, Haeckel's version of heterochrony is essentially evidence against the notion that ontogeny recapitulates phylogeny, the basic tenet of Haeckel's biogenetic law (Spicer 2006). Though Haeckel himself was aware that heterochrony explained exceptions to the biogenetic law, the definition of heterochrony itself has evolved over the past century. The study of heterochrony is now mostly centered on understanding the timing of developmental events and the underlying mechanisms that result in an alteration in developmental timing *with respect to the evolution of species* (Spicer 2006). Because of this focus, heterochrony is thought to be a major driving force in evolution though natural selection, and its influence is well-illustrated in differences in the timing of ontogeny that generate diversity of adaptive significance (De Beer 1958; Keyte and Smith 2014; Yanai et al. 2011).

2.2.1.1 Timing Mechanisms

Evolution of vertebrate body plans through heterochrony is well-illustrated by the relationship between the number of body segments that arise during embryogenesis

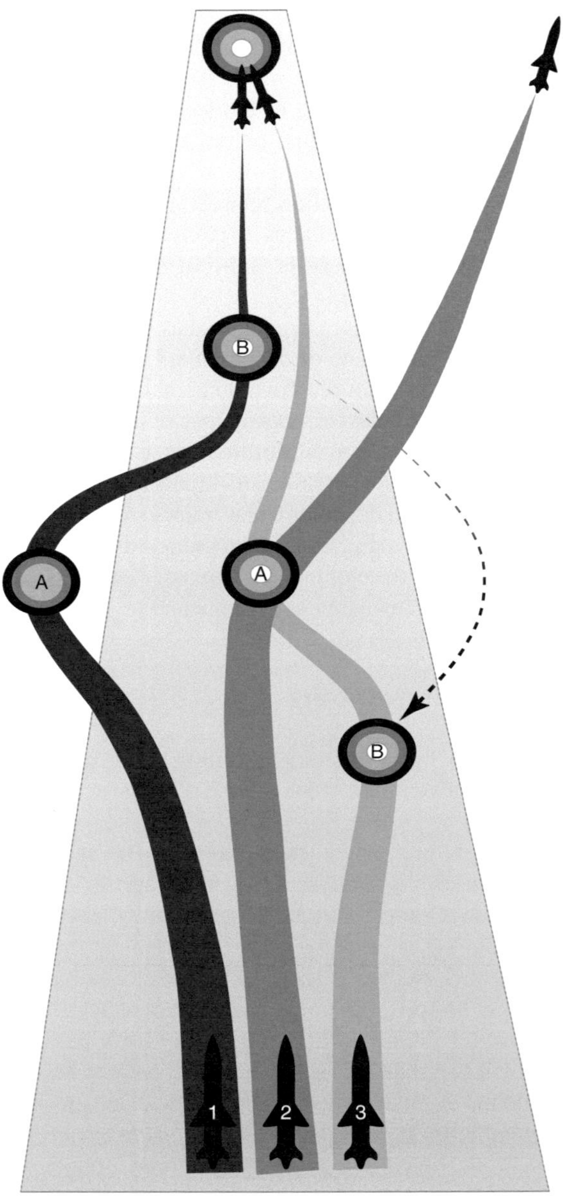

Fig. 2.2 Alteration of developmental trajectory. The innate trajectory of a developing organism can become altered. Here the analogy of a guided missile is used to visualize three scenarios where developing organisms (missiles) may have altered trajectories to developmental landmarks required for survival (A and B). In scenario 1, the organism may be pushed off course by a stressor but undergo self-repair and still be able to reach important landmarks of development required for survival. In scenario 2, the organism may be irreparably pushed off track, such that the organism will not reach the landmarks required for developmental success. In scenario 3, the stressor evokes change in the sequence of events but is still able to successfully complete development

and their speed of formation (Keyte and Smith 2014; Tollis et al. 2014). These segments are bilateral sections of tissue called somites, which arise in pairs along the axis of the embryo and eventually differentiate into various skeletal, endothelial, and epithelial structures. For example, the number of somites formed in the embryo corresponds to the number of vertebrae formed in the adult. The process of somite formation is called somitogenesis. Timing of somite formation during somitogenesis is a venerable model system for studying heterochrony since somites can be directly observed as they arise in order, from cranial to caudal, at a defined interval and with rhythm (Webb and Oates 2016). As a result, the rate of somitogenesis, the total number of somites, and the underlying molecular mechanisms that guide somite formation have been recorded and characterized in numerous embryonic model species. The timing mechanism of somitogenesis occurs through the clock and wavefront model, where a rhythmic *wavefront* of molecular signals moves through the tissue, causing the cells in the presomatic (unsegmented) mesodermal (PSM) tissue of the embryo to oscillate between either permissive or nonpermissive states where a permissive signal causes differentiation (Bertrand et al. 2015; Webb and Oates 2016). The molecular components of this "segmentation clock" (Notch, FGF and Wnt signaling pathways) can be visualized and quantified as molecular markers are expressed in the cells as the wavefront of permissive/nonpermissive signal moves through the tissue (Keyte and Smith 2014; Webb and Oates 2016). As the embryo elongates, the demarcation of where the wavefront of signal begins and ends determines where a segment boundary is formed, such that the cells of the PSM which undergo one wave of the molecular signal become the next somite (Bertrand et al. 2015). As mentioned above, the clock and wavefront have been characterized as the interval timer of somitogenesis in several metazoans. However, the pacemaker that sets the rhythm of the segmentation clock is still unknown.

2.2.1.2 Examples of Alteration of Somitogenesis

Study of the characteristics of timing of somite formation between species and the adaptive significance of the resulting body plans indeed show that alterations in the speed of somitogenesis can drive evolution (Gomez and Pourquie 2009; Keyte and Smith 2014). In vertebrates, the speed of somite formation ultimately determines the number of segments, resulting in variation in body plans (e.g., number of vertebrae, pattern of skeletogenesis). In snakes, an increased oscillation of the segmentation clock causes a shorter interval between somite formation that ultimately results in an increased number of smaller somites and thus more vertebrae in reptile species compared to other amniotes (Gomez et al. 2008; Gomez and Pourquie 2009; Keyte and Smith 2012, 2014; Tollis et al. 2014). Indeed, that increase in speed of development can also be altered by intrinsic factors, considering the variation in developmental trajectories across species and phyla and extrinsic factors (e.g., temperature).

Heterochrony is also evident in marsupial and placental mammals where alteration in the rate of somitogenesis during embryonic development causes a unique phenotype at hatch (Keyte and Smith 2012). In the opossum, somitogenesis rate decreases nearly fourfold as the somite formation progresses from rostral to caudal, indicating that clock speed of the wavefront interval becomes progressively longer as somitogenesis proceeds (Keyte and Smith 2012, 2014). The progressive slowing of somitogenesis rate in the opossum is linked to both underdeveloped hind limbs and an anterior-posterior gradient of chondrification of the axial skeleton in these altricial mammals, who have less-developed posterior skeletal structures, compared to mammals that are more precocial at birth (i.e., mouse) (Keyte and Smith 2012).

In the case of the opossum and other marsupials that are born in a more "embryonic" state, well-developed forelimbs pose an adaptive advantage for these altricial mammals, which can effectively climb from the birth canal to the teat without assistance (Keyte and Smith 2010). This is exemplary of how ontogenic rate can vary at different times throughout development, where different systems have adapted to proceed along different trajectories.

2.2.1.3 Heterochronic Genes and Disease

Heterochronic genes control the timing of events during developmental transitions and have been identified in several species (for an introduction into the literature, see Faunes and Larrain 2016). In *C. elegans*, heterochronic genes are under control of microRNAs (miRNA) that act as switches that time the transition through the discrete developmental stages of these organisms (Harandi and Ambros 2015; Moss 2007; Nimmo and Slack 2009). For example, the *lin-4 and let-7* miRNA families target most heterochronic genes, including Lin-14, Lin-28, and Lin-41, and are well conserved in other species, including mammals (Nimmo and Slack 2009). In zebrafish and Japanese flounder, isoforms of *let*-7 miRNA are highest during embryo to larval transitions, and in *Xenopus* at pro-metamorphosis, suggesting that this may partially be under the control of thyroid hormone (Faunes and Larrain 2016), the master regulator of metamorphosis in vertebrates (Laudet 2011). Importantly, the heterochronic gene program modulates cell division and cell differentiation, highlighting stem cells as key players in developmental transitions, including those that occur during postembryonic development (Fig. 2.3) (Faunes and Larrain 2016). Several heterochronic genes, including *lin-28, lin-41*, and the miRNAs *let-7 and miR-125* regulate stem cell renewal and differentiation and are highly conserved in mammalians (Faunes and Larrain 2016). The *let-7* miRNAs are *also* implicated as tumor suppressors, linking developmental timing to

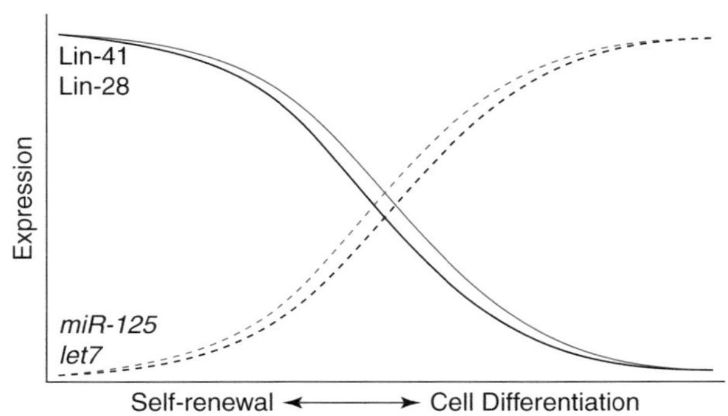

Fig. 2.3 Heterochronic genes and modulators. Heterochronic genes modulate cell division and cell differentiation. As expression of heterochronic genes (e.g., Lin-41 and Lin-28) decreases, active cell differentiation increases. Several miRNAs (e.g., *let-7* and *miR-125*) modulate the expression of heterochronic genes, ushering cells from differentiation toward self-renewal. Modified from Faunes and Larrain (2016)

oncogenesis (Nimmo and Slack 2009). This is not surprising in terms of heterochrony, since alteration of developmental sequence of a cell lineage, by halting cell differentiation while allowing the division of essentially immature cells, is a major tenant of the cancer stem cell hypothesis (Bielecka et al. 2017; Visvader and Lindeman 2008). Such alteration in gene expression by miRNAs is also epigenetic since it can alter gene transcription independent of genomic guidance (Tammen et al. 2013). It follows that since miRNAs may influence selection during development by alteration of developmental timing mechanisms, epigenetic modulation can be an impetus for non-genomic inheritance of novel phenotypes (see Chaps. 7 and 17 for more information on epigenetic inheritance) (Peterson and Muller 2016). Few studies have analyzed heterochronic gene isoforms and tissue-specific expression during development (Faunes and Larrain 2016). However, understanding the role of heterochronic genes in timing of physiological capacities during development will provide functional knowledge that might explain how heterochrony works to drive the evolution of species.

2.2.1.4 Physiological Heterochrony

The above examples of heterochrony illustrate that alterations in the timing of cellular events during development can drive diversity of species and are under molecular and hormonal control. Although the majority of the study of heterochrony has been directed toward morphological characteristics (Smith 2002), the less-studied occurrence of *physiological heterochrony* may be a more likely target of selective pressure due to environmental stressors. Physiological heterochrony is the alteration of the timing of emergence of a physiological trait that offers a selective advantage for the evolution of a species. Below, we illustrate how physiological heterochrony can take the form of alterations in sequence and speed of development. Further we show how physiological heterochrony is not restricted to species-level evolution, since it can occur during microevolution of adaptive traits in populations.

2.2.1.5 Sequence Heterochrony

Alteration of *sequence* of emergence of physiological phenotypes has been observed in several vertebrate (and invertebrate) models (Chap. 4; Mueller et al. 2015). For example, Spicer (2006) describes several examples such as how salinity tolerance in salmonid fish first emerges at different times during ontogeny, triggered by different habitat-specific cues in related species, how thermoregulation in birds and mammals emerges at different timelines along a continuum of altricial to precocial depending on the species and niche, and how the emergence of various respiratory and cardiovascular landmarks appears earlier in ontogeny in more evolved species; and several other examples are present in literature (Spicer 2006). Further, Sirsat et al. (2016) showed that the developmental trajectory of metabolism, growth, and mitochondrial function in birds, compared to related ectothermic archosaurs, was a primary driver of the evolution of endothermy (Sirsat et al. 2016). Though there was no mention of heterochrony in Sirsat et al., the work illustrates that the sequence of morphological, physiological, and cellular events is likely a common target of selective pressure in the evolution of species. It further shows that although the overall rate of development may be increased in examples of heterochrony, this is not a requirement, and alteration of speed of development is not definitive of heterochrony.

A further example of sequence heterochrony at the whole-animal level is illustrated in the sequence of formation of the dermal skeleton in amniotes. The development of osteoderms (dermal bones) in mammals and reptiles forms through a delayed-onset mechanism where ossification does not begin until after birth (Dubansky and Dubansky 2017; Vickaryous et al. 2015; Vickaryous and Sire 2009). Indeed, the boney matrix of the osteoderm of the American alligator (*Alligator mississippiensis*) develops within fully differentiated tissue, instead of embryonic mesenchyme where the rest of the skeleton forms during embryogenesis (Dubansky and Dubansky 2017). Here, sequence heterochrony is illustrated in the exclusion of osteoderm development from the normal sequence of embryonic skeletogenesis. Nearly a year after hatch, osteogenesis resumes and finishes the formation of the skeleton after osteoderm development is complete. Several human pathologies that involve heterotopic bone formation occur via mechanisms similar to alligator osteoderm development and may expose links between human disease and heterochronic genes, where alteration of sequence and order of normal development results in adverse outcomes like in the pathology of heterotopic ossification following trauma (Dubansky and Dubansky 2017). Indeed, it has been shown that the heterochronic miRNA *let-7*, described above as both a switch for timing progression through early stages and as a oncosuppressor, also promotes bone growth in stem cells (Wei et al. 2014). Although there has been virtually no molecular characterization of osteoderm development, *let-7* has been proposed as a key regulator of bone formation in osteogenic lineages of mesenchymal stem cells and suggested as a regulator of ectopic bone formation in humans (Wei et al. 2014). As such, it is likely that heterochronic genes are key regulators of both the pathologic conditions in humans and the heterochronic alteration of skeletogenesis in alligators.

To evaluate sequence heterochrony between species and phyla, several methodologies have been suggested based on the appearance of landmark developmental events (Bininda-Emonds et al. 2007). Indeed, a comparison of the ordering of emergence of a few traits (i.e., A, B, C, D versus A, C, B, D) can show alteration in sequence both within and between species (Mueller et al. 2015; Tills et al. 2010) and can be used to create phylogenetic comparisons (Smirthwaite et al. 2007). Several new methods are available to analyze multiple events in multiple taxa in a phylogenetic context (Keyte and Smith 2014). For example, parsimony-based ontogenic sequence analysis (OSA) uses algorithms to account for variability of sequence of traits as that arise within a species (Bininda-Emonds et al. 2007; Colbert and Rowe 2008). This technique then maps the hierarchy of the sequence pattern against multiple species into network diagrams or phylogenetic trees. Such techniques will likely become more useful as such work expands to include more physiological indices in addition to morphological phenotypes, and can evaluate the ordering of landmark events within the context of developmental time.

2.2.1.6 Population-Level Heterochrony

Early definitions of heterochrony make less reference to population-level alterations (Spicer 2006), though it is well understood that populations can evolve gene-level adaptations that make them physiologically distinct from other populations of the same species (Reid et al. 2016; Reid and Whitehead 2016). Further, population differences in timing, if under selective pressure, could lead to speciation (Spicer et al. 2011). Here we posit that physiological heterochrony can act on a population. However, it should first be clarified that alteration in the order or timing of developmental events caused by novel environmental stressors within an individual or population is considered heterokairy (discussed below and in detail in Chap. 4), not heterochrony. Although stressors may exert selective pressure, heterokairy is an agent of plasticity (Mueller et al. 2015), not evolution. We see a clear yet thin line between heterokairy and heterochrony well illustrated with a new example of physiological heterochrony within populations of Gulf killifish (*Fundulus grandis*) found near the Houston Ship Channel (HSC) in Texas (see Chap. 14 for an in-depth review of HSC killifish). These populations of Gulf killifish in the HSC have heterochronic adaptations associated with evolved resistance to dioxin-like compounds (DLCs) that are ubiquitous in the home range of these fish populations. DLCs, such as polychlorinated dibenzodioxin (PCDD), polychlorinated biphenyls (PCBs), and polycyclic aromatic hydrocarbons (PAHs), are strongly carcinogenic and teratogenic (Oziolor et al. 2014, 2016a, b). These HSC populations of Gulf killifish exhibit a collective phenotype that suggests that along with resistance to the teratogenicity of DLCs, these unique fish populations have an altered developmental trajectory compared to typical Gulf killifish populations.

When *any* species of fish is exposed to DLCs during embryogenesis, a suite of characteristic developmental defects is manifested. These distinctive pathological features include decreased size, craniofacial defects, ascites, hemorrhage, spinal defects, cardiac dysmorphogenesis, and altered heart rate and arrhythmia (see Chap. 10 for detailed review of effects of DLCs on development) (Burggren et al. 2017; Dubansky et al. 2013; Incardona 2017; Lema et al. 2007; Weis and Weis 1974). This suite of defects can also be induced experimentally in fish embryos by exposure to several other stressors unrelated to DLCs including heat (Burggren and Bagatto 2008), insecticides (Clark and Di Giulio 2012; Oziolor et al. 2016b; Weis and Weis 1974, 1977), metabolic waste and nitrogenous compounds (Wolf 1957), and several other toxicants besides those found in crude oil (Clark and Di Giulio 2012; Oziolor et al. 2016b; Weis and Weis 1997). Although this pathology has been recognized for over a century as a collective response to stressor exposure (Wolf 1954, 1957), this characteristic response to stressors has been used extensively to classify the effects of DLCs on embryonic fish development (Incardona 2017; Incardona et al. 2011, 2013). However, the physiological mechanisms that cause this diverse suite of defects are still uncertain.

In vertebrates, the AHR pathway is, to a large degree, responsible for the homeostatic response to DLCs (Dubansky et al. 2013, 2017; Oziolor et al. 2016b; Whitehead et al. 2012). Paradoxically, inactivity of the AHR in the HSC fish is associated with resistance to teratogenicity from exposure to these toxicants

(Oziolor et al. 2014, 2016b). However, the summative phenotype of these resistant fish is clearly beyond xenobiotic metabolism via the AHR pathway (Bickham 2011; Oziolor et al. 2016b). Interestingly, HSC Gulf killifish also possess several novel physiological phenotypes that are not typically associated with metabolism of xenobiotics, such as an increased metabolic rate at hatch and variation in response to hypoxia between populations (Oziolor et al. 2016b). Here, we highlight a few new observations in Gulf killifish embryos and larvae from the HSC, including an increased growth rate (Fig. 2.4) in larvae, and indications of increased rate of development in embryos, which may be adaptive for survival in the heavily polluted HSC (see below). Since the altered developmental trajectory is heritable, unique to HSC populations, and has adaptive value, it is hypothesized here that *the accelerated development observed in HSC Gulf killifish populations is an example of population heterochrony*.

HSC fish, in addition to an increased rate of growth, have an increase in the rate of pigmentation development during embryogenesis (Fig. 2.5). The development of pigmentation is driven by the central nervous system signal for production of melanophore-stimulating hormone (MSH) (Bolker and Hill 2000). Although MSH regulates much of the function of melanocytes, pigmentation can be profoundly affected by blocking the conversion of thyroid hormones thyroxin (T4) to

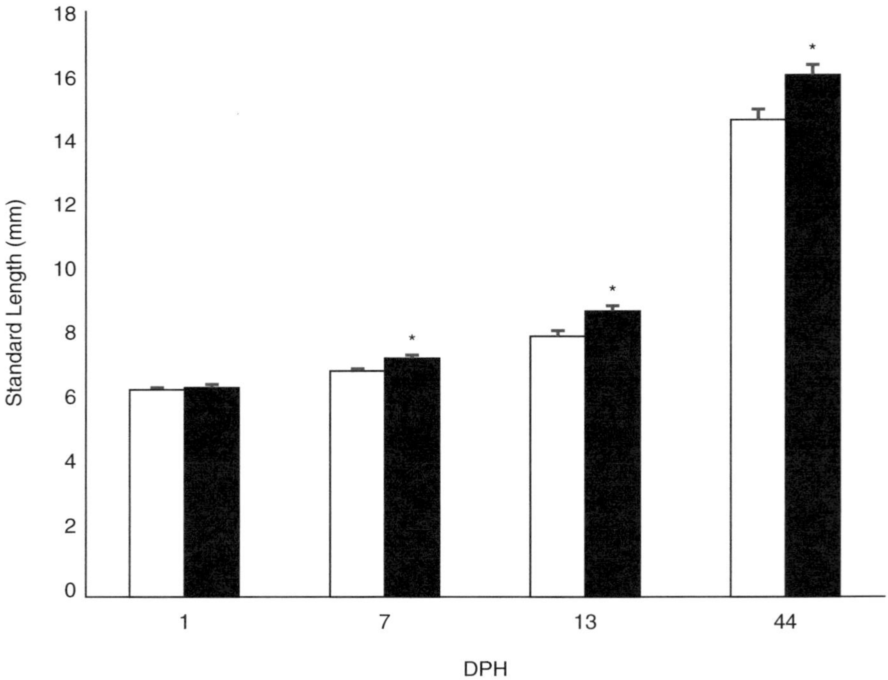

Fig. 2.4 Enhanced growth rate in DLC-resistant populations of Gulf killifish (*Fundulus grandis*). Gulf killifish populations that are resistant to dioxin-like compounds have an increased growth rate

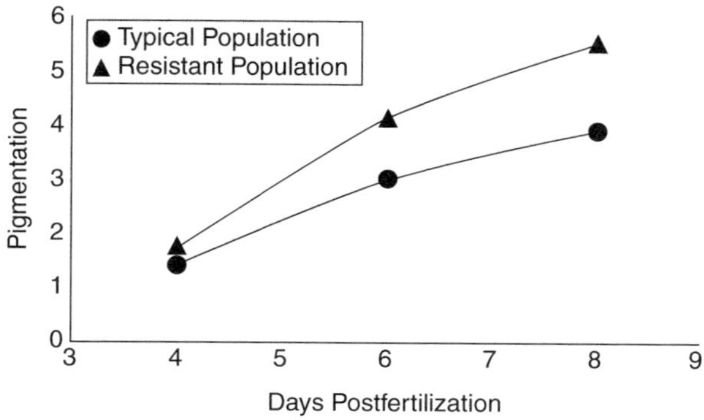

Fig. 2.5 Development of pigmentation in DLC-resistant and typical populations of Gulf killifish (*Fundulus grandis*). Gulf killifish populations that are resistant to dioxin-like compounds have an increased rate of development, illustrated by an increased rate of pigmentation development, a widely used marker for developmental progression

triiodothyronine (T3) (Li et al. 2012). Further, alteration in thyroid hormones also affects several physiological processes including heart rate and metabolism. Since thyroid hormone is a master regulator of vertebrate metamorphosis (reviewed in Laudet 2011), pigmentation is often used as a developmental marker for early life stages (Chauvigne et al. 2016; Colman et al. 2005; Ditty et al. 2005; Quigley and Parichy 2002; Urho 2002). Further, alteration in pigmentation is associated with exposure to several toxicants, including DLCs in fish (Frantzen et al. 2012; Meier et al. 2010) (discussed more below). Although it is still unclear what mechanisms are responsible for the increased rate of development and early growth observed in these HSC killifish, several physiological mechanisms, including those that increase the rate of pigmentation development, are likely enhanced in DLC-resistant fish from the HSC. In addition, it appears that the increase in overall growth and development seen during embryogenesis continues at least to the juvenile stage (Fig. 2.4).

Since a strong selective pressure is often associated with fitness costs, it is possible that the evolution of resistance to DLCs in HSC fish is concomitant with physiological deficiency, as a trade-off for rapid acquisition of their toxicant-resistant phenotype. However, it is also possible that genomic regions, beyond xenobiotic metabolic pathways, were selected for along with the resistant phenotype, which underlay coincidentally advantageous phenotypes (i.e., spandrels) (reviewed in Gould and Lewontin 1979). Such coadaptations are seen in strains of fish species with high growth rates artificially selected for use in the aquaculture industry. In that example, comparison of growth rates between farmed fish shows dramatically enhanced growth rates compared to wild fish of the same species. However, selecting for enhanced growth occurs at the expense of several traits that would likely preclude survival outside of the aquaculture setting (Sibly et al. 2015). Still, such examples further illustrate that heterochrony can be induced and observed at the population level within a species, when altered developmental trajectories become sustained and heritable. Furthermore, since smaller species generally reach reproductive age faster than larger species (Sibly et al. 2015), altered growth rates observed between larger and smaller species could also be a result of trade-offs of maturity for size. At present, adult size distribution has not been studied in HSC Gulf killifish.

In the absence of a size difference between adult HSC Gulf killifish and typical adult killifish populations, it could be reasoned that accelerated growth seen in early stage HSC is physiological heterochrony, as part of the collective adaptive phenotype of DLC-resistant fish. An accelerated rate toward reproductive age may be of selective advantage for HSC fish, considering the carcinogenic potential of the DLCs in their environment and the slow onset of cancer manifestation. Most importantly, the contaminants at the HSC are highly teratogenic compounds that would otherwise cause severe developmental deformities in cardiovascular and morphological endpoints in nonresistant fish at the concentrations present in the HSC. It is likely that the collective phenotype seen in the HSC killifish is due to a combination of heritable mechanisms of developmental plasticity that have permanently altered the ontogenic program of HSC fish. This altered ontogenic program allows these HSC killifish to proceed through embryogenesis and larval development, while essentially ignoring otherwise toxic doses of teratogenic stressors, providing protection (at least) until these fish reach reproductive age.

2.2.2 Heterokairy

Developmental plasticity may allow for altered trajectory of development during self-repair or lead to permanent alterations, thus altering the timing of when physiological capacities arise (Burggren 2018). With heterokairy, the occurrence of such *alterations in timing* of development *of an individual* is in response to an environmental cue or stressor (Chap. 4). In contrast to heterochrony, heterokairy is generally considered to *not* create heritable phenotypes that are the target of natural selection but rather a response to environmental stressors. However, the fixing within the genome of otherwise temporary phenotypes is of considerable interest to those studying epigenetic inheritance and climate change (Burggren 2014, 2015, 2016, 2017; Dickins and Rahman 2012). Environmental stressors that evoke heterokairy can be either perceived directly, such as exposure to a physical or chemical stressor, or may be an indirect effect of parental exposure to a stressor (Chaps. 4 and 15) (Mueller et al. 2015; Spicer and Burggren 2003; Spicer et al. 2011). In this we note that exposure history has potential to alter the downstream physiology of an individual or offspring. Here, *physiological heterokairy is* an alteration in the sequence or timing of emergence of physiological capacities (Spicer and Burggren 2003).

Physiological heterokairy can be observed as alteration to the developmental trajectory via alteration of sequence or speed of developmental rate (i.e., interval or rhythm), either in association with self-repair or as a consequence of adverse effects. Self-repair mechanisms could alter the developmental trajectory of an individual if the physiological system comes online at a delayed or enhanced schedule (as in developmental trajectory Fig. 2.2) (Chap. 4). Likewise, if the timing of emergence of a critical physiological function is altered, one could imagine a number of scenarios where such a change could affect survival (e.g., osmoregulatory capacity, cardiovascular regulation, hypoxia tolerance, etc.). As such, though heterokairy occurs in an individual, the *ability* to perform this sort of plasticity can be of adaptive advantage (Mueller et al. 2015).

Following the example above for physiological heterochrony at the population level in DLC-resistant Gulf killifish, *physiological heterokairy* can also be seen during exposures of *typical* Gulf killifish embryos to toxic DLCs from crude oil. In this scenario, heterokairy is an effect of embryonic exposure of typical Gulf killifish to DLCs. It should be reiterated that DLCs are teratogenic to *any* typical fish species (and all other vertebrates). However, depending on toxicant dose and innate physiological capability, dynamic alterations in physiological phenotype can occur during development, which alters developmental trajectory toward self-repair *or* retardation. This is illustrated in the development of pigmentation in the skin versus eye and head size in typical populations of Gulf killifish (Fig. 2.6). As discussed above, several molecular pathways under physiological regulation contribute to the development of pigmentation, and as such pigmentation has been used as a developmental marker in control fish. Further, there is also evidence that the pathway chiefly responsible for metabolizing DLCs is involved in the regulation of pigmentation in the skin (Esser et al. 2013). So, when exposed to lower doses of DLCs, typical Gulf killifish will have a slower development of pigmentation during earlier stages of development, but by the time they hatch, they will have the same pigmentation as controls, indicating self-repair mechanisms (Fig. 2.6a). At higher doses, the larvae will be less pigmented at all stages up to hatch. This trend of self-repair at low doses

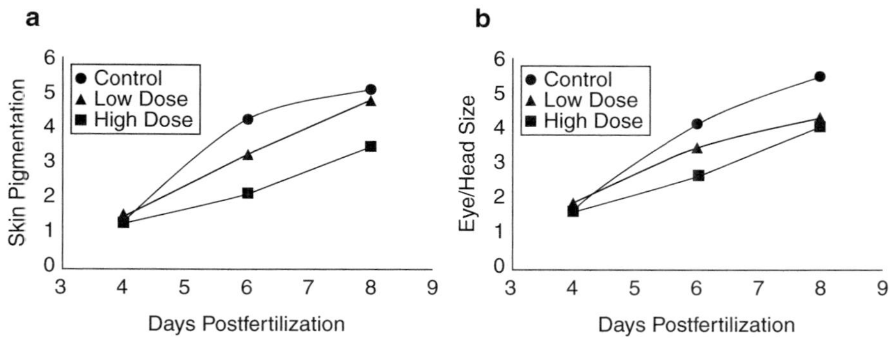

Fig. 2.6 Self-repair during development. Alterations in physiological phenotype associated with exposure of typical Gulf killifish to DLCs from crude oil indicate that the developmental trajectory can become altered toward self-repair *or* retardation, depending on toxicant dose. (**a**) Low-dose exposure to airborne PAHs from crude oil affects pigmentation at early stages, though recovery to control levels of pigmentation occurs prior to hatch indicating self-repair. (**b**) During early stages of embryogenesis, head and eye size are significantly smaller in killifish exposed to lower doses of airborne PAHs, though this effect is magnified at higher doses. However, upon hatch, both low and high doses had similar effects on head and eye size

is seen in several other physiological variables (see below), though morphological indices such as eye and head size may be less plastic (Fig. 2.6b).

It is also likely that physiological processes can be brought online in an altered order in response to stressors. Spicer and Burggren (2003) proposed that alteration of the timing of when physiological phenotypes arise is a hallmark of physiological heterokairy. Additional examples of physiological heterokairy are well described in Chap. 4. Here, we emphasize that physiological heterokairy is most certainly a phenomenon that has received little attention, and as a result, alterations in timing of developmental events are likely common occurrences in experimental developmental biology.

2.3 Developmental Time: Practical Considerations

Ontogenic rate, sequence, and trajectory can become altered due to environmental stressors, and such alterations can be of selective advantage. As mentioned above, developmental time is not only set by extrinsic astrological phenomena but is also set by some intrinsic oscillators (e.g., somite clock) and cues (e.g., hormones) and also is ordered as a sequence based on other events during development. Further, there is great variability in developmental plans between and within species according to both the mode of development (i.e., altricial, precocial, parental care, etc.) and chronological timing of developmental events. This presents a major hurdle for comparison between species and phyla. So, with the multitude organisms under study, how can one relate developmental progression of, say, a zebrafish that hatches in a few days to a rainbow trout that may take a month or more to reach an equivalent life stage?

There is, in fact, no unit for developmental time. Chronological time is set by the earth's revolution and rotation as the oscillators. It is measured in hours and years and further subdivided into months, weeks, hours, minutes, seconds, etc. However, variation in the rates and sequence of development obscures a comparison along a chronological time scale (Jeffery et al. 2005). So, it is nearly impossible to make a comparison between species with respect to the age of the organisms, based on chronological time. Comparison between organisms of the same species with respect to chronological time is somewhat murky, as well, without the context of identical environmental conditions, as mentioned above. However, several methods have attempted to rectify this by comparing early landmarks in development with respect to common landmarks that occur later in development.

2.3.1 Developmental Rate and Chronological Time

Since it is difficult to compare ontogenic rates with respect to chronological time, several techniques have been suggested to standardize comparisons between species and experimental groups. During portions of ontogeny where intervals of developmental time are set by a known internal oscillator (i.e., heterochronic genes, clock and wavefront model, etc.), it is possible to compare relative ontogenic rates between experimentally induced *and* inherent variation between species and groups. This relative comparison can then be superimposed on the oscillatory rhythm of chronological time. For example, since intrinsic oscillators drive gene expression to set a rhythmic interval of somite formation, and one somite is formed per oscillation, somitogenesis rate can be standardized between species, in relation to chronological time (Gomez et al. 2008; Keyte and Smith 2014). At species optimal development temperatures, corn snakes form one somite every 100 min; house snakes, every 60 min; zebrafish, every 30 min; chickens, every 90 min; and mouse somites, every 120 min. (Gomez et al. 2008). Like the methods for comparison of the sequence of appearance of landmark traits, comparison of the time it takes to reach those landmark traits can be used to show relative rates of development. Gomez et al. (2008) proposed that somite formation time can be framed within the context of relative rate of development to measure somite clock speed, relative to the time it takes to reach a common developmental landmark. The authors reasoned that since the cells of the somites come from the PSM, somite formation reduces the size of the PSM by a fraction proportional to the somite during an interval defined by the clock and wavefront signal (Gomez et al. 2008). Since each somite is the result of one oscillation of the clock and wavefront signal along the PSM, the relationship between the somite clock, the size of the somite, and the growth reduction of the PSM is proportional within an animal. So, since the size of the PSM is relative to the size and total number of somites, the interval of oscillation of the segmentation clock can be compared relative to the chronological time between two developmental landmarks (Gomez et al. 2008). In this way, rate of somitogenesis between experimental groups *or species* can be standardized for comparison.

Several investigators have proposed other methods to standardize developmental time between species. Since growth and development are heavily tied to food resources and environmental conditions, only comparisons between individuals held in identical conditions can be compared based on age within chronological time. Fuiman (2002) posited that for comparison between ontogenic state of fish larvae, size is a more reliable indicator of developmental stage of fish after hatch, since size integrates the total environmental experience of the individual (Fuiman 2002). Fuiman further noted that as development proceeds, less and less ontogenic changes occur, and growth increases, because development follows a logarithmic scale with respect to larval size. This logarithmic growth occurs in all species studied during the larval stage when the length of the fish increases quickly, until reaching a near-isometric growth rate, when metamorphosis to the juvenile stage is complete. So, just after hatching, a small change in length will be accompanied by a more drastic increase in development compared to the same change in length nearer metamorphosis (Fuiman 2002). If the average size at metamorphosis to juvenile stage is known for a species, the

Ontogenic Index of the larvae at a given length (O_L) is equal to the logarithm of the larval length ($\log_L$) divided by the logarithm of the length at juvenile metamorphosis (L_{juvenile}):

$$O_L = \frac{\log L}{\log L_{\text{juvenile}}} \times 100$$

The Ontogenic Index thus converts length to a percentage of the ontogenic state of the larva, such that comparison of relative ontogeny can be compared between species or treatment groups. Such a comparison could be used to account for ontogenic rate as an effect of experimental variables. Importantly, the size at metamorphosis can be substituted for the size at which an organism obtains a landmark trait. As in the above example for comparison of somitogenesis rate, such a conserved landmark could also be used as the comparative reference point for estimating Ontogenic Index (Fuiman et al. 1998). The Ontogenic Index is well suited for early development of fishes, and similar methods could be applied to other organisms. However, additional considerations are likely necessary for other clades, developing animals with indeterminate size, or those that undergo molting.

2.3.2 Developmental Time and Experimental Planning

Laboratory studies on live organisms generally aim to control for variables by creating a stable environment, though experimental variables introduce stochasticity. We have shown here how environmental stressors can exert effects in developing organisms, beyond what would be observed in adults. Indeed, developmental timing can become altered by environmental stressors that can be either adaptive or maladaptive. Since we know, to some extent, how this works, it is important to consider that manipulation of experimental variables often alters developmental timing. Such an alteration in developmental time could conceivably masquerade as a physiological effect, when really the observed effect is just an alteration in the timing of ontogeny of physiological phenotype. It follows that the interpretation of experimental data could be well served with an assessment of relative developmental progression.

Characterization of the ontogeny of model organisms is a valuable context for explaining effects that occur within developmental time. With characterization of the ontogeny of a model species, the choice of measurement times and endpoints can further provide a higher resolution within the context of the ontogenic plan. For example, the measurement of heart rate can be best interpreted with respect to the ontogeny of cardiovascular regulation. As mentioned above, early life stages are considered more sensitive to environmental stressors than adults. This is in large part because developing organisms do not have the compensatory abilities afforded by the adult physiology. Here it is hypothesized that the alteration of developmental timing is a common and primary effect of environmental stress, manifesting as variation in physiology.

2.3.3 Developmental Time and Physiology

Early work of this author measured heart rates of Gulf killifish (*Fundulus grandis*), in embryos exposed to contaminants from crude oil at a single time point at 75% of embryogenesis. At this time point, it was reasoned that the heart is chambered, and heart beat is statistically stable within a day or so before and after that time *in control fish*, depending on incubation conditions. Thus, 75% of embryogenesis is the perfect time to assess effects on cardiovascular physiology. Dubansky et al. (2013) found that when embryos were exposed to PAHs from sediments contaminated with crude oil, average heart rate and hatching success was significantly depressed compared to controls at this time point (Fig. 2.7a) (Dubansky et al. 2013). However, at the conclusion of the experiment, many exposed embryos failed to hatch, and those that did hatch were significantly smaller and listless. Indeed, when looking back at the ancillary data collected from that experiment, it is apparent that the exposed larvae appear to be less developed than control larvae (see below in Fig. 2.8). Indeed, retarded embryonic development has been observed by this author for nearly all fish species, when exposed to higher doses of PAHs. The resultant conclusion is that if the development of exposed killifish is retarded, the alteration of heart rate is more likely a function of the altered ontogeny of the cardiovascular system rather than bradycardia, which would indicate an alteration of cardiovascular regulation. However, no physiological explanation that accounts for *developmental influence* has been supplied for the alteration of heart rate in fish embryos exposed to PAHs; however molecular alterations between exposed and unexposed early life stage fish have been proposed (Chap. 10). It is likely that a mechanistic explanation for acute cardiovascular effects of PAHs is explanatory for the decrease in heart rate seen in embryonic and larval fish exposed to PAHs. Here, we highlight that in the author's experiments (and all others known to the author), time was not measured in accordance with a method that standardized developmental time between treatment groups, in lieu of using chronological time as a proxy. Since hatching was delayed at high-dose treatment groups, and those that did hatch had a less-developed phenotype (Fig. 2.8), no common developmental landmark was available to account for developmental time comparison. So, in retrospect, it could be hypothesized that the timing of emergence of physiological phenotypes is altered in embryos exposed to PAHs, thus affecting the heart rate at the time of measurement.

More recently, exposure of embryonic Gulf killifish to low concentrations of PAHs from crude oil suggested that significant alteration in heart rate during embryogenesis may be of little consequence in some scenarios. Exposure to low concentrations of PAHs from crude oil again caused significant decrease in heart rate at time points bracketing 75% of embryonic development (Fig. 2.7b). However, at early time points, and at time points approaching hatch, low concentrations of PAHs caused no difference in hatching success or timing, and the ontogenic state at hatch for all larvae was indistinguishable in controls and experimental groups. The absence of observable effect at hatch, and daily data collection throughout the experiment, allowed direct comparison of effects within the context of developmental time and along a chronological timeline, using hatching as a common

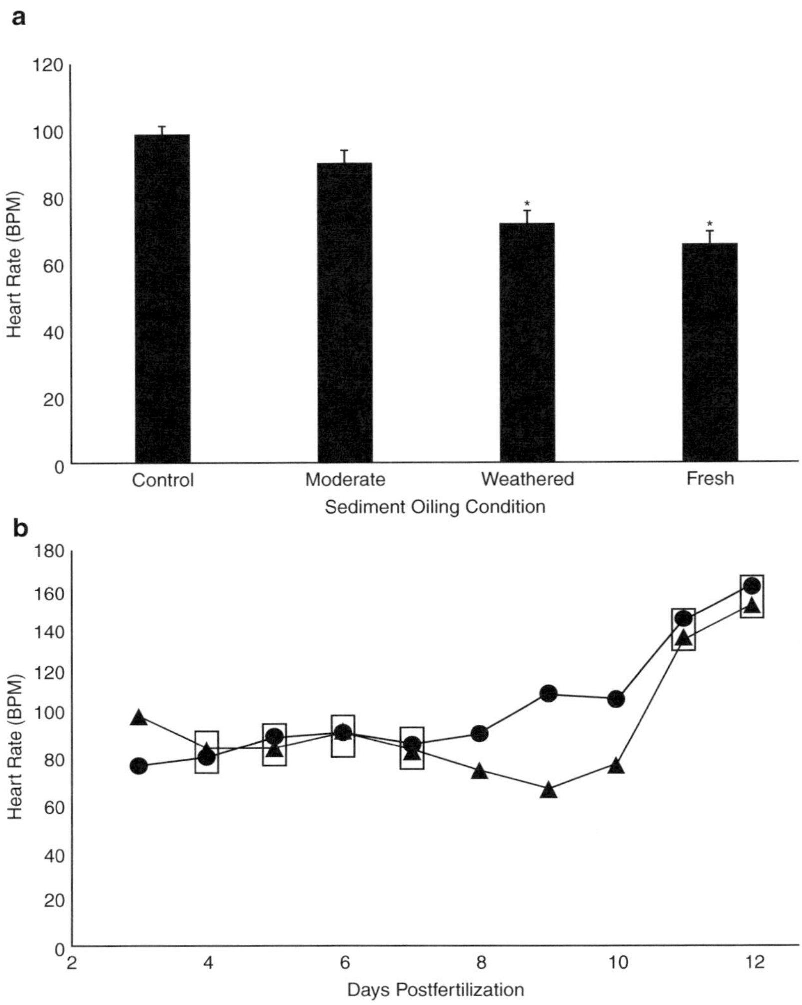

Fig. 2.7 Development of cardiovascular regulation. (**a**) At 75% of embryonic development, elevated levels of oil in sediment cause a decrease in heart rate. (**b**)At 75% of embryonic development, elevated toxicity from airborne oil vapors causes a decrease in heart rate, although at early time points and at time points nearing hatch, the heart rate in exposed embryos is the same as control embryos

developmental landmark. In this experiment, the apparent and significant effects to the cardiovascular system in the oil-exposed fish were abolished prior to hatch (Fig. 2.7b). This could be reflective that there is an alteration of the timing of cardiovascular regulatory systems. For example, heart rate steadily increases as the stimulatory beta-adrenergic tone of the autonomic nervous system comes online. In Gulf killifish, beta-adrenergic tone is online prior to the time points where heart rate

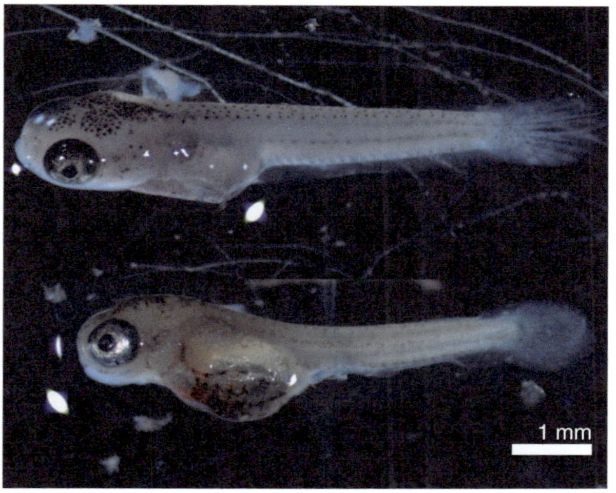

Fig. 2.8 Control versus PAH-exposed larvae. At hatch, control larvae (top) are larger than PAH-exposed larvae (below), which retain a less-developed phenotype

is significantly decreased by PAHs in Fig. 2.7. A possible explanation for the observed effects could be that beta-adrenergic tone was slower to develop in the oil-exposed embryos, although the system came online a few days later, allowing the heart rate to increase in time for hatch. Here, hatching success and timing, morphology, and heart rate were similar at hatch. As such, the significantly lower heart rates of the oil-exposed embryos were more likely due to alteration in the timing of development of the cardiovascular regulation, rather than a mechanistic effect that caused bradycardia. However, this supposition could be predictive of downstream cardiovascular effects that are more nuanced (and perhaps more insidious) than depression of heart rate.

"Development waits for no (wo)man," *anonymous*, and this is all too true when following an experiment that measures variables in a quickly developing embryo. Indeed, a fast-developing organism might relegate 50% of the available observation times to the normal sleeping schedule of the observer. For example, imagine your organism completes embryogenesis at 48-h post-fertilization. During that 48-h window, cardiovascular regulation is actively developing, altering the heart rate considerably as it proceeds through embryonic development in just a couple days. Despite the researcher's creed to work around the clock, let's consider that the heart beat may not be readily observable until 24-h post-fertilization. This leaves a window of 1 day to measure the development of cardiovascular regulation. Now, take into account that there is also upward of 25% variation in the time to hatch, reflecting an inherent variability of ontogenic rate. In chronological time, that is half of a day—and half of your observation time window! Now consider adding a stressor, noting the above references that purport that alteration of an organism's ontogenic rate and/or sequence is nearly inevitable when a developing organism is exposed to environmental stressors. Though this case is generalized, this example's timeline is quite realistic for some fish species. In such a case, it is hypothesized here that variation in heart rate between experimental groups, at any single time point, does not account for the developmental state of the cardiovascular system, and thus not indicative of a *lasting* physiological effect.

2.3.4 Developmental Time and Model Organisms

It is common to assign stages to development defined by the appearance of morphological and physiological landmarks that occur along an expected developmental timeline. However, individual variation of developmental timing and environmental influences provide a wide range of altered, yet normal development along a continuum. Buds form before limbs, cells migrate before the heart forms, and receptors arise before innervation of nerves, which grow as cells are added. Indeed, the appearance of a phenotype during development is not a digital event that is either there, or not there. In effect, development is non-saltatory, where structure and function emerge along a continuum, as phenotypes emerge slowly, while all of the necessary components become arranged and begin to attain functionality. Here, the author concedes that assigning stages to early development, though often necessary for communicating context, is overly simplistic in the experimental setting and not always in accordance with developmental or chronological time.

While time is of little worry for a primary investigator overlooking the labor of laboratory studies, choosing appropriate model organisms can maximize time spent in the lab. Indeed, a quick embryonic development may be an attribute in some situations, such as when *larval physiology* is the focus, or when hatching, mortality, or other such discrete time points are the primary interest. So, when time is short, and a quick turnaround is desired, a quick embryonic stage may be a desired attribute for early life stage studies. In the environmental sciences, a number of fish species have emerged as models, including zebrafish, salmonid species, herring, killifish, sheepshead minnows, a number of other freshwater species, and several demersal and pelagic marine fish species. However, we contend that the concepts here can be applied to any species, with respect to the conflicts between developmental and chronological time. As such, there is a large existing pool of species with a venerable background for use in culturing and holding experimental organisms. However, if the goal is to study the effects of the environment on developmental physiology, a model organism with a shorter embryonic period may not be supportive of that goal with respect to chronological time. Though it is easy to miss the action, since development waits for no man (or woman), there is a simple solution to this conundrum. We can, quite comfortably and within the confines of our personal life, *wait on development*, providing we use the right model species.

It is evident that consideration of alteration of ontogenic rate and sequence provides a holistic context for data interpretation that is not otherwise possible, as illustrated in the above examples. This chapter has often focused on Gulf killifish, an ideal model organism for addressing effects on the environment and for increasing our understanding of developmental timing. Several of the author's ongoing projects compare ontogenic rate as a variable for comparison between Gulf killifish populations and treatment groups. In contrast to fish models that hatch within a few days into altricial larvae, Gulf killifish can take 10–14 days to hatch into precocial larvae with considerable physiological ability at hatch in comparison to altricial fish species. This allows study of several physiological attributes, including development of cardiovascular regulation during the embryonic stage. This is a highly desirable attribute, since during the embryonic stage, fish are immobile such that they do not require anesthesia or restraint. This opens considerably more possibilities for studying developmental physiology that are not nearly as feasible in an altricial species that may have little ability to regulate cardiovascular physiology until sometime after hatch. Other species with a protracted embryonic stage may also offer platforms that allow a higher resolution of developmental time, within our self-imposed confines of chronological time. That is, since we generally work during the day, sometimes take weekends off, and have kids and relationships outside of our work, organisms with a protracted development can be attractive to better enable researchers to align personal schedules with developmental time! Most importantly, however, since a protracted development essentially offers a stretched out, smoother version of development, a likewise smoother and more comprehensive dataset can be obtained offering higher resolution for observing the emergence of phenotypes over time.

Conclusion

Though it is mostly unavoidable, utilizing calendars and clocks to plan experiments may be placing constraints on understanding the developmental physiology of organisms. However, the stability and familiarity of chronological time is helpful for conceptualizing ontogeny. Developmental time is the temporal framework inherent to a species' developmental program and is an integral part of a species' phenotype. Importantly, developmental time can be altered in response to the environment as a function of developmental plasticity. Alteration in developmental timing can also be adaptive for a species or a population. Indeed, variation of developmental timing and trajectory is a widespread effect that is likely often overlooked. However, with the increasing use of developing organisms in environmental sciences, we call to attention that sensitive early life stages will be more affected by environmental stressors than adults. This sensitivity of developing organisms is an attribute for gaining an integral understanding how stressors may affect populations. However, since such developmental effects depend on the stressor intensity, duration, or dosage, we can easily predict that effects will be eminent (i.e., altered heart rate example in Fig. 2.7a, b). When this is the circumstance, we are essentially casefinding results by identifying effects that we already know will occur! Although this is a fundamental tenant of toxicology, it is not necessarily supportive of increasing the understanding of the developmental physiology of the organisms studied, *unless* such studies are placed within the context of developmental time.

References

Albertson RC, Cresko W, Detrich HW 3rd, Postlethwait JH (2009) Evolutionary mutant models for human disease. Trends Genet 25:74–81

Amaral FG, Castrucci AM, Cipolla-Neto J, Poletini MO, Mendez N, Richter HG, Sellix MT (2014) Environmental control of biological rhythms: effects on development, fertility and metabolism. J Neuroendocrinol 26:603–612

Bertrand S, Aldea D, Oulion S, Subirana L, de Lera AR, Somorjai I, Escriva H (2015) Evolution of the role of RA and FGF signals in the control of somitogenesis in chordates. PLoS One 10:e0136587

Bickham JW (2011) The four cornerstones of evolutionary toxicology. Ecotoxicology 20:497–502

Bielecka ZF, Maliszewska-Olejniczak K, Safir IJ, Szczylik C, Czarnecka AM (2017) Three-dimensional cell culture model utilization in cancer stem cell research. Biol Rev 92:1505–1520

Bininda-Emonds ORP, Jeffery JE, Sanchez-Villagra MR, Hanken J, Colbert M, Pieau C, Selwood L, ten Cate C, Raynaud A, Osabutey CK, Richardson MK (2007) Forelimb-hindlimb developmental timing changes across tetrapod phylogeny. BMC Evol Biol 7:182

Bolker JA, Hill CR (2000) Pigmentation development in hatchery-reared flatfishes. J Fish Biol 56:1029–1052

Burggren W (2016) Epigenetic inheritance and its role in evolutionary biology: re-evaluation and new perspectives. Biology 5:24

Burggren W (2018) Developmental phenotypic plasticity helps bridge stochastic weather events associated with climate change. J Exp Biol 221. https://doi.org/10.1242/jeb.161984

Burggren W, Bagatto B (2008) Cardiovascular anatomy and physiology of larval fishes. In: Finn N, Kapoor BG (eds) Fish larval physiology. Oxford & IBH Publishing Co. Pvt. Ltd., New Delhi, pp 119–161

Burggren W, Dubansky B, Martinez-Bautista N (2017) Cardiovascular development of embryonic and larval fishes. In: Gamperl AK, Gillis TE (eds) Fish physiology. Elsevier, Amsterdam

Burggren WW (2014) Epigenetics in comparative animal physiology - or - Lamarck is lookin' pretty good these days. J Exp Biol 217:682–689

Burggren WW (2015) Dynamics of epigenetic phenomena: intergenerational and intragenerational phenotype 'washout'. J Exp Biol 218:80–87

Burggren WW, Reyna KS (2011) Developmental trajectories, critical windows and phenotypic alteration during cardio-respiratory development. Respir Physiol Neurobiol 178:13–21

Burggren WW, Santin JF, Antich MR (2016) Cardio-respiratory development in bird embryos: new insights from a venerable animal model. Rev Bras Zootec 45:709–728

Burggren WW, Warburton S (2007) Amphibians as animal models for laboratory research in physiology. ILAR J 48:260–269

Carlos-Wallace FM, Zhang L, Smith MT, Rader G, Steinmaus C (2016) Parental, in utero, and early-life exposure to benzene and the risk of childhood leukemia: a meta-analysis. Am J Epidemiol 183:1–14

Chauvigne F, Fjelldal PG, Cerda J, Finn RN (2016) Auto-adhesion potential of extraocular aqp0 during teleost development. PLoS One 11:e0154592

Christ E, Korf HW, von Gall C (2012) When does it start ticking? Ontogenetic development of the mammalian circadian system. In: Kalsbeek A, Merrow M, Roenneberg T, Foster RG (eds) Neurobiology of circadian timing, pp 105–118

Clark BW, Di Giulio RT (2012) Fundulus heteroclitus adapted to PAHs are cross-resistant to multiple insecticides. Ecotoxicology 21:465–474

Colbert MW, Rowe T (2008) Ontogenetic sequence analysis: using parsimony to characterize developmental sequences and sequence polymorphism. J Exp Zool B Mol Dev Evol 310:398–416

Colman JR, Twiner MJ, Hess P, McMahon T, Satake M, Yasumoto T, Doucette GJ, Ramsdell JS (2005) Teratogenic effects of azaspiracid-1 identified by microinjection of Japanese medaka (Oryzias latipes) embryos. Toxicon 45:881–890

De Beer G (1958) Embryos and ancestors. Clarendon Press, Oxford

Dickins TE, Rahman Q (2012) The extended evolutionary synthesis and the role of soft inheritance in evolution. Proc R Soc B-Biol Sci 279:2913–2921

Ditty JG, Shaw RF, Fuiman LA (2005) Larval development of five species of blenny (Teleostei: Blenniidae) from the Western Central North Atlantic, with a synopsis of blennioid family characters. J Fish Biol 66:1261–1284

Dubansky B, Rice CD, Galvez F (2017) Biomarkers of aryl-hydrocarbon receptor activity in gulf killifish (*Fundulus grandis*) from the norther Gulf of Mexico following the Deepwater Horizon oil spill. Arch Environ Contam Toxicol 73(1):63–75

Dubansky B, Whitehead A, Miller J, Rice CD, Galvez F (2013) Multi-tissue molecular, genomic, and developmental effects of the Deepwater Horizon oil spill on resident gulf killifish (*Fundulus grandis*). Environ Sci Technol 47:5074–5082

Dubansky B, Whitehead A, Rice CD, Galvez F (2014) Response to comment on "multi-tissue molecular, genomic, and developmental effects of the Deepwater Horizon oil spill on resident gulf killifish (*Fundulus grandis*)". Environ Sci Technol 14(13):7679–7680

Dubansky BH, Dubansky B (2017) Natural development of dermal ectopic bone in the American alligator (*Alligator mississippiensis*) resembles heterotopic ossification disorders in humans. Anat Rec (Hoboken) 301(1):56–76

Duffy TA, Childress W, Portier R, Chesney EJ (2016) Responses of bay anchovy (*Anchoa mitchilli*) larvae under lethal and sublethal scenarios of crude oil exposure. Ecotoxicol Environ Saf 134:264–272

Esser C, Bargen I, Weighardt H, Haarmann-Stemmann T, Krutmann J (2013) Functions of the aryl hydrocarbon receptor in the skin. Semin Immunopathol 35:677–691

Faunes F, Larrain J (2016) Conservation in the involvement of heterochronic genes and hormones during developmental transitions. Dev Biol 416:3–17

Felter SP, Daston GP, Euling SY, Piersma AH, Tassinari MS (2015) Assessment of health risks resulting from early-life exposures: are current chemical toxicity testing protocols and risk assessment methods adequate? Crit Rev Toxicol 45:219–244

Frantzen M, Falk-Petersen I-B, Nahrgang J, Smith TJ, Olsen GH, Hangstad TA, Camus L (2012) Toxicity of crude oil and pyrene to the embryos of beach spawning capelin (*Mallotus villosus*). Aquat Toxicol 108:42–52

Fuiman LA (2002) Special considerations of fish eggs and larvae. In: Fuiman LA, Werner RG (eds) Fishery science: the unique contributions of early life stages. Blackwell Science, Oxford, pp 1–32

Fuiman LA, Poling KR, Higgs DM (1998) Quantifying developmental progress for comparative studies of larval fishes. Copeia 1998:602–611

Garcia GR, Noyes PD, Tanguay RL (2016) Advancements in zebrafish applications for 21st century toxicology. Pharmacol Ther 161:11–21

Goldberg J, Barrasso DA, Agostini MG, Quinzio S (2016) Vocal sac development and accelerated sexual maturity in the lesser swimming frog, Pseudis minuta (Anura, Hylidae). Zoology (Jena) 119:489–499

Gomez C, Ozbudak EM, Wunderlich J, Baumann D, Lewis J, Pourquie O (2008) Control of segment number in vertebrate embryos. Nature 454:335–339

Gomez C, Pourquie O (2009) Developmental control of segment numbers in vertebrates. J Exp Zool 312B:533–544

Gomula A, Koziel S (2015) Post-migration adaptation and age at menarche in the second generation of migrants. Anthropol Anz 72:245–255

Gould SJ (1977) Ontogeny and phylogeny. Belknap Press, Cambridge

Gould SJ, Lewontin RC (1979) Spandrels of San-Marco and the panglossian paradigm–a critique of the adaptationist program. Proc R Soc Lond Ser B-Biol Sci 205:581–598

Grandjean P (2016) Paracelsus revisited: the dose concept in a complex world. Basic Clin Pharmacol Toxicol 119:126–132

Harandi OF, Ambros VR (2015) Control of stem cell self-renewal and differentiation by the heterochronic genes and the cellular asymmetry machinery in *Caenorhabditis elegans*. Proc Natl Acad Sci U S A 112:E287–E296

Heindel JJ, Balbus J, Birnbaum L, Brune-Drisse MN, Grandjean P, Gray K, Landrigan PJ, Sly PD, Suk W, Cory Slechta D, Thompson C, Hanson M (2015) Developmental origins of health and disease: integrating environmental influences. Endocrinology 156:3416–3421

Incardona JP (2017) Molecular mechanisms of crude oil developmental toxicity in fish. Arch Environ Contam Toxicol 73(1):19–32

Incardona JP, Collier TK, Scholz NL (2011) Oil spills and fish health: exposing the heart of the matter. J Expo Sci Environ Epidemiol 21:3–4

Incardona JP, Swarts TL, Edmunds RC, Linbo TL, Aquilina-Beck A, Sloan CA, Gardner LD, Block BA, Scholz NL (2013) Exxon Valdez to Deepwater Horizon: comparable toxicity of both crude oils to fish early life stages. Aquat Toxicol 142:303–316

Jeffery JE, Bininda-Emonds OR, Coates MI, Richardson MK (2005) A new technique for identifying sequence heterochrony. Syst Biol 54:230–240

Jolma IW, Laerum OD, Lillo C, Ruoff P (2010) Circadian oscillators in eukaryotes. Wiley Interdiscipl Rev Syst Biol Med 2:533–549

Keyte A, Smith KK (2012) Heterochrony in somitogenesis rate in a model marsupial, Monodelphis domestica. Evol Dev 14:93–103

Keyte AL, Smith KK (2010) Developmental origins of precocial forelimbs in marsupial neonates. Development 137:4283–4294

Keyte AL, Smith KK (2014) Heterochrony and developmental timing mechanisms: changing ontogenies in evolution. Semin Cell Dev Biol 34:99–107

Laudet V (2011) The origins and evolution of vertebrate metamorphosis. Curr Biol 21:R726–R737

Lema SC, Schultz IR, Scholz NL, Incardona JP, Swanson P (2007) Neural defects and cardiac arrhythmia in fish larvae following embryonic exposure to 2,2′,4,4′-tetrabromodiphenyl ether (PBDE 47). Aquat Toxicol 82:296–307

Lewis PD, Danisman R, Gous RM (2008) Illuminance, sexual maturation, and early egg production in female broiler breeders. Br Poult Sci 49:649–653

Li ZR, Ptak D, Zhang LY, Walls EK, Zhong WX, Leung YF (2012) Phenylthiourea specifically reduces zebrafish eye size. PLoS One 7:14

Lotufo GR, Farrar JD, Biedenbach JM, Laird JG, Krasnec MO, Lay C, Morris JM, Gielazyn ML (2016) Effects of sediment amended with Deepwater Horizon incident slick oil on the infaunal amphipod leptocheirus plumulosus. Mar Pollut Bull 109:253–258

Love JW, Rees BB (2002) Seasonal differences in hypoxia tolerance in gulf killifish, *Fundulus grandis* (Fundulidae). Environ Biol Fish 63:103–115

Mahler GJ, Butcher JT (2011) Cardiac developmental toxicity. Birth Defects Res C Embryo Today 93:291–297

McDonald MD, Gilmour KM, Walsh PJ, Perry SF (2010) Cardiovascular and respiratory reflexes of the gulf toadfish (*Opsanus beta*) during acute hypoxia. Respir Physiol Neurobiol 170:59–66

Meier S, Morton HC, Nyhammer G, Grosvik BE, Makhotin V, Geffen A, Boitsov S, Kvestad KA, Bohne-Kjersem A, Goksoyr A, Folkvord A, Klungsoyr J, Svardal A (2010) Development of Atlantic cod (*Gadus morhua*) exposed to produced water during early life stages effects on embryos, larvae, and juvenile fish. Mar Environ Res 70:383–394

Monsalve GC, Frand AR (2012) Toward a unified model of developmental timing: a "molting" approach. WormBook 1:221–230

Moss EG (2007) Heterochronic genes and the nature of developmental time. Curr Biol 17:R425–R434

Mueller CA, Eme J, Burggren WW, Roghair RD, Rundle SD (2015) Challenges and opportunities in developmental integrative physiology. Comp Biochem Physiol A-Mol Integr Physiol 184:113–124

Nimmo RA, Slack FJ (2009) An elegant miRror: microRNAs in stem cells, developmental timing and cancer. Chromosoma 118:405–418

Oziolor EM, Bigorgne E, Aguilar L, Usenko S, Matson CW (2014) Evolved resistance to PCB- and PAH-induced cardiac teratogenesis, and reduced CYP1A activity in gulf killifish (*Fundulus grandis*) populations from the Houston Ship Channel, Texas. Aquat Toxicol 150:210–219

Oziolor EM, De Schamphelaere K, Matson CW (2016a) Evolutionary toxicology: meta-analysis of evolutionary events in response to chemical stressors. Ecotoxicology 25:1858–1866

Oziolor EM, Dubansky B, Burggren WW, Matson CW (2016b) Cross-resistance in gulf killifish (*Fundulus grandis*) populations resistant to dioxin-like compounds. Aquat Toxicol 175:222–231

Pasparakis C, Mager EM, Stieglitz JD, Benetti D, Grosell M (2016) Effects of Deepwater Horizon crude oil exposure, temperature and developmental stage on oxygen consumption of embryonic and larval mahi-mahi (*Coryphaena hippurus*). Aquat Toxicol 181:113–123

Peterson T, Muller GB (2016) Phenotypic novelty in evodevo: the distinction between continuous and discontinuous variation and its importance in evolutionary theory. Evol Biol 43:314–335

Phillips JB, Westerfield M (2014) Zebrafish models in translational research: tipping the scales toward advancements in human health. Dis Model Mech 7:739–743

Pie HV, Schott EJ, Mitchelmore CL (2015) Investigating physiological, cellular and molecular effects in juvenile blue crab, *Callinectes sapidus*, exposed to field-collected sediments contaminated by oil from the Deepwater Horizon incident. Sci Total Environ 532:528–539

Pourquie O (2011) Vertebrate segmentation: from cyclic gene networks to scoliosis. Cell 145:650–663

Quigley IK, Parichy DM (2002) Pigment pattern formation in zebrafish: a model for developmental genetics and the evolution of form. Microsc Res Tech 58:442–455

Ramakrishnan V (2007) Salinity, pH, temperature, desiccation and hypoxia tolerance in the invasive freshwater apple snail Pomacea insularum. ProQuest dissertations and Theses, The University of Texas at Arlington, Ann Arbor, p 262

Ransom JT, Filbrun JE, Hernandez FJ (2016) Condition of larval Spanish mackerel *Scomberomorus maculatus* in relation to the Deepwater Horizon oil spill. Mar Ecol Prog Ser 558:143–152

Reid NM, Proestou DA, Clark BW, Warren WC, Colbourne JK, Shaw JR, Karchner SI, Hahn ME, Nacci D, Oleksiak MF, Crawford DL, Whitehead A (2016) The genomic landscape of rapid repeated evolutionary adaptation to toxic pollution in wild fish. Science 354:1305–1308

Reid NM, Whitehead A (2016) Functional genomics to assess biological responses to marine pollution at physiological and evolutionary timescales: toward a vision of predictive ecotoxicology. Brief Funct Genomics 15:358–364

Schartl M (2014) Beyond the zebrafish: diverse fish species for modeling human disease. Dis Model Mech 7:181–192

Sibly RM, Baker J, Grady JM, Luna SM, Kodric-Brown A, Venditti C, Brown JH (2015) Fundamental insights into ontogenetic growth from theory and fish. Proc Natl Acad Sci U S A 112:13934–13939

Sirsat SKG, Sirsat TS, Price ER, Dzialowski EM (2016) Post-hatching development of mitochondrial function, organ mass and metabolic rate in two ectotherms, the American alligator (*Alligator mississippiensis*) and the common snapping turtle (*Chelydra serpentina*). Biol Open 5:443–451

Smirthwaite JJ, Rundle SD, Bininda-Emonds OR, Spicer JI (2007) An integrative approach identifies developmental sequence heterochronies in freshwater basommatophoran snails. Evol Dev 9:122–130

Smith KK (2002) Sequence heterochrony and the evolution of development. J Morphol 252:82–97

Spicer JI (2006) A physiological approach to heterochrony. In: Warburton S, Burggren W, Pelster B, Reiber C, Spicer JI (eds) Comparative developmental physiology. Oxford University Press, New York, pp 191–202

Spicer JI, Burggren WW (2003) Development of physiological regulatory systems: altering the timing of crucial events. Zoology 106:91–99

Spicer JI, Rundle SD, Tills O (2011) Studying the altered timing of physiological events during development: It's about time … or is it? Respir Physiol Neurobiol 178:3–12

Stefansson ES, Langdon CJ, Pargee SM, Blunt SM, Gage SJ, Stubblefield WA (2016) Acute effects of non-weathered and weathered crude oil and dispersant associated with the Deepwater Horizon incident on the development of marine bivalve and echinoderm larvae. Environ Toxicol Chem 35:2016–2028

Tammen SA, Friso S, Choi SW (2013) Epigenetics: the link between nature and nurture. Mol Asp Med 34:753–764
Tills O, Spicer JI, Rundle SD (2010) Salinity-induced heterokairy in an upper-estuarine population of the snail Radix balthica (Mollusca: Pulmonata). Aquat Biol 9:95–105
Tollis M, Hutchins ED, Kusumi K (2014) Reptile genomes open the frontier for comparative analysis of amniote development and regeneration. Int J Dev Biol 58:863–871
Tsai TY-C, Choi YS, Ma W, Pomerening JR, Tang C, Ferrell JE (2008) Robust, tunable biological oscillations from interlinked positive and negative feedback loops. Science (New York, NY) 321:126–129
Tsai TYC, Theriot JA, Ferrell JE (2014) Changes in oscillatory dynamics in the cell cycle of early *Xenopus laevis* embryos. PLoS Biol 12:e1001788
Urho L (2002) Characters of larvae–what are they? Folia Zool 51:161–186
Vickaryous MK, Meldrum G, Russell AP (2015) Armored geckos: a histological investigation of osteoderm development in Tarentola (Phyllodactylidae) and Gekko (Gekkonidae) with comments on their regeneration and inferred function. J Morphol 276:1345–1357
Vickaryous MK, Sire JY (2009) The integumentary skeleton of tetrapods: origin, evolution, and development. J Anat 214:441–464
Visvader JE, Lindeman GJ (2008) Cancer stem cells in solid tumours: accumulating evidence and unresolved questions. Nat Rev Cancer 8:755–768
Vornanen M, Hassinen M (2016) Zebrafish heart as a model for human cardiac electrophysiology. Channels 10:101–110
Wankowska M, Polkowska J (2010) The pituitary endocrine mechanisms involved in mammalian maturation: maternal and photoperiodic influences. Reprod Biol 10:3–18
Warkman AS, Krieg PA (2007) Xenopus as a model system for vertebrate heart development. Semin Cell Dev Biol 18:46–53
Warner DA (2014) Fitness consequences of maternal and embryonic responses to environmental variation: using reptiles as models for studies of developmental plasticity. Integr Comp Biol 54:757–773
Webb AB, Oates AC (2016) Timing by rhythms: daily clocks and developmental rulers. Dev Growth Diff 58:43–58
Webb SE, Miller AL (2012) Ca^{2+} signaling during activation and fertilization in the eggs of teleost fish. Cell Calcium 53:24–31
Wei J, Li H, Wang S, Li T, Fan J, Liang X, Li J, Han Q, Zhu L, Fan L, Zhao RC (2014) Let-7 enhances osteogenesis and bone formation while repressing adipogenesis of human stromal/mesenchymal stem cells by regulating HMGA2. Stem Cells Dev 23:1452–1463
Weis JS, Weis P (1997) Aquatic testing with early life stages of killifish. In: Wells PG, Lee K, Blaise C (eds) Microscale testing in aquatic toxicology. CRC Press, Boca Roton, pp 479–490
Weis P, Weis JS (1974) Cardiac-malformation and other effects due to insecticides in embryos of killifish, *Fundulus heteroclitus*. Teratology 10:263–267
Weis P, Weis JS (1977) Methylmercury teratogenesis in killifish, *Fundulus heteroclitus*. Teratology 16:317–325
Whitehead A, Dubansky B, Bodinier C, Garcia TI, Miles S, Pilley C, Raghunathan V, Roach JL, Walker N, Walter RB, Rice CD, Galvez F (2012) Genomic and physiological footprint of the Deepwater Horizon oil spill on resident marsh fishes. Proc Natl Acad Sci 109:20298–20302
Whitworth KW, Symanski E, Coker AL (2008) Childhood lymphohematopoietic cancer incidence and hazardous air pollutants in Southeast Texas, 1995–2004. Environ Health Perspect 116:1576–1580
Whitworth KW, Symanski E, Lai D, Coker AL (2011) Kriged and modeled ambient air levels of benzene in an urban environment: an exposure assessment study. Environ Health 10:21–21
Wilkinson JL, Hooda PS, Barker J, Barton S, Swinden J (2016) Ecotoxic pharmaceuticals, personal care products, and other emerging contaminants: a review of environmental, receptor-mediated, developmental, and epigenetic toxicity with discussion of proposed toxicity to humans. Crit Rev Environ Sci Technol 46:336–381

Wilson GN, Optiz JM, Reynolds JF (1988) Heterochrony and human malformation. Am J Med Genet 29:311–321

Wittig J, Münsterberg A (2016) The early stages of heart development: insights from chicken embryos. J Cardiovasc Dev Dis 3:12

Wolf K (1954) Progress report on blue-sac disease. Prog Fish-Culturist 16:51–59

Wolf K (1957) Experimental induction of blue-sac disease. Trans Am Fish Soc 86:61–70

Yanai I, Peshkin L, Jorgensen P, Kirschner MW (2011) Mapping gene expression in two xenopus species: evolutionary constraints and developmental flexibility. Dev Cell 20:483–496

3 Critical Windows in Animal Development: Interactions Between Environment, Phenotype, and Time

Casey A. Mueller

Abstract

Observable phenotypic traits of an animal are a result of the interaction between the genome and environment. Differences in phenotypic traits between individuals induced by the environment, an indicator of phenotypic plasticity, may have immediate and long-term consequences for individuals, populations, and species. During development, animals are often most responsive or susceptible to changes in their environment, and phenotypic plasticity can be particularly prevalent. It is increasingly apparent that the way in which the environment influences an animal's physiology may differ not just across a species' lifetime but also within a species' ontogeny. Periods of development during which an animal may show greater likelihood of phenotypic changes are termed "critical windows" or "sensitive periods." Across animal taxa, experiments utilize exposures to particular environmental, chemical, or pharmacological stressors at certain time points of development to detect and understand critical windows during development. This chapter examines the emergence of critical windows as an important physiological concept using examples from the literature that span model and non-model invertebrates and vertebrates exposed to a range of environmental conditions. This chapter also outlines considerations for the continued search for critical windows. Critical window experimental designs can range in complexity, and variables such as the timing of exposures, if a single or multiple doses of a stressor are used, and when endpoints are assessed should be considered. A continued focus on critical windows will no doubt contribute to our growing knowledge of the interaction between the environment and physiology during animal development.

C. A. Mueller
Department of Biological Sciences, California State University San Marcos, San Marcos, CA, USA
e-mail: cmueller@csusm.edu

© Springer International Publishing AG, part of Springer Nature 2018

W. Burggren, B. Dubansky (eds.), *Development and Environment*,
https://doi.org/10.1007/978-3-319-75935-7_3

3.1 Introduction

3.1.1 Definition and History of Developmental Critical Windows

Phenotype is the result of the interaction between the genome and environment. Differences in phenotypic traits between individuals induced by the environment, an indicator of phenotypic plasticity, may have immediate and long-term consequences for individuals, populations, and species. During development, animals are often most responsive or susceptible to changes in their environment, and phenotypic plasticity can be particularly prevalent. Recognizing and understanding both the potential positive and negative consequences of plasticity is at the core of developmental physiological research. Chronic exposure to certain environmental conditions throughout development is the classic approach to understanding how environment and physiology interact. However, it is apparent that the way in which the environment influences an animal's physiology may differ not just across a species' lifetime but also within a species' ontogeny. Periods of development during which an animal may show greater likelihood of phenotypic changes are termed developmental "critical windows" or "sensitive periods."

The terms "sensitive period" or "critical period" first appeared in the medical literature in the 1940s and 1950s and became prevalent in the 1970s and beyond, particularly in relation to the required developmental processes that occur at certain times during human development (Vito et al. 1979; Colombo 1982; Johnson and Newport 1989). The idea of sensitive periods has been discussed extensively in relation to sensory development, with critical windows defined as periods when developing neural circuits are particularly sensitive to stimuli and may need signals for normal development to occur (Rice and Barone Jr 2000; Andersen 2003; Hensch 2004; Knudsen 2004; Uylings 2006). Similarly, sensitive periods in cardiovascular, endocrine, reproductive, respiratory, and immune development are discussed in relation to the developmental trajectories of these systems (Barr Jr et al. 2000; Dietert et al. 2000; Pryor et al. 2000; Selevan et al. 2000; Andersen et al. 2006). More recently, mammalian studies are concerned with identifying critical windows as they relate to developmental origins of diabetes (Reece et al. 1996; Lumey et al. 2015), hypertension (Lusk et al. 2015; Svitok et al. 2015), and neurodevelopmental disorders (Meredith 2015).

Much of the earliest work on critical windows occurred in humans, or with a human focus, but critical windows have also become a fundamental concept of comparative physiology and animal toxicology. Across animal taxa, experiments that utilize exposure to particular environmental, chemical, or pharmacological stressors at certain time points of development are used as a means to detect and understand critical windows during development. Detecting such periods is vital for understanding how the environment may influence the developmental phenotype both in the short-term (during embryonic or larval development) and long-term (mature life stages).

3.1.2 Critical Windows Are Central to the Interaction Between Development, Physiology, and Environment

The environment may exert larger impacts on physiological systems at particular time points due to the developmental status of the animal. Exposure to a stressor that itself may play a key role in developmental processes can be used to uncover the developmental trajectory of an animal or system. Retinoic acid, for example, plays a key role in axis formation and limb patterning, and thus, its application during certain stages of development can be used to infer the series of developmental events that constitute these important developmental processes. Embryos of the African clawed frog (*Xenopus laevis*) are most sensitive to retinoic acid in early gastrulation stages, with significant truncation of the body axis. This sensitivity is related to the disruption of the expression of cement-gland-specific genes that are normally expressed in late gastrula and early neurula stages (Sive et al. 1990). Thus, retinoic acid exposure illustrates that early in progression of the body axis patterning the process is somewhat plastic, and this may be important for subsequent development.

Periods of developmental plasticity or susceptibility can be defined by piecing together the findings of numerous studies that cover multiple developmental stages. Again using the example of retinoic acid, separate studies in rodents have examined the production of morphological abnormalities following retinoic acid exposure at gastrulation (Vickers 1985; Sulik et al. 1995) and organogenesis (Kochhar et al. 1984). Assessment of exposures and doses of retinoic acid used in these studies, and the resultant effects, indicates that retinoic acid sensitivity decreases as development proceeds in rodents. In some instances, an individual researcher or research lab has pieced together changes in developmental responses to a stressor across various published works. An excellent example of this is in a series of papers from 1956 to 1971 that examined the hatchability of embryonic chickens (*Gallus gallus*) following exposure to hypoxia, hyperoxia, and hypercapnia at particular development ages (Taylor et al. 1956, 1971; Taylor and Kreutziger 1965, 1966, 1969). After examining these different exposure studies, a number of conclusions on the stage-specific effects of various respiratory gas exposures on chicken embryo hatchability can be made. Embryos show a general trend for increased tolerance with later hypercapnic exposures, but no significant sensitive period for the effect of hypoxia on hatchability (hypoxia does exert stage-specific morphological and physiology effects on chicken embryos, discussed in more detail in Sect. 3.2.1). Hatchability is also sensitive to hyperoxia exposure during days 5–8 and particularly during days 17–21. Thus, the eventual hatchability of embryos is influenced most by early hypercapnia and late hyperoxia exposure.

With recognition of the importance of critical windows, a more systematic approach has emerged in which individual studies perform multiple exposures during particular, distinct periods of development to assess variability in sensitivity across development. The timing of exposures is often determined based on significant developmental events, such as hatching, birth, metamorphosis, and molting. Thus, many studies examine the sensitivity of an animal during the embryo versus larval period (e.g., Fent and Meier 1994; Bridges 2000; Greulich and Pflugmacher 2003). Yet, within these periods, differential susceptibility may also occur as an animal develops and their physiological status progresses so that the extent of their cellular differentiation, organogenesis, and enzymatic activity influences how sensitive or responsive they may be to environmental stressors. For example, animals may show varied responses to environmental toxicants due to the stage of maturation of the immune system and the developmental status of immune cells and organs (Dietert et al. 2002). In light of this, developmental milestones within an individual physiological system are now being considered as a means for dividing up development into different windows of exposure (Dietert et al. 2000; Landreth 2002).

The majority of critical window studies use a design in which a subset of animals are raised in control conditions, a subset of animals are chronically exposed to the stressor of interest, and a subset of animals are exposed to the stressor during distinct, separate windows (Fig. 3.1). These windows may be chosen based on developmental events (Aronzon et al. 2011; Eme et al. 2015; Mueller et al. 2015b), or they may be arbitrary divisions of development (Dzialowski et al. 2002; Chan and Burggren 2005; Oxendine et al. 2006; Hanlon and Parris 2014). In either case, this approach is a tried and trusted method for detecting periods of susceptibility or plasticity in the physiology, morphology, and biochemistry of an animal.

This chapter examines how critical windows are detected using examples from the literature that span model and non-model invertebrates and vertebrates exposed to a range of environmental conditions. The examples reflect areas of research in which critical windows have received the most attention, as well as areas in which there is an opportunity to undertake a search for critical windows. Exposures to important naturally occurring environment variables, such as hypoxia and temperature, during distinct developmental periods have been undertaken across animal

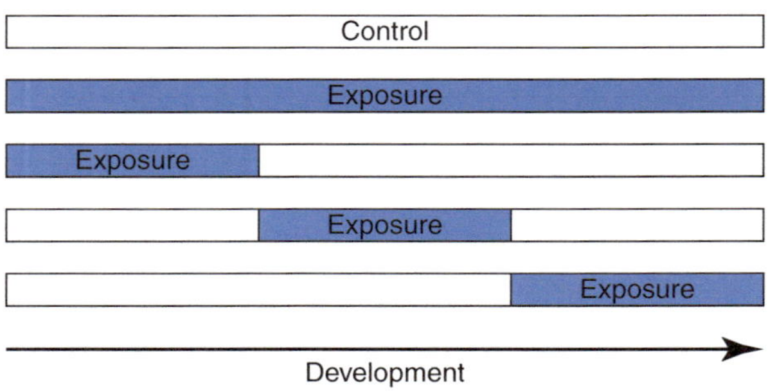

Fig. 3.1 Schematic of a simple critical window experimental design. The design includes a control treatment, in which animals are exposed to control conditions throughout development; a chronic exposure treatment, in which animals are exposed to an environmental stressor throughout development; and treatments with exposures during distinct, separate periods of development. In this instance the windows of exposure cover the first, middle, and last third of development

groups. The field of environmental toxicology also has a focus on understanding periods of susceptibility, particularly exposures to heavy metals, pesticides, and endocrine-disrupting chemicals in developing invertebrates and aquatic vertebrates. The concept of critical windows is central to appreciating the importance of the environment during development, and this chapter outlines considerations for the continued search for developmental periods of sensitivity or plasticity that will ensure critical window research remains central to the field of developmental biology.

3.2 Stage-Specific Sensitivity to Naturally Occurring Environmental Stressors

3.2.1 Respiratory Gases

Hypoxia and hypercapnia are naturally occurring environmental stressors for many developing animals, both in aquatic and terrestrial environments. Experimental manipulation of oxygen and carbon dioxide levels can reveal ecological implications for animals (Petranka et al. 1982; Rombough 1988; Latham and Just 1989; Kam 1993; Mills and Barnhart 1999; Seymour et al. 2000; Mueller et al. 2011a) and is also very useful for assessing the physiology of the developing respiratory and cardiovascular systems (Tazawa 1981; Tazawa et al. 1992; Dzialowski et al. 2002; Bavis 2005; Crossley II and Altimiras 2005; Liu et al. 2006; Bavis and Mitchell 2008; Ferner and Mortola 2009; Eme et al. 2011a, b, 2013, 2014; Bavis et al. 2013).

Hypoxia has been used to examine the development of respiratory control in rats during the first 3 weeks of postnatal development (Wong-Riley and Liu 2005; Liu et al. 2006, 2009; Liu and Wong-Riley 2010). The hypoxic ventilatory response following 5 min of exposure to hypoxia (10% oxygen) is blunted on days 12–16, and particularly on day 13. Measurement of respiratory variables, including minute ventilation, breathing frequency, and tidal volume throughout prenatal development, indicates that respiratory control undergoes a considerable shift at this time, with respiratory frequency peaking on day 13 (Liu et al. 2006). Additionally at this time, body temperature abruptly increases, metabolic rate is heightened in normoxia but comparatively reduced in hypoxia (Liu et al. 2009), the brain stem respiratory nuclei demonstrate a transient dominance of inhibitory over excitatory neurotransmission (Wong-Riley and Liu 2005), and serotonin transmission decreases (Liu and Wong-Riley 2010). In light of these findings, days 12–16 most likely represent a critical window during which numerous physiological and neurochemical changes occur simultaneously, which may result in a reduction in respiratory modulation that causes animals to be less responsive to respiratory stressors. This period in rats is comparable to 2–4 months postnatal development in humans, a time represented by the highest incidence of sudden infant death syndrome (SIDS) (Hakeem et al. 2015). Thus, understanding the developmental changes that occur in rodents during this time may shed light on the physiological mechanisms that lead to SIDS.

Avian and reptile embryos are often used as substitutes for mammalian fetuses for understanding developmental physiology, particularly as they are separate from maternal influences. In ovo hypoxia exposure is easy to undertake and ecologically relevant in many instances. Many chronic exposure studies in bird and reptile embryos have demonstrated changes in morphological and physiological phenotype of embryos, particularly in response to hypoxia (Wangensteen et al. 1974; McCutcheon et al. 1982; Crossley II et al. 2003; Crossley II and Altimiras 2005; Copeland and Dzialowski 2009; Eme et al. 2011a, b, 2013, 2014). However, in recent years studies have examined if there are particular critical windows for hypoxia sensitivity (Dzialowski et al. 2002; Chan and Burggren 2005; Tate et al. 2015). The physiology of chicken embryos is generally more hypoxia sensitive as development progresses (Grabowski and Paar 1958), and hyperoxia during days 14–18 of the 21-day incubation period produces greater decreases in body mass, hematocrit, and lung mass compared to embryos exposed during days 7–18 (Xu and Mortola 1989). Thus, later stage chicken embryos are more sensitive to both low and high environmental oxygen. This finding is not surprising considering the increase in metabolic activity as embryos approach hatching (Romanoff 1967).

Hypoxia exposure during distinct windows of chicken development reveals some interesting time-specific phenotypic changes. For example, embryos exposed to 15% oxygen during days 1–6 have reduced body mass and a lower oxygen consumption rate on day 12 compared to normoxic embryos, but body mass and metabolism recover by hatching. Likewise, embryos incubated in hypoxia during days 12–18 also have reduced mass, with only a lower dry mass persisting at hatch. Embryos exposed during days 6–12 also have reduced mass on day 12 and 18 but recover by hatching. However, these embryos show an altered respiratory phenotype that persists to hatching. They are initially able to cope well with hypoxia but show an eventual decrease in oxygen consumption at hatching compared to normoxic embryos (Dzialowski et al. 2002). Thus, the middle third of embryonic development of chickens appears to be a critical window for oxygen consumption; however, hypoxic-induced alterations in organ and size occur throughout embryonic development, and in some instances hatchlings display normal morphological sizes following return to normoxia. As the chorioallantoic membrane, the main gas exchange organ of these embryos, increases to cover a significantly larger portion of the inner eggshell during this time (Ackerman and Rahn 1981), it is likely to be impacted by hypoxia and may alter the metabolic phenotype of the embryos. While the middle third of development is important for oxygen consumption, the last third of incubation is a critical window during which hypoxia blunts ventilation of chicken embryos. This is thought to be due to hypoxia influencing the normal development of the carotid bodies that become functional during this time (Ferner and Mortola 2009).

Different windows for hypoxia sensitivity for oxygen consumption and ventilation in chicken embryos indicate how different components of a system may show critical windows that correspond, overlap, or occur at separate times of development. Exposure to hypoxia (10% O_2) between 50 and 70% of development in the common snapping turtle (*Chelydra serpentina*) enlarges heart size relative to body size. Yet, the critical window for baseline mean arterial pressure is broader, with mean arterial pressure decreasing when hypoxia exposure occurs from 20 to 70% of development (Tate et al. 2015). In this instance, the sensitivity of the physiological function of a system is more extensive, and morphological effects of a stressor are confined to a smaller proportion of development. It is not known if the critical windows of these two components influence each other.

3.2.2 Temperature

Temperature has pervasive effects on all biological processes. Animals that develop in utero (mammals) or with parental care (many birds) are somewhat protected from variations in environmental temperature. For ectothermic animals, however, including invertebrates, fishes, amphibians, and reptiles, temperature can drive survival, development times, growth, metabolism, and sex. These temperature effects have been assessed by incubating developing animals in different constant temperatures and measuring physiological functions such as growth rate (Sweeney and Schnack 1977; Angilletta et al. 2004), development rate (McLaren and Cooley 1972; Herzig and Winkler 1986; Rombough 2003), and oxygen consumption rate (Kuramoto 1975; Feder 1985; Kamler et al. 1998; Gillooly et al. 2001; Mueller et al. 2011b).

Periods of increased thermal sensitivity during fish development, when physiological variables are particularly plastic, are of increased interest due to predicted higher temperatures due to climate change. Atlantic salmon (*Salmo salar*) embryos raised to a larval feeding stage in chronically heated water, 4.6 °C above ambient (2.6 or 7.2 °C), display higher maximum growth rates compared to control fish when both groups are raised after feeding at common temperatures (Finstad and Jonsson 2012). Zebrafish (*Danio rerio*) raised from hatching to adulthood at 27 °C, but incubated as embryos at 22 °C, 27 °C (control), or 32 °C, show increased thermal sensitivity to exercise performance at temperatures different than respective embryonic incubation temperatures. Furthermore, both high and low temperature incubation groups display better exercise performance than control fish at 16 °C (Scott and Johnston 2012). The effects of temperature changes during embryonic development are also of interest. Oxygen consumption, heart rate, and survival of lake whitefish (*Coregonus clupeaformis*) embryos are reduced following a temperature shift at the end of gastrulation compared to embryos in constant temperatures. In comparison, when the temperature shift occurs at the end of organogenesis, the embryos show no change in metabolism or heart rate (Eme et al. 2015; Mueller et al. 2015b). Thus, lake whitefish embryos show greater plasticity in these variables with a temperature change during organogenesis than later in development before hatch. The thermal environment in embryogenesis up to the eyed embryo stage also persistently altered the muscle phenotype of Atlantic salmon (Macqueen et al. 2008). Thus, the processes of gastrulation and early organogenesis may be particularly important periods of thermal exposure that induces metabolic or cardiac phenotypic plasticity.

3.2.2.1 Temperature-Dependent Sex Determination

An excellent example of how genome-environment interaction can determine the phenotype of an animal is environmental sex determination. Temperature during embryogenesis or larval development is the prevailing environmental factor that is involved in environmental sex determination in ectothermic vertebrates, such as reptiles, amphibians, and fishes (Hillman 1977; Conover 1984; Korpelainen 1990; Baroiller and D'cotta 2001; Sarre et al. 2004).

The period during development in which temperature-dependent sex determination occurs in turtles provides some of the first examples of critical window studies in comparative physiology (e.g., Yntema 1979; Bull and Vogt 1981; Pieau and Dorizzi 1981; Yntema and Mrosovsky 1982). The critical window for sex determination is assessed by examining resultant sex ratios after shifting eggs between male- and female-producing temperatures during certain stages of development. In the common snapping turtle, the male producing temperature is 26 °C, while the female producing temperatures are 20 and 30 °C (Yntema 1976). Yntema (1979) demonstrated that developmental stages 14–19 (stages defined by cranial, neck, and forelimb formation; Yntema 1968) are temperature sensitive for female determination at 30 °C, whereas the window for female determination at 20 °C is during stages 14–16. Thus, while the critical window at 20 °C covers less developmental time, the total chronological time for the critical window at 20 °C (21 d) is greater than at 30 °C (12 d) due to a relatively slower development rate at 20 °C. Sensitive stages of male determination at 26 °C in the common snapping turtle are influenced by incubation temperature prior and subsequent to the thermally sensitive windows (stages 14–19 or 14–16). When embryos are incubated at 30 °C prior to stage 14, stages 14–19 are the male-producing critical window for embryos shifted to 26 °C, whereas embryos incubated at 20 °C prior to stage 14, stages 14–16 are the male-producing critical window for embryos shifted to 26 °C (Yntema 1979). The critical thermal window for sex determination occurs between stages 14–20 in the loggerhead sea turtle (*Caretta caretta*), stages 16–20 in the red-eared slider (*Trachemys scripta elegans*), and stages 16–22 in the map turtle (*Graptemys ouachitensis*), painted turtle (*Chrysemys picta*), and European pond turtle (*Emys orbicularis*) (Bull and Vogt 1981; Pieau and Dorizzi 1981; Yntema and Mrosovsky 1982; Wibbels et al. 1991). These critical windows represent approximately 15–20% of total incubation (Pieau and Dorizzi 1981) and occur approximately during the middle third of incubation prior to sexual differentiation of the gonads (Wibbels et al. 1991). The similarity in the sensitive stages across turtles indicates that the window for temperature-dependent sex determination is conserved across species in this clade. All extant crocodilians studied also show temperature-dependent sex determination, and in the American alligator (*Alligator mississippiensis*), similar to the turtles discussed above, the critical window for temperature-dependent sex determination covers 20% of incubation, during weeks 2 and 3 of the 10-week incubation period (Ferguson and Joanen 1982). This period is earlier in incubation compared to the turtles, but alligators are laid at a more advanced embryonic stage and so the critical window is quite similar (Ferguson and Joanen 1982).

Interestingly, Pieau and Dorizzi (1981) found the length of exposure depends upon the stage within the critical period at which the temperature shift occurs. That is, if the temperature shift is late in the critical window, a longer exposure to that temperature is required. Additionally, different incubation temperatures, irrespective of if male- or female-producing (Wibbels et al. 1991), can have differing influence on sex differentiation, and this may influence the exact width of the critical window. There is some suggestion that critical windows for sexual differentiation are not so important, and instead temperature drives the expression of certain genes that will produce variable ratios of one sex over another (Deeming et al. 1988). However, recent work has found that genes that may be involved in sex differentiation have temperature-dependent expression, and this alteration in regulation often occurs during the critical window for sex determination (Kettlewell et al. 2000; Murdock and Wibbels 2002).

Fish also show temperature-dependent sex determination. The mangrove killifish (*Rivulus marmoratus*) is the only self-fertilizing hermaphrodite vertebrate, and hermaphrodites are the predominant form in nature (Turner et al. 1992). However, while hermaphrodites are produced at temperature of 25 °C or above, exposures to 20 °C or under during a small window prior to hatching will produce males (Harrington 1968). As males are scarce in nature, however, it seems unlikely that killifish are often exposed to temperatures below 20 °C during this critical window for male production. In Atlantic silverside (*Menidia menidia*), temperature-dependent sex determination is correlated with body size, and, as found for the common snapping turtle, the timing and length of the critical window is dependent on temperature (Conover and Fleisher 1986). When shifted between 15 and 21°C, the sex ratio of Atlantic siverside larvae ranging in size from 8–21 mm is intermediate between that observed at constant 15°C (1:1, male:female) and constant 21°C incubation (9:1, male:female). At 21 °C the window for sex determination is shorter both in terms of total development time and in relation to the developmental stage of the larvae (Conover and Fleisher 1986). Thus, the sex ratio of silverside is determined by a combination of the overall incubation temperature and if larvae experience a temperature change during the critical window.

3.2.3 Salinity

Salinity is increasing in many freshwater systems as a result of clearance of native vegetation, agricultural irrigation, and rising groundwater (Williams 2001), and this has led to a focus on salinity tolerance for a range of freshwater aquatic organisms (Kefford et al. 2003; Karraker et al. 2010). However, whether these organisms are particularly sensitive to increased salinity at particular developmental time points is still largely unstudied. A study examining the effect of salinity during certain windows of exposure on survival, growth, and development of the salt-adapted brine shrimp (*Artemia franciscana*) is an example of a search for a critical window for salinity tolerance. Brine shrimp survival is highest following exposure to low salinity (10 and 20 ppt) compared to high salinity (40 and 50 ppt) early in development, but this difference disappears when animals are exposed later in development (Burggren and Mueller 2015). Similarly, differences in growth and maturation between salinities are greatest following exposure during days 1–6 of the 15-day development period (Mueller et al. 2016). Thus, the earliest brine shrimp instar stages are most sensitive to changes in salinity. This early sensitivity may be related to an immature osmoregulatory ability due to the absence of adult osmoregulatory organs. Early instars have a neck organ and renal maxillary gland that are thought to serve as the main osmoregulatory organs (Croghan 1958; Conte 1984), but the absence of full development swimming appendages and branchiae may result in increased salinity sensitivity (Mueller et al. 2016). Similarly, the earliest instars of *A. franciscana* also show greater sensitivity to an acidic environment compared with adults (Doyle and McMahon 1995), and thus, morphological and physiological changes during this period may be particularly important in brine shrimp environmental tolerance.

Studies that relate tolerance to environmental stressors to developmental status and physiological function provide an opportunity to understand the mechanism behind periods of sensitivity. Tolerance of low salinity in American lobster (*Homarus americanus*) and Japanese tiger shrimp (*Penaeus japonicus*) decreases during larval development, is lowest at metamorphosis, and increases during postlarval development (Charmantier et al. 1988). It is possible that the late larval stages, and the process of metamorphosis, are most sensitive to salinity changes. Charmantier et al. (1988) correlated high salinity sensitivity to low osmoregulatory ability, particularly during metamorphosis, when both species shift from osmoconformers to adult type osmoregulators. Thus, the shift in osmoregulatory ability defines the critical window for salinity tolerance in these species.

3.2.4 Pathogens

The sensitivity of an animal to pathogens is related to its resistance (ability to resist infection following exposure) and tolerance (ability to survive damage following infection). Resistance and tolerance may change throughout development resulting in critical windows of high pathogen susceptibility. For example, rainbow trout (*Oncorhynchus mykiss*) fry challenged with the myxozoan parasite *Myxobolus cerebralis* exhibit at least a 95% infection rate following exposure at 1, 3, 5, 7, or 9 weeks post hatch. In comparison, the prevalence of infection in Chinook salmon (*O. tshawytscha*) decreases from 100 to 90% in 1- and 3-week old fry, respectively, to just 1.7% in 9-week old fry (Sollid et al. 2003). Despite differences in infection rates across fry development between the two species, the disease states, such as whirling behavior and blacktail, exhibited by both species decreases as fry development proceeds (Sollid et al. 2003). Thus, early fry stages have both low resistance and tolerance to disease, the tolerance of rainbow trout increases during development, and both resistance and tolerance increases in Chinook salmon.

Disease is a common cited cause of worldwide amphibian declines (Stuart et al. 2004). Thus, assessment of critical windows of disease susceptibility will contribute to understanding the effect of disease on amphibian populations, particularly as frog tadpoles demonstrate stage-specificity sensitivity to pathogens. Mortality and abnormalities in the Pacific chorus frog (*Pseudacris regilla*) following infection with the nematode parasite (*Ribeiroia ondatrae*) is greatest in pre-limb (stages 23–24; Gosner 1960) and early limb (Gosner stages 26–30) tadpoles, and this sensitivity steadily decreases until metamorphosis (Johnson et al. 2011). Similarly, mortality of northern leopard frog (*Lithobates pipiens*) tadpoles exposed to the same parasite is also highest following exposure during pre-limb stages (Gosner stage 24–25), and abnormalities are highest following exposure during limb-bud stages (Gosner stage 27–28) (Schotthoefer et al. 2003). Additionally, the type of abnormalities displayed by the Pacific chorus frog is stage-specific. When infection occurs prior to limb development, the predominant abnormality is missing limbs while later exposure results in extra limbs or abnormal limb projections (Johnson et al. 2011). The abnormalities from such stage-specific exposures can be related to abnormalities seen in the wild, which may have implications for conservation efforts.

3.3 Critical Windows for Exposure to Environmental Contaminants

3.3.1 Ethanol

While not an environmental contaminant in the traditional sense, the effect of ethanol on mammalian development is of interest when examining the consequences of alcohol consumption during human pregnancy. Fetal alcohol syndrome (FAS) in mammals includes growth deficiency, cognitive impairment, and craniofacial abnormalities following alcohol consumption by the mother (Mattson and Riley 1998). Whether FAS is stage-specific has been examined in rats and primates. Short, early exposures produce neurobehavioral and morphological abnormalities in rhesus monkeys (*Macaca mulatta*) and the southern pig-tail macaque (*Macaca nemestrina*), indicating early embryogenesis may represent a critical window for FAS (Clarren et al. 1992; Schneider et al. 2001). In contrast, hyperactivity of rats is highest following ethanol exposure during the later half of gestation (Tran et al. 2000), but learning does not appear to be affected by ethanol exposure at various time points throughout development (Cronise et al. 2001). Thus, specific critical windows for certain neural and morphological effects of FAS may be species-specific in mammals.

The search for critical windows of exposure for mammalian-relevant stressors, such as ethanol, is often undertaken in lower vertebrates in which exposure occurs in ovo. For example, zebrafish mortality is highest after ethanol exposure during gastrulation, while morphological malformations and a reduction in larval swimming performance are most severe after ethanol exposure during subsequent embryonic organogenesis (Ali et al. 2011). Likewise, different endpoints in Japanese medaka (*Oryzias latipes*) embryos display different potential critical windows for ethanol exposure. Embryos exposed to ethanol in the first and last third of development (0–3 and 6–9 dpf (days post fertilization)) have the highest incidence of reduced head width and body length. However, the incidence of apoptosis decreases with early exposure and increases with late exposure, suggesting 6–9 dpf may be when medaka are most sensitive to ethanol (Oxendine et al. 2006).

3.3.2 Heavy Metals

The lethal and sublethal effects of heavy metals in the environment have been studied across a range of animal groups (Pérez-Coll and Herkovits 1990; Mariño-Balsa et al. 2000; Bunn et al. 2001; Lee et al. 2001). Exposure of heavy metals during development of aquatic invertebrates in particular has implications not only for the invertebrates themselves but for higher trophic levels (Lavolpe et al. 2004). Metals such as mercury, copper, zinc, and lead are toxic during development and alter developmental variables such as survival, growth, morphological development, swimming, feeding behavior, and osmoregulation (DeCoursey and Vernberg 1972; Wong et al. 1993; Bambang et al. 1995; Itow et al. 1998; Lavolpe et al. 2004). Lead is most toxic during the first half of embryonic development of the estuarine crab (*Chasmagnathus granulatus*). However, exposure to copper and zinc during the second half of embryonic development of *C. granulatus* produces a higher incidence of eye abnormalities and chromophore hypopigmentation (Lavolpe et al. 2004). Likewise, exposure of embryos of the grass shrimp (*Palaemonetes pugio*) to copper late in embryonic development increases mortality and eye abnormalities compared to exposure during early embryo stages (López Greco et al. 2002). The increased sensitivity to copper and zinc in later embryonic development is thought to be related to the increased permeability of the egg envelope as the embryo approaches hatching (Glas et al. 1997; López Greco et al. 2002; Lavolpe et al. 2004). However, exposure during post hatching stages has not been undertaken in these species to assess if removal of the egg envelope increases sensitivity. Zinc exposure during gastrulation stages of the embryonic development of American horseshoe crabs (*Limulus polyphemus*) produces the highest percentage of malformations (17%), including double embryos and segment-defective embryos, with the lowest incidence of malformations (1–2%) occurring prior to hatch (Itow et al. 1998). Likewise, embryonic American horseshoe crabs are also highly sensitive to an organotin compound, a tin-containing substance used in biocides and antifouling paints, with the incidence of abnormalities decreasing from 6–11 to 3% at the end of gastrulation. A second peak in abnormalities also occurred prior to first embryo molt before decreasing again prior to hatch (Itow et al. 1998). Japanese horseshoe crabs (*Tachypleus tridentatus*) also show the same two peaks in sensitivity to organotin compounds as their American counterparts (Itow et al. 1998). Thus, early embryonic development until gastrulation and the period around the first embryo molt are likely critical windows for toxicant sensitivity in this group of crabs. Metal toxicity also varies during marine invertebrate larval development. Larvae of the common prawn (*Palaemon serratus*) and fiddler crab (*Uca pugilator*) are increasingly sensitive to mercury exposure (DeCoursey and Vernberg 1972; Mariño-Balsa et al. 2000), whereas copper toxicity decreases during larval development of the Japanese tiger shrimp (Bambang et al. 1995).

With the increasing occurrence of deformed individuals in the environment, developmental stages of amphibians are also used extensively to assess developmental toxicity of a number of aquatic contaminants (Degitz et al. 2000). These studies highlight that amphibians often exhibit stage-specific sensitivity to a range of toxicants. Embryos of the South American toad (*Rhinella arenarum*) have a reduced body size, curved bodies, and tail, fin, and gill deformities following lead exposure at any stage of embryogenesis. However, exposure in the middle of embryonic development during neuromuscular activity results in a significant decrease in survival and higher incidence of malformations than at other stages of exposure (Pérez-Coll and Herkovits 1990). Similarly, western-clawed frogs (*Xenopus tropicalis*) are most sensitive to organotin compounds in the middle stages of embryonic development. During this time, normal fin formation is disrupted and other abnormalities include an enlarged proctodaeum and pigment loss in melanophores (Yuan et al. 2011).

Both bird and mammal embryos show increased immunotoxicity in response to lead following exposure during particular periods in ovo or in utero. Rat offspring have higher levels of lead in their blood and bone after the mother is given water-containing lead during gestational days 15–21 compared to days 3–6 (Bunn et al. 2001). Despite no differences in consumption, mothers have higher blood lead levels when exposed late in their pregnancy, which, together with a potential higher transfer between mother and fetus, leads to greater suppression of the cellular immune response in offspring following late gestation exposure. In chicken embryos exposed to sublethal lead levels, stage-specific immunotoxic effects occur (Lee et al. 2001). Lead exposure on day 5, 7, and 9 of incubation affects primarily macrophages, while exposure on day 12 results in a loss of T-cell immune function. This pattern may be due to lead interrupting the migration of T-cells from the bone marrow to the thymus, a process that only occurs from day 12 (Lee et al. 2001).

3.3.3 Endocrine Disrupting Chemicals

Both natural and man-made chemicals that mimic hormones can disrupt normal development and reproduction in a range of animals (Tyler et al. 1998). This has been of particular focus in organisms that are often used as indicators of environmental health and in toxicological assays, such as aquatic invertebrates, amphibians, and fish. Water fleas (*Daphnia magna*) are parthenogenetic, and females will produce either all female broods during times of abundant resources or male broods when resources become limited. Male offspring will sexually reproduce with females to produce eggs that can enter diapause until conditions improve. The switch from producing female to male broods is responsive to multiple environmental cues (Stross and Hill 1965; Carvalho and Hughes 1983) but is also under endocrine control by the natural crustacean hormone, methyl farnesoate (Olmstead and Leblanc 2002). Chronic exposure to methyl farnesoate causes male brood production, but exposure must occur during ovarian egg maturation, and the 60–72 h period of ovarian egg development in particular produces the most male broods (Olmstead and Leblanc 2002).

Endocrine disruption during development can also affect single-sex dimorphisms, and help reveals how such dimorphisms are regulated. Two critical windows have been detected for the development of male dimorphism in dung beetles (*Onthophagus taurus*). Dung beetles exhibit male horn length dimorphism, in which larvae that experience favorable conditions grow large and produce horns while larvae in poor conditions do not (Hunt and Simmons 1997). Exposure to a juvenile hormone analog, methoprene, late in the third instar stage, once larvae have ceased feeding prior to metamorphosis, can switch the trajectory of male development. That is, exposure to methoprene during this window can lead to small males producing horns (Emlen and Nijhout 1999). Furthermore, exposure to methoprene earlier in development, during the late feeding stage, increases the body size threshold at which horns are developed so that fewer males develop horns (Emlen and Nijhout 2001). Exposure to methoprene during these two periods suggests the first period is the window during which the fate of the males, whether horned or hornless, is determined, while the second period represents the time during which horn growth is regulated so that it matches body size (Emlen and Nijhout 2001).

Fish in particular are often used to infer toxic effects in higher animal groups, as the molecular processes involved in responses to toxic substances are highly conserved across vertebrates (Ankley and Johnson 2004). The effects of endocrine-disrupting chemicals on fish development are of particular interest from a toxicological perspective, especially considering the plasticity in fish gonadal development (van Aerle et al. 2002). The effects of hormones on developing stages are also useful for sexual reversal techniques used to optimize commercial production of farmed species (Hunter and Donaldson 1983; Chiasson et al. 2008). Exposure to estrogens and androgens during early larval development of zebrafish, medaka, fathead minnow (*Pimephales promelas*), Mozambique tilapia (*Tilapia mossambica*), and Egyptian mouthbrooder (*Hemihaplochromis multicolor*) influences sexual differentiation and gonadal development during a period in which the reproductive system is particularly plastic (Nakamura and Takahashi 1973; Hackmann and Reinboth 1974; Koger et al. 2000; van Aerle et al. 2002; Ankley and Johnson 2004; Maack and Segner 2004). When exactly the critical window for endocrine disruption or sex differentiation occurs is species-specific. Fathead minnow are most sensitive to ethinylestradiol exposure during days 10–15 of their 100-day post-hatch development (van Aerle et al. 2002). Likewise, zebrafish are most sensitive to ethinylestradiol during 43–71 dpf of the 190 dpf development, a time during which the hermaphrodite gonad differentiates into testes or ovary (Maack and Segner 2004). Another study found zebrafish are most sensitive to ethinylestradiol during a similar window from approximately 22–62 dpf (Andersen et al. 2003). In some instances, such as in the common carp (*Cyprinus carpio*), the period prior to sexual differentiation is also a sensitive period for endocrine disruption (Gimeno et al. 1997). In comparison, sexual differentiation in marine species occurs after metamorphosis during juvenile development (Hendry et al. 2002; Chiasson et al. 2008). Thus, these fish are likely to display a later critical window for hormone sensitivity.

Amphibians exhibit heightened sensitivity to endocrine disruptors during metamorphosis, a period that is characteristic of large changes in the synthesis and action of endogenous hormones, such as thyroid hormone (Kikuyama et al. 1993; Hayes 1997). The northern leopard frog is developmentally delayed following ethinylestradiol exposure during the middle of metamorphosis (Gosner stages 30–36; Gosner 1960), and this delayed development persists until metamorphic climax (Hogan et al. 2008). This sensitivity corresponds to the onset of thyroid gland function and a rise in thyroid hormone levels; thus, estrogens may modulate thyroid action during this metamorphic critical window. In comparison, ethinylestradiol exposure during the early period of metamorphosis (stages 27–30) produces a female-biased sex ratio compared to unexposed tadpoles, and this is related to the period of gonadal differentiation (Hogan et al. 2008). Thus, ethinylestradiol can induce different windows of sensitivity for different physiological parameters, and this is dependent upon the trajectory of certain developmental processes. Examining critical windows for endocrine disruption during fish and amphibian development allows straightforward toxicological assays to be completed in a relatively short time frame. Such research can provide a basis for quantifying the physiological and ecological impacts of endocrine disrupters (Ankley and Johnson 2004).

3.3.4 Sensitivity of Embryo Versus Larval Stages

In light of the many environmental stressors, animal models, physiological systems, and stages of development assessed across critical window studies, it is difficult to make a broad statement about when during development animals are most sensitive to environmental stressors. However, critical windows have been frequently demonstrated across vertebrate and invertebrate taxa, suggesting that more complex experimental designs (Figs. 3.2, 3.3, and 3.4) may better represent the effects of environmental stressors on physiological phenotypic plasticity, compared to classic experimental designs (Fig. 3.1). It can be argued that later developmental stages will show increased tolerance to stressors due to advancement of their physiological systems. Embryos, as the earliest developmental stage, may be considered the most sensitive to stress. However, for those species that develop in ovo, the egg itself may provide some protective benefits, resulting in embryos being more tolerant than post-hatch stages (Van Leeuwen et al. 1985; Green et al. 1986; Fent and Meier 1994).

The egg chorion of invertebrates, fish, and amphibians, also known as the capsule or envelope, is thought to serve as a potential barrier against environmental contaminants. For example, the chorion of fish eggs can absorb up to 98% of cadmium and 70% of zinc due to binding between the metals and proteins in the chorion-containing sulfhydryl groups (Wedemeyer 1968; Blaxter 1969; Beattie and

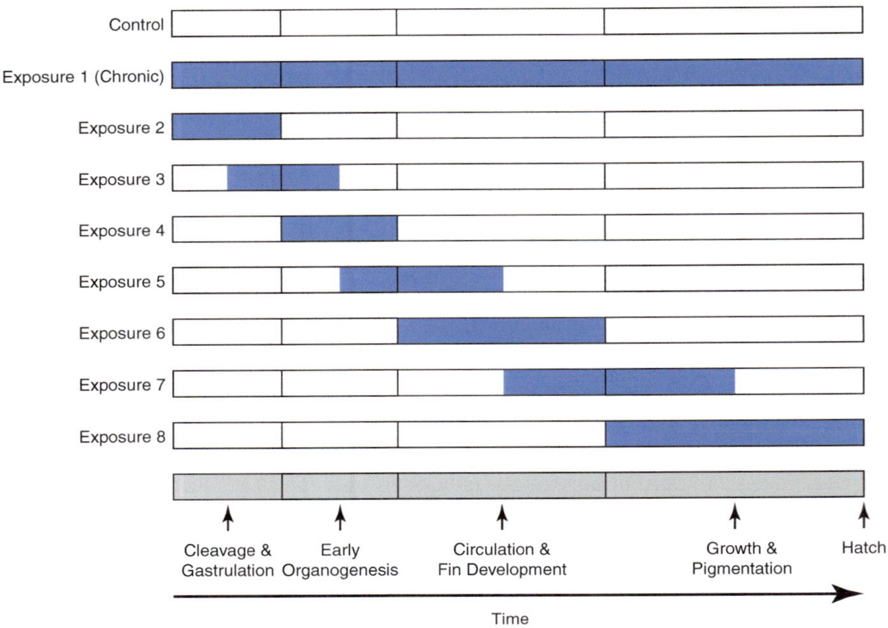

Fig. 3.2 Schematic of a critical window experimental design during the embryonic development of a fish, modeled on lake whitefish (*Coregonus clupeaformis*) (Price 1934a, b, 1935; Eme et al. 2015). Cleavage and gastrulation, during which division of early cells and formation of the germ layers occur, cover approximately the first 15% of development. Early organogenesis, which represents the first formation of organs from the ectoderm, mesoderm, and endoderm and ends at the onset of the heartbeat, covers approximately 16–30% of development. Circulation and fin development, which represents continued differentiation of tissues, cover 31–60% of development. Growth and pigmentation, during which pigmentation increases and embryos grow and become progressively active, occur during 61% hatch. The design includes a control and a chronic exposure to an environmental stressor (exposure 1). Windows of exposure also occur during each developmental phase (exposures 2, 4, 6, and 8) or overlap these phases (exposures 3, 5, and 7) to create a series of overlapping exposure windows

Pascoe 1978). The egg capsule of amphibians can also provide a protective barrier against pesticides. While early tadpole stages represent a critical window for parasite infection in the Pacific chorus frog, the embryos are unaffected by exposure, indicating the function of the egg capsule to prevent parasite infection (Johnson et al. 2011).

The protective nature of the egg, if present, may change during development. The insecticide cypermethrin is most lethal to medaka embryos late in embryonic development (stage 34 of 39; Iwamatsu 2004), however, a high incidence of edema occurs earlier in embryonic development (González-Doncel et al. 2004). As evident by the morphological abnormalities, cypermethrin must be able to enter medaka eggs at all stages of embryogenesis, but the higher lethality later in development suggests that changes in the egg chorion structure, a natural process of hatching,

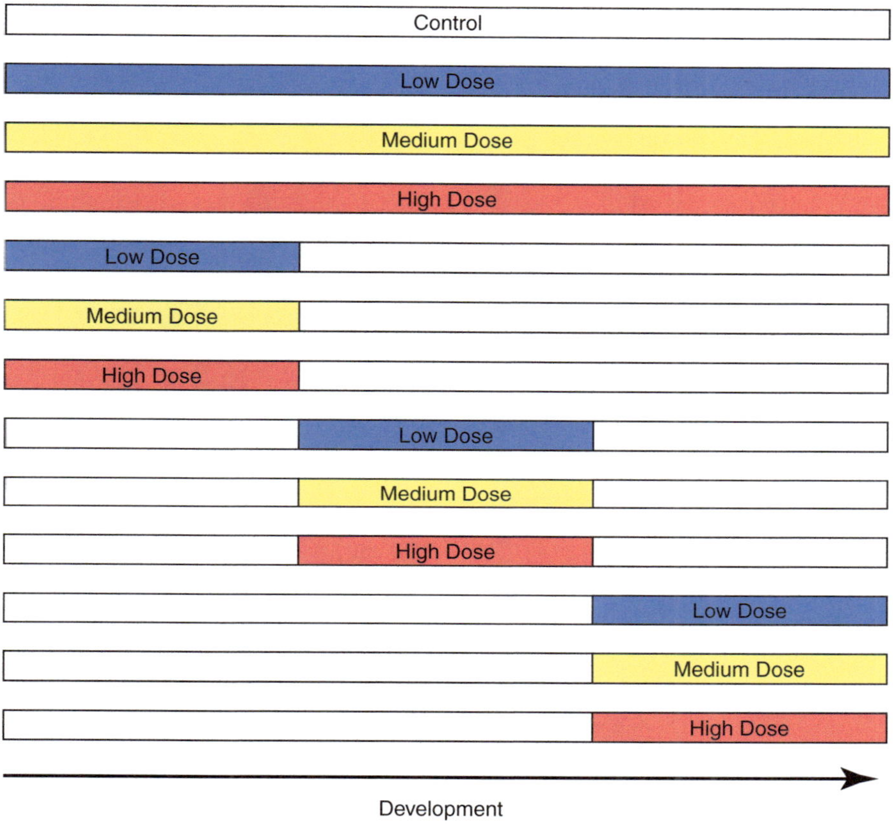

Fig. 3.3 Schematic of a critical window experimental design with multiple exposure windows and stressor doses. The design is similar to that presented in Fig. 3.1 but includes chronic exposure treatments for each dose and treatments with an exposure to each dose during three distinct, separate periods of development. This design can be used to build a 3-D critical window, which illustrates the interaction between exposure window, dose, and phenotypic modification (Fig. 3.4)

may decrease the protective nature of the chorion. In fact, when medaka embryos are exposed to the herbicide thiobencarb with and without the chorion intact, the LC50 (lethal concentration, where 50% of the population is killed by the contaminant) of dechorionated embryos is 2.0 mg/l of thiobencarb compared to 3.9 mg/l in chlorinated embryos, illustrating the protective nature of the egg structure.

The chorion may also be directly affected by a toxicant. For example, the egg chorion of medaka develops fractures and holes following exposure to silver nanoparticles (Wu and Zhou 2012). The loss of chorion integrity compromises osmotic balance within the egg, resulting in edema and morphological abnormalities of medaka embryos following silver nanoparticle exposure (Wu and Zhou 2012). In this instance the egg chorion may actually facilitate the toxicity of the nanoparticles (Villalobos et al. 2000).

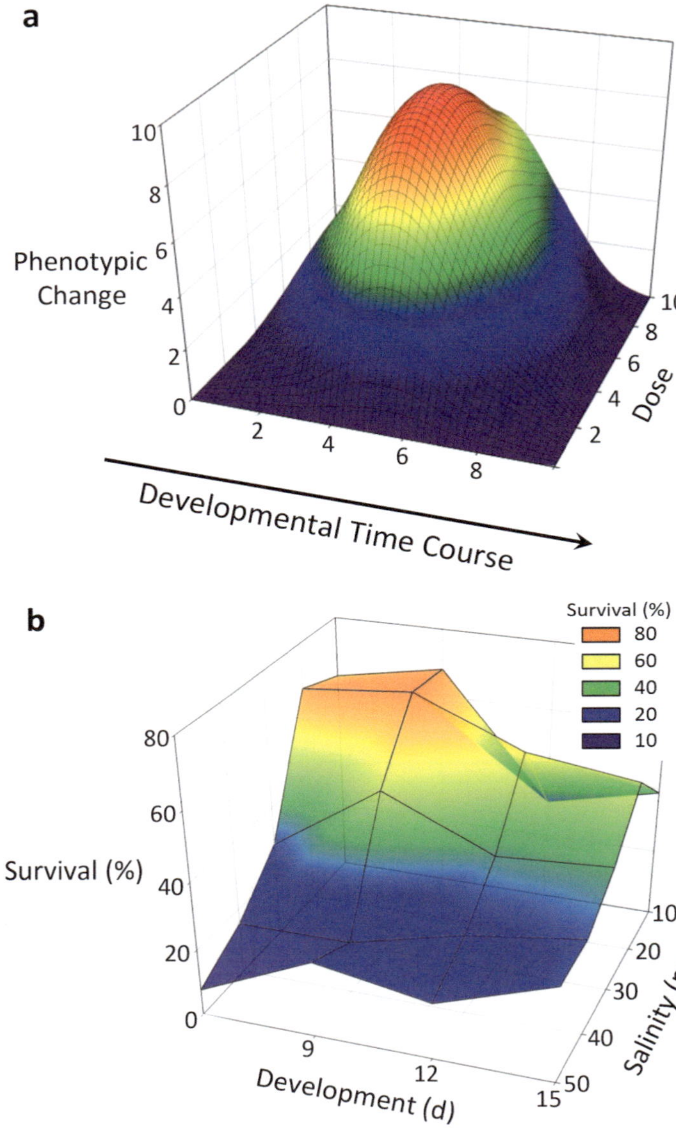

Fig. 3.4 (**a**) A critical window as a three-dimensional construct illustrating the interaction between exposure time, stressor dose, and phenotypic modification. This three-dimensional window can be constructed using an experimental design as illustrated in Fig. 3.3. (**b**) Relationship between salinity (ppt), time of exposure in development (d), and mean survival (%) of developing brine shrimp. Modified from Burggren and Mueller (2015) with permission. © 2015 The University of Chicago Press

3.4 Considerations for Critical Window Experimental Approaches

As our knowledge of developmental phenotypic plasticity and critical windows grows, how we approach their study must also evolve. Experiments can range from those that expose a developing animal to a single stressor dose during a few windows of development (Fig. 3.1) to those that utilize multiple exposure windows and stressor doses to examine the extent of phenotypic change (Figs. 3.3 and 3.4). The latter designs increase in complexity, which can result in large treatment and animal numbers that are not always possible or manageable. Species that are relatively easy to obtain, house, and care for in large numbers lend themselves to complex critical window studies. Irrespective of the animal model, there are a number of factors to consider when designing an experiment aimed at detecting critical windows, and they will be discussed below.

3.4.1 Length of Exposure and Overlapping Exposures

Often, the length of the exposure to an environmental stressor can impact the phenotypic changes induced in the developing animal. Furthermore, the length of exposure may vary considerably across species. For example, common prawn and spider crab (*Maja squinado*) larval mortality is limited following 24 h exposure to mercury, copper, or cadmium but increases following 72 h exposure. In comparison, mortality caused by heavy metal exposure in the larvae of lobster (*Homarus gammarus*) is evident after just 24 h (Mariño-Balsa et al. 2000). The way in which exposure times affect species is dependent upon how the length of exposure is related to overall development time. That is, the same exposure time in two species with different developmental trajectories will more than likely result in different effects in each. Within a species, an increase in an effect with a longer exposure time may indicate the accumulative effect of the stressor. However, it may also indicate that the longer exposure time covers more of a critical window of sensitivity during development. For Japanese tiger shrimp nauplii, the ammonia concentration that produces 50% mortality is much lower in late nauplii after 48 h of exposure compared to early nauplii after 24 h exposure (Lin et al. 1993). This is because post-molt late nauplii are no longer relying on endogenous reserves and are feeding, increasing their sensitivity to the environment. Even in the more ammonia-tolerant zoea and mysis developmental stages, ammonia-induced mortality increases with exposure time due to increased sensitivity during molting (Lin et al. 1993).

Whether the researcher is purposely investigating the presence of a critical window or not, the fact that experimental animals are constantly developing must be taken into account when designing exposure times. This is particularly relevant if comparing across species, as the exposure times must be relative to their developmental trajectories. Figure 3.2 demonstrates an experimental design in which exposures occur during well-defined periods of fish embryogenesis. This design is perhaps an improvement on experiments that use arbitrary exposures that divide development into even exposure windows (Fig. 3.1). In this instance, phenotypic changes that may occur during exposures can be more easily related to what is happening developmentally.

Furthermore, Fig. 3.2 shows a design that incorporates overlapping exposures, which can be useful for defining the boundaries of a detected critical window. For example, using exposures 2, 4, 6, and 8 only in Fig. 3.2, if a phenotypic change occurs following exposure 6, then the critical window may be defined as occurring during circulation and fin development. However, in an instance when all overlapping exposures in Fig. 3.2 are used, a phenotypic change may be greatest following exposures 5 and 6 but not 7. This suggests that exposure must occur during the first half of circulation and fin development (as in exposure 5 and 6), and this period is the true critical window. A limited number of studies have employed the approach of overlapping exposure windows (Olmstead and Leblanc 2002; Hogan et al. 2008; Tate et al. 2015), and they demonstrate the usefulness of overlapping exposures. An example analogous to that described in Fig. 3.2 is an evident in an experiment using juvenile hormone exposure to find the critical window for sex determination in *Daphnia* (Olmstead and Leblanc 2002). Exposure to juvenile hormone during 36–60 h and 72–96 h of development produces a low percentage (<40%) of males, while exposure during 48–72 h and 60–84 h produces a high percentage (>80%) of males. Therefore, the critical window for producing male broods is defined as 60–72 h, as this is the common period shared by the two exposure windows that produce the highest percentage of males. In comparison, if exposures had only occurred during separate, nonoverlapping 24 h exposures (0–24, 24–48, 48–72, and 72–96 h), the window would be defined as wider (48–72 h) than what actually exists.

3.4.2 Stressor Doses

The dose or level of a stressor used is a vital consideration in critical window designs. In chronic studies, a very high dose of an environmental toxicant that causes mortality is useful for defining lethality only. A slightly lower dose, however, may provide more information about the morphological, physiological, and biochemical effects of a toxicant, particularly if the dose is environmentally relevant. In critical window studies, a dose that produces mortality following exposure during certain periods can be used to define a critical window for mortality. However, doses that are sublethal at all stages of development may provide more useful information about phenotypic traits that are affected by the stressor and if phenotypic changes are more prevalent during certain critical windows.

Another aspect of critical window experiments that requires attention is if a single dose or level of a stressor is used, or if multiple doses are used (as depicted in Fig. 3.3). Of the critical window studies cited in this chapter, less than 40% use more than a single dose or level of a stressor. Many studies undertake preliminary tests to determine the best dose to use before undertaking exposures during multiple windows (e.g., López Greco et al. 2002; Ali et al. 2011) or select a single environmentally relevant dose (e.g., Clarren et al. 1992; Boone et al. 2013). However, this approach may limit the information that can be obtained from the study and may mask the subtleties of stressor-induced effects. This is particularly relevant when attempting to define the extent of a critical window. For example, a higher dose may induce a phenotypic change slightly earlier in development, and it may persist slightly longer as the critical period closes. Thus, the critical window defined from the higher dose will be wider than the same window defined from a lower dose (Mueller et al. 2015a).

Experimental designs that employ multiple exposure windows and stressor doses are useful for defining the extent of the critical window. However, by examining the phenotypic modification induced by all possible combinations of exposure time and dose allows critical windows to be considered using a three-dimensional construct (Fig. 3.4a; Burggren and Mueller 2015). In this construct, phenotypic changes can represent continuous reaction norms that are dependent upon the timing of exposure and the dose. This approach allows the effect of a stressor to be examined in significant detail as the magnitude of the phenotypic change, not just whether a phenotypic change is present or not, becomes the central focus. Studies employing multiple exposure windows and stressor doses that have presented data in a three-dimensional manner include examination of salinity tolerance in developing brine shrimp (Fig. 3.4b; Burggren and Mueller 2015) and pesticide sensitivity in the southern leopard frog (Bridges 2000). This approach may be particularly useful for assessing how a phenotypic trait is modified by different stressor doses within the critical window once the potential window has been detected using a more simple exposure design (such as depicted in Fig. 3.1). The study examining salinity tolerance in brine shrimp (detailed in Sect. 3.2.3) used this approach to visualize the interaction between exposure time, dose, and magnitude of phenotypic modification of maturation and morphology (Mueller et al. 2016). Such an approach may be useful to pinpoint the mechanism of stressor action and examine phenotype modification at multiple levels of the organism.

3.4.3 Systems Approach to Critical Windows

Critical window studies often focus on changes or abnormalities in morphological characteristics, as they are often the easiest to observe and measure. Studies examining sex determination and endocrine disruption, for example, tend to examine the morphology of the developing gonads. Yet, the critical window for endocrine disruption may be influenced by other physiological aspects, such as the onset of hormonal regulation due to the presence of endogenous hormones and receptors. For instance, the critical window for endocrine disruption in coho salmon (*Oncorhynchus kisutch*) is just after hatching (Piferrer and Donaldson 1989), and this occurs in concert with a peak in endogenous testosterone (Feist et al. 1990).

The absence of traditional indicators of phenotypic modification during critical window exposure to a stressor is not always indicative of a noneffect. For example, 3-day exposure to carbaryl, a broad-spectrum insecticide, late in tadpole development of the green frog (*Lithobates clamitans*) does not alter survival or time to and size metamorphosis but alters mRNA levels of thyroid-regulated genes in the brain 4 months later, following metamorphosis (Boone et al. 2013). Thus, assessing one or two aspects of a system, in this instance growth and development rate, may fail to uncover other potential stressor effects, such as transcriptome changes. Taking a system approach to studying critical windows and understanding that different aspects of a system, be it morphological, physiological, or biochemical, may have different periods of sensitivity are key to understanding the overall system effect of an environmental stressor.

3.4.4 Combined Stressors

Understanding the impacts of natural and anthropogenic environmental stressors on animals has moved from the study of individual stressors alone to combined effects, as this is more representative of what animals are likely to experience in their natural environment. The consequences of combined stressors are also worth consideration in the context of critical windows as the combination of stressors may influence the timing of susceptibility during development. The concept of combined stressors is particularly relevant in the context of environmental contamination in a situation in which a species is already under some natural environmental stress. Combined stressors may have an additive or synergistic effect (e.g., Relyea and Mills 2001; Jones et al. 2011), or one stressor may lessen the effect of the other. An example of the latter situation is pesticide exposure in frog tadpoles naturally exposed to pathogens. The effect of fungal infection on tadpoles, including reduced survival and growth, is mitigated by the broad-spectrum herbicide glyphosate in the gray tree frog (*Hyla versicolor*) and by the fungicide thiophanate-methyl in the southern leopard frog (*Lithobates sphenocephalus*) (Hanlon et al. 2012; Hanlon and Parris 2014). Hanlon and Parris (2014) exposed gray tree frog tadpoles to glyphosate and the fungus during early, mid, or late tadpole development but did not find a stage-specific interaction between the two stressors. However, this study is an example of how the concept of critical windows can be incorporated when examining the potential effects of combined stressors in the environment.

3.4.5 Timing of Endpoint Measurement

The search for critical windows is often concerned with the immediate, or developmental, impacts on the animal. However, the long-term consequences of an environmental change or stressor exposure during a certain period of development are also of importance. Phenotypic changes induced during critical window exposure may persist into mature life stages, influencing the adult phenotype. In some instances, immediate developmental effects may not be evident but may materialize at a later stage, such as the transcriptome changes that appear months after insecticide exposure in green frog tadpoles discussed in Sect. 3.4.3. Often the appearance of an effect is specific to the endpoint. For example, transient morphological abnormalities, such as edema, occur in medaka embryos exposed to the insecticide cypermethrin early in embryonic development, with embryos recovering prior to hatch. However, following late embryonic exposures, mortality is high and longer term effects, such as spinal curvature, an inability to inflate the swim bladder and behavioral changes, manifest in the surviving larvae (González-Doncel et al. 2004).

While some effects may not appear until later in life, developmental effects may also disappear later in life, often evident of self-repair in the animal (Burggren and Reyna 2011). Morphological changes, such as decreased body mass and shorter toes and beaks, following hypoxia exposure during the windows of day 1–6, 6–12, or 12–18 of chicken embryo development are present on day 12 and 18, but almost all changes disappear by hatching (Dzialowski et al. 2002; Chan and Burggren 2005). Thus, given time back in normoxia, chicken embryos show a remarkable ability to recover to the normal morphological phenotype. Likewise, zebrafish exposed to estrogens during the period for sex differentiation show an initial skew in the sex ratio and kidney pathology, but after recovery in clean water until the adult stage, the sex ratio, gonad morphology, and kidney histology return to normal (Hill Jr and Janz 2003; Weber et al. 2003). However, the exposure may still have long-term ecological implications as viable egg production and hatchability are reduced. In comparison, exposure to an estrogen alters sex ratios and increases the incidence of intersex in the northern leopard frog 2–3 months after tadpole exposure (Hogan et al. 2008). Thus, the ability to recover from endocrine disruption during critical windows of reproductive development is variable among fish and amphibian species and may be related to their mode of sex differentiation (Hogan et al. 2008).

The endpoints chosen in a critical window study will depend on the particular biological question of interest. If examining developmental toxicity of an environmental contaminant, then endpoints measured shortly after exposure during development should be the focus. However, if potential long-term consequences are of interest, then endpoints measured at the end of development or in an adult stage will be most useful. In many cases, multiple endpoints across time will provide the best picture of phenotypic changes induced by an environmental stressor. Endpoints measured at different life stages can also shed light on the mechanism of self-repair and if such a process occurs following exposure during certain distinct windows of exposure. Beyond endpoints measured during the life history of a single individual, the potential transgenerational effect of exposure to a stressor during a critical window of development may also be of interest. Transgenerational transfer and epigenetic effects are of increasing focus in developmental physiology (Ho and Burggren 2010; Burggren 2014; Burggren and Crews 2014; Mueller et al. 2015a). The interplay between critical windows, the resultant phenotype of an individual at all life stages, and potential transgenerational effects should be of focus due to their ability to link the developmental environment to physiological traits at all life stages and across generations.

Conclusions

The concept of critical windows has become ubiquitous in the medical, comparative physiology, and environmental toxicology fields, as it recognizes that the environment may influence the biology of an animal differently during ontogeny. The study of critical windows has the ability to inform us about normal developmental progression of different physiological systems while also demonstrating how animals may respond via phenotypic modification in response to environmental stressors. For example, exposure to hypoxia during distinct periods of development can inform us about the ontogeny of the cardiovascular and respiratory systems, while exposure to different salinities during specific time points can inform us about osmoregulation. At the same time, exposure to these environmental variables during specific windows of development can indicate how an animal may be able to respond to changes in their natural environment. This is particularly relevant for appreciating the potential challenges, such as global climate change, faced by animals during development. Likewise, understanding if there are critical windows during which developing animals are particularly sensitive to environmental contaminants, and how that sensitivity relates to developmental status, can be used to understand the mechanism of toxicity and may be useful in conservation efforts.

As our knowledge of developmental phenotypic plasticity and critical windows grows, we must continue to assess and improve how we approach their study. A search for a critical window may begin broadly, covering as much of development as possible and using a relatively simple experimental design. Once a period of plasticity or susceptibility has been detected, then that time period can be of closer focus, with more detailed experimental approaches used to define the window with as much accuracy as possible. The considerations for critical window experimental designs discussed in this chapter need not be incorporated into every study, as experiments would no doubt become unmanageable in terms of the number of treatments and animals required. However, variables such as the timing of exposures and if a single or multiple doses of the stressor or combined stressors are used should be considered in the context of the study species and questions of interest. Likewise, the form of the endpoints examined, whether morphological, physiological, or biochemical, and when those endpoints are measured are important decisions that can help understand how the developmental environment influences animals at all stages of life. A continued focus on critical windows, and advancements in the approaches used to detect and study them, will no doubt contribute to our growing knowledge of the interaction between development, environment, and physiology.

References

Ackerman RA, Rahn H (1981) In vivo O_2 and water vapor permeability of the hen's eggshell during early development. Respir Physiol 45(1):1–8

Ali S, Champagne DL, Alia A, Richardson MK (2011) Large-scale analysis of acute ethanol exposure in zebrafish development: a critical time window and resilience. PLoS One 6(5):e20037

Andersen SL (2003) Trajectories of brain development: point of vulnerability or window of opportunity? Neurosci Biobehav Rev 27(1):3–18
Andersen L, Holbech H, Gessbo Å, Norrgren L, Petersen GI (2003) Effects of exposure to 17α-ethinylestradiol during early development on sexual differentiation and induction of vitellogenin in zebrafish (*Danio rerio*). Comp Biochem Physiol C Toxicol Pharmacol 134(3):365–374
Andersen HS, Gambling L, Holtrop G, McArdle HJ (2006) Maternal iron deficiency identifies critical windows for growth and cardiovascular development in the rat postimplantation embryo. J Nutr 136(5):1171–1177
Angilletta MJ, Steury TD, Sears MW (2004) Temperature, growth rate, and body size in ectotherms: fitting pieces of a life-history puzzle. Integr Comp Biol 44(6):498–509
Ankley GT, Johnson RD (2004) Small fish models for identifying and assessing the effects of endocrine-disrupting chemicals. ILAR J 45(4):469–483
Aronzon CM, Sandoval MT, Herkovits J, Pérez-Coll CS (2011) Stage-dependent toxicity of 2,4-dichlorophenoxyacetic on the embryonic development of a South American toad, *Rhinella arenarum*. Environ Toxicol 26(4):373–381
Bambang Y, Thuet P, Charmantier-Daures M, Trilles J-P, Charmantier G (1995) Effect of copper on survival and osmoregulation of various developmental stages of the shrimp *Penaeus japonicus* bate (Crustacea, Decapoda). Aquat Toxicol 33(2):125–139
Baroiller J, D'cotta H (2001) Environment and sex determination in farmed fish. Comp Biochem Physiol C Toxicol Pharmacol 130(4):399–409
Barr M Jr, DeSesso JM, Lau CS, Osmond C, Ozanne SE, Sadler TW, Simmons RA, Sonawane BR (2000) Workshop to identify critical windows of exposure for children's health: cardiovascular and endocrine work group summary. Environ Health Perspect 108(Suppl 3):569
Bavis RW (2005) Developmental plasticity of the hypoxic ventilatory response after perinatal hyperoxia and hypoxia. Respir Physiol Neurobiol 149(1):287–299
Bavis RW, Mitchell GS (2008) Long-term effects of the perinatal environment on respiratory control. J Appl Physiol 104(4):1220–1229
Bavis RW, Fallon SC, Dmitrieff EF (2013) Chronic hyperoxia and the development of the carotid body. Respir Physiol Neurobiol 185(1):94–104
Beattie J, Pascoe D (1978) Cadmium uptake by rainbow trout, *Salmo gairdneri* eggs and alevins. J Fish Biol 13(5):631–637
Blaxter J (1969) Development: eggs and larvae. In: Hoar WS, Randall DJ (eds) Fish physiology, vol 3. Academic Press, San Diego, pp 177–252
Boone MD, Hammond SA, Veldhoen N, Youngquist M, Helbing CC (2013) Specific time of exposure during tadpole development influences biological effects of the insecticide carbaryl in green frogs (*Lithobates clamitans*). Aquat Toxicol 130:139–148
Bridges C (2000) Long-term effects of pesticide exposure at various life stages of the southern leopard frog (*Rana sphenocephala*). Arch Environ Con Tox 39(1):91–96
Bull J, Vogt R (1981) Temperature-sensitive periods of sex determination in emydid turtles. J Exp Zool 218(3):435–440
Bunn T, Parsons P, Kao E, Dietert R (2001) Exposure to lead during critical windows of embryonic development: differential immunotoxic outcome based on stage of exposure and gender. Toxicol Sci 64(1):57–66
Burggren WW (2014) Epigenetics as a source of variation in comparative animal physiology—or—Lamarck is lookin' pretty good these days. J Exp Biol 217(5):682–689
Burggren WW, Crews D (2014) Epigenetics in comparative biology: why we should pay attention. Integr Comp Biol 54(1):7–20
Burggren WW, Mueller CA (2015) Developmental critical windows and sensitive periods as 3-D constructs in time and space. Physiol Biochem Zool 88:91–102
Burggren WW, Reyna KS (2011) Developmental trajectories, critical windows and phenotypic alteration during cardio-respiratory development. Respir Physiol Neurobiol 178:13–21
Carvalho GR, Hughes RN (1983) The effect of food availability, female culture-density and photoperiod on ephippia production in *Daphnia magna* Straus (Crustacea: Cladocera). Freshwater Biol 13(1):37–46

Chan T, Burggren W (2005) Hypoxic incubation creates differential morphological effects during specific developmental critical windows in the embryo of the chicken (*Gallus gallus*). Respir Physiol Neurobiol 145(2–3):251–263
Charmantier G, Charmantier-Daures M, Bouaricha N, Thuet P, Trilles J-P, Aiken D (1988) Ontogeny of osmoregulation and salinity tolerance in two decapod crustaceans: *Homarus americanus* and *Penaeus japonicus*. Biol Bull 175(1):102–110
Chiasson M, Benfey TJ, Martin-Robichaud DJ (2008) Gonadal differentiation in Atlantic cod, *Gadus morhua* L., and haddock, *Melanogrammus aeglefinus* (L.) Acta Ichtyol Piscat 38(2):127–133
Clarren SK, Astley SJ, Gunderson VM, Spellman D (1992) Cognitive and behavioral deficits in nonhuman primates associated with very early embryonic binge exposures to ethanol. J Pediatr 121(5):789–796
Colombo J (1982) The critical period concept: research, methodology, and theoretical issues. Psychol Bull 91(2):260
Conover DO (1984) Adaptive significance of temperature-dependent sex determination in a fish. Am Nat 123:297–313
Conover DO, Fleisher MH (1986) Temperature-sensitive period of sex determination in the Atlantic silverside, *Menidia menidia*. Can J Fish Aquat Sci 43(3):514–520
Conte FP (1984) Structure and function of the crustacean larval salt gland. Int Rev Cytol 91:45–106
Copeland J, Dzialowski EM (2009) Effects of hypoxic and hyperoxic incubation on the reactivity of the chicken embryo (*Gallus gallus*) ductus arteriosi in response to catecholamines and oxygen. Exp Physiol 94(1):152–161
Croghan P (1958) The mechanism of osmotic regulation in *Artemia salina* (L.): the physiology of the branchiae. J Exp Biol 35(1):234–242
Cronise K, Marino MD, Tran TD, Kelly SJ (2001) Critical periods for the effects of alcohol exposure on learning in rats. Behav Neurosci 115(1):138
Crossley DA II, Altimiras J (2005) Cardiovascular development in embryos of the American alligator *Alligator mississippiensis*: effects of chronic and acute hypoxia. J Exp Biol 208(Pt 1):31–39
Crossley DA II, Burggren WW, Altimiras J (2003) Cardiovascular regulation during hypoxia in embryos of the domestic chicken *Gallus gallus*. Am J Physiol Regul Integr Comp Physiol 284(1):R219–R226
DeCoursey PJ, Vernberg WB (1972) Effect of mercury on survival, metabolism and behaviour of larval *Uca pugilator* (Brachyura). Oikos 23:241–247
Deeming D, Ferguson M, Mittwoch U, Wolf U, Dorizzi M, Zaborski P, Sharma H (1988) Environmental regulation of sex determination in reptiles. Philos T Roy Soc B 322(1208):19–39
Degitz SJ, Kosian PA, Makynen EA, Jensen KM, Ankley GT (2000) Stage-and species-specific developmental toxicity of all-trans retinoic acid in four native North American ranids and *Xenopus laevis*. Toxicol Sci 57(2):264–274
Dietert RR, Etzel RA, Chen D, Halonen M, Holladay SD, Jarabek AM, Landreth K, Peden DB, Pinkerton K, Smialowicz RJ (2000) Workshop to identify critical windows of exposure for children's health: immune and respiratory systems work group summary. Environ Health Perspect 108(Suppl 3):483
Dietert R, Lee J, Bunn T (2002) Developmental immunotoxicology: emerging issues. Hum Exp Toxicol 21(9–10):479–485
Doyle JE, McMahon BR (1995) Effects of acid exposure in the brine shrimp *Artemia franciscana* during development in seawater. Comp Biochem Physiol A 112:123–129
Dzialowski EM, von Plettenberg D, Elmonoufy NA, Burggren WW (2002) Chronic hypoxia alters the physiological and morphological trajectories of developing chicken embryos. Comp Biochem Physiol A 131(4):713–724
Eme J, Altimiras J, Hicks JW, Crossley DA II (2011a) Hypoxic alligator embryos: chronic hypoxia, catecholamine levels and autonomic responses of in ovo alligators. Comp Biochem Physiol A 160(3):412–420
Eme J, Hicks JW, Crossley DA II (2011b) Chronic hypoxic incubation blunts a cardiovascular reflex loop in embryonic American alligator (*Alligator mississippiensis*). J Comp Physiol B 181(7):981–990

Eme J, Rhen T, Tate K, Gruchalla K, Kohl Z, Slay C, Crossley D II (2013) Plasticity of cardiovascular function in snapping turtle embryos (*Chelydra serpentina*): chronic hypoxia alters autonomic regulation and gene expression. Am J Physiol Regul Integr Comp Physiol 304:R966–R979

Eme J, Rhen T, Crossley DA II (2014) Adjustments in cholinergic, adrenergic and purinergic control of cardiovascular function in snapping turtle embryos (*Chelydra serpentina*) incubated in chronic hypoxia. J Comp Physiol B 184(7):891–902

Eme J, Mueller CA, Manzon RG, Somers CM, Boreham DR, Wilson JY (2015) Critical windows in embryonic development: shifting incubation temperatures alter heart rate and oxygen consumption of Lake whitefish (*Coregonus clupeaformis*) embryos and hatchlings. Comp Biochem Physiol A 179:71–80

Emlen DJ, Nijhout HF (1999) Hormonal control of male horn length dimorphism in the dung beetle *Onthophagus taurus* (Coleoptera: Scarabaeidae). J Insect Physiol 45(1):45–53

Emlen D, Nijhout H (2001) Hormonal control of male horn length dimorphism in *Onthophagus taurus* (Coleoptera: Scarabaeidae): a second critical period of sensitivity to juvenile hormone. J Insect Physiol 47(9):1045–1054

Feder ME (1985) Thermal acclimation of oxygen consumption and cardiorespiratory frequencies in frog larvae. Physiol Zool 58(3):303–311

Feist G, Schreck CB, Fitzpatrick MS, Redding JM (1990) Sex steroid profiles of coho salmon (*Oncorhynchus kisutch*) during early development and sexual differentiation. Gen Comp Endocrinol 80(2):299–313

Fent K, Meier W (1994) Effects of triphenyltin on fish early life stages. Arch Environ Con Tox 27(2):224–231

Ferguson MW, Joanen T (1982) Temperature of egg incubation determines sex in *Alligator mississippiensis*. Nature 296:850–853

Ferner K, Mortola JP (2009) Ventilatory response to hypoxia in chicken hatchlings: a developmental window of sensitivity to embryonic hypoxia. Respir Physiol Neurobiol 165(1):49–53

Finstad AG, Jonsson B (2012) Effect of incubation temperature on growth performance in Atlantic salmon. Mar Ecol-Prog Ser 454:75–82

Gillooly JF, Brown JH, West GB, Savage VM, Charnov EL (2001) Effects of size and temperature on metabolic rate. Science 293(5538):2248–2251

Gimeno S, Komen H, Venderbosch PW, Bowmer T (1997) Disruption of sexual differentiation in genetic male common carp (*Cyprinus carpio*) exposed to an alkylphenol during different life stages. Environ Sci Technol 31(10):2884–2890

Glas PS, Courtney LA, Rayburn JR, Fisher WS (1997) Embryonic coat of the grass shrimp *Palaemonetes pugio*. Biol Bull 192(2):231–242

González-Doncel M, Fernández-Torija C, Hinton D, Tarazona J (2004) Stage-specific toxicity of cypermethrin to medaka (*Oryzias latipes*) eggs and embryos using a refined methodology for an in vitro fertilization bioassay. Arch Environ Con Tox 48(1):87–98

Gosner KL (1960) A simplified table for staging anuran embryos and larvae with notes on identification. Herpetologica 16:183–190

Grabowski CT, Paar JA (1958) The teratogenic effects of graded doses of hypoxia on the chick embryo. Am J Anat 103(3):313–347

Green DW, Williams KA, Pascoe D (1986) The acute and chronic toxicity of cadmium to different life history stages of the freshwater crustacean *Asellus aquaticus* (L). Arch Environ Con Tox 15(5):465–471

Greulich K, Pflugmacher S (2003) Differences in susceptibility of various life stages of amphibians to pesticide exposure. Aquat Toxicol 65(3):329–336

Hackmann E, Reinboth R (1974) Delimitation of the critical stage of hormone-influenced sex differentiation in *Hemihaplochromis multicolor* (Hilgendorf)(Cichlidae). Gen Comp Endocrinol 22(1):42–53

Hakeem GF, Oddy L, Holcroft CA, Abenhaim HA (2015) Incidence and determinants of sudden infant death syndrome: a population-based study on 37 million births. World J Pediatr 11:41–47

Hanlon SM, Parris MJ (2014) The interactive effects of chytrid fungus, pesticides, and exposure timing on gray treefrog (*Hyla versicolor*) larvae. Environ Toxicol Chem 33(1):216–222

Hanlon SM, Kerby JL, Parris MJ (2012) Unlikely remedy: fungicide clears infection from pathogenic fungus in larval southern leopard frogs (*Lithobates sphenocephalus*). PLoS One 7(8):e43573
Harrington RW (1968) Delimitation of the thermolabile phenocritical period of sex determination and differentiation in the ontogeny of the normally hermaphroditic fish *Rivulus marmoratus* Poey. Physiol Zool 41:447–460
Hayes TB (1997) Steroids as potential modulators of thyroid hormone activity in anuran metamorphosis. Am Zool 37(2):185–194
Hendry C, Martin-Robichaud D, Benfey T (2002) Gonadal sex differentiation in Atlantic halibut. J Fish Biol 60(6):1431–1442
Hensch TK (2004) Critical period regulation. Annu Rev Neurosci 27(1):549–579
Herzig A, Winkler H (1986) The influence of temperature on the embryonic development of three cyprinid fishes, *Abramis brama*, *Chalcalburnus chalcoides mento* and *Vimba vimba*. J Fish Biol 28:171–181
Hill RL Jr, Janz DM (2003) Developmental estrogenic exposure in zebrafish (*Danio rerio*): I. Effects on sex ratio and breeding success. Aquat Toxicol 63(4):417–429
Hillman R (1977) Polygenic control of *Drosophila* morphogenesis during the stages of determination and specification of adult structures. Am Zool 17(3):521–533
Ho DH, Burggren WW (2010) Epigenetics and transgenerational transfer: a physiological perspective. J Exp Biol 213(1):3–16
Hogan NS, Duarte P, Wade MG, Lean DRS, Trudeau VL (2008) Estrogenic exposure affects metamorphosis and alters sex ratios in the northern leopard frog (*Rana pipiens*): identifying critically vulnerable periods of development. Gen Comp Endocrinol 156(3):515–523
Hunt J, Simmons LW (1997) Patterns of fluctuating asymmetry in beetle horns: an experimental examination of the honest signalling hypothesis. Behav Ecol Sociobiol 41(2):109–114
Hunter GA, Donaldson EM (1983) Hormonal sex control and its application to fish culture. In: Hoar WS, Randall DJ, Donaldson EM (eds) Fish physiology, vol 9. Academic Press, New York, pp 223–303
Itow T, Loveland R, Botton M (1998) Developmental abnormalities in horseshoe crab embryos caused by exposure to heavy metals. Arch Environ Con Tox 35(1):33–40
Iwamatsu T (2004) Stages of normal development in the medaka *Oryzias latipes*. Mech Develop 121(7):605–618
Johnson JS, Newport EL (1989) Critical period effects in second language learning: the influence of maturational state on the acquisition of English as a second language. Cogn Psychol 21:60–99
Johnson PT, Kellermanns E, Bowerman J (2011) Critical windows of disease risk: amphibian pathology driven by developmental changes in host resistance and tolerance. Funct Ecol 25(3):726–734
Jones DK, Hammond JI, Relyea RA (2011) Competitive stress can make the herbicide roundup® more deadly to larval amphibians. Environ Toxicol Chem 30(2):446–454
Kam YC (1993) Physiological effects of hypoxia on metabolism and growth of turtle embryos. Respir Physiol 92:127–138
Kamler E, Keckeis H, Bauer-Nemeschkal E (1998) Temperature-induced changes of survival, development and yolk partitioning in *Chondrostoma nasus*. J Fish Biol 53:658–682
Karraker NE, Arrigoni J, Dudgeon D (2010) Effects of increased salinity and an introduced predator on lowland amphibians in southern China: species identity matters. Biol Conserv 143(5):1079–1086
Kefford BJ, Papas PJ, Nugegoda D (2003) Relative salinity tolerance of macroinvertebrates from the Barwon River, Victoria, Australia. Mar Freshw Res 54(6):755–765
Kettlewell JR, Raymond CS, Zarkower D (2000) Temperature-dependent expression of turtle Dmrt 1 prior to sexual differentiation. Genesis 26(3):174–178
Kikuyama S, Kawamura K, Tanaka S, Yamamoto K (1993) Aspects of amphibian metamorphosis: hormonal control. In: Jeon KW, Jarvik J (eds) International review of cytology, vol 145. Academic Press, San Diego, pp 105–105

Knudsen E (2004) Sensitive periods in the development of the brain and behavior. J Cognitive Neurosci 16(8):1412–1425

Kochhar D, Penner JD, Tellone CI (1984) Comparative teratogenic activities of two retinoids: effects on palate and limb development. Teratogen Carcin Mut 4(4):377–387

Koger CS, Teh SJ, Hinton DE (2000) Determining the sensitive developmental stages of intersex induction in medaka (*Oryzias latipes*) exposed to 17β-estradiol or testosterone. Mar Environ Res 50(1–5):201–206

Korpelainen H (1990) Sex ratios and conditions required for environmental sex determination in animals. Biol Rev 65(2):147–184

Kuramoto M (1975) Adaptive significance in oxygen consumption of frog embryos in relation to the environmental temperatures. Comp Biochem Physiol A 52(1):59–62

Landreth K (2002) Critical windows in development of the rodent immune system. Hum Exp Toxicol 21(9–10):493–498

Latham KE, Just JJ (1989) Oxygen availability provides a signal for hatching in the rainbow trout (*Salmo gairdneri*) embryo. Can J Fish Aquat Sci 46:55–58

Lavolpe M, Greco LL, Kesselman D, Rodríguez E (2004) Differential toxicity of copper, zinc, and lead during the embryonic development of *Chasmagnathus granulatus* (Brachyura, Varunidae). Environ Toxicol Chem 23(4):960–967

Lee J-E, Chen S, Golemboski KA, Parsons PJ, Dietert RR (2001) Developmental windows of differential lead-induced immunotoxicity in chickens. Toxicology 156(2):161–170

Lin H-P, Thuet P, Trilles JP, Mounet-Guillaume R, Charmantier G (1993) Effects of ammonia on survival and osmoregulation of various development stages of the shrimp *Penaeus japonicus*. Mar Biol 117(4):591–598

Liu Q, Wong-Riley MT (2010) Postnatal changes in the expressions of serotonin 1A, 1B, and 2A receptors in ten brain stem nuclei of the rat: implication for a sensitive period. Neuroscience 165(1):61–78

Liu Q, Lowry TF, Wong-Riley MT (2006) Postnatal changes in ventilation during normoxia and acute hypoxia in the rat: implication for a sensitive period. J Physiol 577(3):957–970

Liu Q, Fehring C, Lowry TF, Wong-Riley MT (2009) Postnatal development of metabolic rate during normoxia and acute hypoxia in rats: implication for a sensitive period. J Appl Physiol 106(4):1212–1222

López Greco L, Rodriguez E, Bolaños J, Hernandez G, Fingerman M (2002) Toxicity of copper sulphate during early and late embryonic development of a palaemonid shrimp (Crustacea). Invertebr Reprod Dev 41(1–3):165–170

Lumey L, Khalangot MD, Vaiserman AM (2015) Association between type 2 diabetes and prenatal exposure to the Ukraine famine of 1932–33: a retrospective cohort study. Lancet Diabetes Endocrinol 3(10):787–794

Lusk LA, Wai KC, Moon-Grady AJ, Steurer MA, Keller RL (2015) Persistence of pulmonary hypertension by echocardiography predicts short-term outcomes in congenital diaphragmatic hernia. J Pediatr 166(2):251–256. e251

Maack G, Segner H (2004) Life-stage-dependent sensitivity of zebrafish (*Danio rerio*) to estrogen exposure. Comp Biochem Physiol C Toxicol Pharmacol 139(1–3):47–55

Macqueen DJ, Robb DH, Olsen T, Melstveit L, Paxton CG, Johnston IA (2008) Temperature until the 'eyed stage'of embryogenesis programmes the growth trajectory and muscle phenotype of adult Atlantic salmon. Biol Lett 4:294–298

Mariño-Balsa J, Poza E, Vázquez E, Beiras R (2000) Comparative toxicity of dissolved metals to early larval stages of *Palaemon serratus*, *Maja squinado*, and *Homarus gammarus* (Crustacea: Decapoda). Arch Environ Con Tox 39(3):345–351

Mattson SN, Riley EP (1998) A review of the neurobehavioral deficits in children with fetal alcohol syndrome or prenatal exposure to alcohol. Alcohol Clin Exp Res 22(2):279–294

McCutcheon IE, Metcalfe J, Metzenberg AB, Ettinger T (1982) Organ growth in hyperoxic and hypoxic chick embryos. Respir Physiol 50(2):153–163

McLaren IA, Cooley JM (1972) Temperature adaptation of embryonic development rate among frogs. Physiol Zool 45:223–228

Meredith R (2015) Sensitive and critical periods during neurotypical and aberrant neurodevelopment: a framework for neurodevelopmental disorders. Neurosci Biobehav R 50:180–188
Mills NE, Barnhart MC (1999) Effects of hypoxia on embryonic development in two *Ambystoma* and two *Rana* species. Physiol Biochem Zool 72:179–188
Mueller CA, Joss JMP, Seymour RS (2011a) Effects of environmental oxygen on development and respiration of Australian lungfish (*Neoceratodus forsteri*) embryos. J Comp Physiol B 181:941–952
Mueller CA, Joss JMP, Seymour RS (2011b) The energy cost of embryonic development in fishes and amphibians, with emphasis on new data from the Australian lungfish, *Neoceratodus forsteri*. J Comp Physiol B 181:43–52
Mueller CA, Eme J, Burggren WW, Rundle SD, Roghair RD (2015a) Challenges and opportunities in integrative developmental physiology. Comp Biochem Physiol A 184:113–124
Mueller CA, Eme J, Manzon RG, Somers CM, Boreham DR, Wilson JY (2015b) Embryonic critical windows: changes in incubation temperature alter survival, hatchling phenotype and cost of development in Lake whitefish (*Coregonus clupeaformis*). J Comp Physiol B 185:315–331
Mueller CA, Willis E, Burggren WW (2016) Salt sensitivity of the morphometry of Artemia franciscana during development: a demonstration of 3D critical windows. J Exp Biol 219(4):571–581
Murdock C, Wibbels T (2002) Expression of Dmrt1 in a turtle with temperature-dependent sex determination. Cytogenet Genome Res 101(3–4):302–308
Nakamura M, Takahashi H (1973) Gonadal sex differentiation in *Tilapia mossambica*, with special regard to the time of estrogen treatment effective in inducing complete feminization of genetic males. Bull Fac Fish Hokkaido Univ 24(1):1–13
Olmstead AW, Leblanc GA (2002) Juvenoid hormone methyl farnesoate is a sex determinant in the crustacean *Daphnia magna*. J Exp Zool 293(7):736–739
Oxendine SL, Cowden J, Hinton DE, Padilla S (2006) Vulnerable windows for developmental ethanol toxicity in the Japanese medaka fish (*Oryzias latipes*). Aquat Toxicol 80(4):396–404
Pérez-Coll C, Herkovits J (1990) Stage dependent susceptibility to lead in *Bufo arenarum* embryos. Environ Pollut 63(3):239–245
Petranka JW, Just JJ, Crawford EC (1982) Hatching of amphibian embryos: the physiological trigger. Science 217:257–259
Pieau C, Dorizzi M (1981) Determination of temperature sensitive stages for sexual differentiation of the gonads in embryos of the turtle, *Emys orbicularis*. J Morphol 170(3):373–382
Piferrer F, Donaldson EM (1989) Gonadal differentiation in coho salmon, *Oncorhynchus kisutch*, after a single treatment with androgen or estrogen at different stages during ontogenesis. Aquaculture 77(2):251–262
Price JW (1934a) The embryology of the whitefish *Coregonus Clupeaformis*, (Mitchill). Part I. Ohio J Sci 34:287–305
Price JW (1934b) The embryology of the whitefish *Coregonus Clupeaformis*, (Mitchill). Part II. Organogenesis. Ohio J Sci 34:399–414
Price JW (1935) The embryology of the whitefish *Coregonus Clupeaformis*, (Mitchill). Part III. The second half of the incubation period. Ohio J Sci 35:40–53
Pryor JL, Hughes C, Foster W, Hales BF, Robaire B (2000) Critical windows of exposure for children's health: the reproductive system in animals and humans. Environ Health Perspect 108(Suppl 3):491
Reece EA, Wiznitzer A, Homko CJ, Hagay Z, Wu YK (1996) Synchronization of the factors critical for diabetic teratogenesis: an in vitro model. Am J Obstet Gynecol 174(4):1284–1288
Relyea RA, Mills N (2001) Predator-induced stress makes the pesticide carbaryl more deadly to gray treefrog tadpoles (*Hyla versicolor*). P Natl Acad Sci USA 98(5):2491–2496
Rice D, Barone S Jr (2000) Critical periods of vulnerability for the developing nervous system: evidence from humans and animal models. Environ Health Perspect 108(Suppl 3):511
Romanoff AL (1967) Biochemistry of the avian embryo. Wiley, New York
Rombough PJ (1988) Respiratory gas exchange, aerobic metabolism, and effects of hypoxia during early life. In: Hoar WS, Randall DJ (eds) The physiology of developing fish, Fish physiology, vol 11. Academic Press, San Diego, pp 59–161

Rombough PJ (2003) Development rate: modelling developmental time and temperature. Nature 424(6946):268–269
Sarre SD, Georges A, Quinn A (2004) The ends of a continuum: genetic and temperature-dependent sex determination in reptiles. BioEssays 26(6):639–645
Schneider ML, Moore CF, Becker EF (2001) Timing of moderate alcohol exposure during pregnancy and neonatal outcome in rhesus monkeys (*Macaca mulatta*). Alcohol Clin Exp Res 25(8):1238–1245
Schotthoefer A, Koehler A, Meteyer C, Cole R (2003) Influence of *Ribeiroia* infection on limb development and survival of northern leopard frogs: effects of host-stage and parasite exposure level. Can J Zool 81:1144–1153
Scott GR, Johnston IA (2012) Temperature during embryonic development has persistent effects on thermal acclimation capacity in zebrafish. P Natl Acad Sci USA 109(35):14247–14252
Selevan SG, Kimmel CA, Mendola P (2000) Identifying critical windows of exposure for children's health. Environ Health Perspect 108(Suppl 3):451
Seymour RS, Roberts JD, Mitchell NJ, Blaylock AJ (2000) Influence of environmental oxygen on development and hatching of aquatic eggs of the Australian frog, *Crinia georgiana*. Physiol Biochem Zool 73:501–507
Sive H, Draper B, Harland R, Weintraub H (1990) Identification of a retinoic acid-sensitive period during primary axis formation in *Xenopus laevis*. Genes Dev 4:932–942
Sollid SA, Lorz HV, Stevens DG, Bartholomew JL (2003) Age-dependent susceptibility of Chinook salmon to *Myxobolus cerebralis* and effects of sustained parasite challenges. J Aquat Anim Health 15(2):136–146
Stross RG, Hill JC (1965) Diapause induction in *Daphnia* requires two stimuli. Science 150(3702):1462–1464
Stuart SN, Chanson JS, Cox NA, Young BE, Rodrigues ASL, Fischman DL, Waller RW (2004) Status and trends of amphibian declines and extinctions worldwide. Science 306(5702):1783–1786
Sulik KK, Dehart DB, Rogers JM, Chernoff N (1995) Teratogenicity of low doses of all-trans retinoic acid in presomite mouse embryos. Teratology 51(6):398–403
Svitok P, Molcan L, Vesela A, Kruzliak P, Moravcik R, Zeman M (2015) Increased salt intake during early ontogenesis lead to development of arterial hypertension in salt-resistant Wistar rats. Clin Exp Hypertens 37(2):142–147
Sweeney BW, Schnack JA (1977) Egg development, growth and metabolism of *Sigara alternata* (say) (Hemiptera: Corixidae) in fluctuating thermal environments. Ecology 58:265–277
Tate KB, Kohl ZF, Eme J, Rhen T, Crossley DA (2015) Critical windows of cardiovascular susceptibility to developmental hypoxia in common snapping turtle (*Chelydra serpentina*) embryos. Physiol Biochem Zool 88:103–115
Taylor LW, Kreutziger GO (1965) The gaseous environment of the chick embryo in relation to its development and hatchability 2. Effect of carbon dioxide and oxygen levels during the period of the fifth through the eighth days of incubation. Poult Sci 44(1):98–106
Taylor LW, Kreutziger GO (1966) The gaseous environment of the chick embryo in relation to its development and hatchability 3. Effect of carbon dioxide and oxygen levels during the period of the ninth through the twelfth days of incubation. Poult Sci 45(5):867–884
Taylor LW, Kreutziger GO (1969) The gaseous environment of the chick embryo in relation to its development and hatchability 4. Effect of carbon dioxide and oxygen levels during the period of the thirteenth through the sixteenth days of incubation. Poult Sci 48(3):871–877
Taylor LW, Sjodin RA, Gunns C (1956) The gaseous environment of the chick embryo in relation to its development and hatchability 1. Effect of carbon dioxide and oxygen levels during the first four days of incubation upon hatchability. Poult Sci 35(6):1206–1215
Taylor LW, Kreutzige GO, Abercrombie GL (1971) The gaseous environment of the chick embryo in relation to its development and hatchability 5. Effect of carbon dioxide and oxygen levels during the terminal days of incubation. Poult Sci 50(1):66–78
Tazawa H (1981) Effect of O_2 and CO_2 in N_2, He, and SF_6 on chick embryo blood pressure and heart rate. J Appl Physiol 51(4):1017–1022

Tazawa H, Hashimoto Y, Nakazawa S, Whittow GC (1992) Metabolic responses of chicken embryos and hatchlings to altered O_2 environments. Respir Physiol 88:37–50
Tran TD, Cronise K, Marino MD, Jenkins WJ, Kelly SJ (2000) Critical periods for the effects of alcohol exposure on brain weight, body weight, activity and investigation. Behav Brain Res 116(1):99–110
Turner BJ, Davis WP, Taylor D (1992) Abundant males in populations of a selfing hermaphrodite fish, *Rivulus marmoratus*, from some Belize cays. J Fish Biol 40(2):307–310
Tyler C, Jobling S, Sumpter J (1998) Endocrine disruption in wildlife: a critical review of the evidence. CRC Cr Rev Toxicol 28(4):319–361
Uylings HB (2006) Development of the human cortex and the concept of "critical" or "sensitive" periods. Lang Learn 56(s1):59–90
van Aerle R, Pounds N, Hutchinson TH, Maddix S, Tyler CR (2002) Window of sensitivity for the estrogenic effects of ethinylestradiol in early life-stages of fathead minnow, *Pimephales promelas*. Ecotoxicology 11(6):423–434
Van Leeuwen C, Griffioen P, Vergouw W, Maas-Diepeveen J (1985) Differences in susceptibility of early life stages of rainbow trout (*Salmo gairdneri*) to environmental pollutants. Aquat Toxicol 7(1):59–78
Vickers T (1985) Embryolethality in rats caused by retinoic acid. Teratology 31(1):19–33
Villalobos SA, Hamm JT, Teh SJ, Hinton DE (2000) Thiobencarb-induced embryotoxicity in medaka (*Oryzias latipes*): stage-specific toxicity and the protective role of chorion. Aquat Toxicol 48(2):309–326
Vito CC, Wieland SJ, Fox TO (1979) Androgen receptors exist throughout the 'critical period'of brain sexual differentiation. Nature 282:308–310
Wangensteen OD, Rahn H, Burton RR, Smith AH (1974) Respiratory gas exchange of high altitude adapted chick embryos. Respir Physiol 21(1):61–70
Weber LP, Hill RL Jr, Janz DM (2003) Developmental estrogenic exposure in zebrafish (*Danio rerio*): II. Histological evaluation of gametogenesis and organ toxicity. Aquat Toxicol 63(4):431–446
Wedemeyer G (1968) Uptake and distribution of Zn^{65} in the coho salmon egg (*Oncorhynchus kisutch*). Comp Biochem Physiol 26(1):271–279
Wibbels T, Bull JJ, Crews D (1991) Chronology and morphology of temperature-dependent sex determination. J Exp Zool 260(3):371–381
Williams W (2001) Anthropogenic salinisation of inland waters. Hydrobiologia 466:329–337
Wong C, Chu K, Tang K, Tam T, Wong L (1993) Effects of chromium, copper and nickel on survival and feeding behaviour of *Metapenaeus ensis* larvae and postlarvae (Decapoda: Penaeidae). Mar Environ Res 36(2):63–78
Wong-Riley MT, Liu Q (2005) Neurochemical development of brain stem nuclei involved in the control of respiration. Respir Physiol Neurobiol 149(1):83–98
Wu Y, Zhou Q (2012) Dose- and time-related changes in aerobic metabolism, chorionic disruption, and oxidative stress in embryonic medaka (*Oryzias latipes*): underlying mechanisms for silver nanoparticle developmental toxicity. Aquat Toxicol 124–125(0):238–246
Xu L, Mortola JP (1989) Effects of hypoxia or hyperoxia on the lung of the chick embryo. Can J Physiol Pharmacol 67(5):515–519
Yntema C (1968) A series of stages in the embryonic development of *Chelydra serpentina*. J Morphol 125(2):219–251
Yntema C (1976) Effects of incubation temperatures on sexual differentiation in the turtle, *Chelydra serpentina*. J Morphol 150(2):453–461
Yntema C (1979) Temperature levels and periods of sex determination during incubation of eggs of *Chelydra serpentina*. J Morphol 159(1):17–27
Yntema C, Mrosovsky N (1982) Critical periods and pivotal temperatures for sexual differentiation in loggerhead sea turtles. Can J Zool 60(5):1012–1016
Yuan J, Zhang X, Yu L, Sun Z, Zhu P, Wang X, Shi H (2011) Stage-specific malformations and phenotypic changes induced in embryos of amphibian (*Xenopus tropicalis*) by triphenyltin. Ecotoxcol Environ Safe 74(7):1960–1966

Developmental Plasticity and Heterokairy

4

John I. Spicer, Oliver Tills, Manuela Truebano, and Simon D. Rundle

Abstract

There is a resurgence of interest in using phenotypic plasticity, 'the environmentally sensitive production of alternative phenotypes by given genotypes', as a framework in the study of evolutionary biology. The term developmental plasticity describes a more specific strand of investigation dealing with how alterations to developmental processes and outcomes shape such environmentally induced variation. Nested within developmental plasticity is the notion of heterokairy, the potential of a single genotype to alter the timing of a developmental event (e.g. onset of a particular structure, function or components of that function), in response to an environmental signal or influence. Here we make a case for using the term heterokairy as a way of focusing on altered timing across different biological disciplines, and we suggest a road map for such an approach. Heterokairy as an interdisciplinary term could be used to (a) bring together the substantial knowledge currently available of environmentally sensitive, genetic and hormonal control of the timing of developmental transitions, (b) embed the study of altered timing of developmental events within developmental plasticity and (c) highlight the role that plasticity can play in adaptive evolution, particularly in response to global environmental change.

J. I. Spicer (✉) · O. Tills · M. Truebano · S. D. Rundle
Marine Biology and Ecology Research Centre, School of Biological and Marine Sciences, University of Plymouth, Plymouth, UK
e-mail: J.I.Spicer@plymouth.ac.uk

© Springer International Publishing AG, part of Springer Nature 2018

W. Burggren, B. Dubansky (eds.), *Development and Environment*,
https://doi.org/10.1007/978-3-319-75935-7_4

4.1 Introduction

> …the heart is the first thing that lives and the last that dies, but its auricles…have life before the ventricles and die after them…whence it is that that which in living creatures is made last fails first, and that which is first made fails last.
>
> William Harvey (1628)

The notion of general immutability and predictability in the itinerary of successful development, as voiced by the founder of modern physiology William Harvey in *De Mortu Cordis* (Sloan 1978), has prevailed until recent times. Investigations of cell lineages in the nematode worm *Caenorhabditis elegans*, for example, show little interindividual variation (Sulston and Horvitz 1977), facilitating a genetic analysis of development by screening for mutants (Sulston and Horovitz 1981). Deviations from either 'the' (now defunct) genetic blueprint (Pigliucci 2010) or from a fixed, tightly controlled developmental process seem to fly in the face of Darwin's (1859) suggestion in the *Origin* that natural selection, as a major mechanism underlying evolutionary change, could act on intraspecific variation during development. At the same time, more recent attention to phenotypic and developmental plasticity (DeWitt and Scheiner 2004; West-Eberhard 1989, 2003, 2005; Pigliucci 1998, 2001) has given greater prominence to the environment in driving, as well as selecting for, phenotypic variation (Laland et al. 2014).

Studying phenotypic plasticity has seen greater prominence given to the roles of development and the environment in organismal biology. Interestingly, plasticity in physiological function, and in particular plasticity in the development of that function, has not figured prominently in that research agenda. This lack of integration of physiology into the study of plasticity may result partly from the perception that physiological variation comes more within the context of phenotypic flexibility (e.g. Piersma and Drent 2003), with physiological alterations to the phenotype being more likely to be reversible and of short duration (Table 4.1). The term heterokairy, first proposed by Spicer and Burggren (2003), gave focus to the investigation of plasticity within a physiological context, by stressing the importance of environmentally induced variation in the timing of the onset of physiological function within a species. The notion of heterokairy has subsequently been extended to

Table 4.1 Mutually exclusive definitions of the four most commonly used categories of phenotypic plasticity and how heterokairy compares with these definitions

Plasticity category	Phenotypic change reversible?	Variation within a single individual	Phenotypic change seasonally cyclic
Developmental plasticity	No	No	No
Polyphenism	No	Yes	Yes
Phenotypic flexibility	Yes	Yes	No
Life cycle staging	Yes	Yes	Yes
Heterokairy	No	No	No (but could be)

Modified from Piersma and Drent (2003)

reinforce the idea that intraspecific variation in the timing of development of other types of trait could provide an important focus for studying plastic responses within behavioural, ecological and evolutionary contexts (Muratori 2010; Muratori et al. 2010; Warkentin 2007, 2011; Moravek and Martin 2011; Rundle and Spicer 2017). There is also the potential for this term to embrace the substantial work on environmental control of developmental timing that has focused on the hormonal and genetic influence on major developmental transitions in animals and plants.

The main aims of this chapter are to attempt to embed heterokairy explicitly within the context of developmental plasticity and to explore its future utility as a unifying term. To do this we firstly outline how heterokairy relates to phenotypic and developmental plasticity. We then provide a brief, critical review of the literature which investigates heterokairy, either explicitly or implicitly. Finally we produce a prospectus outlining what we see the term heterokairy being 'good for', focusing on the role of altered timing during an organism's life cycle and how this is informed and shaped by the notion of developmental plasticity. We hope that this chapter will spark debate and act as a catalyst for interaction between those from different biological communities interested in the mechanisms and importance of altered timing in an organism's life cycle.

4.2 Phenotypic Plasticity, Developmental Plasticity and Heterokairy

4.2.1 Phenotypic Plasticity

Phenotypic plasticity is central to the long-standing nature versus nurture debate (Pigliucci 2001). Its study can be traced back to Woltereck's use of the term 'Reaktionsnorms' (reaction norms—see below) to highlight the importance of the environment in genotype-phenotype interactions (Sarkar 2004). Despite the subsequent emphasis placed by the likes of Waddington and Dobzhansky in the middle of the twentieth century on a potential role for the environment in generating selectable phenotypic variation, the prominence of genetic mutations in providing the raw material for selection dominated evolutionary thought (see Sarkar 2004 for a comprehensive view of this history). Arguably it wasn't until Bradshaw's review of plasticity in plants (Bradshaw 1965) that the environment began to figure more prominently in biological thought although, ironically, Bradshaw emphasized the fact that such plasticity might be under genetic control. Since then, the interest in phenotypic plasticity as a biological phenomenon has burgeoned albeit in parallel with, rather than embedded within, the study of ecology and evolution (DeWitt and Scheiner 2004; Forsman 2015; Pigliucci 2001; De Witt 2016; see Dickins Chap. 7, this volume).

As studies of phenotypic plasticity have increased over the past quarter century, so too has the number of definitions of plasticity. 'Plasticity' can now mean so many different things that some have counselled caution when inferring or generalizing from studies in the vast literature (Stearns 1989; Forsman 2015). For example, plasticity may be referred to as 'active' or 'passive' depending, respectively, on whether organisms respond to the environment in a very specific, tightly coordinated way (e.g. induced defences in response to predators) or in an unregulated way (e.g. effects of temperature on enzyme function). Plasticity can also be classified as 'adaptive' or 'non-adaptive' (see below) and can be described as transgenerational (Bateson et al. 2004). Here we follow DeWitt and Scheiner (2004, p. 2) with their flexible (but relatively common) definition of phenotypic plasticity as 'the environmentally sensitive production of alternative phenotypes by given genotypes'.

Phenotypic plasticity is commonly investigated using an analysis of variance approach for investigating the relative importance of the genotype (G), environment (E) and genotype × environment interaction (G×E) for the expression of the phenotype (P), which is usually a single trait. These relationships are commonly depicted graphically as reaction norm plots that display the value of the trait of interest in two or more environments (Fig. 4.1). This approach is central to the testing of the

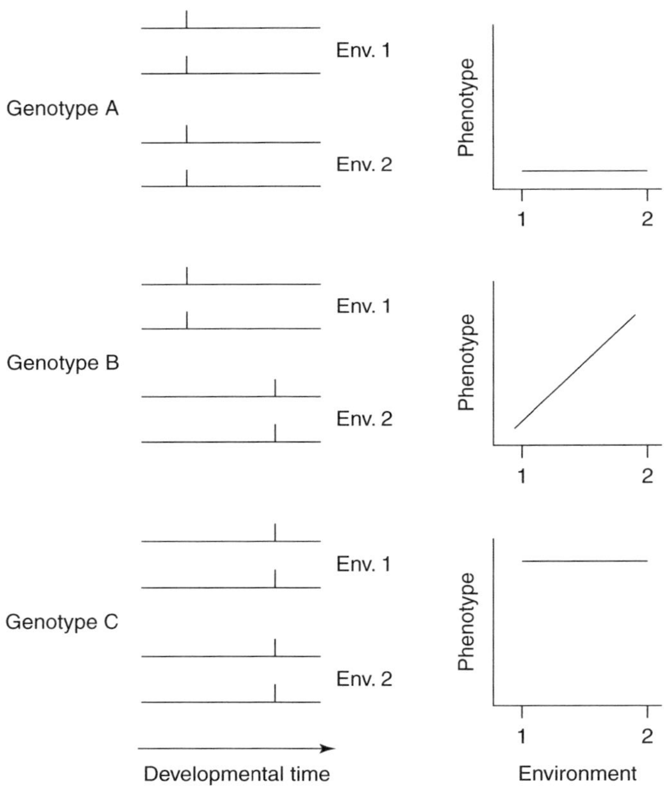

Fig. 4.1 Heterokairy and plasticity. Left-hand plots show the timing of onset of a developmental event for two individuals from each of three genotypes exposed to the same two environments. Right-hand plots depict reaction norms for the phenotypes (event timings). Genotype A shows an early timing but no plasticity of this event. Genotype B shows plasticity in the timing of this event. If this plasticity allows genotype B to colonize a new habitat (environment 2), the new timing of this event could be selected for and become fixed, leading to a loss of plasticity and potentially giving rise to a new, nonplastic genotype—genotype C

evolutionary potential of phenotypic (and as we shall see later, developmental) plasticity, and for inferring whether plasticity gives rise to optimal or suboptimal phenotypes in focal environments (see below).

4.2.2 Developmental Plasticity

The term developmental plasticity delineates a more specific strand of phenotypic plasticity that focuses on how alterations to developmental processes and outcomes shape such environmentally induced variation (West-Eberhard 1989, 2003, 2005; Fusco and Minelli 2010; Bateson and Gluckman 2011; Moczek et al. 2011; Mateus et al. 2014). However, it might also be argued that if development is how the genotype is translated into the phenotype, then all expressions of phenotypic plasticity are by definition developmental. Currently, however, plasticity is commonly examined with respect to the generation of morphologically different adult phenotypes, often ignoring earlier developmental stages (Spicer and Rundle 2007); hence, our understanding of developmental plasticity lags behind that of phenotypic plasticity more generally.

4.2.3 Heterokairy

The term heterokairy was originally coined within the context of comparative physiological development. Its original definition focused on intraspecific 'plasticity in the timing of the onset of physiological regulatory systems or their components' (Spicer and Burggren 2003) and has since been expanded to include intraspecific differences in the timing of morphological, behavioural and life history traits (see Spicer and Rundle 2007; Rundle and Spicer 2017 and references therein). The topic of timing has also been incorporated previously into the study of phenotypic plasticity, and specifically developmental plasticity, through the use of ontogenetic reaction norms (ORNs). ORNs add an additional time axis to the traditional reaction norm approach and can be used to test for significant genotype × environment × time interactions (Fig. 4.2) (Pigliucci et al. 1996; Pigliucci and Schlichting 1995; Schlichting and Pigliucci 1998). While ORNs enabled some investigation of the importance of variation in plasticity through developmental time, they have focused on changes in size and shape, rather than altered timing, through ontogeny. Hence, they have been used primarily to investigate changes in morphological traits (e.g. Hjelm et al. 2001; Corse et al. 2012; Arnqvist and Johansson 1998). In contrast, by fixing on the time of onset of developmental events, heterokairy can embrace all traits, i.e. morphological, physiological, behavioural, life history. Thus time of onset is the defining feature of heterokairy, and the results of altered timing can readily be expressed using the reaction norm approach that characterizes much plasticity research (Fig. 4.1).

We propose using heterokairy, a subset of developmental plasticity (Table 4.1), as an umbrella term that could provide an important way of bringing together previously disparate disciplines and groups which study altered timing, to share

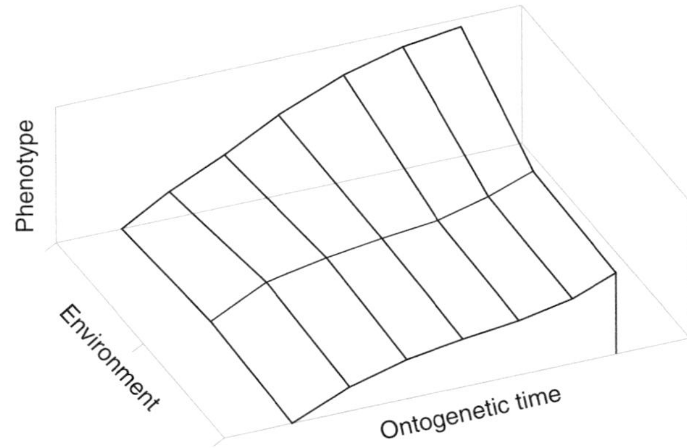

Fig. 4.2 Concept of developmental phenotypic plasticity illustrated using the developmental reaction norm. This is an extension of the genotypic reaction norm with the extra dimension of developmental time included. The graph above is taken from Pigliucci and Schlichting (1995) and shows plasticity (as the change in a hypothetical, multivariant phenotype). In this example, there is an interactive effect between ontogeny (time) and environment

knowledge and understanding. In the following sections, we set out an agenda which indicates what heterokairy is 'good for' outlining how this interdisciplinary term is best used to (a) incorporate substantial knowledge of (environmentally sensitive) genetic and hormonal control of the timing of developmental transitions within a more general biological framework (Sect. 4.2.3.1), (b) embed physiological heterokairy/plasticity within the field of developmental plasticity (Sects. 4.2.3.2 and 4.2.3.3), (c) contribute to the study of developmental plasticity in adaptive evolution (Sect. 4.2.3.3) and (d) aid understanding of the potential role of developmental plasticity in species' responses to environmental change (Sect. 4.2.3.4).

4.2.3.1 Is Heterokairy a Way of Integrating Disparate but Parallel Literatures that Investigate Altered Timing During Development?

While the term heterokairy gives prominence to the use of time as a way of framing questions on developmental plasticity, this is not to claim that timing has not figured strongly in other research within development and biology more generally. Indeed, work which involved altered timing of events (including physiological events) during development was being carried out long before the advent of the term heterokairy and continues to be carried out without reference to it (e.g. Bäurle and Dean 2006). The question is, how possible, or even desirable, is integration of these different and disparate strands to produce an understanding of heterokairy?

Below we review three main strands of research activity (the control of timing in plant and animal development; ecological transitions and how they are influenced by external factors; and ideas in evo-devo) that address this question before formulating an answer to the question that makes up the subheading of this section.

Developmental Transitions in Plants and Animals

The control of the timing of developmental transitions, and their effectors, during plant ontogeny has a long and distinguished history (Fig. 4.3). The environmental triggers, hormonal and molecular basis of three main transitions, germination, shift from juvenile to an adult vegetative phase and floral transition have received considerable attention (Huijser and Schmid 2011; Wagner 2016, 2017).

Particular interest has been drawn to parallels in the molecular basis of the vegetative phase change and the transition to flowering. This said, the identity of the floral promotor 'florigen' still alludes researchers although some of its components have recently come to light (Huijser and Schmid 2011). Work here has concentrated on the role of trans-acting small interfering RNA (tasiRNA), which when inhibited shortens the juvenile phase (Rougvie 2005). One of the main general take-home messages is that there is a striking commonality in the environmental effectors (i.e. day length, light and temperature change) that regulate the timing of the three main developmental transitions (Sun et al. 2014). However, the physiological evidence points to multiple pathways that affect timing changes; commonality in the identity of the environmental triggers belies the complexity of the mechanisms underpinning responses. Studying the control of developmental transitions in animals is also a major undertaking and has understandably attracted considerable financial and intellectual support. Pursuing this research agenda at different levels of organization has only been made tractable by concentrating on a few model organisms. While the fruit fly *Drosophila melanogaster* and the frog *Xenopus laevis* have been the focus of the role of hormones in metamorphosis, *Caenorhabditis elegans* has been key in understanding the genetic control of such transitions (Faunes and Larraín 2016). Explicit attempts are now underway to link the function of heterochronic genes (i.e. those that alter the timing of developmental processes), hormones and environmental cues in the onset and modification of developmental transitions to look for commonality of function. Such a comparative approach has also started to take place between kingdoms (Drost et al. 2017). Faunes and Larraìn (2016) erected a

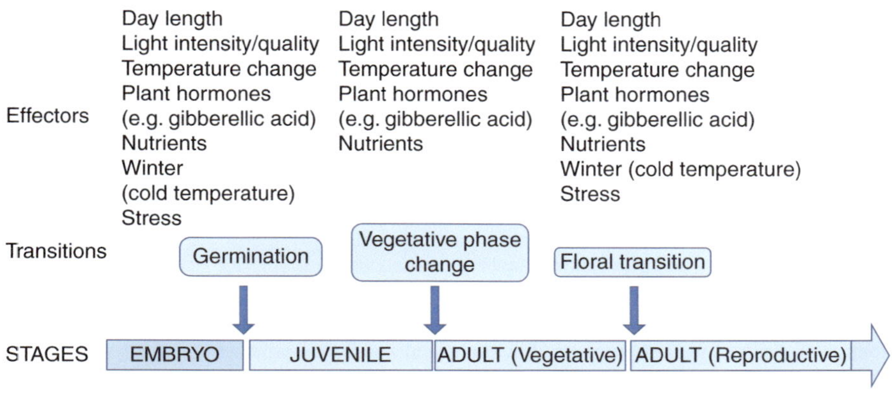

Fig. 4.3 Brief summary of key developmental transitions and their effectors during plant ontogeny (information from Fig. 1 in Bäurle and Dean (2006))

hypothesis that animal developmental transitions are regulated by an evolutionary conserved mechanism where heterochronic genes and hormones, open to environmental modulation, interact to determine developmental timing, via the control of stem/progenitor cell proliferation, cell cycle exit, quiescence and differentiation (Fig. 4.4). Interestingly, their integrative model does not include the ontogeny of physiological regulations or functions (perhaps with the exception of hormonal responses?) and how they interact with more 'conventional' developmental mechanisms. Nor, arguably, does it focus on altered timing of events although they are clearly included in their assessment. That said, they do hypothesize that hormones and genetic networks could be linked.

Environmental Modification of Developmental Transitions

A common assumption across different areas of predominantly ecological investigations is that environmental modulation of developmental transitions aligns development with favourable, environmental conditions. Thus a number of research foci have emerged, sometimes involving sections of the communities introduced above, sometimes not. These foci centre on environmental modulation and plasticity of key life history transitions, in vertebrates and invertebrates, including hatching (Dimichele and Taylor 1980; Martin 1999; Martin et al. 2011; Oyarzun and

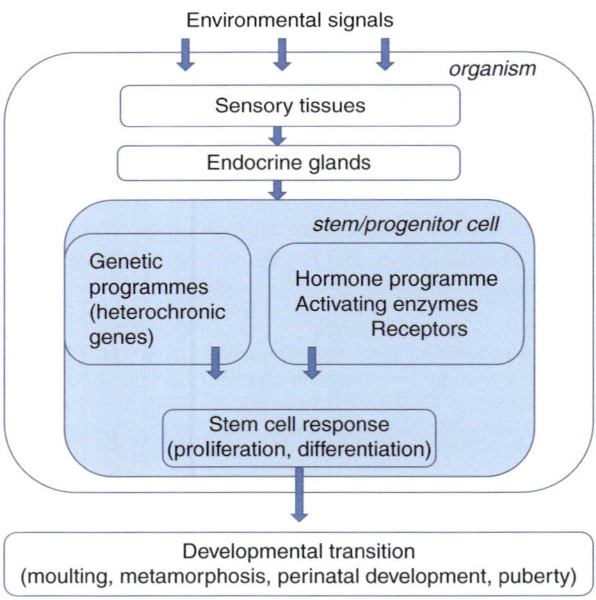

Fig. 4.4 Faunes and Larraín's (2016) attempt to integrate environmental, hormonal and genetic control of developmental transitions. Environmental conditions produce hormonal signals (via sensors), which transform environmental signals into specific hormones which act on stem and progenitor cells (whose timing of expression are controlled by heterochronic genes). They hypothesize that hormones, and genetic networks may be linked, thereby synchronizing physiological signals with intrinsic cellular programmes during developmental transitions

Strathmann 2011; Warkentin 2011; Collin et al. 2016), metamorphosis (Pechenik 1990; Forward et al. 2001; Voss and Smith 2005; Gebauer et al. 2003; Benard 2015), diapause (Podrabsky et al. 2016) and spawning (Loman 2009). That such ecological transitions are at the centre of heterokairy is clear, but it is worth noting that many such studies deal with the consequences rather than the mechanisms of altered timing.

Testing Ideas from the Evo-Devo Community

As well as potentially drawing on the substantial body of knowledge on the control of developmental transitions, the investigation of heterokairy could also provide additional opportunities to test some of the ideas arising from the evo-devo community. Quantifying plasticity in timing of multiple traits in response to different environmental stressors could be used to test for developmental modules (i.e. groups of traits that show similar relative timing in different environments) and levels of dissociation in different environments (i.e. environment-dependent shifts in timing of developmental modules) (Poe 2004; Raff and Raff 2000). Recent studies of heterokairy in the pond snail *Radix balthica* found that, under predation stress, both the timing of the onset of mantle muscle flexing and crawling during development had changed (Rundle et al. 2011) suggesting that these two developmental events might be part of the same developmental module. Hypersaline stress altered timing in different developmental events, showing that the sensitivity of different developmental modules is probably context-dependent in this species (Tills et al. 2010). Selection pressures could drive the dissociation of traits from the same developmental module if the optimum timing for each differed in a novel environment. It would also be informative to test how costs of plasticity (Auld et al. 2010; Murren et al. 2015) vary through development and quantify trade-offs between traits with different timing, e.g. earlier expression of a 'costly' trait could affect traits occurring later in development (Monaghan and Haussmann 2015) with implications for life history trajectories. Access to inbred lines which show heritable developmental events (Tills et al. 2013a) is invaluable, as is employing model species such as *C. elegans* for which the function of heterochronic genes has been most fully elucidated (Moss 2007; Faunes and Larraìn 2016). Finally, the approach of factoring sequence changes into the study of developmental plasticity could also be extended to include tests of quantitative genetic models for evolutionary change, including the potential role of epigenetic effects (Atchley and Zhu 1997).

Overall, the evo-devo communities have produced a wealth of empirical research and ideas which could be used even more widely than they have, to investigate patterns, mechanisms and consequences of the altered timings of developmental transitions, with an emphasis on the altered timing within an environmental framework and not solely the ontogeny of structures and functions.

Heterokairy Is a Way of Integrating Disparate but Parallel Research Interests

It is clear from the text above that there are three (and perhaps more) distinct strands, each with their own set of research questions and scientific communities, that could feed directly into our understanding of heterokairy. However, without a common focus that complements the interests of these communities, it is unlikely that there will be any cross-fertilization of ideas. We suggest that the notion of heterokairy is a focal point, providing a common direction and platform, enabling and marshalling a greater understanding of the mechanisms underlying, and the implications of, environmentally altered timing in the onset of events during development.

4.2.3.2 Evaluating Current Explicit Investigations of Heterokairy

Reintegration of Development, Evolution and Physiology

The (re)integration of development with evolution, pushed forward in seminal books by Gould (1977), Riedl (1978) and Raff and Kaufman (1983), has certainly shaped, informed and challenged the study of developmental plasticity. Initially the integration seemed to have little impact on comparative animal physiology (CAP). At the time CAP was only just moving away from adaptationist thinking (Gould and Lewontin 1979) towards a more rigorous and concerted approach to evolutionary physiology (Feder et al. 1987, 2000; Burggren and Bemis 1990; Garland and Carter 1994). In terms of integrating development with CAP, there are important, although until recently largely ignored, antecedents (e.g. Needham 1933; Adolph 1968). However, recent proponents, pre-eminently Burggren and his collaborators, have championed the integration of CAP with both development and evolution (Burggren 1992, 1998, 2000, 2005, 2006; Burggren and Just 1992; Burggren and Reyna 2011) and have established what is essentially an interdisciplinary convergence termed comparative developmental physiology (Burggren and Warburton 2005; Warburton et al. 2006) or, more recently, developmental integrative physiology (Mueller et al. 2015). This convergence has developmental plasticity at the heart of its interests and agenda; the investigation of inter- and intraspecific variation in the timing of physiological development, epigenetics, critical windows and the fitness consequences of such plasticity are key themes highlighted (Mueller et al. 2015). While many of these key themes deal with irreversible plasticity, that is not to say that more labile components have been excluded. What is clear is that with obvious common ground, there is great potential for interaction between this convergence and current plasticity research priorities.

Explicit Studies of Heterokairy

Current studies which explicitly set out to study physiological heterokairy involve a relatively wide range of taxa and environmental effectors. Nevertheless the two main take-home messages, while clear, are quite restricted in scope and often only provide small pieces of what is a very large puzzle. Firstly, the ontogeny of a particular regulation is affected by altered internal or external effectors, e.g. hyperoxia resulted in earlier maturation of respiratory phenotype in rats (Bavis et al. 2010), hypoxia delayed the onset of air breathing in blue gourami but advanced it in Siamese fighting fish (Mendez-Sanchez and Burggren 2012, 2014), and hypercapnia switched on enzyme activity of carbonic anhydrase earlier in chickens (De Smit et al. 2006).

Secondly, the ontogeny of certain components of a physiological regulation is affected by altered internal or external effectors. Here there are several examples: both hypoxia and hormone treatment resulted in an earlier onset and accelerated rate of lung surfactant (Blacker et al. 2004; Orgeig and Daniels 2009); hypothermia delayed the onset of thermoregulatory ability (Black and Burggren 2004); gestation of deer mice at altitude resulted in earlier rapid growth of organs that contribute to aerobic performance (Russell et al. 2008); hypoxia delayed the onset of cardiovascular function in zebrafish (Bagatto 2005); culture in hypersaline conditions resulted in a later onset of cardiac activity in estuarine snail (Tills et al. 2010); hypoxia accelerated appearance of functional larval and adult hearts in gastropod embryos (Rudin-Bitterli et al. 2016); alarm cues accelerated onset of muscle contraction and heart function in zebrafish (Mourabit et al. 2010); and exposure to predator cues resulted in delayed body flexing and earlier appearance of the eyes and heart in the snail *Radix auricularia* and late body flexing and earlier crawling/attachment to egg capsule wall in the congeneric *R. balthica* (Rundle et al. 2011).

Studies of altered timing in one or more physiological regulations, and a detailed study of the ontogeny of the constituent parts of each of those regulations at different hierarchical levels, as originally envisaged by Spicer and Burggren (2003) are, understandably, rarer in the literature because of the huge amount of work such studies entail. Most of what is already published that manages to integrate different hierarchical approach, though not (always) labelled as heterokairy, has been achieved by focusing on (1) the altered timing in salinity tolerance during ontogeny of anandromous salmonid fish and plasticity in the ontogeny of physiological regulations during marine-freshwater transitions (McCormick and Saunders 1987; Sakamoto et al. 1993; McCormick 2001, 2009, 2013; Varsamos et al. 2005; Evans 2008); (2) the role of hypoxia in the development of respiratory regulation, and its constituent parts, in brine shrimp (Spicer and El-Gamal 1999); (3) environmental and hormonal basis for the developmental transition from water to air by amphibian physiology (Burggren and Just 1992); and (4) physiological transitions at birth in higher mammals, including humans (Polin et al. 2011; Longo 2013).

Nested Hierarchy Hypothesis

It is with data from the type of studies presented in the previous section perhaps that we currently have the most extensive raw material to test a hypothesis of a nested hierarchy of patterns of physiological heterokairy as envisaged by Spicer and Burggren (2003) (Fig. 4.5). Implicit in this nested hierarchy hypothesis is the prediction that changes in the timing of onset at one hierarchical level will impact, or even determine, changes in the timing of onset at a higher hierarchical level. In other words, the hypothesis is that heterokairy at the level of physiological regulation is driven by heterokairy at the level of the tissues, which, in turn, is driven by heterokairy at the level of molecules. Spicer and Burggren (2003) flagged the possibility that heterokairy may be considered as a process as well as a description of pattern. But they, and subsequent papers (Spicer and Rundle 2007; Spicer et al. 2011), have cautioned that altered timing in a pattern during development does not necessarily have to reflect altered timing of processes or mechanisms that underpin that pattern.

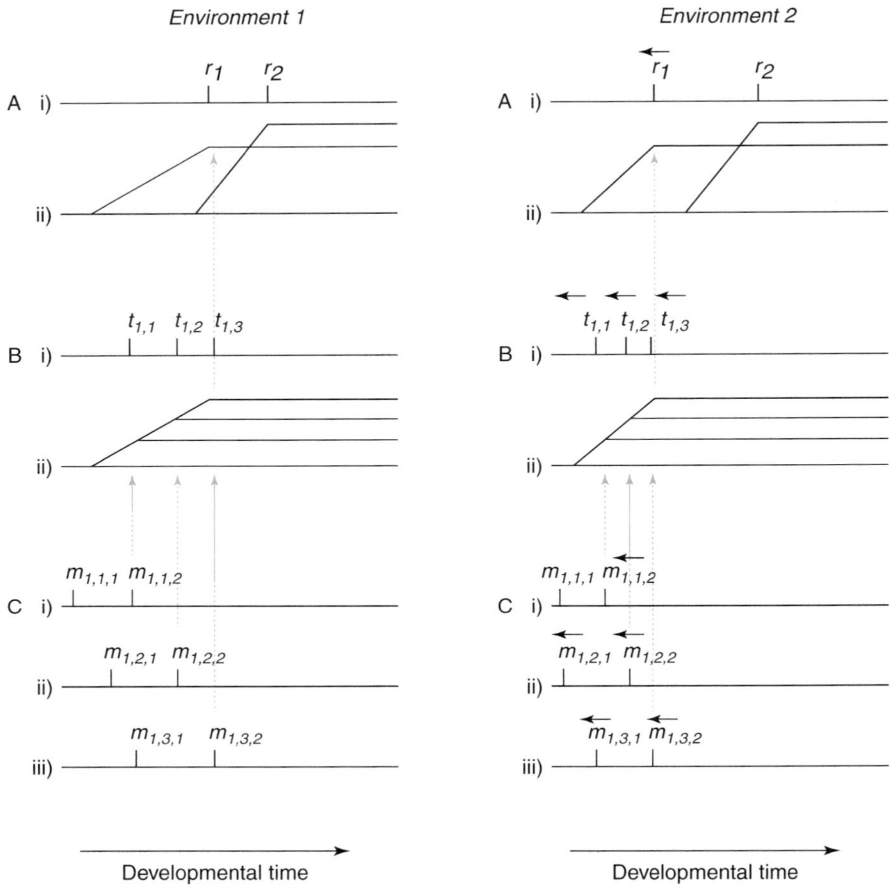

Fig. 4.5 The nested hierarchy hypothesis. As envisaged by Spicer and Burggren (2003), physiological heterokairy could be present at different functional levels, from the organismal (i.e. physiological regulation) to the molecular level. The lower levels are nested within the higher levels as depicted in this theoretical example. (**a**) The time of onset of two types of physiological variation (r_1 and r_2) with nominal values is depicted in two environments. r_1 shows heterokairy—it occurs earlier in environment 2 than in environment 1. (**b**) This regulation r_1 is dependent on the formation of three tissue types ($t_{1,1}$, $t_{1,2}$ and $t_{1,3}$) all of which have an earlier development in environment 2. (**c**) These tissues, in turn, are dependent on two molecular-level processes, and heterokairy occurs by one (i) or two (ii) of these showing early expressions. Note that the number of all of these levels could vary

To address this issue, the nested hierarchy hypothesis is erected as a testable model of the relationship between heterokairy as a pattern and as a process (Fig. 4.5).

Currently there are no explicit tests of this hypothesis. However, there are studies which allow us, to some extent, to test the prediction that changes in the timing of onset at one hierarchical level will determine changes in the timing of onset at a higher hierarchical level. Consider two examples which support the hypothesis by

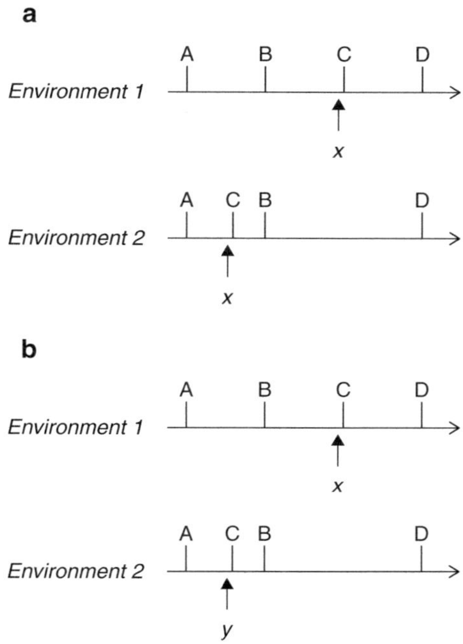

Fig. 4.6 A suggested mechanistic basis for patterns of physiological heterokairy. Heterokairy as a pattern may be due to (**a**) an homologous physiological mechanism appearing at a different time in development (in this case the mechanism x appears earlier in development in environment 1 compared with in environment 2) and (**b**) an analogous physiological mechanism appearing at a different time in development (in this case, in environment 2, the mechanism y occurs earlier in development leading to the same functional physiological feature earlier than that due to mechanism x in environment 1) (After Spicer and Rundle 2007)

upholding this prediction. First, the environmentally sensitive proliferation of gill Na/K ATPase and chloride cells which determines osmoregulation and salinity tolerance appears to provide support for the hypothesis, i.e. heterokairy at the molecular and cellular levels, and drives heterokairy at the level of the tissues and then the regulation which those tissues produce (McCormick 2013). Second, the early production of a respiratory pigment with a high O_2 affinity, as a result of exposure to reduced environmental O_2, brings forward the onset of respiratory regulation that normally takes place during the transition from planktonic larva to benthic juvenile in Norway lobster *Nephrops norvegicus* (Spicer 1995; Spicer and Eriksson 2003). Heterokairy in the production of a respiratory pigment with a high O_2 affinity is what drives the heterokairy observed at the level of whole animal respiratory regulation.

However, we should be cautious, as we also seem to have cases where the prediction is not upheld. First, Pedersen et al. (2013) seem to show that the genes involved in early smolting of the Atlantic salmon *Salmo salar* are different from those that 'normally' control smolting. In other words, at the gene expression level, early smolting is not merely switching the same genes on earlier, i.e. heterokairy in function may not be the result of heterokairy at the molecular level. Second, the early onset of an adult pattern of respiratory regulation in brine shrimp initiated by culture under reduced O_2 does not result from altered timing of the appearance of the heart and gills which accompanies the appearance of this regulation under normoxic conditions (Spicer and El-Gamal 1999). Instead, it is the hypoxia-induced production of large concentrations of a high affinity haemoglobin pigment which produces the early onset of the adult pattern of respiratory regulation. Therefore, the heterokairy involved in bringing the adult pattern of respiratory regulation earlier involves a novel mechanism (production of a haemoglobin with a high O_2 affinity) and not a change in timing of the appearance of the gas exchange surfaces and cardiovascular system which are invariant in their ontogeny.

In summary we have evidence that supports, and evidence that seems to reject, the nested hierarchy hypothesis. But more targeted studies crafted to its predictions are still urgently required.

Adaptive Plasticity and Flexible Stem Hypotheses

There are near analogues of the nested hierarchy hypothesis, applied above the level of the species, which have been investigated in the phenotypic plasticity literature. These are the adaptive plasticity hypothesis, which uses phylogenetically controlled analyses to test for correlations between plasticity and environmental tolerance amongst species (Van Buskirk 2002) and the flexible stem hypothesis, which investigates plasticity in ancestral species as a source for speciation (West-Eberhard 2003; Gomez-Mestre and Buchholz 2006; Wund et al. 2008). These hypotheses have not been formally tested in the context of heterokairy and its microevolutionary potential, but there is no reason why they shouldn't be. That heterokairy is an obvious candidate mechanism for the evolution of species-specific differences in timing (i.e. heterochrony) has been highlighted on numerous occasions (Raff 1992, p. 211; Raff and Wray 1989; Strathmann et al. 1992; Spicer et al. 2011; Heyland and Hodin 2004; Keyte and Smith 2014; Poullet et al. 2016; Spicer 2006; Spicer and Rundle 2006). Recent studies have demonstrated considerable standing variation in event timing exists (Tills et al. 2013a, b; Rundle et al. 2011; De Jong et al. 2009). Some of this standing variation seems to have some genetic basis, i.e. it is heritable (Tills et al. 2011), and a parent-offspring comparison also revealed heritability in the timing of some specific developmental events (Tills et al. 2013a).

To conclude, we have made good progress in investigating heterokairy explicitly and both the nested hierarchy hypothesis and the flexible stem hypothesis show the benefits of closer integration of heterokairy and the developmental plasticity literature. However, in terms of producing more fully nested and complete studies of heterokairy, the construction of a nested hierarchy of physiological function, going from molecular mechanisms to whole organism function as envisaged by Spicer and Burggren (2003) (Fig. 4.5), still seems far off at present although we have no reason yet to believe that it is not obtainable.

4.2.3.3 Heterokairy and Adaptive Evolution

Recent calls for a reappraisal of the evolutionary synthesis (Laland et al. 2014) suggest that phenotypic plasticity should be given more prominence as a potential driver of evolutionary change. The adaptive potential of plasticity has been hypothesized to occur through mechanisms such as phenotypic and genetic accommodation and genetic assimilation (Pigliucci et al. 2006; Suzuki and Nijhout 2006; Crispo 2007; Sikkink et al. 2014), and there has been increased emphasis on investigating the potential for plasticity to drive evolution. If adaptive plasticity produces a phenotype close to the optimal for a particular habitat, it should be subject to stabilizing but not directional selection. As such it is unlikely to evolve. However, if the phenotype produced is suboptimal for the new, or changed, habitat but still survives, then it is likely to be subject to directional selection and evolution is more likely (Price et al. 2003), particularly if there are costs of such plasticity (Murren et al. 2015).

That adaptive plasticity could result in evolutionary change is perhaps unsurprising. However, recently it has been suggested that non-adaptive plasticity could play a role in evolution and may be more likely to give rise to evolutionary change than adaptive plasticity (Ghalambor et al. 2015). When a population encounters a new environment, the manifest plastic response of its constituent individuals, in many cases, is the production of a maladaptive phenotype (i.e. one that diverges from the optimal phenotype) that does not persist in the new environment. However, if the response includes an increase in the variation of phenotypes expressed revealed through cryptic genetic variation (Le Rouzic and Carlborg 2008), some of the phenotypic variants may be close to the optimum for the new environment and so be subject to selection (Ghalambor et al. 2007) (see Fig. 4.7).

The study of heterokairy can be seen to fit within this interest in the adaptive potential of phenotypic plasticity. Indeed there are several examples of studies on heterokairy that illustrate it can be non-adaptive. Some species alter their developmental itinerary in response to altered environmental conditions, and this has associated costs (Spicer and Rundle 2007; Spicer et al. 2011). Brine shrimp *Artemia franciscana* responded to culture under reduced O_2 (50% air saturation) by bringing forward in chronological and developmental time the adult ability to oxyregulate over a wide range of external O_2 tensions. This heterokairy enabled individuals to survive to become reproductive adults and so was adaptive within this sense, but there were costs associated with it in the form of reduced reproductive output (Spicer and El-Gamal 1999), an example of a cost of plasticity that could drive selection in the new environment. A recent study also showed that those individuals of the intertidal snail, *Littorina obtusata*, that survived embryonic development in chronically low O_2 displayed an earlier onset of their adult heart beat than those that died during culture under these conditions (Rudin-Bitterli et al. 2016). In this example, the large variation in the new environment was, on average, non-adaptive, but for the proportion of individuals that did survive, the timing of heart development was key and, again, might be selected for (Fig. 4.7). These examples of where stress increases phenotypic variance and where clear measures of fitness are possible offer useful models for exploring the role of heterokairy in non-adaptive plasticity.

4.2.3.4 Heterokairy and Global Environmental Change

Our planet is experiencing unprecedented changes in environmental conditions (IPCC 2014), and so understanding species responses to these altered conditions is one of the greatest challenges biologists face in the twenty-first century (Pecl et al. 2017; Wade et al. 2017). Phenotypic plasticity is viewed as an important way that species can survive (and adapt—see above) to changing/new environments and developmental timing, and hence, heterokairy is a critically important component of such responses.

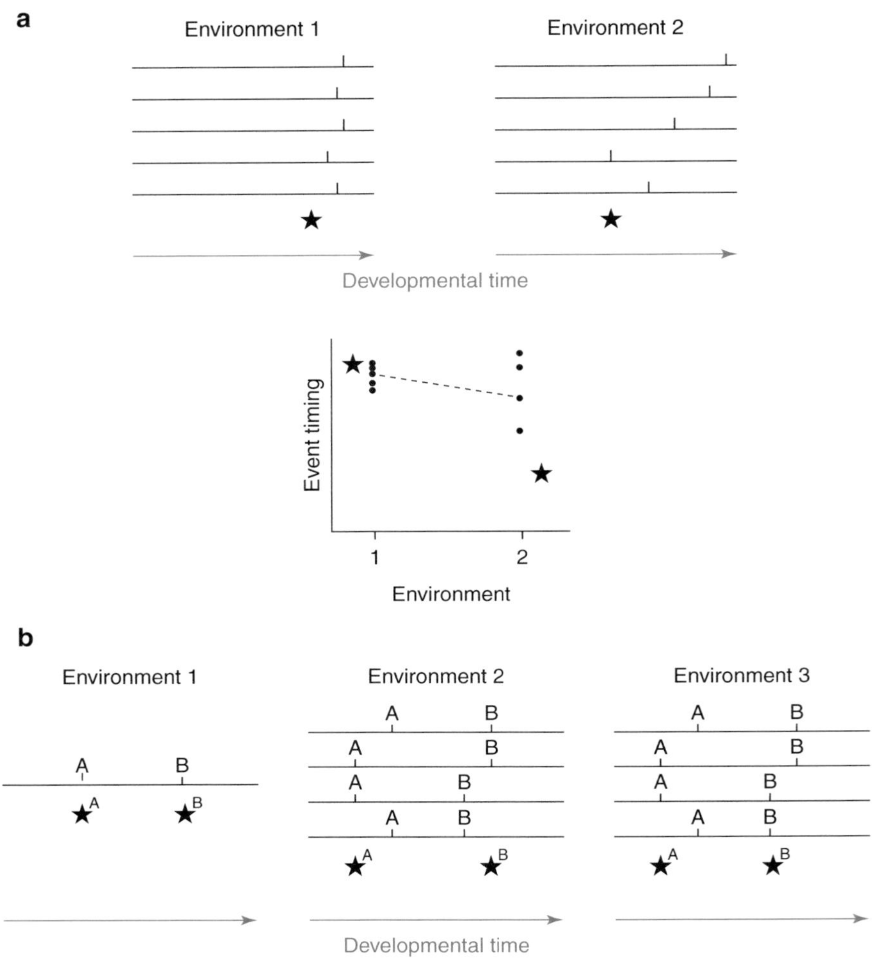

Fig. 4.7 (**a**) Non-adaptive heterokairy. The timing of a single developmental event is shown for ten individuals of the same species, five in each of two different environments. These event timings are projected onto a reaction norm plot. The optimal timing for this event at each site is indicated by a black star—at site 2, an earlier expression of the event is optimal. The reaction norm (dashed line), based on the mean event timing at each site, shows that the plasticity exhibited is non-adaptive as the mean falls short of the (earlier) optimal event timing value for site 2. However, the high phenotypic variation in site 2 means that one individual has an event timing that is close to the optimum. This could lead to directional selection in site 2 (After Rundle and Spicer 2017). (**b**) Heterokairy in multiple events. This plot depicts the timing of two developmental events (A and B) in an ancestral (environment 1) and two descendent (environments 2 and 3) environments. Four possible scenarios for the relative timing of two traits are shown for each of two 'descendent' environments. The optimal event timing for each trait in each environment is indicated by a star: in environment 2 the optimal timing has shifted early for event A, but not event B; in environment 3 the optimal timing has shifted earlier for both events. Note that where the timing of trait B is associated with trait A (suggesting they might be part of the same developmental module), the timing of trait B is suboptimal in environment 2 but optimal in environment 3, whereas dissociation between events A and B leads to optimal timing of event B in environment 2 but suboptimal timing in environment 3 (After Rundle and Spicer 2017)

A powerful approach in this regard is where species are exposed to novel environments and can persist under the new conditions through plasticity, allowing a comparison of ancestral and derived populations (e.g. Scoville and Pfrender 2010). There are currently a large number of ecosystems that are undergoing substantial changes to their environmental conditions through biotic (e.g. introduced species) or abiotic drivers (e.g. increasing air and sea temperatures and levels of hypoxia in the sea). Several of the examples of physiological heterokairy outlined here, and in previous reviews of heterokairy (Spicer and Rundle 2007; Spicer et al. 2011), concerned the effects of such drivers associated with anthropogenic activities, suggesting that the ability of species to respond to environmental change may, in part, relate to their ability to show plasticity in the timing of developmental events such as physiological variation. Again, this strengthens the case for the inclusion of heterokairy within a developmental plasticity framework.

As well as species showing altered timing of developmental events in response to altered environmental conditions, major climate-driven changes in life history strategies might also impact developmental timing (Rudolf and Singh 2013). Such phenological variation affects many aspects of a species' ecology, such as the duration of life stages and the extent of temporal overlap with competitors, predators and mutualists (Forrest and Miller-Rushing 2010; Benard 2015). For many species, the timing of breeding, a key life history event, has shifted earlier in the year in response to changing climate (Root et al. 2003; Parmesan 2006). Shifts in the date of breeding can be large, and the average rate of change in breeding date for many taxonomic groups is approximately 5–7 days earlier per decade (Root et al. 2003). Such phenological shifts may affect how species persist in the face of climate change, and thus understanding the consequences of variation in the timing of breeding is critical for understanding the effects of climate change (Bellard et al. 2012; Cleland et al. 2012). However, few studies have evaluated the extent to which breeding early affects the growth, development and survival of individuals (Miller-Rushing et al. 2010; Yang and Rudolf 2010), i.e. we still know little of the effect of environmental change on heterokairy. Temperature and perhaps more importantly light have been highlighted as playing key roles in controlling the timing of developmental events in many biological systems, determining fitness in seasonal environments, and initiating physiological changes which prepare organisms to develop, reproduce, enter or come out of dormancy/hibernation and initiate migration (Bradshaw and Holzapfel 2010). As organisms must rely on current environmental cues as predictors of future environmental conditions, the rapid rate of current environmental change could distort the quality of information taken from cues to make developmental decisions (Bradshaw and Holzapfel 2010). Thus scope for heterokairy may be vital if such organisms, through restricted plasticity, are not to be caught in developmental traps where rapid environmental change forces them into potentially maladaptive developmental decisions (The lost generation hypothesis; Van Dyck et al. 2015).

4.3 Heterokairy: The 'Big Picture'

Clearly heterokairy is closely related to the notions of phenotypic plasticity and developmental plasticity presented above, although perhaps this was not originally made as explicit as it could have been. Here we have made the case that it is a subset of developmental plasticity. Spicer and Burggren (2003) suggested the term because it focused specifically on plasticity in the *timing* of onset, not primarily on the *nature of the change* at the time of onset. The notion expressed by Heraclitus (540–480 B.C.), 'You could not step twice into the same river, for other waters are ever flowing on to you' (Jones 1931), is central to the notion of heterokairy. This is also why heterokairy is not just phenotypic flexibility—for although the processes and mechanisms involved could be reversible, the timing of their appearance is not (Table 4.1). The notion of heterokairy is one way in which disparate literatures (e.g. comparative animal physiology, evo-devo, phenology) and the multitude of different scientific communities interested (explicitly or implicitly) in altered timing in development (or indeed at any point in an organism's life cycle) could extract a common currency of exchange and thereby construct a 'bigger picture' of heterokairy.

The emphasis on the influence of the environment inherent in studying plasticity and timing in particular has been an exciting, welcome and all pervasive change in biology. But the implications are still emerging. Evolutionary biology has recently given more credence to the roles of developmental bias and noise, epigenetics and niche construction in evolutionary change (Laland et al. 2014; Noble et al. 2014; Pigliucci and Müller 2010; Arthur 2004; West-Eberhard 2003), and it should also be so for heterokairy.

It is conceivable that at least some of the variation in event timing between individuals is produced by 'noise'. This noise could be extrinsic, originating from fluctuations in the environment mediated to the organism, or intrinsic, perturbations that arise from random fluctuations at the molecular/cellular level ('developmental noise'), whose causes are still incompletely known. Such intrinsic variation is often perceived as negative. However, Woods (2014) suggested that such intrinsic noise could produce, 'simultaneous, spatially distributed plasticity' (mosaic physiology) which could be adaptive in combination with phenotypic plasticity. Such a possibility is exciting and has yet to be explored within the context of heterokairy.

References

Adolph EF (1968) Origins of physiological variation. Academic Press, New York

Arnqvist G, Johansson F (1998) Ontogenetic reaction norms of predator-induced defensive morphology in dragonfly larvae. Ecology 79:1847–1858

Arthur W (2004) Biased embryos and evolution. Cambridge University Press, Cambridge

Atchley WR, Zhu J (1997) Developmental quantitative genetics, conditional epigenetic variability and growth in mice. Genetics 147:765–776

Auld JR, Agrawal AA, Relyea RA (2010) Re-evaluating the costs and limits of adaptive phenotypic plasticity. Proc R Soc B Biol Sci 277:503–511

Bagatto B (2005) Ontogeny of cardiovascular control in zebrafish (*Danio rerio*): effects of developmental environment. Comp Biochem Physiol A Mol Integr Physiol 141:391–400

Bateson P, Barker D, Clutton-Brock T, Deb D, D'Udine B, Foley RA, Gluckman P, Godfrey K, Kirkwood T, Lahr MM, McNamara J, Metcalfe NB, Monaghan P, Spencer HG, Sultan SE (2004) Developmental plasticity and human health. Nature 430:419–421

Bateson P, Gluckman P (2011) Plasticity, robustness, development and evolution. Cambridge University Press, Cambridge

Bäurle I, Dean C (2006) The timing of developmental transitions in plants. Cell 125:655–664
Bavis RW, Young KM, Barry KJ, Boller MR, Klein PM, Ovrutsky AR, Rampersad DA (2010) Chronic hyperoxia alters the early and late phases of the hypoxic ventilatory response in neonatal rats. J Appl Physiol 109:796–803
Bellard C, Bertelsmeier C, Leadley P, Thuiller W, Courchamp F (2012) Impacts of climate change on the future of biodiversity. Ecol Lett 15:365–377
Benard MF (2015) Warmer winters reduce frog fecundity and shift breeding phenology, which consequently alters larval development and metamorphic timing. Glob Chang Biol 21:1058–1065
Black JL, Burggren WW (2004) Acclimation to hypothermic incubation in developing chicken embryos (*Gallus domesticus*). J Exp Biol 207:1543–1552
Blacker HA, Orgeig S, Daniels CB (2004) Hypoxia control of the development of the surfactant system in the chicken: evidence for physiological heterokairy. Am J Physiol Regul Integr Comp Physiol 287:R403–R410
Bradshaw WE (1965) Evolutionary significance of phenotypic plasticity in plants. Adv Genet 13:115–155
Bradshaw WE, Holzapfel CM (2010) Light, time, and the physiology of biotic response to rapid climate change in animals. Annu Rev Physiol 72:147–166
Burggren WW (1992) The importance of an ontogenic perspective in physiological studies: amphibian cardiology as a case study. In: Wood SC, Weber R, Hargens A, Millard R (eds) Physiological adaptations in vertebrates: respiration, circulation and metabolism. Dekker, New York, pp 235–253
Burggren WW (1998) Studying physiological development: past, present and future. Biol Bull Nat Taiwan Normal Univ 33:71–84
Burggren WW (2000) Developmental physiology, animal models, and the August Krogh principle. Zoology 102:148–156
Burggren WW (2005) Developing animals flout assumptions of ecological physiology. Comp Biochem Physiol A Mol Integr Physiol 141:430–439
Burggren WW (2006) Complexity change during physiological development. In: Warburton S, Burggren W, Pelster B, Reiber CL, Spicer JI (eds) Comparative developmental physiology: contributions, tools and trends. Oxford University Press, Oxford, pp 174–190
Burggren WW, Bemis T (1990) Studying physiological evolution: paradigms and pitfalls. In: Nltecki MH (ed) Evolutionary innovations: patterns and processes. Oxford University Press, Oxford, pp 191–228
Burggren WW, Just JJ (1992) Developmental changes in amphibian physiological systems. In: Feder ME, Burggren WW (eds) Environmental physiology of the Amphibia. University of Chicago Press, Chicago, pp 467–530
Burggren WW, Reyna KS (2011) Developmental trajectories, critical windows and phenotypic alteration during cardio-respiratory development. Respir Physiol Neurobiol 178:13–21
Burggren WW, Warburton S (2005) Comparative developmental physiology: an interdisciplinary convergence. Annu Rev Physiol 67:203–223
Cleland EE, Allen JM, Crimmin TM, Dunne JA, Pau S, Travers SE, Zavaleta ES, Wolkovich EM (2012) Phenological tracking enables positive species responses to climate change. Ecology 93:1765–1771
Collin R, Roof KE, Spangler A (2016) Hatching plasticity in the tropical gastropod *Nerita scabricosta*. Invertebr Biol 135:87–96
Corse E, Neve G, Sinama M, Pech N, Costedoat C, Chappaz R, Gilles A (2012) Plasticity of ontogenetic trajectories in cyprinids: a source of evolutionary novelties. Biol J Linn Soc 106:342–355
Crispo E (2007) The Baldwin effect and genetic assimilation: revisiting two mechanisms of evolutionary change mediated by phenotypic plasticity. Evolution 61:2469–2479
Darwin C (1859) On the origin of species by means of natural selection, or the preservation of favoured races in the struggle for life. John Murray, London
De Jong IML, Colbert MW, Witte F, Richardson MK (2009) Polymorphism in developmental timing: intraspecific heterochrony in a Lake Victoria cichlid. Evol Dev 11:625–635

De Smit L, Bruggeman DV, Tona JK, Debonne M, Onagbesan O, Arckens L, De Baerdemaeker J, Decuypere E (2006) Embryonic developmental plasticity of the chick: increased CO_2 during early stages of incubation changes the developmental trajectories during prenatal and postnatal growth. Comp Biochem Physiol A Mol Integr Physiol 145:166–175
De Witt TJ (2016) Expanding the phenotypic plasticity paradigm to broader views of trait space and ecological function. Curr Zool 62:463–473
DeWitt TJ, Scheiner SM (eds) (2004) Phenotypic plasticity: functional and conceptual approaches. Oxford University Press, Oxford
Dimichele L, Taylor MH (1980) The environmental control of hatching in *Fundulus heteroclitus*. J Exp Zool 214:181–187
Drost HG, Janitza P, Grosse I, Quint M (2017) Cross-kingdom comparison of the developmental hourglass. Curr Opin Genet Dev 45:69–75
Evans DH (2008) Teleost fish osmoregulation: what have we learned since August Krogh, Homer Smith and Ancel Keys. Am J Physiol Regul Integr Comp Physiol 295:R704–R713
Faunes F, Larraín J (2016) Conservation in the involvement of heterochronic genes and hormones during developmental transitions. Dev Biol 416:3–17
Feder MA, Bennett AF, Burggren WW, Huey RB (eds) (1987) New directions in ecological physiology. Cambridge University Press, New York
Feder MA, Bennett AF, Huey RB (2000) Evolutionary physiology. Annu Rev Ecol Syst 31:315–341
Forrest J, Miller-Rushing AJ (2010) Towards a synthetic understanding of the role of phenology in ecology and evolution. Philos Trans R Soc B 365:3101–3112
Forsman A (2015) Rethinking phenotypic plasticity and its consequences for individuals, populations and species. Heredity 115:276–284
Forward RB, Tankersley RA, Rittschof D (2001) Cues for metamorphosis of brachyuran crabs: an overview. Am Zool 41:1108–1122
Fusco G, Minelli A (2010) Phenotypic plasticity in development and evolution: facts and concepts. Philos Trans R Soc B 365:547–556
Garland T Jr, Carter PA (1994) Evolutionary physiology. Annu Rev Physiol 56:579–621
Gebauer P, Paschke K, Anger K (2003) Delayed metamorphosis in decapod crustaceans: evidence and consequences. Rev Chil Hist Nat 76:169–175
Ghalambor CK, Hoke KL, Ruell EW, Fischer EK, Reznick DN, Hughes KA (2015) Non-adaptive plasticity potentiates rapid adaptive evolution of gene expression in nature. Nature 525:372–375
Ghalambor CK, McKay JK, Carroll SP, Reznick DN (2007) Adaptive versus non-adaptive phenotypic plasticity and the potential for contemporary adaptation in new environments. Funct Ecol 21:394–407
Gomez-Mestre I, Buchholz DR (2006) Developmental plasticity mirrors differences among taxa in spadefoot toads linking plasticity and diversity. Proc Natl Acad Sci U S A 103:19021–19026
Gould SJ (1977) Ontogeny and phylogeny. Belknap, Cambridge
Gould SJ, Lewontin RC (1979) The spandrels of San Marco and the Panglossian paradigm: a critique of the adaptationist programme. Proc R Soc Lond B Biol Sci 205:581–598
Harvey W (1628) An anatomical disputation concerning the movement of the heart and blood in living creatures. Translated by Whitteridge G (1976). Blackwell, Oxford
Heyland A, Hodin J (2004) Heterochronic developmental shift caused by thyroid hormone in larval sand dollars and its implications for phenotypic plasticity and the evolution of nonfeeding development. Evolution 58:524–538
Hjelm J, Svanbäck R, Byström P, Persson L, Wahlström E (2001) Diet-dependent body morphology and ontogenetic reaction norms in Eurasian perch. Oikos 95:311–323
Huijser P, Schmid M (2011) The control of developmental phase transitions in plants. Development 138:4117–4129
IPCC (2014) Climate change 2014: synthesis report. Contribution of working groups I, II and III to the Fifth Assessment Report of the Intergovernmental Panel on Climate Change [Core Writing Team, Pachauri RK, Meyer LA (eds.)]. IPCC, Geneva, p 151
Jones WHS (1931) Hippocrates volume IV on the Universe by Heraclitus. Loebe Classical Library

Keyte AL, Smith KK (2014) Heterochrony and developmental timing mechanisms: changing ontogenies in evolution. Semin Cell Dev Biol 34:99–107
Laland K, Uller T, Feldman M, Sterelny K, Müller GB, Moczek A, Jablonka E, Odling-Smee J (2014) Does evolutionary theory need a rethink? Nature 514:161–164
Le Rouzic A, Carlborg Ö (2008) Evolutionary potential of hidden genetic variation. Trends Ecol Evol 23:33–37
Loman J (2009) Primary and secondary phenology. Does it pay a frog to spawn early? J Zool (Lond) 279:64–70
Longo LD (2013) The rise of fetal and neonatal physiology. Basic science to clinical care. Perspectives in Physiology, vol 1. Springer, New York
Martin K, Bailey K, Moravek C, Carlson K (2011) Taking the plunge: Californian grunion embryos emerge rapidly with environmentally cued hatching. Integr Comp Biol 51:26–37
Martin KL (1999) Ready and waiting: delayed hatching and extended incubation of anamniotic vertebrate terrestrial eggs. Am Zool 39:279–288
Mateus ARA, Narques-Pita M, Oostra V, Lafuente E, Brakefield PM, Zwaan BJ, Beldade P (2014) Adaptive developmental plasticity: compartmentalized responses to environmental cues and to corresponding internal signals provide phenotypic flexibility. BMC Biol 12:97
McCormick SD (2001) Endocrine control of osmoregulation in teleost fish. Am Zool 41:781–794
McCormick SD (2009) Evolution of the hormonal control of animal performance: insights from the seaward migration of salmon. Integr Comp Biol 49:408–422
McCormick SD (2013) Smolt physiology and endocrinology. Fish Physiol 32:199–251
McCormick SD, Saunders RI (1987) Preparatory physiological adaptations for marine life in salmonids: osmoregulation, growth and metabolism. Am Fish Soc Symp 1:211–229
Mendez-Sachez JF, Burggren W (2012) Modulation of the onset of air-breathing of the Siamese fighting fish and the blue gourami. FASEB J 31:1071–1079
Mendez-Sanchez JF, Burggren W (2014) Environmental modulation of the onset of air-breathing and survival of *Betta splendens* and *Trichopodus trichopterus*. J Fish Biol 84:794–807
Miller-Rushing AJ, Høye TT, Inouye DW, Post E (2010) The effects of phenological mismatches on demography. Philos Trans R Soc B 365:3177–3186
Moczek AP, Sultan S, Foster S, Ledón-Rettig C, Dworkin I, Nijhout HF, Abouheif E, Pfennig DW (2011) The role of developmental plasticity in evolutionary innovation. Proc R Soc Lond B Biol Sci 278:2705–2713
Monaghan P, Haussmann MF (2015) The positive and negative consequences of stressors during early life. Early Hum Dev 91:643–647
Moravek CL, Martin K (2011) Life goes on: delayed hatching, extended incubation, and heterokairy in development of embryonic California grunion, *Leuresthes tenuis*. Copeia 2011:308–314
Moss EG (2007) Heterochronic genes and the nature of developmental time. Curr Biol 17:R425–R434
Mourabit S, Rundle SD, Spicer JI, Sloman KA (2010) Alarm substance from adult zebrafish alters early embryonic development in offspring. Biol Lett 6:525–528
Mueller CA, Eme J, Burggren WW, Roghair RD, Rundle SD (2015) Challenges and opportunities in developmental integrative physiology. Comp Biochem Physiol A Mol Integr Physiol 184:113–124
Muratori FB (2010) Heterokairy as an anti-predator strategy for parasitic species. Commun Integr Biol 3:309–312
Muratori FB, Borlee S, Messing RH (2010) Induced niche shift as an anti-predator response for an endoparasitoid. Proc R Soc Lond B Biol Sci 277:1475–1480
Murren CJ, Auld JR, Callahan H, Ghalambor CK, Handelsman CA, Heskel MA, Kingsolver JG, Maclean HJ, Masel J, Maughan H et al (2015) Constraints on the evolution of phenotypic plasticity: limits and costs of phenotype and plasticity. Heredity 115:293–301
Needham J (1933) On the dissociability of the fundamental processes in ontogenesis. Biol Rev 8:180–223
Noble D, Jablonka E, Joyner MJ, Müller GB, Omholt SW (2014) Evolution evolves: physiology returns to centre stage. J Physiol 592:2237–2244

Orgeig S, Daniels CB (2009) Environmental selection pressures shaping the pulmonary surfactant system of adult and developing lungs. In: Glass ML, Wood SC (eds) Cardio-respiratory control in vertebrates. Springer, Berlin, pp 205–239
Oyarzun FX, Strathmann RR (2011) Plasticity of hatching and the duration of planktonic development in marine invertebrates. Integr Comp Biol 51:81–90
Parmesan C (2006) Ecological and evolutionary responses to recent climate change. Annu Rev Ecol Evol Syst 37:637–669
Pechenik JA (1990) Delayed metamorphosis by larvae of benthic marine invertebrates: does it occur? Is there a price to pay? Ophelia 32:63–94
Pecl GT, Araujo MB, Bell J, Blanchard J, Bonebrake TC, Chen I, Clark TD, Colwell RK, Danielsen F, Evengard B, Robinson S et al (2017) Biodiversity redistribution under climate change: impacts on ecosystems and human well-being. Science 355:1–9
Pedersen S, Berg PR, Culling M, Danzmann RG, Glebe B, Leadbetter S, Lien S, Moen T, Vandersteen W, Boulding EG (2013) Quantitative trait loci for precocious parr maturation, early smoltification, and adult maturation in double-backcrossed trans-Atlantic salmon (*Salmo salar*). Aquaculture 410-411:164–171
Piersma T, Drent J (2003) Phenotypic flexibility and the evolution of organismal design. Trends Ecol Evol 18:228–233
Pigliucci M (1998) Developmental phenotypic plasticity: where internal programming meets external environment. Curr Opin Plant Biol 1:87–91
Pigliucci M (2001) Phenotypic plasticity: beyond nature and nurture. Johns Hopkins University Press, Baltimore
Pigliucci M (2010) Genotype-phenotype mapping and the end of the 'genes as blueprint' metaphor. Philos Trans R Soc B 365:557–566
Pigliucci M, Müller GB (2010) Evolution: the extended synthesis. MIT Press, Cambridge
Pigliucci M, Murren CJ, Schlichting CD (2006) Phenotypic plasticity and evolution by genetic assimilation. J Exp Biol 209:2362–2367
Pigliucci M, Schlichting CD (1995) Ontogenetic reaction norms in *Lobelia siphilitica* (Lobeliaceae): response to shading. Ecology 76:2134–2144
Pilgliucci M, Schlichting CD, Jones CS, Schwenk K (1996) Developmental reaction norms: the interactions among allometry, ontogeny and plasticity. Plant Species Biol 11:69–85
Podrabsky JE, Romney AL, Culpepper KM (2016) Alternative developmental pathways. In: Berois N, García G, de Sá RO (eds) Annual fishes. Life history strategy, diversity and evolution. CRC Press, Taylor & Francis Group, Boca Raton, pp 63–74
Poe S (2004) A test for patterns of modularity in sequences of developmental events. Evolution 58:1852–1855
Polin RA, Fox WW, Abman SH (2011) Fetal and neonatal physiology (2 volumes). Saunders, Elsevier, Philadelphia
Poullet N, Vielle A, Gimond C, Carvalho S, Teotónio H, Braendle C (2016) Complex heterochrony underlies the evolution of *Caenorhabditis elegans* hermaphrodite sex allocation. Evolution 70:2357–2369
Price TC, Qvarnström A, Irwin DE (2003) The role of phenotypic plasticity in driving genetic evolution. Proc R Soc B 270:1433–1440
Raff EC, Raff RA (2000) Dissociability, modularity, evolvability. Evol Dev 2:235–237
Raff RA, Kaufman TC (1983) Embryos, genes and evolution. MacMillan, New York
Raff RA, Wray GA (1989) Heterochrony: developmental mechanisms and evolutionary results. J Evol Biol 2:409–434
Raff RR (1992) Direct-Developing Sea urchins and the evolutionary reorganisation of early development. BioEssays 14:211–218
Riedl R (1978) Order in living systems: a systems analysis of evolution. Wiley, New York
Root TL, Price JT, Hall KR, Schneider SH, Rosenzweig C, Pounds JA (2003) Fingerprints of global warming on wild animals and plants. Nature 421:57–60
Rougvie AE (2005) Intrinsic and extrinsic regulators of developmental timing: from miRNAs to nutritional cues. Development 132:3787–3798

Rudin-Bitterli TS, Spicer JI, Rundle SD (2016) Differences in the timing of cardio-respiratory development determine whether marine gastropod embryos survive or die in hypoxia. J Exp Biol 219:1076–1085

Rudolf VHW, Singh M (2013) Disentangling climate change effects on species interactions: effect of temperature, phenological shifts, and body size. Oecologia 173:1043–1052

Rundle SD, Smirthwaite JJ, Colbert MW, Spicer JI (2011) Predator cues alter the timing of developmental events in gastropod embryos. Biol Lett 7:285–287

Rundle SD, Spicer JI (2017) Heterokairy: a significant form of developmental plasticity? Biol Lett 12:20160509

Russell GA, Rezende EL, Hammond KA (2008) Development partly determines the aerobic performance of adult deer mice, *Peromyscus maniculatus*. J Exp Biol 211:35–41

Sakamoto T, McCormick SD, Hirano T (1993) Osmoregulatory actions of growth hormone and its mode of action in salmonids: a review. Fish Physiol Biochem 11:155–164

Sarkar S (2004) From the *Reaktionsnorm* to the evolution of adaptive plasticity: a historical sketch, 1909-1999. In: DeWitt TJ, Scheiner SM (eds) Phenotypic plasticity: functional and conceptual approaches. Oxford University Press, Oxford, pp 10–30

Schlichting CD, Pigliucci M (1998) Phenotypic evolution: a reaction norm perspective. Sinauer Associates, Sunderland

Scoville AG, Pfrender ME (2010) Phenotypic plasticity facilitates recurrent rapid adaptation to introduced predators. Proc Natl Acad Sci U S A 107:4260–4263

Sikkink KL, Reynolds RM, Tuarte CM, Cresko WA, Phillips C (2014) Rapid evolution of phenotypic plasticity and shifting thresholds of genetic assimilation in the nematode *Caenorhabditis remanei*. Genes Genomes Genet 4:1103–1112

Sloan AW (1978) William Harvey, physician and scientist. S Afr Med J 54:247–252

Spicer JI (1995) Ontogeny of respiratory function in crustaceans exhibiting either direct or indirect development. J Exp Zool 272:413–418

Spicer JI (2006) A physiological approach to heterochrony. In: Warburton S, Burggren W, Pelster B, Reiber CL, Spicer JI (eds) Comparative developmental physiology: contributions, tools and trends. Oxford University Press, New York, pp 191–202

Spicer JI, Burggren WW (2003) Development of physiological regulatory systems: altering the timing of crucial events. Zoology 106:91–99

Spicer JI, El-Gamal MM (1999) Hypoxia accelerates the development of respiratory regulation in brine shrimp – but at a cost. J Exp Biol 202:3637–3646

Spicer JI, Eriksson SP (2003) Does the development of respiratory regulation always accompany the transition from pelagic larvae to benthic fossorial postlarvae in the Norway lobster *Nephrops norvegicus* (L)? J Exp Mar Biol Ecol 295:219–243

Spicer JI, Rundle SD (2006) Out of place and out of time – towards a more integrated approach to heterochrony. Anim Biol 56:487–502

Spicer JI, Rundle SD (2007) Plasticity in the timing of physiological development: physiological heterokairy – what is it, how frequent is it, and does it matter? Comp Biochem Physiol A Mol Integr Physiol 148:712–719

Spicer JI, Rundle SD, Tills O (2011) Studying the altered timing of physiological events during development: it's about time…or is it? Respir Physiol Neurobiol 178:3–12

Stearns SC (1989) The evolutionary significance of phenotypic plasticity. Bioscience 39:436–445

Strathmann RR, Fenaux L, Strathmann MF (1992) Heterochronic developmental plasticity in larval sea urchins and its implications for evolution of nonfeeding larvae. Evolution 46:972–986

Sulston JE, Horvitz HR (1977) Post-embryonic cell lineages of the nematode *Caenorhabditis elegans*. Dev Biol 56:110–156

Sulston JE, Horvitz HR (1981) Abnormal cell lineages in mutants of the nematode *Caenorhabditis elegans*. Dev Biol 82:41–55

Sun LD, Ye M, Hao H, Wang NT, Wang YQ, Cheng TR, Zhang QX, Wu RL (2014) A model framework for identifying genes that guide the evolution of heterochrony. Mol Biol Evol 31:2238–2247

Suzuki Y, Nijhout HF (2006) Evolution of a polyphenism by genetic accommodation. Science 311:650–652
Tills O, Rundle SD, Salinger M, Haun T, Pfenninger M, Spicer JI (2011) A genetic basis for intra-specific differences in developmental timing? Evol Dev 13:542–548
Tills O, Rundle SD, Spicer JI (2013a) Parent-offspring similarity in the timing of developmental events: a potential link between ontogeny and phylogeny. Proc R Soc Lond B Biol Sci 280:2013479
Tills O, Rundle SD, Spicer JI (2013b) Variance in developmental event timing occurs predominantly at low biological levels: implications for heterochrony. Biol J Linn Soc 110:581–590
Tills O, Spicer JI, Rundle SD (2010) Salinity-induced heterokairy in an upper-estuarine population of the snail *Radix balthica* (Mollusca: Pulmonata). Aquat Biol 9:95–105
Van Buskirk J (2002) A comparative test of the adaptive plasticity hypothesis: relationships between habitat and phenotype in anuran larvae. Am Nat 160:87–102
Van Dyck H, Bonte D, Puls R, Gotthard K, Maes D (2015) The lost generation hypothesis: could climate change drive ectotherms into a developmental trap? Oikos 124:54–61
Varsamos S, Nebel C, Charmantier G (2005) Ontogeny of osmoregulation in postembryonic fish: a review. Comp Biochem Physiol A Mol Integr Physiol 141:401–429
Voss SR, Smith JJ (2005) Evolution of salamander life cycles: a major-effect quantitative trait locus contributes to discrete and continuous variation for metamorphic timing. Genetica 170:275–281
Wade AA, Hand BK, Kovach RP, Muhlfeld CC, Waples RS, Luikart G (2017) Assessments of species' vulnerability to climate change: from pseudo to science. Biodivers Conserv 26:223–229
Wagner D (2016) Making flowers at the right time. Dev Cell 37:208–210
Wagner D (2017) Key developmental transitions during flower morphogenesis and their regulation. Curr Opin Genet Dev 45:44. https://doi.org/10.1016/j.gde.2017.01.018
Warburton SJ, Burggren WW, Pelster B, Reiber CL, Spicer JI (eds) (2006) Comparative developmental physiology. Oxford University Press, New York
Warkentin KM (2007) Oxygen, gills, and embryo behaviour: mechanisms of adaptive plasticity in hatching. Comp Biochem Physiol A Mol Integr Physiol 148:720–731
Warkentin KM (2011) Environmentally cued hatching across taxa: embryos respond to risk and opportunity. Integr Comp Biol 51:14–25
West-Eberhard MJ (1989) Phenotypic plasticity and the origins of diversity. Annu Rev Ecol Syst 20:249–278
West-Eberhard MJ (2003) Developmental plasticity and evolution. Oxford University Press, Oxford
West-Eberhard MJ (2005) Developmental plasticity and the origin of species differences. Proc Natl Acad Sci U S A 102:6543–6549
Wood HA (2014) Mosaic physiology from developmental noise: within organism physiological diversity as an alternative to phenotypic plasticity and phenotypic flexibility. J Exp Biol 217:35–45
Wund MA, Baker JA, Clancy B, Golub JL, Fosterk SA (2008) A test of the "flexible stem" model of evolution: ancestral plasticity, genetic accommodation and morphological divergence in the three-spine stickleback radiation. Am Nat 172:449–462
Yang LH, Rudolf VHW (2010) Phenology, ontogeny and the effects of climate change on the timing of species interactions. Ecol Lett 13:1–10

Responses to Environmental Stressors in Developing Animals: Costs and Benefits of Phenotypic Plasticity

5

Bernd Pelster and Warren W. Burggren

Abstract

The phenotype of an animal results from interaction between genetic information and environmental influences. Embryos and larvae of most animals are freely exposed to the environment, so environmental influences may start affecting development as soon as egg deposition. A response to the environment requires information about the environment, which then triggers modifications in cell and organ function. Even in the earliest developmental stages, receptors may be functional, and a modification of receptor activity may induce physiological responses or modifications in cell proliferation. In embryos, operation of these control loops, which in adults typically are humoral or neuronal loops, is hampered by the somewhat delayed development of the nervous system. In addition, so-called critical windows may severely restrict the time of responsiveness to certain signals or stressors during development. Nevertheless, modifications in heart rate, ventilation, or metabolic activity demonstrate the existence of physiological plasticity in earliest developmental stages. In addition, differences between individuals or populations in cell number and organ size reveal a remarkable plasticity in structural development. Flexibility and plasticity certainly are beneficial because aerobic metabolism, for example, can be adjusted to changes in oxygen availability, ensuring an optimal outcome especially in an environment with variable oxygen tensions, like aquatic environments. The timing of developmental milestones like hatching can be adjusted to optimal envi-

B. Pelster (✉)
Institute of Zoology, University of Innsbruck, Innsbruck, Austria

Center for Molecular Biosciences, University of Innsbruck, Innsbruck, Austria
e-mail: Bernd.Pelster@uibk.ac.at

W. W. Burggren
Department of Biological Sciences, University of North Texas, Denton, TX, USA

© Springer International Publishing AG, part of Springer Nature 2018

W. Burggren, B. Dubansky (eds.), *Development and Environment*,
https://doi.org/10.1007/978-3-319-75935-7_5

ronmental conditions in terms of oxygen availability, temperature, or humidity or in response to biotic cues from conspecifics or predators. On the other hand, any response takes time and additional energy and thus may slow down development and growth rate and thus increase the risk of predation in vulnerable larval stages. In species with thermal sex determination, the environmental influence may result in an unbalanced sex ratio of a population, which obviously may be very disadvantageous for the propagation of a population. Organ size and function may be adjusted to the current situation encountered during development and cannot be reversed if environmental conditions return to previous settings. In this case reduced fitness may be the consequence. Our discussion of these options reveals that a certain degree of phenotypic plasticity is essential and required for a species to prosper in a variable environment, and certainly many ecosystems are changing, or predicted to change, in a dramatic fashion. This flexibility derived from phenotypic plasticity does, however, come with a cost and may occasionally result in reduced fitness and disadvantageous phenotypes in the less plastic adult animals.

5.1 Introduction

Phenotypic plasticity is the ability of an organism with given genetic background to react to an environmental input with a change not only in behavior and physiology, but also in form and state (West-Eberhard 2003; Beaman et al. 2016). While a change in behavior, activity status, physiological performance, or structure is considered as active plasticity, the transient deformation of the head of a mammal during birth or the reduction in swim bladder size induced by increasing hydrostatic pressure may be considered passive plasticity (West-Eberhard 2003). Passive plasticity defined in this way may be considered as tissue compliance, and it does not include any kind of metabolic or behavioral modification and does not involve any change in gene expression. In this study we therefore would like to focus our attention on active plasticity.

The genotype and the coordinated expression of genes comprising the genotype represent crucial determinants for the phenotype of an individual, but even with an identical genotype, quite different phenotypes may develop (West-Eberhard 2003). For an entry in to the extensive literature (see Moczek 2010; Lillycrop and Burdge 2014; Schlichting and Wund 2014; Rundle and Spicer 2016; Turcotte and Levine 2016; Wright and Turko 2016; Corona et al. 2016). Clearly, there is remarkable plasticity in the development of a phenotype, and environmental influences are responsible for this variation in phenotype over time and topology. Plasticity is especially obvious during development and has been shown for many species of both animals and plants. In fact, phenotypic plasticity has been regarded as a key mechanism enabling developing organisms to survive in the face of environmental change (Traevis 1994; Murren et al. 2015). Yet, a modified phenotype resulting from plasticity that is expressed during early development may not transfer to the adult. For example, larvae of the gray tree frog *Hyla versicolor* can develop resistance to the insecticide Sevin, but this resistance was not retained at later developmental stages (Jones and Relyea 2015). Thus, phenotypic plasticity should be viewed as a dynamic, even reversible process rather than a switch from one static state to another.

5.2 Is Phenotypic Plasticity Always Advantageous?

Phenotypic plasticity is usually neutrally defined – i.e., some variation on "the ability of one genotype to produce multiple phenotypes," without any suggestion that phenotypic plasticity can be disadvantageous or advantageous. Yet, there is a subtle, implicit bias in which phenotypic plasticity is often discussed in the context of the *advantages* that it confers on organisms. Thus, a typical discussion might consider an environmental stressor (e.g., temperature, nutrition, predation, oxygen) and whether the organism exhibits a phenotypic adaptation that mitigates the effects of the environmental stressor and allows the organism to survive, thrive, and ultimately reproduce. If the organism modifies its phenotype in a way that helps the organism, it is usually regarded as an example of phenotypic plasticity. Interestingly, however, an induction of a *disadvantageous* phenotype, which may result from exactly the same mechanism and take exactly the same time to develop, is usually labeled simply as a "negative response" to the environment, as the inability of the organism to cope with the stressor, or even as "just" a pathology.

In truth, few examples of modified phenotypes resulting from phenotypic plasticity that carry demonstrated advantages are free of disadvantages. For example, the copepod *Eurytemora affinis* shows plasticity in the extent of pigmentation it carries. More pigmented individuals have higher antioxidant capacity, higher RNA:DNA ratios and reproductive output (advantages) but are also more visible and so experience higher rates of predation (disadvantage) (Gorokhova et al. 2013). As we will discuss below, advantageously modified phenotypes often come at an energetic cost (a disadvantage), so phenotypic plasticity is best viewed as having components that may be both advantageous and disadvantageous (and thus a trade-off is present), rather than as a "good" phenotypic modification.

5.3 What Is the Time Course of Phenotypic Plasticity?

The conventional definition of phenotypic plasticity does not have embedded in it a "time course," i.e., a time frame for how long the phenotypic modification should take to arise. And, indeed, the time courses of various phenotypic modifications as a result of phenotypic plasticity can span minutes in very short-lived organisms to years in longer-lived organisms. Additionally, developing animals are generally regarded as more "plastic" because of the ongoing processes of differentiation, organogenesis, and growth (Del Giudice 2015; Moczek 2015; Beaman et al. 2016; Bavis and MacFarlane 2016; Rundle and Spicer 2016).

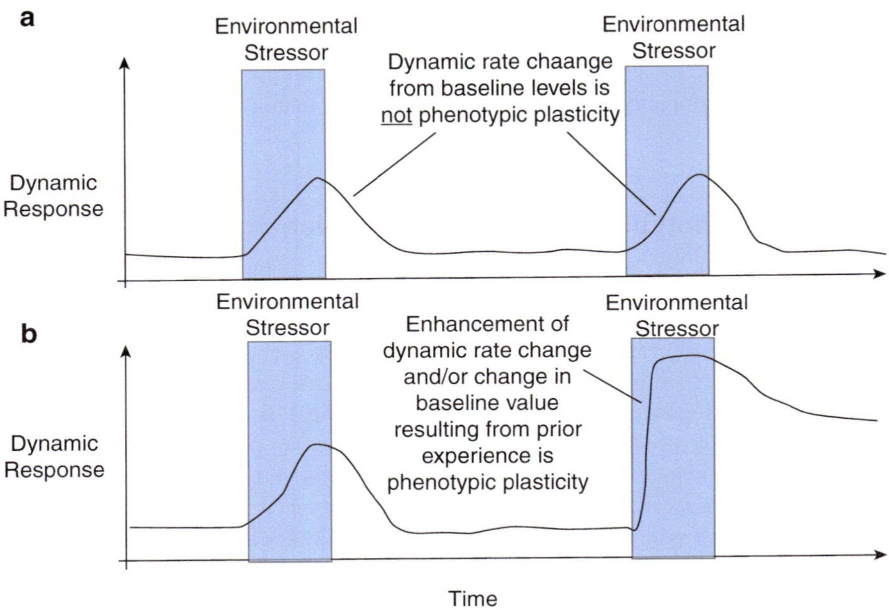

Fig. 5.1 Potential responses to an environmental stressor. (**a**) The stressor invariably produces the same homeostatic dynamic response each time an organism experiences it. (**b**) The dynamic response to the stressor is modified because of previous exposures, with changes in magnitude of the response and/or changes in baseline

A key issue in considering the time course of phenotypic plasticity, however, revolves around whether an observed change is an actual change in phenotype or more simply a change in performance or a homeostatic adjustment. Unfortunately, some have considered an acute response in behavior, biochemistry, or physiology as plasticity. However, the purpose of an organism's regulatory system is to respond dynamically to environmental change (Fig. 5.1). Thus, the dynamic response per se is not an example of phenotypic plasticity, but if that response changes over time because of previous prior experience of the stressor, then this *is* an example of phenotypic plasticity. Phenotypic plasticity may of course also be observed without prior experience, as obvious when discussing, for example, predator-induced plasticity (see Sect. 5.4). Similarly, if we consider environmental changes like the currently discussed climate change, phenotypic plasticity not related to prior experience will be crucial for a species to survive and thrive under the new environmental settings.

5.4 Advantageous Phenotypic Plasticity

What are the advantages conferred by phenotypic plasticity, particularly in developing animals? A well-known and intensely studied example for plasticity is predator-induced plasticity, i.e., the development of a modified phenotype that makes it more difficult for a predator to overwhelm its prey (Benard 2004; Relyea and Auld 2005).

As an example, in a predator environment, larval anurans (tadpoles) typically forage less and develop larger tails and small bodies, and these larvae are more resistant to predation (Relyea and Auld 2005).

The benefit of developing this modified phenotype for the prey organism is obvious – a better chance of survival in the presence of the predator. On the other hand, this phenotype is not developed in the absence of the predator, suggesting that there is some disadvantage connected to it (possibly a significant energetic cost or a change in developmental timing), which is avoided by retaining the "conventional" phenotype in the absence of predation pressure. Unfortunately, the nature of such disadvantages to developing an alternative phenotype is not easily identified, although it is generally accepted that there must be some disadvantage to the predator-induced phenotype, given that it is not universally found in the population. Even in an experimentally controlled condition – for example, presence or absence of predation – identification and measurement of costs of phenotypic plasticity is not an easy task. Indeed, an accurate description of plasticity and in particular of its benefits or costs is even more difficult in nature (Pigliucci 2005), where environmental parameters affecting an organism vary spatially and temporally between populations.

In addition to understanding the cost of phenotypic plasticity, the extent to which a phenotype developed in response to an environmental signal is reversible or persistent is key to understanding long-term effects. No organism can be infinitely or ideally plastic (Murren et al. 2015). There obviously are limits to plasticity, and if a parent animal responded to environmental changes for better survival, this does not necessarily mean the offspring will similarly show the same plasticity and survive. Thus, two additional questions arise: *Can a beneficial trait developed in response to environmental changes be genetically (or epigenetically) inherited, and thus can plasticity contribute to speciation?* Additionally, *can plasticity itself be considered an inheritable trait that can be conveyed to offspring?*

In this review we will first discuss some examples for phenotypic plasticity, in particular developmental plasticity, and consider the various levels at which such plasticity can be observed. Following a discussion of benefits of plasticity, we will attempt to delineate disadvantages or costs of plasticity. To do so it is essential to first define benefits and costs for an organism. We will see that the definition of costs in particular may render a proper determination of the actual costs encountered with development of a particular phenotype very difficult or even impossible. Finally, we will direct our attention toward limits of plasticity, which obviously must exist, because no performance or trait can be infinitely good.

5.5 Benefits and Costs of Phenotypic Plasticity

5.5.1 Phenotypic Plasticity Observed in Nature and Its Benefits

Not all organismal responses to phenotypic plasticity occur at the same rate, nor do they necessarily have the same impact. For example, behavioral and physiological adjustments in response to environmental changes are arguably the most highly responsive aspects of a phenotype in terms of speed of response, as well as reversibility. Important to note is the fact that responding behaviorally or physiologically does not in itself constitute phenotypical plasticity (see Fig. 5.1). Consider, for example, that an animal responding to heat stress may both pant (physiological response) and seek shade (behavioral response). These responses do not comprise plasticity. However, an animal (or population) that over time pants at greater rates or seeks shade at lower temperatures than others in the population is showing plasticity.

Being able to pant at greater rates or seeking shade at lower temperatures can be considered specializations, and organisms may in fact facultatively switch among specializations, producing various polymorphisms, polyphenisms, and different behavioral and morphological traits (West-Eberhard 2003; Pfennig et al. 2010; Murren et al. 2015). Such responses are achieved through differential expression of different sets of genes in response to changing environmental signals. Thus, these genes determine which of the possible alternate programs for functional and morphological development will be selected from the repertoire of a single genome. For example, some fish species may move from freshwater to seawater, and this change in osmolarity is possible by reprogramming ion movements in gill cells, in particular gill ionocytes. While in freshwater ions are taken up from the water, in seawater ions are released into the water (Evans 2008; Hsu et al. 2014), and this change is possible by large-scale transcriptomic changes in these ionocytes (Leguen et al. 2015).

To account for the fact that phenotypic plasticity allows a given genotype to produce different, alternative phenotypes, the concept of a reaction norm has been introduced, which describes the range of phenotypes that can be produced from a given genotype (Pigliucci 2005; Callahan et al. 2008). Again, the remarkable salinity tolerance of euryhaline fish may serve as an example, assuming that a gradual transition from the freshwater, ion-resorbing type, to the seawater, ion-secreting type, is observed during gradual salinity change in brackish water. The stenohaline fish, in contrast, is not able to tolerate these severe changes in external osmolarity.

5.5.1.1 Examples of Phenotypic Plasticity

Classic examples of advantageous phenotypic plasticity are evident in the changes in ion regulation observed in catadromous and anadromous fishes (see above). The mechanisms for ion uptake or ion secretion at the gills become highly modified when switching from freshwater to seawater and vice versa (Evans 2008; Hiroi and McCormick 2012; Leguen et al. 2015). This morphological and, indeed, molecular plasticity allows the adult fish to exploit multiple habitats with very different salinities.

Pseudocrenilabrus multicolor victoriae is a cichlid fish living in the Nile river system and in Lake Victoria basin. Fish from the population in Lake Manywa exposed to stable hypoxic conditions have a 29% greater total gill surface area than fish living in well-oxygenated Kayanja water. A greater gill surface area is advantageous for oxygen uptake, and thus, oxygen requirements can be met even in a habitat with low oxygen availability. Offspring from the same parents living either under hypoxic or normoxic conditions similarly show an 18% greater total gill surface area if parents lived in hypoxia (Chapman et al. 2000). This observation indicates that there may be genetic (or epigenetic) assimilation of the physiological adaptations induced by a changing environment. Phenotypic plasticity of gas exchange organs during their development can also be evoked experimentally by chronic hypoxic exposure in amphibians (e.g., Burggren and Mwalukoma 1983) and insects (e.g., Mortimer and Moberg 2009; Wong et al. 2014), to mention just a few of the animals exhibiting respiratory phenotypic plasticity, in particular during development.

As another example, the plasticity of skeletal muscle tissue in response to different workloads may result in a complete remodeling of muscle fiber size and metabolism (Pette 2001; Pette and Staron 2001; Flück and Hoppeler 2003; Brook et al. 2016). As a consequence, muscle performance is optimized to the type of workload frequently encountered, resulting in a better performance of the animal. Similarly, cardiac muscle can be remodeled by a variety of work-related stressors (Tessier and Storey 2016) and also by development itself (Zhang et al. 2013).

Juvenile instars of the water flea *Daphnia longicephala* or of *Daphnia cucullata* are textbook examples for cyclomorphosis. In the presence of a predator like *Leptodora kindtii*, a larger water flea, the helmet almost doubles in size, and similar results are observed in the presence of another predator, the larvae of the midge *Chaoborus flavicans* (Agrawal et al. 1999; Barry 2000; Weiss et al. 2015). The large helmet size makes *Daphnia* a more difficult prey to catch and therefore increases survival in the presence of a predator. Offspring of mothers treated with kairomone to mimic the presence of a predator always had larger helmets than offspring of mothers kept under control conditions. In these experiments kairomone treatment often is preferred to a real predator. If a predator actually starts feeding on its prey, the predator-prey ratio changes, which may have significant influence on the outcome of the proceeding experiment.

Predator-induced plasticity is also well known for plants, which start producing secondary metabolites difficult to digest and occasionally even poisonous for herbivores (Agrawal et al. 1999; Van Kleunen and Fischer 2005; Mitchell et al. 2016). For example, in wild radish plants, the induction of defenses by the specialist caterpillar *Pieris rapae* persists in the progeny of the damaged plants, indicating that the environmentally induced trait can be transferred to progeny (Agrawal et al. 1999).

Another prominent and well-studied example for plasticity is the influence of diet on mouth and head morphology, quantified for fish, grasshoppers, and larval amphibians, or on the claw size of crabs, for example (West-Eberhard 2003; Relyea and Auld 2005). The benefit of diet-induced plasticity is that a modified phenotype related to diet may make accessible formerly unavailable food resources. Indeed, Charles Darwin recognized that diet-induced plasticity can contribute to speciation.

A fascinating example of extreme developmental plasticity is the two-legged goat. This goat, born with congenital defect on the front legs, developed an upright position and bipedal locomotion using its hind legs alone, which resembles the pattern well known for kangaroos (West-Eberhard 1989, 2005). These changes included significant modifications of the skeleton as well as of leg muscles. Similar phenomena have been observed in various quadrupeds (Pigliucci 2005).

A classic example for developmental origins of beneficial novelties is neoteny in adult salamanders, where a hormonally mediated block of metamorphosis results in the production of sexually mature adults that retain partial larval morphology.

Finally, phenotypic plasticity in developing animals is often viewed as a process occurring in relatively well-formed albeit immature animals. In fact, even the earliest stage of animals can exhibit plasticity, though its effects may show up later in development. For example, the environmental conditions in which fertile but as yet un-incubated chicken eggs are stored can result in changes in survival, hematology, and acid-base balance (Branum et al. 2016).

5.5.1.2 Consequences of Phenotypic Plasticity

The selection of examples of phenotypic plasticity presented above clearly demonstrates that plasticity – whether occurring during development, when plasticity may be highest, or during adulthood – can have obvious benefits. Phenotypic modification may protect the individual from being eaten, expose new food resources, allow exploitation of new habitats, create better mobility, and have many other benefits. Relevant in evolutionary terms, however, is not the advantage for the individual but ultimately its evolutionary fitness, i.e., its reproductive success as aided by the phenotypic modification. Accordingly, the general definition for declaring a phenotypic change as a result of plasticity is that it must lead to an increase in fitness. Moreover, there are several examples suggesting that this plasticity may be inherited by progeny, likely through epigenetic mechanisms (Chap. 7).

Plasticity may include a change in gene expression, either from changes in degree of expression of genes or the awakening/silencing of transcription of different sets of genes. There may also be changes in protein-DNA interaction sites, with alternative splicing, or transposable elements may be involved. A quantitative trait loci analysis (QTL) for plasticity of wild barley revealed that there are genes that directly affect phenotypic plasticity (Elberse et al. 2004). Clearly, there is a genetic background to plasticity, in fact there is indication that plasticity itself has a genetic basis, essentially being a heritable character (Moczek 2008; Espinosa-Soto et al. 2011; Hamilton and Yu 2012; Bajic and Poyatos 2012; Scheiner and Holt 2012; Robinson 2013). Indeed, tools for analyzing the genetic basis of phenotypic plasticity are emerging (Zhai et al. 2014; Zhou et al. 2015). Genes, however, do not interact with the environment, the influence of the environment is on the phenotype, and selection is on the phenotype. Consequently, a phenotype developed in response to environmental changes can only be developed if the available genetic information is suitable for this accommodation, a process known as phenotypic accommodation (West-Eberhard 2003, 2005; Pigliucci et al. 2006; Schlichting and Wund 2014; Ehrenreich and Pfennig 2015). If animals with this capability/plasticity are more successful in the production of offspring, individuals with this set of genes will spread in the population, and thus, plasticity-driven adaptation may occur. More than 60 years ago, Waddington showed that artificial selection of a phenotype that initially appears only in a few organisms after nongenetic perturbations can easily result in the trait's genetic determination (Waddington 1953) and thus in genotypic accommodation (Pigliucci et al. 2006; Lande 2009; Fitzpatrick 2012). Artificial selection can thus turn an alternative phenotype into a native phenotype. In addition, many observations in wild populations suggest that in multiple cases an ancestral alternative phenotype may have facilitated the evolution of new, genetically fixed adaptive traits (Espinosa-Soto et al. 2011). In summary, phenotypic plasticity may be heritable, may contribute to speciation, and may have significant ecological consequences (West-Eberhard 1989, 2005; Rossiter 1996; Pigliucci et al. 2003; Miner et al. 2005; Pigliucci 2005; Moczek 2008; Pfennig et al. 2010; Fitzpatrick 2012; Robinson 2013; Standen et al. 2014; Wright and Turko 2016).

5.5.2 Costs of Phenotypic Plasticity

We have discussed examples of, and possible mechanisms for, phenotypic plasticity that results in advantages to individuals and that may spread over time through a population. Yet, phenotypic modifications, whether during development or in adults, are not without attendant costs. Measurement of such costs, in the wild or in the laboratory, can be problematic. For example, the cost may or may not show up at the time that the phenotypic modification occurs (Van Buskirk and Saxer 2001). Moreover, the cost may or may not scale with the magnitude of the phenotypic changes (Hettyey et al. 2011).

How might the costs of phenotypic plasticity be measured? Potentially such costs will be reflected in increased oxygen consumption, ATP turnover or other metabolic activities, or decreased growth or fecundity. If, as outlined above, the benefit of plasticity is an increase in fitness, then consequently the costs should be defined as a decrease in fitness. There appears to be general agreement that costs of phenotypic plasticity appear as a reduction in fitness (DeWitt et al. 1998; Relyea 2002; Van Kleunen and Fischer 2005; Callahan et al. 2008; Auld et al. 2009; Murren et al. 2015).

As already seen in describing the benefits of plasticity, a cost that leads to a reduction in fitness may have many different facets. Predator-induced phenotypes may have better survival when the predator is present, but growth rates may be reduced as compared to the phenotype developed without the presence of a predator (Relyea 2002). Other components of fitness are presumed to be affected, such as immunity, number of offspring, fitness of offspring, developmental progress, or energy reserves available (Relyea 2007; Callahan et al. 2008).

A seminal paper described parameters or phenomena that contribute to costs of plasticity and tried to identify possible limits to plasticity (DeWitt et al. 1998). According to this analysis, parameters contributing to costs are *maintenance costs* of the sensory and regulatory machinery, *costs for the production* of environmentally responsive characters, *information acquisition costs*, *costs for developmental instability* and broad phenotype distributions, and *genetic costs*. This approach breaks down costs of plasticity into units that may indeed be quantifiable, but in practice some questions about the distinction and tractability of these costs remain. Costs for the maintenance of the sensory machinery, for example, may be the costs for the chemoreception or light reception to detect a predator. On the other hand, these senses most likely are essential for the orientation in the environment and to find food or a mating partner, even in the absence of a predator. So are there in fact additional costs for building or maintaining these sensory organs necessary for everyday life? However, when attention is directed to the possible presence of a predator, other activities like food uptake most likely are hampered, and this certainly is a cost because less energy can be assimilated.

Additional costs of phenotypic plasticity include a broad phenotype distribution and developmental instability (DeWitt et al. 1998). The concept that for each environment and living strategy a more or less ideal environment-phenotype match can be designed implies that less ideal matches also exist. Therefore, imprecise development results in variability of the actual phenotype and necessarily includes more or less severe environment-phenotype mismatches. It is expected that genotypes expressing plasticity resulting in reduced fitness or that are very costly are quickly eliminated by selection (DeWitt et al. 1998; Van Buskirk and Steiner 2009). Therefore, a canalized development with a very consistent phenotype appears to be more ideal in a constant environment (Waddington 1942, 1953; Lande 2009). It has been shown, however, that there is no necessary connection between developmental instability and plasticity (DeWitt 1998).

5.5.2.1 Measuring the Cost of Phenotypic Plasticity

Numerous studies have attempted to quantify the cost of phenotypic plasticity. While overall the costs for the production of environmentally responsive characters appear to be quite reasonable, all of these studies have collectively demonstrated that measuring the cost of phenotypic plasticity is indeed a very difficult task. For example, a meta-analysis compiling datasets of 227 paired estimates of costs of plasticity revealed that in 141 pairs (~60%), no significant costs of phenotypic plasticity could be detected. In the remaining studies, the majority of costs detected were environment-specific (Auld et al. 2009). It was hypothesized that resource limitation in the different environments may affect costs by influencing energy turnover (Auld et al. 2009) and that local selective pressures are important (Pigliucci 2005). Costs for phenotypic plasticity may be related to the environment, and there appears to be general agreement that plasticity is more costly in a stable environment. In such an environment, variable phenotypes provide little advantage, with only one type of specialization providing an optimal environment-phenotype match (Waddington 1953; Robinson 2013). In another meta-analysis of 16 species of plants and 7 of animals, plasticity and canalization were equally frequent, and costs were only mild (Van Buskirk and Steiner 2009). Environment was the only covariate that was identified as marginally positively related to the costs of plasticity.

Even in controlled experiments, it appears to be difficult to disentangle cost of plasticity from local costs of generating a phenotype (Murren et al. 2015). The freshwater snail *Physa heterostropha* produces elongate or rotund shells, depending on the presence of a predator like crayfish or the molluscivorous sunfish, but no indication in production costs of differences in character could be detected (DeWitt 1998), and similar results were obtained in predator experiments using *Daphnia pulex* (Scheiner and Berrigan 1998).

It has repeatedly been proposed that costs of plasticity must be distinguished from costs of phenotype production, which are not always easy to differentiate (Callahan et al. 2008; Murren et al. 2015). The intimate connection between these two types of costs is demonstrated when we consider the process of learning. During development, a variety of phenotypes may be sampled, sometimes under the guidance of parental animals, to establish the best hunting strategy. This implies trials of limited success, and thus, it implies costs, but are these costs for plasticity or costs for phenotype development? And how can these costs be quantified? In addition, costs invested initially for learning may later be offset if the learning resulted in the development of a very efficient trait (Murren et al. 2015).

However, few measurements have given a more proximate cost than successful reproduction. In larvae of the ocellated wrasse (*Symphodus ocellatus*), for example, presumably beneficial metabolic phenotypic plasticity in the form of elevated metabolic rate in the presence of high ambient CO_2 is accompanied by reduced growth rate (Cattano et al. 2016). In another example, dung flies (*Sepsis punctum*) exposed to heat shock showed phenotypic plasticity in their responses to thermal stress that enhanced survival but at an energetic cost (Esperk et al. 2016). As a final example, the estuarine red drum (*Sciaenops ocellatus*) reduces diffusion distances in its gills when exposed to elevated levels of ambient CO_2, mimicking expected increases leading to ocean acidification (Esbaugh et al. 2016). While reduced diffusion distances aided CO_2 elimination, this morphological phenotypic plasticity came at the cost of enhanced diffusive water loss and the necessary greater energetic costs of osmoregulation necessary to maintain blood osmolarity within normal limits.

Phenotypic plasticity may also incur genetic costs such as for the replication of genes required to build the environmental responsive character, or costs for special plasticity related genes. So, we have to consider not only the trait developed by plasticity but that plasticity itself must also be considered a trait (Callahan et al. 2008). Replication of DNA and the production of trinucleotides certainly require a lot of ATP and thus are very expensive. If we, however, take into account that only a small fraction of total DNA is transcribed for the production of proteins, then the relative contribution of these particular genes to the total requirement of energy for the DNA replication must be quite small. Adaptations observed in humans exposed to chronic hypoxia have been used as an example for genetic changes (Schlichting and Wund 2014) related to plasticity in recent studies identifying genes involved in the Hif-signaling pathway that are modified in high altitude Tibetans (Simonson et al. 2010; Yi et al. 2010; Lorenzo et al. 2014). These signaling pathways and these genes are, however, important for hypoxic signaling in many species. In hypoxia-adapted Tibetans, they are expressed in a slightly modified isoform, and the product has a different activity as the form expressed in other humans. That the expression of a modified version of the genes generates measurable genetic costs appears quite questionable. Accordingly, the fitness costs of the regulatory genetic machinery are probably small (Murren et al. 2015).

A common observation in the discussion of the costs of phenotypic plasticity is that such costs appear to be low, if present at all. Why is this so? One explanation proposed in the literature is that genotypes with higher costs are quickly eliminated by natural selection and therefore have no chance to prosper and spread (Van Buskirk and Steiner 2009). Alternatively, costs could seem to be low because they are uniformly present in current populations. Each organism possesses the capability to somehow adaptively respond to one or another environmental challenge, and therefore each genotype bears more or less similar costs for plasticity (Van Buskirk and Steiner 2009). An unfortunate consequence of these two explanations is that we may never be able to accurately quantify these costs of phenotypic plasticity.

A third explanation proposed by van Buskirk and Steiner (2009) is that phenotypic plasticity is costly only under certain ecological contexts. The observations that in studies including stressful environments slightly higher costs of plasticity have been detected indeed is in line with this hypothesis. Availability of resources, predation risk, availability of shelter, and competition all appear to influence costs of plasticity and are very difficult to replicate in experimental setups. This, however, brings us back to the first two explanations. It is conceivable that plasticity is connected to some sort of costs, but to measure and really quantify these costs appear extremely difficult or perhaps even impossible. Not surprisingly, Fitzpatrick (2012) states that we know very little about the prevalence or influence of costs of plasticity.

5.5.3 Limits of Phenotypic Plasticity

No organism can be infinitely plastic; no design and no trait can be infinitely good. Limits of plasticity prevent the expression of an optimal phenotype (Pigliucci 2005; Murren et al. 2015). A limiting factor for plasticity is the reliability of the information obtained from the environment. The quality of the acquired information may be misleading and result in a mismatch of phenotype and environmental requirements (DeWitt et al. 1998). Available time for the response to develop may also be limiting. The band width of a certain trait may be limited, and an epiphenotype problem may be encountered, because an add-on phenotype may be less effective or even ineffective compared with phenotypic elements integrated earlier in development. The new structure or function must be added to the existing phenotype, which may be another problem (DeWitt et al. 1998). Plasticity necessarily starts with a given phenotype, which may be partially reorganized, or a new trait is added to an existing phenotype. Environmental changes also may occur late in ontogeny, and the nearly mature organism may not be able to reorganize its phenotype adequately (Auld et al. 2009; Burggren and Reyna 2011). Alternatively, changes in the environment may occur too early for the developing embryo so that either the sensory system or the regulatory loop organizing a coordinated response may not yet be functional. It is well known that the development of the nervous system is somewhat delayed and initially regulatory loops are mostly controlled by hormonal action, if possible at all (Whittow and Tazawa 1991; Pelster 2002; Müller et al. 2015). Thus, critical windows may cause a problem because the change in the environment may occur when this developmental window is closed (Burggren and Reyna 2011), i.e., the responsiveness to this input is lost (see Chap. 3).

Plasticity history may also be limiting. Thus, the potential for a response to an environmental change may depend on previous responses that occurred earlier in ontogeny (Auld et al. 2009). Ecological limits may also occur, when different environmental influences would require competing modifications to obtain an optimal environment-phenotype match (Auld et al. 2009). Fish frequently exposed to hypoxic water may use the swim bladder for additional aerial gas exchange. To avoid losing oxygen through the gills to the hypoxic water, gill surface area may be reduced. Gills are multifunctional organs, and a reduction in surface area compromises the ion exchange capacity, resulting in a shift of this function to other organs, like the gut, for example (Wood et al. 2016). Strength of selection may also be a limiting factor that influences the evolution of plasticity (Murren et al. 2015). Selection acts on the various traits of a phenotype, and if plasticity itself is a trait, then selective forces have an influence on the propagation of this trait and thus may limit plasticity. Selective forces acting on plasticity may of course be different from the selective forces acting on other trait values, as induced by the environmental changes (Auld et al. 2009).

The costs of plasticity may be limiting to plasticity itself. As outlined above selection is expected to rapidly eliminate plasticity if the costs for this trait are inordinate and result in a reduction in fitness, i.e., when the costs exceed the benefit.

Conclusion

Phenotypic plasticity and in particular developmental phenotypic plasticity are important phenomena allowing organisms to respond to a changing environment. Considering the instability of most environments, this is crucial for the individual organism for survival. This plasticity is based on existing genetic information, which allows for facultative expression of different traits, resulting in modified specializations. The benefit of these specializations ultimately is an increase in fitness, i.e., in the production of offspring. Inheritability of these traits and selection on these modified specializations may thus significantly contribute to speciation. Accordingly, plasticity can be considered one of the driving forces of evolution.

Plasticity clearly comes with costs and with limitations, although these costs are not easily measured and quantified. The trade-off between costs and benefits clearly is to the advantage of benefits, otherwise plasticity would have been eliminated in evolution. Selective forces are apparently not in favor of non-plastic genotypes, a consideration that is supported by the observation that special plasticity genes or loci have been identified, which may assure the perpetuation of plasticity.

References

Agrawal AA, Laforsch C, Tollrian R (1999) Transgenerational induction of defences in animals and plants. Nature 401:60–63

Auld JR, Agrawal AA, Relyea RA (2009) Re-evaluating the costs and limits of adaptive phenotypic plasticity. Proc R Soc Lond B Biol Sci 277:503

Bajic D, Poyatos JF (2012) Balancing noise and plasticity in eukaryotic gene expression. BMC Genomics 13:1–12
Barry JM (2000) Inducible defences in daphnia: responses to two closely related predator species. Oecologia 124:396–401
Bavis RW, Macfarlane PM (2016) Developmental plasticity in the neural control of breathing. Exp Neurol 287:176
Beaman JE, White CR, Seebacher F (2016) Evolution of plasticity: mechanistic link between development and reversible acclimation. Trends Ecol Evol 31:237–249
Benard MF (2004) Predator-induced phenotypic plasticity in organisms with complex life histories. Annu Rev Ecol Evol Syst 35:651–673
Branum SR, Tazawa H, Burggren WW (2016) Phenotypic developmental plasticity induced by preincubation egg storage in chicken embryos (Gallus gallus domesticus). Phys Rep 4:e12712
Brook MS, Wilkinson DJ, Phillips BE, Perez-Schindler J, Philp A, Smith K, Atherton PJ (2016) Skeletal muscle homeostasis and plasticity in youth and ageing: impact of nutrition and exercise. Acta Physiol 216:15–41
Burggren WW, Mwalukoma A (1983) Respiration during chronic hypoxia and hyperoxia in larval and adult bullfrogs (Rana catesbeiana) I. Morphological responses of lungs, skin and gills. J Exp Biol 105:191–203
Burggren WW, Reyna KS (2011) Developmental trajectories, critical windows and phenotypic alteration during cardio-respiratory development. Respir Physiol Neurobiol 178:13–21
Callahan HS, Maughan H, Steiner UK (2008) Phenotypic plasticity, costs of phenotypes, and costs of plasticity. Ann N Y Acad Sci 1133:44–66
Cattano C, Giomi F, Milazzo M (2016) Effects of ocean acidification on embryonic respiration and development of a temperate wrasse living along a natural Co2 gradient. Conserv Physiol 4:cov073
Chapman LG, Galis F, Shinn J (2000) Phenotypic plasticity and the possible role of genetic assimilation: hypoxia-induced trade-offs in the morphological traits of an African cichlid. Ecol Lett 3:387–393
Corona M, Libbrecht R, Wheeler DE (2016) Molecular mechanisms of phenotypic plasticity in social insects. Curr Opin Insect Sci 13:55–60
Del Giudice M (2015) Plasticity as a developing trait: exploring the implications. Front Zool 12:1–11
Dewitt TJ (1998) Costs and limits of phenotypic plasticity: tests with predator-induced morphology and life history in a freshwater snail. J Evol Biol 11:465–480
Dewitt TJ, Sih A, Wilson DS (1998) Costs and limits of phenotypic plasticity. Trends Ecol Evol 13:77–81
Ehrenreich IM, Pfennig DW (2015) Genetic assimilation: a review of its potential proximate causes and evolutionary consequences. Ann Bot 117:769
Elberse IAM, Vanhala TK, Turin JHB, Stam P, Damme JMM, Van Tienderen PH (2004) Quantitative trait loci affecting growth-related traits in wild barley (Hordeum spontaneum) grown under different levels of nutrient supply. Heredity 93:22–33
Esbaugh AJ, Ern R, Nordi WM, Johnson AS (2016) Respiratory plasticity is insufficient to alleviate blood acid-base disturbances after acclimation to ocean acidification in the estuarine red drum, Sciaenops ocellatus. J Comp Physiol B 186:97–109
Esperk T, Kjaersgaard A, Walters RJ, Berger D, Blanckenhorn WU (2016) Plastic and evolutionary responses to heat stress in a temperate dung fly: negative correlation between basal and induced heat tolerance? J Evol Biol 29:900–915
Espinosa-Soto C, Martin OC, Wagner A (2011) Phenotypic plasticity can facilitate adaptive evolution in gene regulatory circuits. BMC Evol Biol 11:1–14
Evans DH (2008) Teleost fish osmoregulation: what have we learned since August Krogh, Homer Smith, and Ancel Keys. Am J Physiol Regul Integr Comp Physiol 295:R704–R713
Fitzpatrick BM (2012) Underappreciated consequences of phenotypic plasticity for ecological speciation. Int J Ecol 2012:1

Flück M, Hoppeler H (2003) Molecular basis of skeletal muscle plasticity-from gene to form and function. Rev Physiol Biochem Pharmacol 146:159–216
Gorokhova E, Lehtiniemi M, Motwani NH (2013) Trade-offs between predation risk and growth benefits in the copepod Eurytemora affinis with contrasting pigmentation. PLoS One 8:E71385
Hamilton BA, Yu BD (2012) Modifier genes and the plasticity of genetic networks in mice. PLoS Genet 8:E1002644
Hettyey A, Vincze K, Zsarnoczai S, Hoi H, Laurila A (2011) Costs and benefits of defences induced by predators differing in dangerousness. J Evol Biol 24:1007–1019
Hiroi J, Mccormick SD (2012) New insights into gill ionocyte and ion transporter function in euryhaline and diadromous fish. Respir Physiol Neurobiol 184:257–268
Hsu HH, Lin LY, Tseng YC, Horng JL, Hwang PP (2014) A new model for fish ion regulation: identification of ionocytes in freshwater- and seawater-acclimated medaka (Oryzias latipes). Cell Tissue Res 357:225–243
Jones DK, Relyea RA (2015) Here today, gone tomorrow: short-term retention of pesticide-induced tolerance in amphibians. Environ Toxicol Chem 34:2295–2301
Lande R (2009) Adaptation to an extraordinary environment by evolution of phenotypic plasticity and genetic assimilation. J Evol Biol 22:1435–1446
Leguen I, Le Cam A, Montfort J, Peron S, Fautrel A (2015) Transcriptomic analysis of trout gill Ionocytes in fresh water and sea water using laser capture microdissection combined with microarray analysis. PLoS One 10:E0139938
Lillycrop KA, Burdge GC (2014) Environmental challenge, epigenetic plasticity and the induction of altered phenotypes in mammals. Epigenomics 6:623–636
Lorenzo FR, Huff C, Myllymaki M, Olenchock B, Swierczek S, Tashi T, Gordeuk V, Wuren T, Ri-Li G, Mcclain DA, Khan TM, Koul PA, Guchhait P, Salama ME, Xing J, Semenza GL, Liberzon E, Wilson A, Simonson TS, Jorde LB, Kaelin WG Jr, Koivunen P, Prchal JT (2014) A genetic mechanism for Tibetan high-altitude adaptation. Nat Genet 46:951–956
Miner BG, Sultan SE, Morgan SG, Padilla DK, Relyea RA (2005) Ecological consequences of phenotypic plasticity. Trends Ecol Evol 20:685–692
Mitchell C, Brennan RM, Graham J, Karley AJ (2016) Plant defense against herbivorous pests: exploiting resistance and tolerance traits for sustainable crop protection. Front Plant Sci 7:1132
Moczek AP (2008) On the origins of novelty in development and evolution. BioEssays 30:432–447
Moczek AP (2010) Phenotypic plasticity and diversity in insects. Philos Trans R Soc B 365:593–603
Moczek AP (2015) Developmental plasticity and evolution - quo vadis? Heredity 115:302–305
Mortimer NT, Moberg KH (2009) Regulation of Drosophila embryonic tracheogenesis by dVHL and hypoxia. Dev Biol 329:294–305
Müller CA, Burggren WW, Tazawa H (2015) The physiology of the avian embryo. In: Scanes CG (ed) Sturkie's avian physiology, vol 6. Elsevier, Amsterdam, pp 739–766
Murren CJ, Auld JR, Callahan H, Ghalambor CK, Handelsman CA, Heskel MA, Kingsolver JG, Maclean HJ, Masel J, Maughan H, Pfennig DW, Relyea RA, Seiter S, Snell-Rood E, Steiner UK, Schlichting CD (2015) Constraints on the evolution of phenotypic plasticity: limits and costs of phenotype and plasticity. Heredity 115:293–301
Pelster B (2002) Developmental plasticity in the cardiovascular system of fish, with special reference to the zebra fish. Comp Biochem Physiol 133a:547–553
Pette D (2001) Plasticity in skeletal, cardiac, and smooth muscle historical perspectives: plasticity of mammalian skeletal muscle. J Appl Physiol 90:1119–1124
Pette D, Staron RS (2001) Transitions of muscle fiber phenotypic plasticity. Histochem Cell Biol 115:359–372
Pfennig DW, Wund MA, Snell-Rood EC, Cruickshank T, Schlichting CD, Moczek AP (2010) Phenotypic plasticity's impacts on diversification and speciation. Trends Ecol Evol 25:459–467
Pigliucci M (2005) Evolution of phenotypic plasticity: where are we going now? Trends Ecol Evol 20:481–486

Pigliucci M, Murren C, Mitton J (2003) Perspective: genetic assimilation and a possible evolutionary paradox: can macroevolution sometimes be so fast as to pass us by? Evolution 57:1455–1464
Pigliucci M, Murren C, Schlichting CD (2006) Phenotypic plasticity and evolution by genetic assimilation. J Exp Biol 209:2362
Relyea RA (2002) Costs of phenotypic plasticity. Am Nat 159:272–282
Relyea RA (2007) Getting out alive: how predators affect the decision to metamorphose. Oecologia 152:389–400
Relyea RA, Auld JR (2005) Predator- and competitor-induced plasticity: how changes in foraging morphology affect phenotypic trade-offs. Ecology 86:1723–1729
Robinson BW (2013) Evolution of growth by genetic accommodation in Icelandic freshwater stickleback. Proc R Soc Lond B Biol Sci 280:20132197
Rossiter M (1996) Incidence and consequences of inherited environmental effects. Annu Rev Ecol Syst 27:451–476
Rundle SD, Spicer JI (2016) Heterokairy: a significant form of developmental plasticity? Biol Lett 12:20160509
Scheiner SM, Berrigan D (1998) The genetics of phenotypic plasticity. VIII. The cost of plasticity in Daphnia pulex. Evolution 52:368–378
Scheiner SM, Holt RD (2012) The genetics of phenotypic plasticity. X. Variation versus uncertainty. Ecol Evol 2:751–767
Schlichting CD, Wund MA (2014) Phenotypic plasticity and epigenetic marking: an assessment of evidence for genetic accommodation. Evolution 68:656–672
Simonson TS, Yang Y, Huff CD, Yun H, Qin G, Witherspoon DJ, Bai Z, Lorenzo FR, Xing J, Jorde LB, Prchal JT, Ge R (2010) Genetic evidence for high-altitude adaptation in Tibet. Science 329:72–75
Standen EM, Du TY, Larsson HCE (2014) Developmental plasticity and the origin of Tetrapods. Nature 513:54–58
Tessier SN, Storey KB (2016) Lessons from mammalian hibernators: molecular insights into striated muscle plasticity and remodeling. Biomol Concepts 7(2):69–92
Traevis J (1994) Size-dependent behavioral variation and its genetic control within and among populations. In: Boake CR (ed) Quantitative genetic studies of behavioral evolution. University Of Chicago Press, Chicago, pp 165–187
Turcotte MM, Levine JM (2016) Phenotypic plasticity and species coexistence. Trends Ecol Evol 31:803–813
Van Buskirk J, Saxer G (2001) Delayed costs of an induced defense in tadpoles? Morphology, hopping, and development rate at metamorphosis. Evolution 55:821–829
Van Buskirk J, Steiner UK (2009) The fitness costs of developmental canalization and plasticity. J Evol Biol 22:852–860
Van Kleunen M, Fischer M (2005) Constraints on the evolution of adaptive phenotypic plasticity in plants. New Phytol 166:49–60
Waddington CH (1942) Canalization of development and the inheritance of acquired characters. Nature 150:563–565
Waddington CH (1953) Genetic assimilation of an acquired character. Evolution 7:118–126
Weiss LC, Leimann J, Tollrian R (2015) Predator-induced defences in Daphnia longicephala: location of kairomone receptors and timeline of sensitive phases to trait formation. J Exp Biol 218:2918
West-Eberhard MJ (1989) Phenotypic plasticity and the origins of diversity. Annu Rev Ecol Syst 20:249–278
West-Eberhard MJ (2003) Developmental plasticity and evolution. Oxford University Press, New York, pp 1–794
West-Eberhard MJ (2005) Developmental plasticity and the origin of species differences. Proc Natl Acad Sci 102:6543–6549
Whittow GC, Tazawa H (1991) The early development of thermoregulation in birds. Physiol Zool 64:1371–1390

Wood CM, Pelster B, Giacomin M, Sadauskas H, Almeida-Val VF, Val AL (2016) The transition from water-breathing to air-breathing is associated with a shift in ion uptake from gills to gut: a study of two closely-related erythrinid teleosts, *Hoplerythrinus unitaeniatus* and *Hoplias malabaricus*. J Comp Physiol B 186:431–445
Wong DM, Shen Z, Owyang KE, Martinez-Agosto JA (2014) Insulin- and warts-dependent regulation of tracheal plasticity modulates systemic larval growth during hypoxia in drosophila melanogaster. PLoS One 9:E115297
Wright PA, Turko AJ (2016) Amphibious fishes: evolution and phenotypic plasticity. J Exp Biol 219:2245–2259
Yi X, Liang Y, Huerta-Sanchez E, Jin X, Cuo ZXP, Pool JE, Xu X, Jiang H, Vinckenbosch N, Korneliussen TS, Zheng H, Liu T, He W, Li K, Luo R, Nie X, Wu H, Zhao M, Cao H, Zou J, Shan Y, Li S, Yang Q, Asan, Ni P, Tian G, Xu J, Liu X, Jiang T, Wu R, Zhou G, Tang M, Qin J, Wang T, Feng S, Li G, Huasang, Luosang J, Wang W, Chen F, Wang Y, Zheng X, Li Z, Bianba Z, Yang G, Wang X, Tang S, Gao G, Chen Y, Luo Z, Gusang L, Cao Z, Zhang Q, Ouyang W, Ren X, Liang H, Zheng H, Huang Y, Li J, Bolund L, Kristiansen K, Li Y, Zhang Y, Zhang X, Li R, Li S, Yang H, Nielsen R, Wang J, Wang J (2010) Sequencing of 50 human exomes reveals adaptation to high altitude. Science 329:75–78
Zhai Y, Lv Y, Li X, Wu W, Bo W, Shen D, Xu F, Pang X, Zheng B, Wu R (2014) A synthetic framework for modeling the genetic basis of phenotypic plasticity and its costs. New Phytol 201:357–365
Zhang L, Allen J, Hu L, Caruthers SD, Wickline SA, Chen J (2013) Cardiomyocyte architectural plasticity in fetal, neonatal, and adult pig hearts delineated with diffusion tensor MRI. Am J Physiol Heart Circ Physiol 304:H246
Zhou T, Lyu Y, Xu F, Bo W, Zhai Y, Zhang J, Pang X, Zheng B, Wu R (2015) A QTL model to map the common genetic basis for correlative phenotypic plasticity. Brief Bioinform 16:24–31

6 The Impact of Early Developmental Exposure to Stressors Related to Individual Fitness in Aquatic Organisms and the Subsequent Reproductive Success and Failure on Populations

Kate Annunziato and Keith R. Cooper

Abstract

The impact of several persistent organic chemical stressors known to primarily impact the early developmental stages of bivalves and finfish. The highest concentrations of these compounds are often present in our waterways, which are nursery areas for shellfish and finfish. It is well documented in aquaculture and aquarium-raised teleosts that even within the same species, strain differences can dramatically alter the dose response and impact fecundity. Such individual variability in reproductive fitness, due to effects on biochemical pathways and cellular physiology, modulates the toxic responses whereby effects could transmit to the population level. Laboratory experiments with field-collected fish demonstrate that exposure history of the specific population from which the individual is obtained can dramatically influence the effective dose response and reproductive success. The sensitivity of the embryonic and early life stages to xenobiotics is due in part to the alteration of synchronization of programed cell movement, the primordial organ formation, and the disruption of programmed biochemical and physiological pathways of growth and development. Even in adult organisms when oogenesis and spermatogenesis are occurring, several compounds can disrupt this process by reducing fecundity and survival of embryos and larvae, by disrupting gonadal structure and function in juveniles and adults, and through interfering with behavior. The impacts

K. Annunziato (✉)
Joint Graduate Program in Toxicology, Rutgers University, New Brunswick, NJ, USA

K. R. Cooper
SEBS, Department of Biochemistry and Microbiology, Rutgers University, New Brunswick, NJ, USA

© Springer International Publishing AG, part of Springer Nature 2018

W. Burggren, B. Dubansky (eds.), *Development and Environment*,
https://doi.org/10.1007/978-3-319-75935-7_6

of xenobiotics on reproductive success, early life stage survival, and altered hormonal cues at the individual level can dramatically increase mortality and decrease the maximum sustainable yield of a population (NJAES 01202, NIEHS T-32-ES007148).

6.1 Introduction

6.1.1 Anthropogenic-Driven Ecosystem Changes

The unprecedented loss of species worldwide has been labeled as the sixth mass extinction by Paul Ehrlich (Ceballos et al. 2015). There are multiple anthropogenic drivers causing ecosystem change worldwide. The discussion below will be limited to the anthropogenic impacts on aquatic ecosystems. Historically, overfishing was considered the major factor, but it is now realized that there are many contributing factors such as loss of nursery areas by anthropogenic activity (physical and chemical), natural disasters, non-native invasive species, and disease epidemics to name a few (FAO 2006, 2012; Kennish 1992; Davis and Kidd 2012; Walther et al. 2002). Regional impacts add additional stressors on local watersheds. Loss of ecological services, such as altered hydrology entering the watershed, due to human activities can dramatically alter the basic physical, hydrological and chemical characteristics of an ecosystem (Davis and Kidd 2012; Kauffman 2016). Fish and shellfish populations from the early 1900s have undergone significant decreases due to local and regional stressors. Both land use management and fisheries management decisions have contributed to the loss of viable nursery areas and fisheries.

Ecological thresholds may be defined as transition points or zones of relatively rapid change between alternate ecosystem states and ecological conditions, often in response to small, continuous changes in one or more stressors. The loss of the cod fishery in the northeast is an example of overfishing that resulted in the collapse of a viable fishery. Indeed, there are many fisheries worldwide, which have suffered the same fate and are attributed to surpassing some ecological threshold (Pinsky and Byler 2015). The identification of ecological thresholds historically has relied on univariant indicators or multivariant reduction in community structure, which lack the robustness to evaluate taxa distributions along an environmental gradient over space and time. Baker and King (2010) have introduced a Threshold Indicator Taxa Analysis (TITAN) approach, which distinguishes both negative and positive taxa responses across multiple taxa using indicator species. TITAN incorporates occurrence, abundance, and decrease or increase in taxa over time and stressor gradient. This, and other newly developed approaches, can be incorporated into management strategies for remediation and for setting resource management guidelines.

The continued expansion of large urban areas, increased sedimentation rates, interruption of freshwater to estuarine hydrology, and the armoring of shorelines have resulted in substantial loss of wetlands reducing the nursery areas for both non-anadromous and anadromous species. The loss and connectivity of nursery areas can dramatically alter regional species diversity and ecological processes. It also can result in genetically isolated populations. In the Chesapeake Bay watershed, the upland drainage from farming, atmospheric input, runoff, and urbanization along the bay illustrates the combined impact on the water quality of this large bay receiving these contaminated waters. Baker and King (2010) demonstrated that sensitive taxa responded as a group to very low levels of development with declines and in some cases extirpation (local extinction). The contaminants in the Chesapeake Bay range from industrial chemicals to nitrates and phosphates from fertilizers to consumer products and drugs from sewage treatment plants. The impact of this type of contamination is also evident along the Gulf Coast where the Mississippi drainage enters the Gulf of Mexico creating a large dead zone (Diaz and Rosenberg 2008). There are many other watersheds that also receive nutrient-rich or chemically contaminated waters from anthropogenic sources that directly impact water quality in local bays and estuaries. As such, intensive studies of one area may model the potential for ecological effects elsewhere.

Ecosystem-level change is seldom the result of a single factor; several stressors usually are linked resulting in a deleterious change. Able et al. (2014) reported that recovery of winter flounder (*Pseudopleuronectes americanus*) in the southern margin of its range was likely not due directly to temperature but may reflect the shift in predator species, which illustrates the complexity of population dynamics. Events at the biochemical level on physiological pathways have the potential to impact reproductive organs and offspring during different life stages. These reproductive and developmental effects coupled with alterations to the food chain can contribute to the complexity with which modification to species composition can lead to community impacts. The relationships between the biological levels of organization are depicted in Fig. 6.1.

There is broad consensus that there are many reasons, especially in the early life history stages, which contribute to fisheries stock and larval recruitment. These include poor egg quality, natural mortality of the larvae, and juveniles due to food availability. Other factors may include predation, adverse dispersal, and habitat degradation to name a few. Because of both rapidly changing anthropogenic and non-anthropogenic stressors in modern aquatic ecosystems, the physical and biological composition of these systems are in flux, and the organisms must evolve or retain appropriate phenotypic plasticity. Historically, selective pressures on multicellular animals have allowed fitter organisms to predominate over centuries. However, in today's world the timeline for adaptation or extinction appears to have been dramatically shortened. More recent studies have shown that species such as the killifish are able to adapt on a diminished timeline to chemical stressors (Reid et al. 2016). The life history of an organism, generation time, and significant variation within the population can influence the time it can take for evolution by natural selection to occur. Taken together this may allow species with these capabilities to survive through rapid changes.

The time scale required to decimate a viable fishery is dependent on many factors that pit recruitment against loss of standing stocks. The time frame is often spread over multiple decades before realization that there is a problem. In targeted

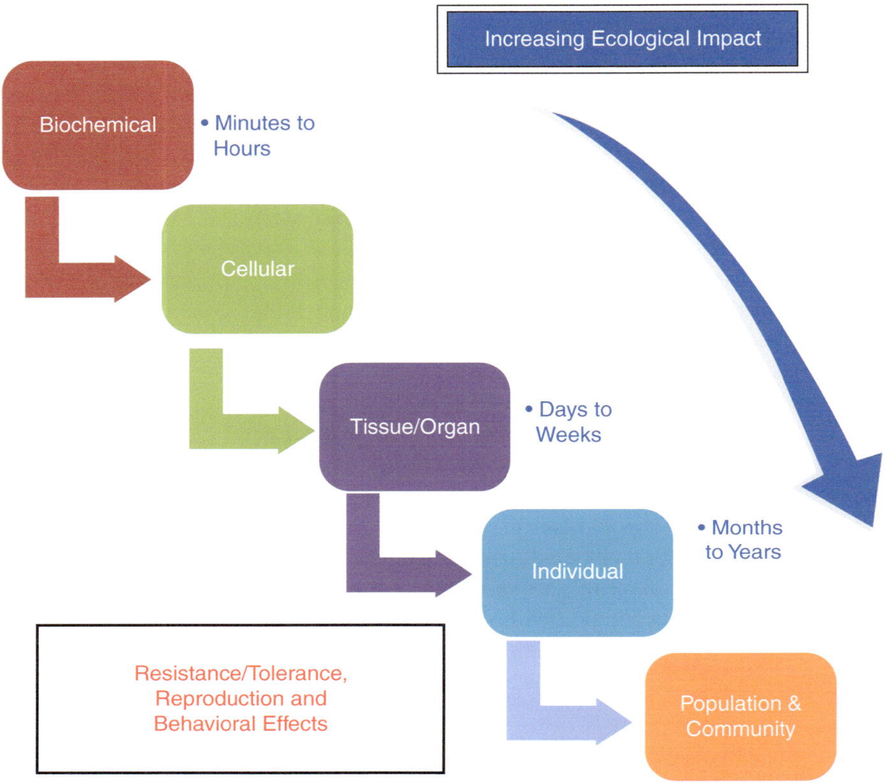

Fig. 6.1 The relationships between distinct levels of organization that can be impacted by chemical and physical stressors and how some changes (arrows) may be transferred to the higher levels of organization are shown. The relative times for events to be manifested range from minutes to hours at the biochemical level and decades to impact the population and community structure

commercial fisheries, there is always an impact of bycatch, and as higher trophic-level species become depleted, lower trophic-level species are commercially substituted. The further depletion of lower trophic-level species has even wider impacts on community structure due to the dependence on energy transfer (FAO 2006, 2012). Influences of biodiversity loss in ocean systems have been shown to result in increased rates of resource collapse and an exponential decrease in recovery potential, stability, and water quality (Worm et al. 2006). It is recognized that there is a much greater appreciation for the complex impacts from terrestrial sources, atmospheric inputs, freshwater watersheds, estuaries, and oceans as interconnected integrated system. Impacts on one component in the system can translate into major changes both locally, such as land utilization, and even on a worldwide basis, such as with changes resulting from global atmospheric changes. For example, Pinsky and Byler (2015) examined the influences of fishing pressures, specific life history traits, and climate variability on the increase in risk of a marine species collapse. Their article pointed to the importance of overfishing but also revealed the importance of life history traits (life expectancy, fecundity, recruiting periodicity, growth, and climate variability for influencing both collapse and significantly depleted stocks). With the continued intense anthropogenic denigration from the inland waters to the estuaries and the open oceans, it is not surprising that fisheries have collapsed around the world and that many more are threatened (FAO 2012). Since the underlying issues are complex and it takes an extended period of time to demonstrate that there is an issue, applying strategies that can predict the future of fisheries by linking biochemical to population-level effects (Fig. 6.1) in risk management is essential.

6.1.2 Linkages Between Biochemical to Population-Level Toxic Effects

In this dynamic field, progress has been made, but with the refinements of established toxicity assays in conjunction with the advent of innovative high-throughput biochemical techniques, behavior studies and physiology protocols have allowed researchers to demonstrate effects at much lower concentrations than previously established and for an increasing number of chemicals (Embry et al. 2010; Kavlock et al. 2012; Padilla et al. 2012; Patyna et al. 1999). Historically the use of aquatic organisms to assess chemical and stressor impacts grew out of the need to regulate effluent discharge toxicity to minimize acute and sub-chronic effects on the receiving waters (Chapman 2000; USEPA 1985, 1993, 1994, 1995). The whole effluent toxicity (WET) paradigm has allowed for the standardization of several toxicity tests which cross large phylogenetic groups found in both fresh- and saltwater environs. Figure 6.1 shows the relationship between the different levels of organization that can be impacted by chemical or physical stressors. Each level is connected to the next illustrating that effects from lower levels can translate into higher-level effects. It is important to also point out that effects may be observed at one level of organization but may not be seen at higher levels of organization, as with alterations in reproduction and behavior, which are generally expressed at the individual level. The timing and dose can also be factors in the assent from lower to higher levels of organization. Different life stages may be more sensitive to disruption than others. For example, in the developing embryo, there are specific stages of development that are more sensitive than others to chemical or other stressors (i.e., first trimester in human fetal development). The dose of a chemical can also impact the progression by overwhelming biochemical and physiological protective processes. The effects at the biochemical to the cellular level can occur in minutes to hours. Impacts at the tissue/organ level may take days to weeks to cause an effect, and at higher levels even more time may be required. Impacts at the individual and population level will generally have ecological impacts on the community and ecosystem level. The selection of individuals that have tolerance to the stressor can be selected, and the trait can be heritable. Importantly, though each organizational level is depicted as a single compartment, there can be any number of sub-compartments interacting as positive or negative feedback systems, and effects can be transmittable through the levels of organization.

To better understand the mechanisms of action (MOA) involved following a toxicant or stressor, exposure requires consideration of multiple endpoints of effect including hormonal, biochemical, and physiological pathways and the integration of multiple interacting systems. The use of aquatic animal models in assays that evaluate reproduction and development along with behavioral assays that assess higher organizational impacts has been recognized as an important criterion for assessing ecological risk. These techniques provide a greater understanding of integrated systems and the potential impacts on wild populations, to allow for better management of aquatic resources. Recent technological developments in omics and meta data analysis have aided in the determination of connections and nodal points that allow for an understanding of complex systems involved in reproduction and behavior. The use of popular model organisms such as Japanese medaka (*Oryzias latipes*), zebra fish (*Danio rerio*), fathead minnow (*Pimephales promelas*), and salmonid species has provided excellent tools for demonstrating impacts from the biochemical through to individual effects and more complex behavioral endpoints. Studies using salmonids and fathead minnows (*Pimephales promelas*) as model organisms have been able to illustrate the importance of chemicals and stressors on the endpoints discussed below as illustrated in Fig. 6.1 (Spitsbergen et al. 1988, 1991; Hornung et al. 1999; Weinberger and Klaper 2014). These complex integrated platforms enable an enhanced assessment of potential MOAs and, in some cases, have aided in identifying specific targeted pathways or networks. Convergence across multiple species and higher vertebrates enables a more robust risk assessment and management decisions that can be applied across species. Additionally, such convergence allows other model organisms to provide similar resolution in determining the effects of compounds on similar biochemical pathways and ultimately higher levels of organization.

Changes in single or multiple pathways controlling integration of complex biological functions can occur with alterations present at the cellular, organ, and individual level, resulting in a decrease in fitness, reproduction, and behavior. Studies of behavior, gonadal development, and survival of offspring incorporate networks of biochemical, endocrine, and cellular interconnected pathways that require proper coordination and expression on a temporal and spatial basis. In order to study these effects, a number of assays have been developed. Visual damage to tissues or organ systems using necropsy and histopathology techniques is useful in assessing an animal's health and identifying organ or cellular specificity and integrates well into such studies. Researchers often focus on specific biomarkers for assessing exposure. Biomarker studies often target a specific component of metabolism or marker of pathway activation. Cytochrome P450 1A1 (CYP1A1) is a common enzyme that has an important function in the metabolism of endogenous and exogenous compounds such that alterations in its levels and activities can be indicative of exposure. There are many biomarkers which have been used to assess exposure or an adverse outcome both in the field and laboratory (metallothionein, vitellogenin, liver function assays, etc.). The development of RNA-Seq and other genomic and proteomic tools is expanding our biomarker capabilities to evaluate pathway analysis. Such examples of biomarkers are all generally assessed by determining if they are present, increased, or decreased in response to specific classes of compounds that cause the specific biological effect. This approach was fashioned after studies carried out principally on higher vertebrates conducted by the pharmaceutical industry, which developed batteries of biomarkers that could be used to screen for efficacy of a drug and anticipate a toxic effect based on the MOA of the compound. Biomarker assays work best for single-compound exposures to naïve animals or in vitro systems but are less informative under environmental scenarios when multiple chemical sources may be present.

6.1.3 Development of Heritable Resistance and Induced Tolerance

Wild populations in estuaries and oceans are exposed to multiple stressors that can alter the organism's response when compared to populations not having been similarly exposed. The population's exposure history may have selected for those individuals which have modified their basal activities to compensate to survive. Chapman (2000) described two types of adaptive changes that can occur in populations: inducible strategies and constitutive strategies. Inducible strategies by the organism likely involve behavior, physiology, and biochemical changes that occur following exposure and require a general lag time. On a morphological basis, this could be hyperplasia (increase in normal epidermal cell numbers) that is induced by a causative agent. When the causative agent is removed, the hyperplasia is reversed with cells returning to normal morphology. In the case of constitutive changes, the modifications are permanent within the population and thought to be irreversible. The use of these two broad categories is likely too simplistic, and there probably exist many intermediate stages depending on the delivered dose, duration, and extent of exposure. For example, a species can develop resistance to a stressor through a heritable adaptation. Van Veld and Nacci (2008) described several biological and physiological factors that could contribute to a heritable resistance to toxicants.

At the population or individual level, the development of resistance or tolerance during a toxicant or stressor exposure is illustrated by comparison between populations that have different exposure histories. A major shift in the dose response curve is evidence of change in the resistance or tolerance of the individual groups. The specific cause for the shift can involve multiple systems either working individually or in combination with several factors or pathways. Intergenerational mechanisms of disease can be somatic mutations, epigenetic effects, or physiological based.

Physiological adaptation usually results in a short-term change in the physiological parameter or biochemical response that can be utilized to maintain homeostasis within the animal. Both the physiological and biochemical responses generally have a limited reserve capacity since long-term or very high exposure levels result in depletion of buffering capacity and/or decreased fitness. For example, organisms that have an increase in tissue damage from endogenous or xenobiotic reactive intermediates can increase cellular glutathione levels to combat the damage only up to a point. In the case where there is somatic genetic damage from a chemical exposure resulting in the transformation of cells, the formation of tumors such as a hepatocarcinoma or epithelial tumors in an individual fish will be self-limiting and not be propagated through to offspring. However, changes to individual fish can have an impact on the population structure and reduce the number of individuals contributing to sustaining the population if the tumors result in death prior to reproductive activity. Intragenerational mechanisms can involve heritable and nonheritable changes that may be neutral under current conditions, advantageous or deleterious. Those mutations on germ cells are heritable. Nonlethal genetic changes in the germ cells of the organism that give an advantage to survival following exposure to a chemical or stressor will be passed along to offspring that will give a fitness advantage over those organisms lacking the genetic alteration. In the case studies discussed below, there are examples where through chronic exposure to specific populations, there is a selection of permanent changes that are distinctive to several oviparous fish species (Nacci et al. 1999; Reid et al. 2016; Wirgin et al. 2011). In these cases, there are essential biochemical metabolic pathways involved in metabolism and elimination of both xenobiotics and endogenous substrates. The selective advantage here is that a group of organisms with a specific and novel trait that allow for survival. The presence of such a phenotype can be observed through carefully chosen biomarker endpoints that reflect the degree of tolerance to novel stressors within the population.

Sublethal disruption of normal homeostasis at the biochemical, cellular, and tissue/organ system level can result in higher morbidity and mortality. The fitness of the organism directly impacts disease resistance and ability to cope with stressors. Adverse effects on swimming ability and behavior following environmental cues can decrease survival from predators, prey capture affecting growth, and courtship affecting reproduction. Reproductive success is one nodal point for any viable population, and stressors or chemicals that prevent adequate recruitment will result in the population's eventual extinction. The rate of extinction is dependent on several factors that relate to the life history of the organisms, population interconnectivity, ecosystem processes, and the severity of the chemical or stressor to impact fecundity. When such measures are paired with biomarker studies, there can be a stronger causal link between the MOA and an observable effect that can translate to population-level alterations.

An underlying principle is that to maintain a population at steady state, it is merely necessary to replace the parental generation (population replacement) within their lifespan. The difficulty with a strict population replacement approach is that the population growth is zero and any increased death rate will result in a population decline. The reproductive strategies of most aquatic species however have evolved to overproduce offspring to maintain a large enough population to survive at an adequate carrying capacity. Therefore, a catastrophic event (i.e., inclement weather, disease, increased fishing pressure, chemical spills, etc.) could temporarily decimate the population, but overtime with decreased selection pressures, the population will likely rebound. In wild populations, there are often large variation in population size from year class to year class, and when the numbers dip below a critical threshold, or critical mass, the population will no longer be sustainable or of sufficient size to remain economically viable such as was the case for the cod industry in the Northeastern USA (Baker and King 2010). Continued steady declines in population size can often go on for decades until the fishery is no longer sustainable. Many invertebrates and teleosts produce large numbers of eggs; however, many eggs are lost due to predation, disease, and other factors. In these populations, there is generally many young of the year (YOY), but in later years the numbers generally decline until reaching sexual maturity and begin spawning. Aquatic organisms are poikilotherms and have requirements for specific ranges of temperature that help define geographical ranges and spawning windows. They also rely on several external physical cues such as temperature and photoperiod to initiate the reproductive cycle both for gonadal maturation and to initiate spawning events. Current changes in seawater temperatures due to climate change are a good example of stressors which can have major impacts on home ranges, gonadal maturation, population, and community composition, but the changes may take decades before a clear trend can be established.

Aquatic communities exist in a balance of populations, where thresholds are in place to maintain community structure. Stressors to a population, such as natural events, the introduction of invasive species, climate change, eutrophication, overfishing, and habitat loss to name a few, can threaten to push a population out of these thresholds. A key stressor that will serve as the basis of discussion of this chapter is persistent organic pollutants (POPs). These compounds can impact biological levels of organization as outline in Fig. 6.1, meaning that a direct interaction of a POP with a biochemical factor or physiological pathway can begin a cascade that has effects on tissue, organ, and individual levels. Many biomarker assays have been developed to examine these effects in a variety of species in both field and laboratory settings. The impacts on the individual, specifically effects on behavior and reproduction, have potential to impact a community. In addition, selection pressures can select for individuals with altered basal activities and traits. The chemical and physical stressors can interfere with key developmental nodal events that alter specific traits that can translate into either adverse (reduced fitness) or beneficial effects (resistance) at the population level. The ecological threshold concept is particularly relevant when anthropogenic environmental stressor gradients are involved because such stressor gradients likely represent novel physical and chemical conditions occurring in contemporary times, which are separate from those experienced by species on an evolutionary time scale (Baker and King 2010; Huggett 2005).

The impacts at the community level are often difficult to assess, because of a deficit in our scientific knowledge about the key roles of many noncommercial and commercial species within a food chain, food web, and the community structure's dependence on sufficient energy transference to the higher trophic levels. In many ecosystems, we have poor data on population sizes, species distribution patterns, migratory patterns, and other essential information on both commercial and noncommercial species that are vital to an ecosystem's health and services. Worm et al. (2006) provide convincing evidence that the establishment of reserves and fishery closures showed increased species diversity of both targeted and nontargeted species with an estimated 23% richness increase. There is a need to incorporate land use restrictions and pollution controls along with fisheries management within any comprehensive strategies to increase species diversity and ecological services (Ravit et al. 2015).

This chapter will concentrate on several classes of chemical stressors that have been shown to result in changes at the biochemical level through the individual level of organization (Fig. 6.1). Case studies of both wild populations and laboratory models will show examples of effects on reproduction, behavior, and the development of resistance in terms of the impacts these changes have on populations and, potentially, the community where these species live. The integration and crosstalk between developmental gene regulatory networks and their impact on higher systems organization involve gene expression, embryonic development, behavior, and reproduction. The sensitivity of the early developing embryonic stages to specific xenobiotics is due in part to the requirement for synchronization of programed cell movement, the primordial organ formation, and the disruption of programmed biochemical and physiological pathways. Even in the adult organisms when oogenesis and spermatogenesis are occurring, several of these compounds can disrupt this process and hormonal regulation reducing the fecundity and survival of the eggs and larvae, disrupt gonadal structure and function, and interfere with behavioral cues. The xenobiotic impacts on reproductive success, larval survival, and altered hormonal cues at the individual level can dramatically increase the natural mortality and decrease the maximum sustainable yield of a population. A stressor or chemical exposure may also select individuals with specific traits that can give them a selective advantage to survive within the stressed environment. Specific chemicals will be used to illustrate the importance of a systems approach interpreting field and laboratory results and the extrapolation of impacts to higher population, community, and ecological levels.

6.2 Effects of Legacy and Emerging Contaminants

The sections below describe specific case studies of reproductive, behavioral, and the development of resistance as a result of chemical exposure on both laboratory and field populations (Table 6.1). In each of the examples, there is an attempt to relate observations made at the various levels of biological organization through the population or community structure (Fig. 6.1). The case studies were selected to illustrate that many different chemical stressors can result in population-level effects on reproduction that directly translates to impacts on the population level.

Table 6.1 Selected studies demonstrating alterations reported in wild fish populations chronically exposed to contaminants that result in individual/population changes

Impacted organism	Stressor/location	Effect compared to reference population	Citation
Atlantic killifish *Fundulus heteroclitus*	PAHs/Elizabeth River, VA	Resistant to PCB126 teratogenicity and modified P4501A induction $F_1 < F_2$	Meyer and Di Giulio (2002), Di Giulio and Clark (2015)
Tomcod *Microgadus tomcod*	PCBs/Hudson River, NY	Resistant to induction via PCB126	Wirgin et al. (2011)
Atlantic killifish *Fundulus heteroclitus*	PCBs/PAH populations, Atlantic Coast	Complex tolerance pathways, log order tolerance differences, AHH-related pathways	Reid et al. (2016)
Atlantic killifish *Fundulus heteroclitus*	PCBs/multiple sites ($N = 24$) VA to MA	Intraspecific variation over four orders of magnitude to PCB126 for induction and teratogenicity	Nacci et al. (2009)
Atlantic killifish *Fundulus heteroclitus*	2,3,7,8-TCDD/ Newark Bay, NJ	Resistant to 2,3,7,8-TCDD compared to reference	Prince and Cooper (1995a, b), Nacci et al. (2000)
Atlantic killifish *Fundulus heteroclitus*	Transplanted Newark Bay, NJ (multiple cont.)	Vitellogenin and ER α induction males, decreased fecundity and embryo survival, smaller egg and yolk volume, ER/AHR crosstalk	Bugel et al. (2011)
Atlantic killifish *Fundulus heteroclitus* and four other species	Mercury/Piles Creek, NJ	Resistant to mercuric chloride, altered morphology, behavior, and reproduction	Weis et al. (2001), Weis and Weis (1977)

In the case studies, both single and multiple chemical stressors are discussed. It is often multiple stressors, working in conjunction with chemical exposure, that lead to these population effects. It has also become evident that as a population's reserve capacity has decreased, sublethal chemical exposure and impacts from chronic legacy chemicals and pharmaceuticals play a larger and larger role in determining the survival of a population. The selective loss of critical populations within a community structure can alter fundamental processes required for maintaining a viable ecosystem.

6.2.1 Case Studies in Wild Fish Populations Exposed to Legacy Contaminants

Alterations in teleosts chronically exposed to anthropogenic compounds show that selection pressures can result in the exposed population to exhibit unique traits when compared to nonexposed populations. Case studies involving fish populations that have been exposed chronically to several persistent organic pollutants (POPs), such as polyaromatic hydrocarbons (PAHs), polychlorinated biphenyls (PCBs), and 2,3,7,8-dibenzo-p-dioxin (2,3,7,8-TCDD), will be discussed. For a more in-depth discussion on the case studies presented below, the reader should consult the cited primary literature listed in Table 6.1.

POPs are legacy contaminants that are persistent in the environment and are generally associated with historical industrial activities. These compounds continue to leach into the system or are associated with contaminated sediments that act as reservoirs for long-term exposure scenarios. Reports have demonstrated that chronically exposed fish populations are resistant to the toxic effects of various POPs and that this resistance is heritable (Bello et al. 2001; Meyer and Di Giulio 2002; Nacci et al. 1999; Prince and Cooper 1995a, b). Two recent studies published by Wirgin et al. (2011) and Reid et al. (2016) in science discussed several populations chronically exposed to POPs as it relates to chemically induced evolutionary adaptation in Atlantic tomcod (*Microgadus tomcod*) and Atlantic killifish (*Fundulus heteroclitus*). One of the major points that these and other researchers have reported within these population is that there are many interconnected biochemical (steroidal, lipid metabolism, hormonal, growth factors) and detoxification pathways (i.e., cytochrome P450s, phase II conjugation, cellular membrane transport) that contribute to the tolerance phenotype. It is thought that local stressors select for individuals with modifications of single or multiple pathways that enhanced survival. Figure 1c in Reid et al. (2016) demonstrated this across several geographically discrete Atlantic killifish populations. The actual fitness cost to the organism and/or the population is often unclear. In many cases where there is an observed higher ecological effect in the field population (i.e., feminization of males), laboratory studies are carried out to enable the researchers to identify the mode of action or a specific mechanism by which the change is caused. What does appear to be evident is that the impacted waters near the resistant populations have less species diversity, which may reflect the resilience for these resistant species to tolerate the exposure compared to more sensitive species. A pathway not directly altered by a chemical can be influenced through crosstalk and the reliance of both pathways on shared limited cofactors that are required for each of the pathways. Differences in species sensitivity has been well documented in teleosts exposed to 2,3,7,8-TCDD as well as other compounds that can also explain the selective loss of species within an ecosystem.

Follow-up studies in the laboratories examining fecundity, embryonic development, and behavioral endpoints contribute to establishing the potential impacts at the population level. Two groups of teleosts, Atlantic killifish and Atlantic tomcod, will be used as case studies of fish populations that have altered metabolism or shifted dose responses to POPs when compared to reference wild fish population (Table 6.1). Killifish have been used extensively for studies that aim to understand how vertebrates respond to environmental stressors. This is because killifish are present in large numbers in tidal salt marshes along the Atlantic Coast, are very tolerant to salinity changes, and have a limited home range allowing for site-specific evaluations in genetically isolated populations. In all the case studies described below and where tolerance has been demonstrated, there have been decades of exposures (>20 years), resulting in selection for survivors with specific traits leading to subpopulations due in part to their limited home range (Brown et al. 2009; Nacci et al. 1999; Lotrich 1975). The large number of embryos spawned each year and exposed from both maternal transfer and direct environmental exposure to POPs may also partially explain the development of tolerant populations, loss of fitness, and resistant embryos (Brown et al. 2009). Atlantic killifish are often one of the few finfish present in both pristine and contaminated locations, which allows for comparisons between chronically exposed and reference fish. Another major advantage is that the eggs can be manually stripped from the female and fertilized in the laboratory to conduct embryo toxicity studies. The eggs have well-defined developmental staging and require about 16–21 days to hatch postfertilization (Armstrong and Child 1965). Embryonic development is highly conserved across oviparous species and is very sensitive to POPs, which allows for many biochemical and morphological endpoints to be examined as it relates to chemical exposure.

The first case study involves a population of Atlantic killifish (*Fundulus heteroclitus)* chronically exposed to a mixture of PAHs and creosote from a wood treatment facility (Atlantic Wood) in the Elizabeth River in Virginia (Huggett 2005). The initial studies were carried out in the early 1990s and have continued until recently when the site underwent remediation and restoration. An extensive review of the history and status of Atlantic killifish living in the Elizabeth River and the effects observed was recently published by Di Giulio and Clark (2015). A central pathway involved in the metabolism of both endogenous compounds and xenobiotics is the aryl hydrocarbon receptor (AHR). The induction of AHR-related genes following binding to DNA response elements can lead to an increase in CYP1A1 which could result in reactive intermediates being formed from PAHs. There are several excellent reviews on this pathway in aquatic organisms and in fish demonstrating the central role it plays (Hahn et al. 1997; Hahn 2003). The initial findings involved the determination that the fish inhabiting this area had increased hepatic tumors (hepatocellular carcinomas) and other lesions (Fournie and Vogelbein 1994). Atlantic killifish living a small distance away, across the river from the site, did not have similar lesions, but as stated above, this species has a small home range that allows for populations to be isolated one from the other. Laboratory studies examining exposure through sediment to the F_1 and F_2 generations demonstrated a dramatic shift in their dose response when exposed to contaminated sediments and pure PAH compounds (Clark et al. 2014). There have been a series of papers examining various characteristics of the fish inhabiting this location from the 1990s through to 2016. Although there appears to have been a viable reproducing population, the surviving population has been reported to have a poor condition index, increased microbial infection, and higher morbidity and mortality when compared to reference populations. Meyer et al. (2003) further reported that F_1 offspring were more sensitive to low oxygen stress and had increased toxicity from photo activation of PAHs. These observations are not surprising considering the potential effects from these compounds on the hematopoietic system and other organ systems in teleosts (Incardona et al. 2004; Faisal and Huggett 1993; Reynaud and Deschaux 2006). As this site is currently being restored, the populations living in this small area will likely be exposed to improving conditions, and following their response will be of great interest. This series of studies also demonstrate that local relatively isolated populations of Atlantic killifish can have different responses to PAHs.

In the case of PCBs, PAHs, and chlorinated dioxins, there have been many studies examining the development of tolerance and resistance in both tomcod and Atlantic killifish. The Hudson River has been the focus of several studies examining the impacts of PCBs released into the river primarily from a general electric facility manufacturing capacitors and transformers which were located upstream at Hudson Falls, NY. Approximately 1.3 million pounds of PCBs were believed to have been released into the Hudson River and 7 million pounds buried in landfills adjacent to the river. Tomcod occur along the Atlantic Coast, and their range is from the Gulf of St Lawrence to Virginia. Tomcod spawn in shallow estuaries, and each female can produce between 6000 and 30,000 eggs that for the first month are associated with the benthos (Klaude et al. 1988). The fish remain in the estuary eating a variety of benthic invertebrates and small fish until they are large enough to migrate to the lower reaches of the river.

The studie*s* examining the prevalence of hepatic tumors in tomcod exposed to PCBs begin in the early 1980s (Dey et al. 1993; Roy and Wirgin 1997). Hepatocellular carcinomas exceeded 50% in 1-year-old fish and over 90% in 2-year-old fish. The population age structure was truncated, and as the Hudson River inputs have declined, so have the prevalence of tumors. Populations from other rivers along the Atlantic seaboard did not show similar pathologies or altered age structures. Several biomarkers such as bile PAH metabolites, CYP1A1 mRNA, and bulky DNA adducts were higher in tomcod obtained from the Hudson River than observed in tomcod from less impacted rivers. It was also shown that induction of CYP1A1 by halogenated aromatic hydrocarbons (HAHs) was inhibited in Hudson River tomcod, while CYP1A1 from fish from other rivers could be induced by both by HAHs and nonhalogenated PAHs. In a similar fashion as to what was observed in the Atlantic killifish at the Atlantic Wood site, chronically exposed populations of fish had developed resistance to PCB126. Wirgin et al. (2011) provided a mechanism that demonstrated how the different populations of tomcod were resistant to PCBs. By examining 8 different geographically isolated tomcod populations, they demonstrated that the AHR2-1 allele, which had a 6-base pair deletion and a nucleotide 3274 substitution, correlated with the resistance. They also demonstrated that the binding to synthesized altered AHR2-1 was approximately one third as effective at binding 2,3,7,8-TCDD and reduced functionality in transfected cells. Reid et al. (2016) reported that the tolerant PAH population was genetically different from the halogenated tolerant populations as it related to duplication of AHR signaling and the involvement of other AHR pathway-related genes: *arntlc*, *hsp90*, *cyp1c1/c2*, *gfrp*, and *gstt1*. This illustrates that depending on the chemical's effect on the cell that other genes may be up- or downregulated and that multiple factors not all fully genetically heritable can confer resistance.

In another series of studies examining predominantly PCB-exposed populations of Atlantic killifish along the Atlantic Coast, a similar pattern of resistant and tolerant population involving both the adults and the sensitivity of embryonic stages to specific halogenated hydrocarbons has emerged (Nacci et al. 1999; Reid et al. 2016). The different populations have all demonstrated resistance compared to reference populations using a prototypical toxicant such as PCB126. This set of studies demonstrates the selective pressures from chronic chemical exposure to select a population with a specific genetic trait. As stated by the authors, there are likely many physiological or chemical genetic modifications that could have impacts on resistance. The selection for viable offspring is likely driven through the divergence between sensitive and tolerant individuals with a beneficial modification in the total aggregate pattern of regulatory linkages or "architecture" controlling gene regulatory network in tolerant individuals. This idea has been elaborated by Levine and Davidson (2005) as it relates to gene regulatory networks involved in morphological development. The AHR regulatory gene battery and their products could be considered an essential biochemical node, which is far more flexible and malleable to moderate chemical stressor effects and allows development of resistant individuals. The continued selection pressure over multiple generations within genetically isolated gene pools would result in a predominantly resistant population of individuals with a trait that is beneficial for survival.

The studies in the Newark Bay populations were principally driven by the high concentrations of 2,3,7,8-TCDD, but there are contributions from many different contaminants including PAHs, PCBs, and metals into the Newark Raritan Bay system from legacy and current inputs (Cooper and Borjan 2010). Studies examining the impact on adult and embryos collected from the Newark Bay complex have been going on since the mid-1980s when very high levels of dioxins were found both on terrestrial sites and in the sediments (Wisk and Cooper 1990a, b; Cooper et al. 1993). Because of the known embryo toxicity associated with this class of compounds, an embryo-larval toxicity approach was developed based on assays from the literature, as well as used in the test for whole effluent toxicity (WET) assessments (Cooper and McGeorge 1991; Cooper 1996; USEPA 1985, 1993, 1994, 1995). The sensitivity and usefulness of examining these life stages were also demonstrated by the Weis's work regarding mercury and other research that focused on 2,3,7,8-TCDD and other halogenated aromatic hydrocarbons (Spitsbergen et al. 1988, 1991; Hornung et al. 1999; Weis and Weis 1977). What became apparent was that populations of Atlantic killifish living in Newark Bay were resistant to exposure to 2,3,7,8-TCDD when compared to Tuckerton, NJ, reference fish. The larvae reared from eggs strip spawned from Newark Bay fish did not show any of the typical dioxin-related lesions including yolk sac and pericardial edema and cardiovascular lesions. Also, the induction of EROD, ethoxyresorufin-O-deethylase, which is a biomarker for CYP1A1 activity in teleosts was not inducible in the wild Newark Bay fish (Nacci et al. 1998). The basal level of CYP1A1 was higher as evidenced through EROD and CYP1A protein levels in Newark Bay fish (Prince and Cooper 1995a, b). Reid et al. (2016) reported that comparison of the tolerant (T3) and sensitive (S3) populations which are from the Newark Bay complex showed that in the T3 population, there was a dramatic duplication in the *CYP1A1* DNA region when compared to the S3 group. Interestingly, the number of individuals for specific deletions is comparable for the Narragansett and Elizabeth River groups as it relates to homozygous deletions, heterozygotes, and homozygotes intact. One possible reason could be that these populations have a large PAH contribution, and this is reflected in the similar patterns observed following mixture exposures (Reid et al. 2016).

In a series of transplant studies of reference Atlantic killifish into Newark Bay and laboratory challenge studies, it has been shown that the vitellogenin and estrogen pathways along with the cytochrome P450 pathways are involved in crosstalk (Bugel et al. 2011, 2013). In the Atlantic killifish populations from Newark Bay, the egg size and number of eggs spawned were reduced, as was the yolk volume. This would suggest that there is a fitness cost to those individuals living within these waters. Within Table 6.2 are summarized several environmental cues and critical pathways that are essential for maintaining a successful reproducing population.

Table 6.2 Cues and pathways that are critical at various stages of teleost development

Gonadal development in adult fish	Gonadal development begins, and it is tightly regulated by light, temperature, and nutrition cues such as kisspeptin. Other factors that regulate development include gonadotropin-releasing hormone (GnRH), thyroid hormones (T3, T4) released from the pituitary, and release of the gonadotropins luteinizing hormone (LH) and follicle-stimulating hormone (FSH)
	Steroid synthesis and metabolism within the organism also regulate the development of gonadal tissue. Steroidal hormone modulators, aromatase activity, and endocrine and exocrine release of hormones such as 17β-estradiol and testosterone create positive and negative feedback networks along with other hormonal metabolites and bioactive conjugates
	Early gonadal development requires the formation of pre-vitellogenin to mature eggs and spermatozoa to sperm
	Maternal development is critical as liver vitellogenin formation and transport follow estrogenic-initiated biochemical pathways. In addition, incorporation of maternal gene products allows early division prior to embryonic metabolic control
Fertilization	Full development of both the sperm and egg is necessary to successfully fertilize the egg and to allow proper early division of the embryo
Embryonic development	Embryonic development is an extremely complicated process that requires the coordination of temporally and spatially choreographed transcription, translation, and expression of vital bioactive compounds. The number of factors contributing to the formation of epidermal, mesodermal, and ectodermal tissues and organs to form an intact fish involves growth factors, exocrine and endocrine hormones, expression of membrane and cytosolic biochemical pathways, metabolizing systems, and an extensive uptake and elimination transport system. Several of these systems are active only during specific windows of development
	The yolk serves as the only source of nutrients for the developing embryo
Eleutheroembryo (yolk-sac larvae)	Embryos break out of the chorion by producing proteases, and the yolk-sac larvae continue utilizing the yolk for its food source prior to active feeding
Juvenile/adult	The juvenile fish will continue to increase in size, and the gonads will develop at the appropriate age. Specific behavioral assays assessing swimming, prey capture, predatory avoidance, and reproductive endpoints can be assessed

Decrease egg number would result in fewer juveniles and adult fish. The reduced yolk volume would provide less nutrients to the eleutheroembryo and a smaller larva. These observations have been observed in the wild fish from Newark Bay when compared to reference populations (Bugel et al. 2011, 2013). In addition, the circulating 17β-estradiol in females was significantly lower than the reference

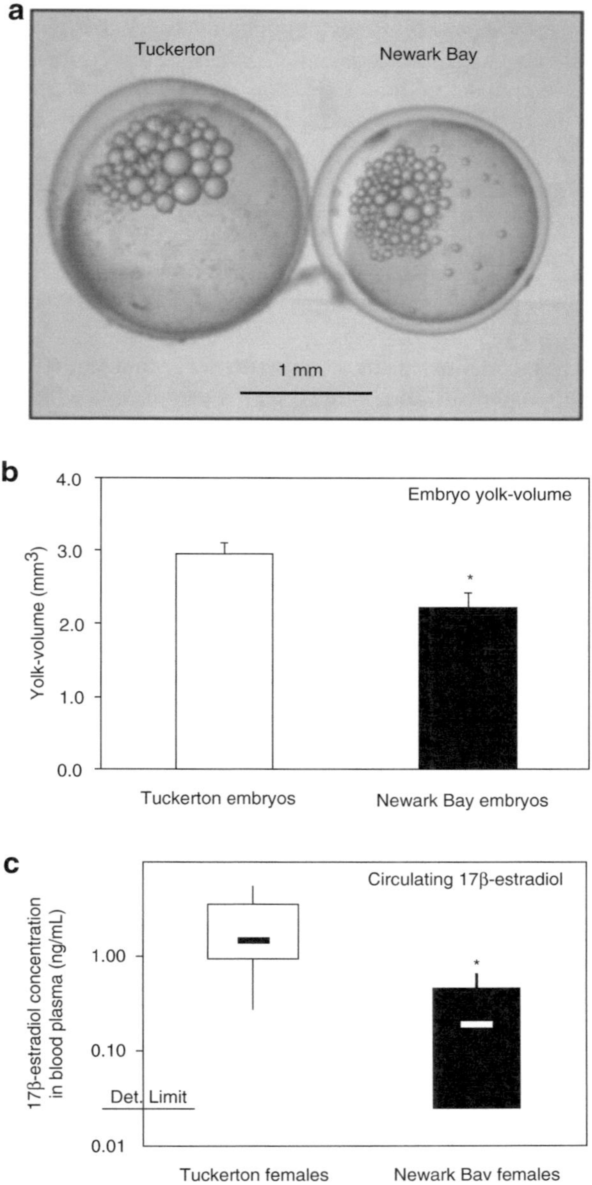

Fig. 6.2 Comparison of Atlantic killifish from reference (Tuckerton, NJ) and contaminated (Newark Bay, NJ) locations in terms of (**a**) egg size, (**b**) embryo yolk volume, and (**c**) circulating 17β-estradiol (Bugel et al. 2011)

populations (Fig. 6.2). These parameters are associated with a decrease in fitness that could affect the embryonic development, juvenile fitness, and reproduction. There is crosstalk between the pathways involved in cytochrome P450 metabolism, conjugation, and steroid and vitellogenin pathways that are likely impacted from chronic xenobiotic exposure. To prevent local extinctions, resident fish populations must be able to adapt by altering gene regulatory pathways that allow survival of tolerant organisms and permit successful reproduction. Although, survival of organisms within these polluted environmental may have a fitness cost.

Although chronic metal contamination (i.e., mercury) will not be discussed in this chapter, there is a similar pattern observed resulting in population effects and development of tolerance. For example, the literature on mercury tolerance in Atlantic killifish reported out of Weis's laboratory paints a very similar picture with impacts on fitness in the exposed adult Atlantic killifish and the development of tolerance in embryos (Weis and Weis 1977, 1989).

6.2.2 Case Study for Bivalve Mollusks in Newark Bay, New Jersey

Bivalve mollusk's adult sessile life stage and ability to filter large volumes of water allow adult bivalve mollusks to accumulate a wide variety of chemicals through both particle associated chemicals and direct absorption across epithelial surfaces (Capuzzo 1996). Bivalves go through seasonal changes that can influence the chemical body burden. For example, during certain periods of the year, a bivalve will accumulate lipids, glycogen, and proteins prior to gonad development and at spawning time will lose up to 30–40% of its body mass (Galtsoff 1964). Since bivalves are poikilotherms, their enzymatic and metabolic rates vary dramatically throughout the year, which can impact their response to stressors. The key roles that mollusks play in an ecosystem as filter feeders, food source, etc. make it important to understand the potential impacts of stressors and chemicals have on their reproductive success and different life stages. There is an extensive literature base on mollusks and specifically bivalves that covers many areas in greater detail. Indeed, there is a long history of using bivalve mollusks (American oyster, *Crassostrea virginica*, and blue mussels, *Mytilus edulis*) as sentinels in both saltwater and freshwater bivalves for assessing chemical accumulation and disease occurrence both in the USA and around the world (i.e., Mussel Watch Program). The next few paragraphs provide only a brief outline of the life history and complex pathways involved in the different life cycle stages of this very diverse group of organisms and their role in remediation efforts. The case study will focus on the American oyster (*C. virginica*), by examining its response to POPs, which unlike in teleosts, resistance has not been demonstrated.

There are many different bivalve species, but the American oyster has both economic and ecological importance along the Gulf Coast and Eastern seaboard in the USA. The American oyster can tolerate wide temperature, salinity, and total suspended solids swings which are some of the reasons for using this organism in biomonitoring studies. The different stages in the life cycle of bivalves, just as with teleosts, have different sensitivities to chemical and stressor impacts. Following the spawning event, the reproductive structures in adults are completely removed by macrophages and replaced with connective tissue. Depending on the location of the bivalve population, they may spawn once or twice a year. After spawning, the bivalve gonadal tissue undergoes reformation based on environmental cues (e.g., temperature, glycogen accumulation). During the senescent period, the bivalve accumulates sufficient energy stores, the gonadal follicles are reformed, and egg and sperm maturation proceeds (Fig. 6.3a). Chemical exposure during the period of reformation can result in poor gonadal development and improper egg and sperm formation.

The spawning of bivalves is primarily triggered by temperature if the gonadal structures are properly formed. Both egg and sperm are released into the water column, and fertilization occurs in the water column (Fig. 6.3b). What is shown in Fig. 6.4b is the progression through different developmental stages that take place over a 2–3 week period. The fertilized egg undergoes cellular division with the formation of ectoderm, mesoderm, and endoderm utilizing the egg's energy stores. Differentiation of primordial organ systems progresses ultimately forming a free-swimming veliger larva with a bivalve shell. The veliger larva filters algae from the water column for food and continues to develop through the straight hinge stage and into the pedal stage where the organism attaches to a substrate. The pedal larvae will settle out of the water column and locate a proper benthic substrate through sensors in the foot, and the spat will attach to the substrate. The development and survival of the pelagic stages, where veligers drift through the water, are likely the most sensitive stages in the oyster life cycle to chemical and other physical stressors. This is due to the rapid cell proliferation and key biochemical signaling between cells and organ development. As with fish development, gene regulatory networks play a critical role in the regulation of expression of critical sets of genes involved in developmental patterns and organ system differentiation. Once attached, the filter feeding oyster will continue to grow laying down shell along the edge of the shell using a specialized shell gland, while the mantle deposits shell nacre along the interior of the shell (Galtsoff 1964). Juvenile bivalves are easily preyed on by many predators and are sensitive to sedimentation and other environmental stressors in part due to their small size and thin shells. In the field, the number of adult bivalves reaching sexual maturity is a small fraction of the spat that initially settled out. Damage to the mantle epithelial cells and the shell gland can disrupt shell formation and result in thin shells. Shell thinning in soft-shell clams (*Mya arenaria*) likely due to POPs was observed in field-collected specimens (Brown et al. 1992, 1994; Cooper and Brown 1995). In laboratory studies, it was shown that calcium deposition to the shell was significantly decreased following exposure to 2,3,7,8-TCDD. Recent studies examining adult oysters collected from the Newark Bay complex also showed thinner shells and hyperplasia and dysplasia of the mantle epithelium and shell glands (Ravit et al. 2015).

The case study described below is primarily based on research examining the effects on *Crassostrea virginica,* the American oyster inhabiting Newark Bay, and

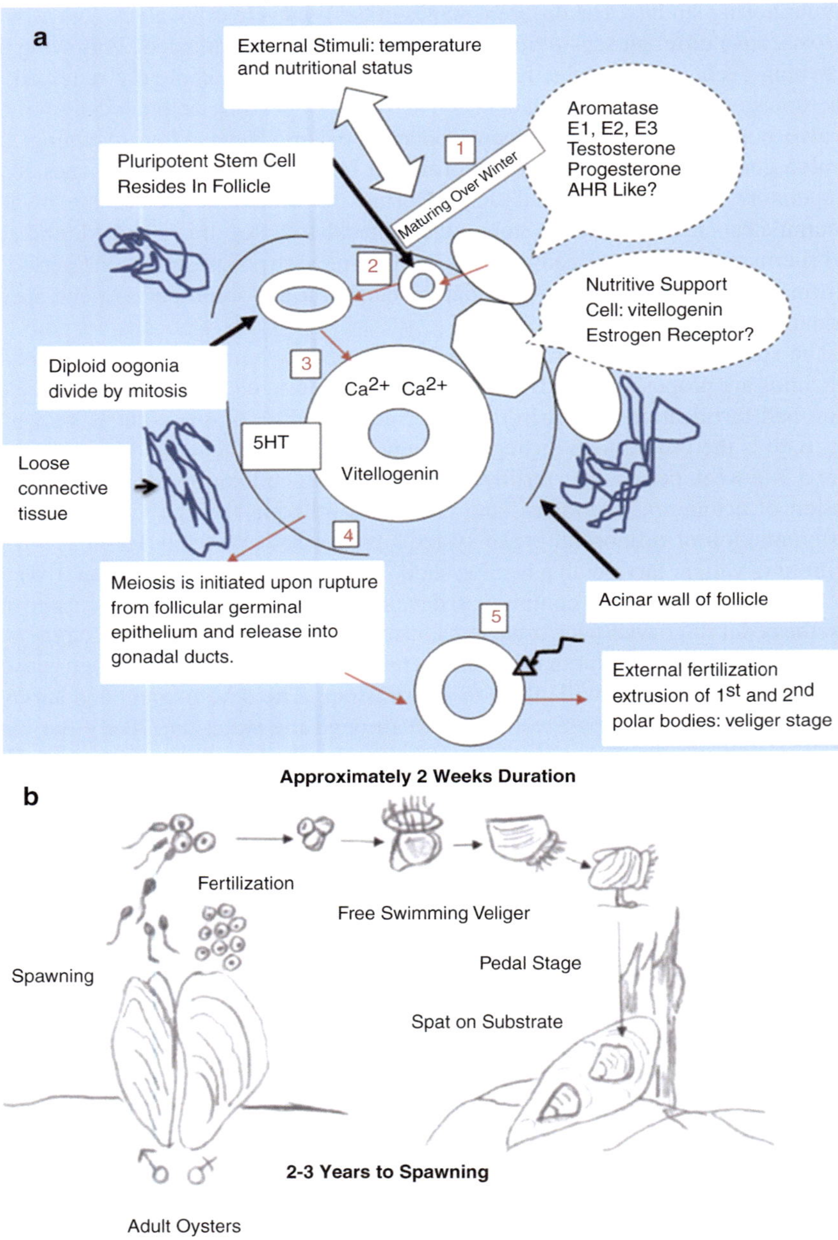

Fig. 6.3 (**a**) The maturation of the female gonadal follicle and egg (red arrows) with the influence of paracrine and autocrine signals from post-spawned follicle with stem cells (1 and 2), early egg development (3), maturation to fully formed vitellogenin egg (4), and release through the gonadal duct and external fertilization (5) (modified from Cooper and Wintermyer 2009). (**b**) The American oyster life cycle from fertilization through the pelagic stages and spat formation on hard substrates

laboratory studies examining impacts of 2,3,7,8-TCDD on bivalve growth, reproduction, and veliger larval development. Within Fig. 6.4a is shown the maturation in the female follicle using a 28-day timeline which recapitulates the overwintering occurring in the field along with the oyster life cycle from spawning until spat formation (Wintermyer and Cooper 2003). The laboratory studies used oysters that had nondifferentiated gonads and were placed into aquaria at 18 °C and fed algae to stimulate gonadal maturation over a 28-day period. That maturation is shown in Fig. 6.4 progressing from undifferentiated to fully developed eggs and sperm. At the start of the experiment, TCDD was injected into the adductor muscle of the oyster where there is a large blood sinus. They were maintained out of the water for several hours and then returned to the treated and non-treated tanks. At the end of the 28-day period, the oysters were strip spawned and assessed for fertilization rates and progression through larval development (Fig. 6.3b).

In a series of studies carried out comparing the impact of 2,3,7,8-TCDD on *C. virginica* using both field and laboratory studies, shell deposition, gonadal development, and veliger development were impaired following transplantation of clean oysters into contaminated waters of Newark Bay, NJ, and the Arthur Kill tributary (September to June) and following controlled dose response laboratory studies. Newark Bay was selected because of the extensive contamination from 2,3,7,8-TCDD and other chemicals. The oysters were placed into oyster bags and hung from a pier at Sandy Hook Bay, NJ, which was considered the reference location, and another set of oysters were hung from a dock in Newark Bay at Newark, NJ, and

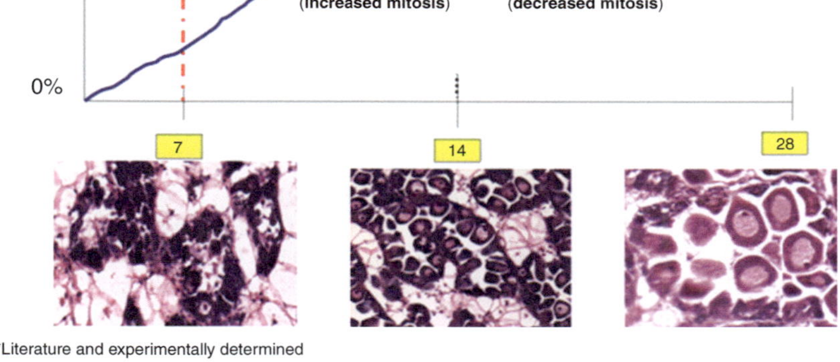

Fig. 6.4 Over a controlled 28-day maturation period, there is steady increase in vitellogenin production as the oocytes are undergoing maturation until they are ready to spawn and photomicrographs show the progression of the eggs within the follicle (Wintermyer and Cooper 2003)

then in the Arthur Kill at Elizabeth, NJ. The field-exposed oysters were assessed for their ability to produce viable eggs and progress through development to straight hinge stage (Fig. 6.3b). Oyster veliger larvae survival from the transplant study for Newark Bay oysters (3.2 pg/g 2,3,7,8-TCDD/2.1 pg/g 2,3,7,8-TCDF), Arthur Kill tributary oysters (1.3 pg/g 2,3,7,8-TCDD/1.7 pg/g 2,3,7,8-TCDF), and Sandy Hook Bay oysters (0.15 pg/g 2,3,7,8-TCDD/2.3 pg/g 2,3,7,8-TCDF) were 3.9%, 7.5%, and 73%, respectively (Wintermyer and Cooper 2003).

Laboratory studies conducted over the 28-day gametogenesis cycle (Fig. 6.4) resulted in control oysters having 80.3% survival (D-stage), while the 2 and 20 pg/g 2,3,7,8-TCDD oyster had 0% survival to the straight hinge stage. Oyster larvae spawned from non-treated oyster (ex vivo) and exposed to rearing solutions dosed with 0, 2, and 10 pg/ml 2,3,7,8-TCDD resulted in 76%, 2.3%, and 1.1% survival to the trochophore and D-stage. 2,3,7,8-TCDD has deleterious effects on the developing larvae both from field-exposed and laboratory-exposed oyster, as well as post-spawned veliger larvae. Within Fig. 6.5 are shown examples of both oocyte development and sperm development in control and 2 pg/g TCDD adductor muscle injected oysters during gametogenesis. What is evident is that even at the lower of the two doses tested (2 and 10 pg/g), both the oocytes and sperm development were disrupted compared to the controls. The spermatids did appear to have been able to develop and produce sperm, but even the sperm showed perinuclear vacuolization, which indicates altered sperm maturation. It is not known whether the sperm quality would interfere with sperm activation of eggs, but based on egg effects, as shown in Fig. 6.5b, fertilization would be dramatically decreased. The poor egg quality would result in a complete failure to reproduce. Based on the above findings, it was concluded that 2,3,7,8-TCDD affected veliger larval survival and proper gonadal maturation in the range from 2.0 to 10 pg/g (Wintermyer and Cooper 2007). The specific mechanism has not been determined, but with recent advances in oyster biology and the sequencing of the genome, new studies may determine the mechanism of action.

These studies demonstrate that in bivalve mollusks, 2,3,7,8-TCDD and related compounds can disrupt normal gonadal development and have a direct impact on the ability to produce viable offspring. As shown in Fig. 6.3, there is likely cross-talk between hormonal and steroidal signaling that can be altered by 2,3,7,8-TCDD and disrupt normal reproductive cycling in the gonad (Cooper and Wintermyer 2009; Ohtake et al. 2008). Studies have not been carried out to determine if resistance or tolerance can be developed in oyster chronically exposed to these contaminants as discussed above in the fish studies. In Newark Bay, the oyster populations have virtually been eliminated except for a few small pockets, and this would indicate that unlike Atlantic killifish, they have not been able to develop resistance. A potential bivalve that does occur in both contaminated and non-contaminated estuaries is the Atlantic ribbed mussel (*Geukensia demissa*), which may have been able to develop some resistance coping mechanisms, but this has not been examined. In the fish studies, there has been a strong emphasis on the cytochrome P450 pathways and related gene regulatory networks, which are less well

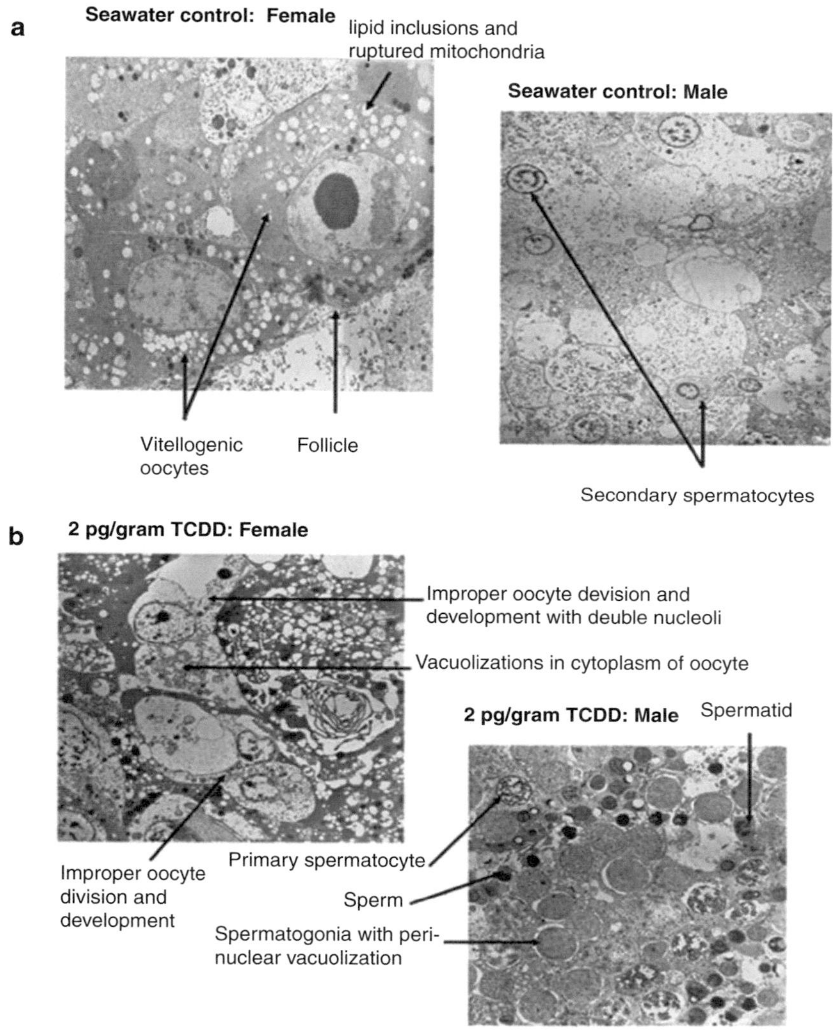

Fig. 6.5 Representative electron micrographs of both male and female gonad development in (**a**) seawater control and (**b**) 2 pg/g 2,3,7,8-TCDD injected into adductor muscle (Wintermyer and Cooper 2007)

defined in the bivalve. The bivalve's PAH metabolic activities are much lower than fish, and they do not have a similar AHR to that of higher vertebrates (Hahn 2003). However, just as in fish, there is a conserved gene regulatory network involved in developmental pattern and growth through the different life stages. Further studies examining these pathways may lead to a better understanding of population effects following chronic exposures.

6.2.3 Case Study in Zebra Fish of Perfluoroalkylated Substances: Emerging Contaminants

The case study discussed below examines perfluoroalkylated substances (PFASs) and uses the zebra fish (*Danio rerio*) to demonstrate how these compounds can result in effects at the biochemical level and manifest effects through to the population.

The perfluoroalkylated substances (PFASs) are a class of compounds known to cause toxic outcomes across many species. Historically, they have been used in a number of industrial processes, including production of nonstick coatings, stain-resistant fabrics, and firefighting foams. These compounds contain a fluorinated carbon backbone and terminal group, and both of these elements have the potential to impact the distribution, metabolism, and interactions with target receptors and, therefore, the relative toxicity of each compound (Fig. 6.6). These toxic outcomes include a myriad of effects such as effects in the liver, reproduction, development, fat metabolism, breast development, hormone receptors, and neuronal toxicity. Each level of biological organization, as characterized in Fig. 6.1, is impacted by interactions with PFASs, and once these observations are strung together, a cascading paradigm of responses becomes apparent. In this way, observations made at the biochemical level can be traced and linked to observations at the individual and ultimately population levels. It is at this point that the full implications of these toxic effects can be determined in terms of ecological relevance and risk management decisions.

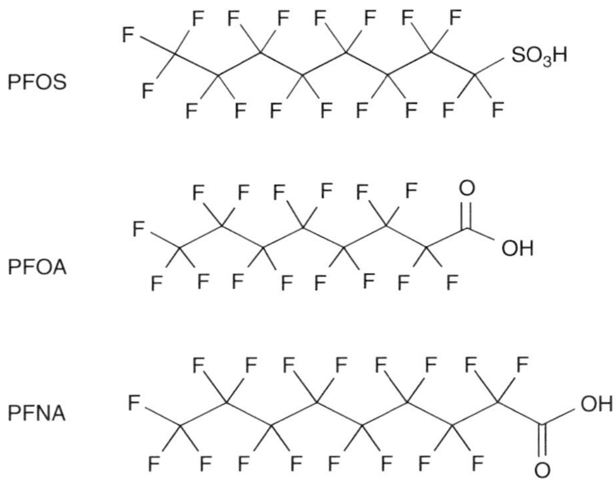

Fig. 6.6 Structural comparison of the non-branched PFASs: perfluorooctanoic sulfonate (PFOS, C8), perfluorooctanoic acid (PFOA, C8), and perfluorononanoic acid (PFNA, C9). The fluorine-carbon bond is resistant to biological degradation resulting in environmental persistence. Increasing carbon length results in higher bioaccumulation, and end groups influence interactions with membrane associated subcellular transporters

Although the cascade begins at the lowest level of biological organization, the biochemical level, the interactions with the PFASs at this level are quite complex. This is especially true when considering the interactions of PFASs with subcellular phase III transporters, which may contribute to the distribution of toxic compounds. Some common classes of transporters include OAT (organic anion transporters), OCT (organic cation transporters), ABC (ATP-binding cassette transporters), and MDR (multidrug resistance ABC transporters). Transporters are enzymatic proteins embedded in cellular membranes and are pivotal in regulating the influx and efflux of endogenous compounds in different organs and tissues. They function in the transfer of substrates, such as amino acids, fatty acids, steroids, hormone derivatives, metabolites, conjugates, and exogenous compounds. Each transporter moves multiple compounds with specific charge or conjugation, and in addition, some contain multiple binding sites, both of which explain the wide range of substrates that can utilize these systems (Klaassen and Aleksunes 2010). Non-endogenous compounds can also make use of this system, as well; the PFASs have been shown to interact with members of the class of organic-anion-transporting polypeptide (oatp) transporters (Weaver et al. 2010; Popovic et al. 2014).

Two PFASs, perfluorooctane sulfonate (PFOS) and perfluorooctanoic acid (PFOA), have been demonstrated to interact with members of the family of organic-anion-transporting polypeptide (oatp) transporters. In cell cultures, PFOS was shown to act as a substrate to the transporter, oatp1d1, and PFOA to act as a substrate to oatp1a1 and an inhibitor to oatp1a1, oatp1a2, and oatp1d1 (Weaver et al. 2010; Yang et al. 2010; Popovic et al. 2014). There are biological implications of such interactions. With the inhibition of a transporter, this enzyme can no longer function in the transfer of critical endogenous compounds across the membrane. This does not necessarily result in an alteration in the levels of these endogenous compounds that use the oatp transporters to maintain homeostasis, as there are multiple members of this transporter family with overlapping substrate capabilities (Shirasaka et al. 2012; Popovic et al. 2013). However, this can lead to competition between endogenous and xenobiotic substrates for these alternative substrate pathways. Because there is great conservation within biological pathways and, specifically, homology across species in terms of the phase III transporters, these data allow for extrapolation across species with high homology.

In particular, one biochemical target following PFAS exposure is the alteration in homeostasis of fatty acids. Following chronic exposure to PFOS, which has also been shown to bioaccumulate, zebra fish were found to have increased fatty acid oxidation and altered serum lipoprotein levels (Cheng et al. 2016). This is possibly confounded due to interactions between PFOS and transporters. In addition, PFOA is a reported ligand of the transcription factor PPARα (peroxisome proliferator-activate receptor alpha) and to some extent PPARγ (Vanden Heuvel et al. 2006). When activated, PPARα leads to the transcription of genes involved with fatty acid transport, binding, and β-oxidation (Tyagi et al. 2011). Taken together, there are multiple mechanisms impacted by PFAS exposure that affect pathways related to fatty acids and lipid homeostasis within the organism. Lipid metabolism, storage, and transport in the liver play an important role in supplying vitellogenin for egg maturation in the gonads of oviparous organisms and development. The proper synchronization and coordination between organ systems (e.g., liver, nervous, gonad) are critical for successful reproduction, and alterations in lipid homeostasis and utilization during critical stages of egg development, eleutheroembryo (yolk-sac larvae) maturation, and juvenile survival can impact the fitness of the teleosts.

At the biochemical level, the PFASs have also been reported to affect a number of neuronal factors. For example, exposure to PFOA, PFOS, and PFNA during early development in zebra fish was reported to alter transcription levels of brain-derived neurotrophic factor, bdnf, which is involved in the regulation of the growth of neurons (Jantzen et al. 2016). Other neuronal pathways have been implicated as well. PFOA exposure in mice has been associated with changes in the profile of neurotransmitters, including glutamate, dopamine, serotonin, and norepinephrine (Yu et al. 2016). Alterations in neurotransmitters and neuronal growth factors can ultimately lead to disrupted function in the developing brain. These effects could be conveyed to higher levels of biological organization that could impact complex behavioral and reproductive capabilities.

The alterations observed at the biochemical level can lead to toxic outcomes at the cellular level, and indeed, many of these pathways targeted at this level are prominent in the associate cell types associated with that function. For example, the theme of dysregulation in fat metabolism is also present at the cellular level. Mitochondrial function is effected in adipocytes following PFOA exposure in mice (Shabalina et al. 2016). Similarly, in vitro treatment of PFASs affects adipocyte growth by increasing cell division rates and triglyceride production but leading to the development of smaller cells (Watkins et al. 2015). These alterations in metabolism at the cellular level in response to PFAS treatment can link back to the biochemical findings regarding the interactions with transporters and PPARα, which are critical in the regulation of fatty acids. The treatment effects of PFASs on neurotransmitters and neuronal growth factors can also be linked to observations at the cellular levels. It has been shown that PFOS exposure reduces the ability of neuronal stem cells to differentiate, as well as disrupts the development of motor neurons in zebra fish (Zhang et al. 2011; Wan Ibrahim et al. 2013). These effects on neuronal development and fat homeostasis will also affect factors further along the cascade.

The endpoints of toxicity observed at the tissue/organ level of biological organization are interconnected to the findings at the cellular and biochemical levels. One such target organ of toxicity is the liver, where fatty liver development and increased liver weight have been observed following exposure to PFASs (Lau et al. 2007). In zebra fish, for example, chronic PFOS exposure has been associated with the development of hepatic steatosis (Cheng et al. 2016). Similarly, effects were observed in the livers of Atlantic salmon and freshwater tilapia (Liu et al. 2007; Krovel et al. 2008). Effects on the liver occur in both lower vertebrates as well as higher vertebrates and are used as a critical endpoint for human risk assessment (Post et al. 2012; NJDWQI 2015, 2017). While these observations in the liver can be associated with responses at lower levels of biological organization, a single pathway cannot account for all the effects. Many of the endpoints of hepatotoxicity are related to the altered fat metabolism and transporter findings reported above, but while interrelated, there is not a single pathway that appears to be targeted. It is not solely the activation of PPARα that leads to the dysregulation of fats and effects in the liver. Even in PPARα-null mice and zebra fish, which contain an inactive PPARα, hepatotoxicity is still observed (Wolf et al. 2008). In addition, rainbow trout (*Oncorhynchus mykiss*) mimic human responses to classic peroxisome proliferators, and treatment with PFOA leads to the formation of hepatic tumors, a classic outcome of PPARα overactivation. Yet, PFOA exposure appears to induce gene expression patterns similar to those observed following estrogen exposure (Tilton et al. 2008). Several of the PFASs were shown to bind to the estrogen receptor and induce vitellogenin, a biomarker of estrogenic activity (Benninghoff et al. 2012). PFOA, PFNA, and to a lesser extent PFOS were all found to act as promoters in the development of hepatic cancer and induce a wide battery of genes that could contribute to their promoting activities (Benninghoff et al. 2012). The fact that these compounds do not interact with DNA to cause mutations directly suggests an alternative mechanism is activated. In the case of trout, this is likely to involve the estrogen receptor, as well as other altered pathways. It is most likely a network of effects at the biochemical level that lead to the observed hepatoxicity in multiple species. As stated above, the ubiquitous nature of the transporters in epithelial tissues and organs throughout eukaryotic vertebrates, their critical roles in toxicokinetics of endogenous compounds and metabolites, and PFASs' direct interaction with the transporters may explain the observed toxicity.

When dealing with observed effects at the organ and higher levels, pinpointing the exact cause is complicated by the increased complexity and feedback mechanisms that can be involved. Effects observed at the individual level contribute to the complete context of PFAS toxicities. As an example, within Fig. 6.7 are summarized some of the effects observed following embryonic (Fig. 6.7a) and chronic (Fig. 6.7b) low-dose PFOA exposure using morphometric, gene expression, behavior, and reproductive endpoints in zebra fish (Jantzen et al. 2016, b). The arrows (Fig. 6.7c) indicate significant changes in the PFOA-exposed fish compared to the controls at different stages of development. In zebra fish larvae exposed to PFOA during their first 5 days of development, increased expression was observed for the genes *cfos* (proto-oncogene), *tcf3a* (transcription factor), *ap1s1* (adaptor protein), and *slco2b1* (oatp transporter). The first three genes are critical in cellular movement and regulation during development and *slco2b1* codes for an oatp transporter shown to interact with PFOA, as discussed previously. In adult zebra fish chronically exposed to PFOA, the expression of multiple oatp transporters is decreased, which may be due to the complex pattern in which these transporters function and are regulated. Morphometric endpoints were also altered at both the eleutheroembryo and larval stage examined in the embryonic exposure study. Sac fry larvae had larger yolk sacs and were significantly smaller, which was also observed in larvae after that same exposure. This demonstrated that these alterations to lipid metabolism and transport can lead to these overall effects on growth. Similar observations were made in Atlantic salmon (*Salmo salar*) where exposure to PFOA increased the body weight of larvae (Arukwe et al. 2013). Effects on an individual in terms of

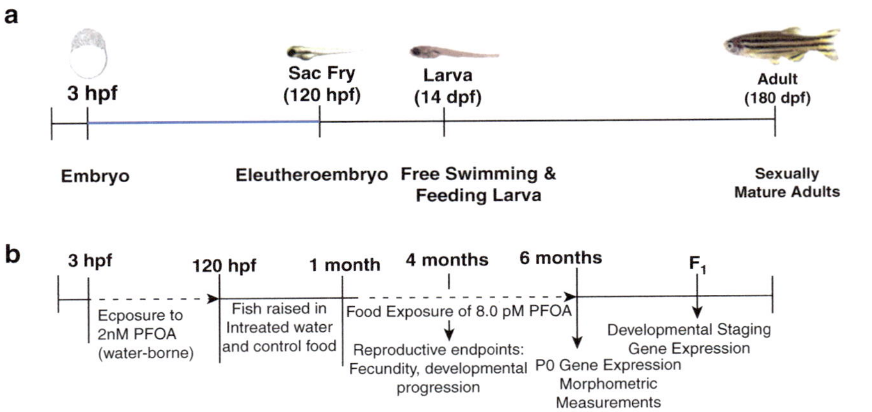

c

PFOA	5 dpf (embryo)	14 dpf (embryo)	Adult (embryo)	Adult (chronic)
Morphometrics	↓ total length ↑ yolk sac size, interoccular distance	↓ total length	NS	↓ total size
Gene Expression	↑ *tcf3a, ap1s1, slco2b1, cfos*	↑*slco2b1, tcf3a*	↓ *slco1d1*	↑ *slo1d1* ↓*tgfb1a, slco3a1, slco2b1,slco4a1*
Behavior		↑distance traveled	↓ time in light	
Reproductive effects				↓ number of eggs, % embryo viability, delayed development

Fig. 6.7 Using PFOA as an example for data that can be collected following exposure ranging from growth to reproductive effects. (**a**) Embryonic exposure study paradigm illustrating the stages examined, from fertilized embryo to adult. Results are based on a 2 μM PFOA exposure from 3 to 120 hpf (hours postfertilization). (**b**) Chronic exposure study paradigm, examining the effects of PFOA in adults and F_1 generations. (**c**) Summary of morphometric, gene expression, behavior, and reproductive effects in both the embryo and chronic exposure paradigms at different time points. The arrows indicate direction of significant effect. *NS* not significant, *blank* not measured. Gene symbols: *slco* organic-anion-transporting polypeptides, *tcf* transcription factor, *tgf* tumor growth factor, *cfos* retrovirus cancer oncogene, *ap1s1* adaptor protein (Jantzen et al. 2016, b)

morphometric endpoints, particularly those related to size, energy stores, and catabolism, will impact the survival of that individual and possibly its offspring.

One way in which toxicity of an individual can be examined is through a battery of behavioral assays. For example, in the Jantzen et al. (2016) study, zebra fish exposed to PFASs following an embryonic exposure (Fig. 6.7a) displayed altered behavior compared to controls. The behavior parameters are determined using infrared cameras designed to be able to monitor both larval fish and adult fish for

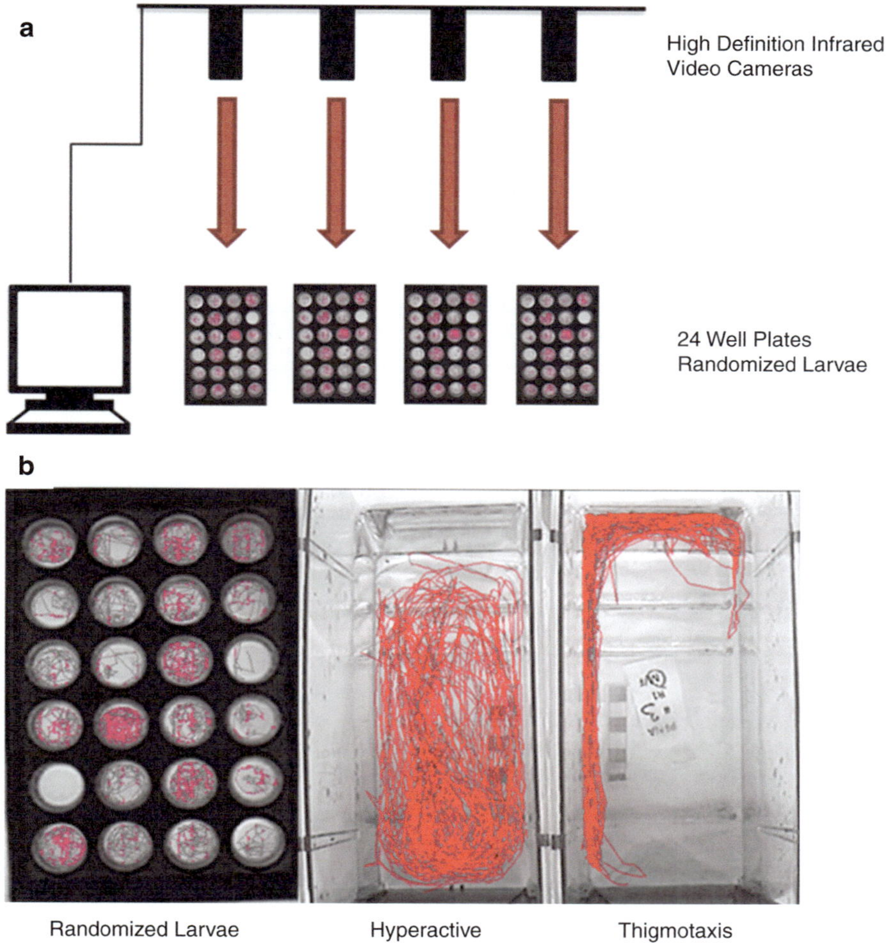

Fig. 6.8 (**a**) Schematic showing the design of the high-definition cameras setup to record movement of either larvae in a 24-well plate or adult fish in larger tanks. (**b**) Representative results from a randomized larval run showing the variability following treatment for swimming behavior, as well as adult fish recorded behavior in a larger swim arena

assessing behavioral traits (Fig. 6.8). Low-dose exposure to PFASs during embryonic development leads to alterations in locomotor activity endpoints in larval fish, including effects on swim speed, swim distance, relative swim time, and thigmotaxis, which is a wall-hugging, stress response-type behavior (Ulhaq et al. 2013; Jantzen et al. 2016). However, the different compounds within this class caused varied responses; summarized in Fig. 6.9 are several of these endpoints with exposure to PFNA, PFOS, and PFOA. Not only were there effects on the swimming behaviors (Fig. 6.9a–c), but there was a significant decrease on prey capture ability

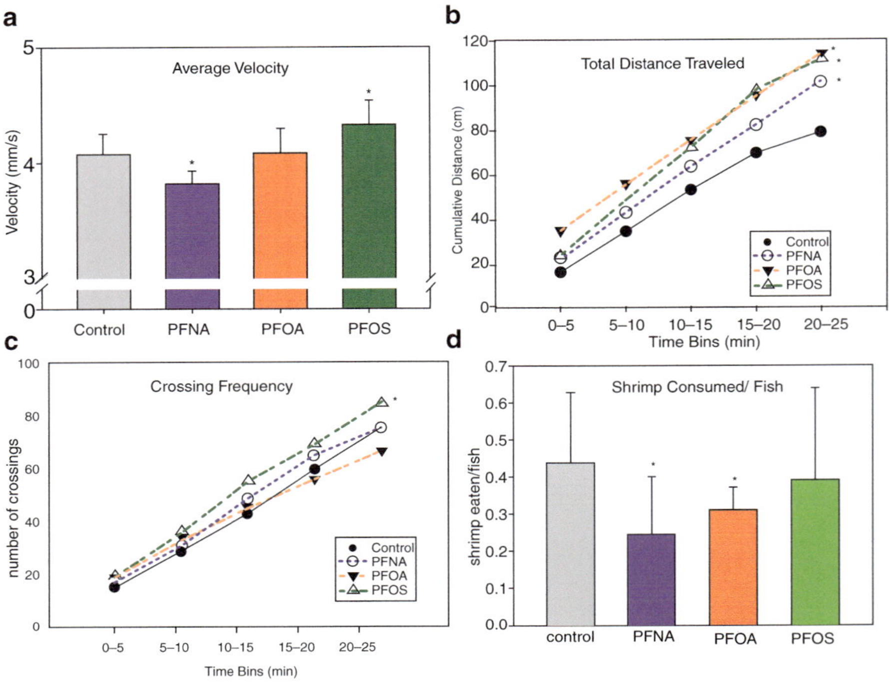

Fig. 6.9 Behavioral assays of larval zebra fish comparing 2 μM PFNA, PFOA, and PFOS to no treatment controls following embryonic exposure (exposure described in Fig. 6.7a). Endpoints examined in larval fish include (**a**) average velocity, (**b**) total distance traveled, (**c**) crossing frequency, and (**d**) shrimp consumed per fish. Asterisks indicate significant difference from control at $p \leq 0.05$ (Jantzen et al. 2016)

for both PFNA- and PFOA-treated larvae (Fig. 6.9d). However, when examining the adult zebra fish, which were again only exposed during embryonic development, for similar responses, behavioral effects were only observed in the males, following a single low-dose developmental exposure to PFNA (Fig. 6.10, Jantzen et al. 2016). Transport and deposition of these compounds vary between males and females, and this difference is likely the cause for the varied responses. Unlike the observations at the larval time point, these fish displayed increased swim velocity, decreased swim distance, as well as decreased time spent immobile and time spent in the middle of the arena, a sign of thigmotaxis. The behaviors observed in these studies mirror what has been reported in the rodent literature, again highlighting the conservation of these pathways (Mariussen 2012). This not only stresses that an exposure during only the development of an organism can have lasting impacts but also the importance of examining multiple time points and that each time point may reveal different response which present different challenges to survival.

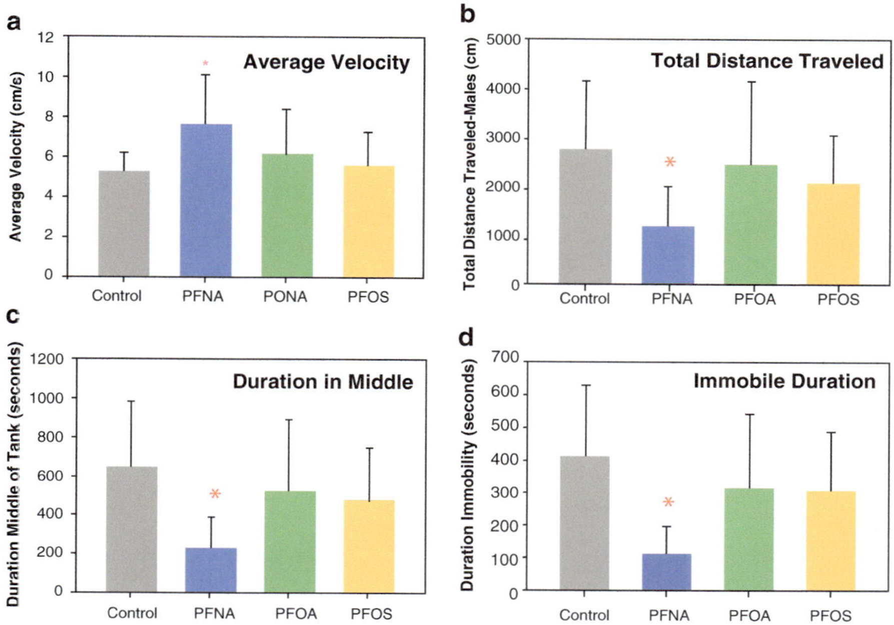

Fig. 6.10 Behavioral assays of male adult zebra fish comparing 2 μM PFNA, PFOA, and PFOS to no treatment controls following embryonic exposure (exposure described in Fig. 6.7a). Endpoints examined in adult fish include (**a**) average velocity, (**b**) total distance traveled, (**c**) duration in the middle of the arena, and (**d**) duration of time immobile. Asterisks indicate significant difference from control at $p \leq 0.05$ (Jantzen et al. 2016)

The culmination of the cascade of biological effects is the impact on the population level. In the zebra fish study, the decreased growth following PFAS exposures was observed at the larval time points, but not in adult fish. However, there is still the potential for this to cause an impact on a population level. As the developmental and larval periods are such sensitive times of growth in the fish, alterations to growth during this time can decrease the likelihood of survival of these fish. The smaller fish will have difficulty competing with larger fish for food and mates, both of which can lead to a decrease in the population. The correlation between size and potential impact on population also exists when considering the stage delay data (Fig. 6.11). Parental fish were chronically exposed to PFOA (Fig. 6.7b), and their embryos were developmentally delayed compared to the embryos reared from control fish. Within this reproductive study, there was also an observed decrease in fecundity and survival of offspring of treated parents (Jantzen et al. 2017). It is clear that decreased reproduction capacity can impact a population and the fish in delayed stages of development are at the same disadvantages as those of a smaller size.

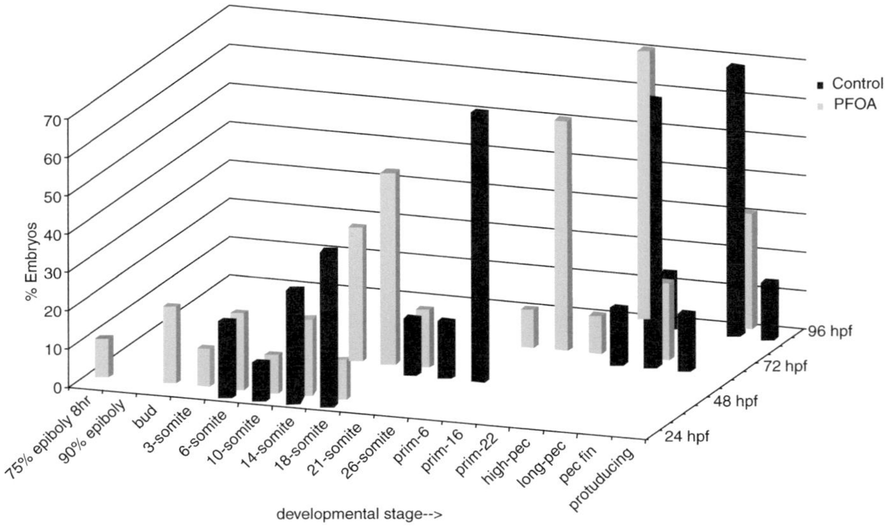

Fig. 6.11 Stage and hatch delay in embryos collected from chronically exposed parental stock and raised in rearing solution free of PFOA (exposure described in Fig. 6.7b). Embryo staging was based on (Kimmel et al. 1995) and recorded at 24, 48, 72, and 96 hpf. There was a significant delay in development when compared to no treatment controls (Jantzen et al. 2017)

The effects on behavior of an individual will also impact the population. As there is homology in responses to PFASs across species, these effects may also contribute to effects in other species as well, potentially altering multiple levels of the ecosystem (Baker and King 2010). In the Jantzen et al. (2016) study, the adult zebra fish, exposed developmentally to PFNA, displayed behaviors in the open-field test consistent with a stress response (i.e., thigmotaxis and increased swim velocity). This type of response is indicative of a decrease in exploratory behavior, which has implications for the population. Similar to the larval findings, this behavior pattern decreases the ability to find food and avoid predators. Alterations in behavior also have reproductive implications for populations as well. The decrease in exploratory behaviors also decreases the ability to find mates. It becomes critical in that the females will explore new surroundings, whereas the males will remain in a familiar environment. The males in this study were more hyperactive which may alter this mating pattern and ultimately decrease reproductive success. In addition to the avoidance behaviors in the male zebra fish exposed to PFNA, they also displayed increased aggression, which has the potential to affect the way in which these male fish interact with females (data not shown). Both observations have the potential to impact the reproductive rates of the population.

Because of the extensive use of this class of compounds in firefighting foams at military bases, airports, and other specific manufacturing facilities, there are wild fish populations that could be examined for these types of effects. As described above, what appear to be subtle effects on a biochemical process have the potential to manifest themselves in ways that impact the growth, development, behavior, and reproductive capabilities of an individual such that effects can transmit to the population. Many the PFAS studies have used laboratory-reared species though it is prudent to include future studies of field populations exposed to PFASs prior to examination over the full spectrum of endpoints from biochemical to population levels (Fig. 6.1). Many fish and wildlife studies have examined tissue burdens, but not biological effects. The considerations made throughout this case study can help bridge the gap between observations in laboratory and field studies, and this highlights the importance of understanding the biology in complex environmental settings.

Conclusion

The early developmental stages in both finfish and bivalves appear to be very sensitive to both PAH and HAH toxicity. This is part due to the ability of the contaminants to accumulate within the organisms and have effects on both the F_0 and F_1 organism as it relates to the mechanism of action of the compound going from biochemical to individual effects (Fig. 6.1). In the case of tolerance being developed, there are several different ways in which this can be achieved in a population. One critical feature is genetic plasticity that allows for selection of a beneficial feature in chronically exposed populations. The tolerance within specific examples provided above requires a population with limited home ranges due to geographic or other reasons, large numbers of progeny, and extended periods of exposure over multiple generations. The crosstalk and interconnections between several critical pathways in early gonadal development, embryo-larval maturation, behavior, and reproduction all seem to play critical roles in viability of these populations. The concept of plasticity in gene regulatory networks involved in early development and reproduction has allowed tolerant populations to evolve mechanisms to survive in chronically contaminated environments. Organisms within a population which are unable to modify their gene regulatory networks are lost from the population and gene pool. There is a fitness cost within several of these populations following exposure which likely contribute to altered population structure.

The bivalve case study demonstrates that chemical challenge during critical periods of gonadal development can result in poor gonadal maturation and loss of viable offspring. This can drastically reduce the population and result in local extinctions due to compounds like 2,3,7,8-TCDD. With the development of the oyster genome being sequenced, future research will be able to examine alterations at the cellular and biochemical levels which have not been available. It also highlights that the time of year of an exposure can have drastic differences in impacts due to the physiological and biochemical state of the bivalves. The interplay between hormonal control involving autocrine and endocrine timing and hormonal levels is critical for proper maturation of the gonads (Fig. 6.3). The proper development in the oyster larvae involves cell migration, tissue development, and organ formation which requires cell-to-cell communication, integrins, heterodimeric receptors, and growth gradients, to name a few, that are conserved and similar to higher vertebrate embryonic development. This type of complexity is not limited to bivalves and reflects phylogenetically conserved pathways across eukaryotic phyla (Ridley et al. 2003).

The work carried out on the zebra fish reinforces the importance of assessing early developmental effects, hatching, juvenile survival, reproductive success, and generational effects. These studies also point out the interconnections between gene effects leading to behavioral effects. The perfluoroalkylated substances (PFASs) are not mutagenic, and yet effects were observed in offspring which had not been exposed. The class of compounds is known to interfere with organic-anion-transporting polypeptides (Fig. 6.7), which may be a key role in many the observed effects. Each of the endpoints from decreased prey capture, altered swimming behavior (Fig. 6.10), and delayed embryonic development (Fig. 6.11) could result in fewer surviving offspring that will successfully reach reproductive age. These subtle changes may reduce the population size over many years and contribute to local extinctions.

This chapter focused on several chemical stressors that have been shown to result in changes from the biochemical level through the individual level of organization (Fig. 6.1). Case studies of both wild populations and laboratory models have demonstrated examples of effects on reproduction, behavior, and the development of resistance in terms of the impacts these changes have on populations and, potentially, the community. There is a need to carry out more experiments in wild populations and laboratory studies to validate these findings and improve the ability to manage aquatic ecosystems as it relates to population dynamics.

References

Able K, Grothues TM, Morson JM, Coleman KE (2014) Temporal variation in winter flounder recruitment at the southern margin of their range: is the decline due to increasing temperatures? ICES J Mar Sci 71(8):2186–2197

Armstrong PB, Child JS (1965) Stages in the normal development of *Fundulus heteroclitus*. Biol Bull 128:143–168

Arukwe A, Cangialosi MV, Letcher RJ, Rocha E, Mortensen AS (2013) Changes in morphometry and association between whole-body fatty acids and steroid hormone profiles in relation to bioaccumulation patterns in salmon larvae exposed to perfluorooctane sulfonic or perfluorooctane carboxylic acids. Aquat Toxicol 130-131:219–230

Baker ME, King RS (2010) A new method for detecting and interpreting biodiversity and ecological community thresholds. Methods Ecol Evol 1:25–37

Bello SM, Franks DG, Stegeman JJ, Hahn ME (2001) Acquired resistance to Ah receptor agonists in a population of Atlantic killifish (*Fundulus heteroclitus*) inhabiting a marine superfund site: *in vivo* and *in vitro* studies on the inducibility of xenobiotic metabolizing enzymes. Toxicol Sci 60:77–91

Benninghoff AD, Orner GA, Buchner CH, Hendricks JD, Duffy AM, Williams DE (2012) Promotion of hepatocarcinogenesis by perfluoroalkyl acids in rainbow trout. Toxicol Sci 125(1):69–78

Brown AR, Hosken DJ, Balloux F, Bickley LK, LePage G, Owen SF, Hertheridge MJ, Tyler CR (2009) Genetic variation, inbreeding and chemical exposure-combined effects in wildlife and critical considerations for ecotoxicology. Philos Trans R Soc B 364:3377–3390

Brown R, Cooper KR, Cristini A (1994) Polychlorinated dibenzo-p-dioxins and dibenzofurans in *Mya arenaria* in the Newark/Raritan Bay estuary. Environ Toxicol Chem 13:523–528

Brown RP, Cristini A, Cooper KR (1992) Histopathological alterations in *Mya arenaria* following a #2 fuel oil spill in the Arthur Kill, Elizabeth, New Jersey. Mar Environ Res 34:65–68

Bugel SM, White LA, Cooper KR (2011) Decreased vitellogenin inducibility and 17β-estradiol levels correlated with reduced egg production in killifish (*Fundulus heteroclitus*) from Newark Bay, NJ. Aquat Toxicol 105(1-2):1–12

Bugel SM, White LA, Cooper KR (2013) Inhibition of vitellogenesis gene induction by 2,3,7,8-tetrachlorodibenzo-p-dioxin is mediated by aryl hydrocarbon receptor 2 (AhR2) in zebrafish (*Danio rerio*). Aquat Toxicol 126:1–8

Capuzzo MDJ (1996) The bioaccumulation and biological effects of lipophilic organic contaminants. In: Kennedy VS, Newell RIE, Eble AF (eds) The eastern oyster *crassostrea virginica*. Maryland Sea Grant College Publication, University of Maryland System, College Park, pp 539–575

Ceballos G, Ehrlich PR, Barnosky AD, Garcia A, Pringle RM, Palmer TM (2015) Accelerated modern human-induced species losses: entering the sixth mass extinction. Sci Adv 1(5):e1400253

Chapman PM (2000) Whole effluent toxicity testing-usefulness, level of protection, and risk assessment. Environ Toxicol Chem 19:3–13

Cheng J, Lv S, Nie S, Liu J, Tong S, Kang N, Xiao Y, Dong Q, Huang C, Yang D (2016) Chronic perfluorooctane sulfonate (PFOS) exposure induces hepatic steatosis in zebrafish. Aquat Toxicol 176:45–52

Clark BW, Bone AJ, Di Giulio RT (2014) Resistance to teratogenesis by F1 and F2 embryos of PAH-adapted *Fundulus heteroclitus* is strongly inherited despite recalcitrance of AHR pathway. Environ Sci Pollut Res Int 21(24):13898–13908

Cooper KR (1996) Effects of polychlorinated dibenzo-p-dioxins and polychlorinated dibenzofurans on aquatic organisms. Aquat Sci 2:227–241

Cooper KR, Borjan M (2010) Northeast coastal and barrier network assessment of contaminant threats: gateway national recreation area. Natural Resource Technical Report NPS/NCBN/NRTR—2010/339. National Park Service, Fort Collins, Colorado, p 392

Cooper KR, Brown R (1995) Toxic effects of 2,3,7,8-tetrachlorodibenzo-p-dioxin (2,3,7,8-TCDD) and related compounds (PCDD/PCDF) on aquatic invertebrate species and specific studies on the soft-shell clam (*Mya arenaria*). Organohalogen Compd 25:339–344

Cooper KR, McGeorge L (1991) Deceased effluent toxicity in the Japanese medaka, *Oryzias latipes*, embryo-larval assay following initiation of toxicant reduction strategy. In: Mayes MA, Barron MG (eds) Aquatic toxicology and risk assessment fourteenth volume, ASTM STP 1124. American Society for Testing and Materials, Philadelphia, pp 67–83

Cooper KR, Schell J, Umbreit T, Gallo M (1993) Fish-embryo toxicity with exposure to soils and sediments contaminated with varying concentrations of dioxins and furans. Mar Environ Res 35:177–180

Cooper KR, Wintermyer M (2009) A critical review: 2,3,7,8-tetrachlorodibenzo-p-dioxin (2,3,7,8-TCDD) effects on gonadal development in bivalve mollusks. J Environ Sci Health C Environ Carcinog Ecotoxicol Rev 27:226–245

Davis J, Kidd IM (2012) Identifying major stressors: the essential precursor to restoring cultural ecosystem services in a degraded estuary. Estuar Coasts 35:1007–1017

Dey WP, Peck TH, Smith CE, Kreamer GL (1993) Epizoology of hepatic neoplasia in Atlantic tomcod (*Microgadus tomcod*) from the Hudson River estuary. Can J Fish Aquat Sci 50:1897–1907

Di Giulio RT, Clark BW (2015) The Elizabeth River: a case study in evolutionary toxicology. J Toxicol Environ Health Part B 18:259–298

Diaz RJ, Rosenberg R (2008) Spreading dead zones and consequences for marine ecosystems. Science 321:926–929

Embry MR, Belanger SE, Braunbeck TA, Galay-Burgos M, Halder M, Hinton DE, Leonard MC, Lillcrap A, Norber-King T, Whale G (2010) The fish embryo toxicity test as an animal alternative method in hazard and risk assessment and scientific research. Aquat Toxicol 97:79–87

Faisal M, Huggett RJ (1993) Effects of aromatic hydrocarbons on the lymphocyte mitogenic responses in spot (*Leiostomus xanthurus*). Mar Environ Res 35:121–124

FAO (2006) World review of fisheries and aquaculture. In: The state of the world fisheries and agriculture. Food and Agriculture Organization of the United Nations, Rome, p 162. ISBN:978-92-5-105568-7
FAO (2012) The state of the world fisheries and agriculture. Food and Agriculture Organization of the United Nations, Rome. ISBN:978-92-5-107225-7
Fournie JW, Vogelbein WK (1994) Exocrine pancreatic neoplasms in the mummichog (*Fundulus heteroclitus*) from a creosote contaminated site. Toxicol Pathol 22(3):237–247
Galtsoff PS (1964) The American oyster Crassostrea virginica (Gmelin). Fish Bull 64:1–456
Hahn ME (2003) Chap. 14: evolutionary and physiological perspectives on Ah receptor function and dioxin toxicity. In: Schecter A, Gasiewicz TA (eds) Dioxins and health, 2nd edn. Wiley, New York, pp 559–602
Hahn ME, Karchner SI, Shapiro MA, Perera SA (1997) Molecular evolution of two vertebrate aryl hydrocarbon (dioxin) receptors (AHR1 and AHR2) and the PAS family. Proc Natl Acad Sci 94:13743–13748
Hornung MW, Spitsbergen JM, Peterson RE (1999) 2,3,7,8-tetrachlorodibenzo-p-dioxin alters cardiovascular and craniofacial development and function in sac fry of rainbow trout (*Oncorhynchus mykiss*). Toxicol Sci 47:40–51
Huggett AJ (2005) The concept and utility of 'ecological thresholds' in biodiversity conservation. Biol Conserv 124:301–310
Incardona JP, Collier TK, Scholz NL (2004) Defects in cardiac function precede morphological abnormalities in fish embryos exposed to polycyclic aromatic hydrocarbons. Toxicol Appl Pharmacol 196:191–205
Jantzen CE, Annunziato KM, Bugel SM, Cooper KR (2016) PFOS, PFNA, and PFOA sub-lethal exposure to embryonic zebrafish have different toxicity profiles in terms of morphometrics, behavior and gene expression. Aquat Toxicol 175:160–170
Jantzen CE, Annunziato KM, Cooper KR (2016) Behavioral, morphometric, and gene expression effects in adult zebrafish (*Danio rerio*) embryonically exposed to PFOA, PFOS, and PFNA. Aquat Toxicol 180:123–130
Jantzen CE, Toor F, Annunziato KM, Cooper KR (2017) Effects of chronic perfluorooctanoic acid (PFOA) at low concentration on morphometrics, gene expression, and fecundity in zebrafish (*Danio rerio*). Reprod Toxicol 69:34–42
Kauffman G (2016) Economic value of nature and ecosystems in the Delaware River Basin. J Contemp Water Res Educ 158:98–119
Kavlock R, Chandler K, Houck K, Hunter S, Judson R, Kleinstreuer N, Knudsen T, Martin M, Padilla S, Reif D, Richard A, Rotroff D, Sipes N, Dix D (2012) Update on EPA's ToxCast program: providing high throughput decision support tools for chemical risk management. Chem Res Toxicol 25:1287–1302
Kennish MJ (1992) Ecology of estuaries: anthropogenic effects. CRC, Boca Raton, p 512
Kimmel CB, Ballard WW, Kimmel SR, Ullmann B, Schilling TF (1995) Stages of embryonic development of the zebrafish. Dev Dyn 203(3):253–310
Klaassen CD, Aleksunes LM (2010) Xenobiotic, bile acid, and cholesterol transporters: function and regulation. Pharmacol Rev 62(1):1–96
Klaude RJ, Moos RE, Schmidt RE (1988) Chap. 8 Life history of Atlantic tomcod, *Microgadus tomcod*, in the Hudson River Estuary, with emphasis on spatio-temporal distribution and movements. In: Smith CL (ed) Fisheries research in the Hudson River. SUNY Press, Albany, pp 219–251
Krovel AV, Softeland L, Torstensen B, Olsvik PA (2008) Transcriptional effects of PFOS in isolated hepatocytes from Atlantic salmon *Salmo salar* L. Comp Biochem Physiol Part C Toxicol Pharmacol 148(1):14–22
Lau C, Anitole K, Hodes C, Lai D, Pfahles-Hutchens A, Seed J (2007) Perfluoroalkyl acids: a review of monitoring and toxicological findings. Toxicol Sci 99(2):366–394
Levine M, Davidson EH (2005) Gene regulatory networks for development. Proc Natl Acad Sci 102(14):4936–4942

Liu C, Yu K, Shi X, Wang J, Lam PK, Wu RS, Zhou B (2007) Induction of oxidative stress and apoptosis by PFOS and PFOA in primary cultured hepatocytes of freshwater tilapia (*Oreochromis niloticus*). Aquat Toxicol 82(2):135–143

Lotrich VA (1975) Home range and movements of *Fundulus heteroclitus* (Pisces: Cyprinodontidae) in a tidal creek. Ecology 56(1):191–198

Mariussen E (2012) Neurotoxic effects of perfluoroalkylated compounds: mechanisms of action and environmental relevance. Arch Toxicol 86(9):1349–1367

Meyer JN, Di Giulio RT (2002) Patterns of heritability of decreased EROD activity and resistance to PCB 126-induced teratogenesis in laboratory-raised offspring of killifish (*Fundulus heteroclitus*) from a creosote-contaminated site in the Elizabeth River, VA, USA. Mar Environ Res 54:621–628

Meyer JN, Smith JD, Winston GW, Di Giulio RT (2003) Antioxidant defenses in killifish (*Fundulus heteroclitus*) exposed to superfund sediments and model prooxidants: short-term and heritable responses. Aquat Toxicol 65:377–395

Nacci D, Coiro L, Champlin D, Jayaraman S, McKinney R, Gleason TR, Munns WR, Specker JL, Cooper KR (1999) Adaptations of wild populations of the estuarine fish *Fundulus heteroclitus* to persistent environmental contaminants. Mar Biol 134:9–17

Nacci D, Coiro L, Champlin D, Jayaraman S, Munns W Jr, Specker J, Cooper KR (2000) Adaptation of wild fish populations to dioxin-like environmental contamination. Mar Environ Res 134:9–17

Nacci D, Coiro L, Champlin D, Munns W Jr, Specker J, Cooper KR (1998) Nondestructive indicator of ethoxyresorufin-o-deethylase activity in embryonic fish. Environ Toxicol Chem 17:2481–2486

Nacci D, Huber M, Champlin D, Jayaraman S, Cohen S, Gauger E, Fong A, Gomez-Chiarri M (2009) Evolution of tolerance to PCBs and susceptibility to a bacterial pathogen (Vibrio harveyi) in Atlantic killifish (*Fundulus heteroclitus*) from New Bedford (MA, USA) harbor. Environ Pollut 157(3):857–864

New Jersey Drinking Water Quality Institute (NJDWQI) (2015) Perfluorononanoic acid. Recommendations for maximum contaminant levels. http://www.nj.gov/dep/watersupply/g_boards_dwqi.html

New Jersey Drinking Water Quality Institute (NJDWQI) (2017) Perfluorooctanoic acid. Recommendations for maximum contaminant levels. http://www.nj.gov/dep/watersupply/g_boards_dwqi.html

Ohtake F, Baba A, Fujii-Kuriyama Y, Kato S (2008) Intrinsic AhR function underlies cross-talk of dioxins with sex hormone signaling. Biochem Biophys Res Commun 370:541–546

Padilla S, Corum D, Padnos B, Hunter DL, Beam A, Houch KA, Sipes N, Kleinstreuer N, Knudson T, Dix DJ, Reif DM (2012) Zebrafish developmental screening of the ToxCast™ phase I chemical library. Reprod Toxicol 33:174–187

Patyna PJ, Davi RA, Parkerton TF, Brown RP, Cooper KR (1999) A proposed multigeneration protocol for Japanese medaka (*Oryzias latipes*) to evaluate effects of endocrine disruptors. Sci Total Environ 233(1-3):211–220

Pinsky ML, Byler D (2015) Fishing, fast growth and climate variability increase the risk of collapse. Proc R Soc B 282:20151053

Popovic M, Zaja R, Fent K, Smital T (2013) Molecular characterization of zebrafish Oatp1d1 (Slco1d1), a novel organic anion-transporting polypeptide. J Biol Chem 288(47):33894–33911

Popovic M, Zaja R, Fent K, Smital T (2014) Interaction of environmental contaminants with zebrafish organic anion transporting polypeptide, Oatp1d1 (Slco1d1). Toxicol Appl Pharmacol 280(1):149–158

Post GA, Cohn PD, Cooper KR (2012) Perfluorooctanoic acid (PFOA), and emerging drinking water contaminant: a critical review of recent literature. Environ Res 116:93–117

Prince R, Cooper KR (1995a) Comparisons of the effects of 2,3,7,8-tetrachlorodibenzo-p-dioxin on chemically impacted and nonimpacted subpopulations of *Fundulus heteroclitus*. I. TCDD toxicity. Environ Toxicol Chem 14:579–587

Prince R, Cooper KR (1995b) Comparisons of the effects of 2,3,7,8-tetrachlorodibenzo-p-dioxin on chemically impacted and nonimpacted subpopulations of *Fundulus heteroclitus*. II. Metabolic considerations. Environ Toxicol Chem 14:589–596

Ravit B, Cooper KR, Buckley B, Comi M, McCandish E (2015) Improving management support tools for reintroducing bivalve species (Eastern oyster *Crassostrea virginica* Gmelin) in urban estuaries. Integr Environ Assess Manag 10(4):555–565

Reid NM, Proestou DA, Clark BW, Warren WC, Colbourne JK, Shaw JR, Karchner SI, Hahn ME, Nacci D, Oleksiak MF, Crawford DL, Whitehead A (2016) The genomic landscape of rapid repeated evolutionary adaptation to toxic pollution in wild fish. Science 354(6317):1305–1308

Reynaud S, Deschaux P (2006) The effects of polycyclic aromatic hydrocarbons on the immune system of fish: a review. Aquat Toxicol 77:229–238

Ridley AJ, Schwartz MA, Burdidge K, Firtle RA, Ginsberg MH, Borisy G, Parson JT, Horwitz AR (2003) Cell migration: integrating signals from front to back. Science 302(5651):1704–1709

Roy NK, Wirgin I (1997) Characterization of the aryl hydrocarbon receptor gene and its expression in Atlantic tomcod. Arch Biochem Biophys 344:373–386

Shabalina IG, Kalinovich AV, Cannon B, Nedergaard J (2016) Metabolically inert perfluorinated fatty acids directly activate uncoupling protein 1 in brown-fat mitochondria. Arch Toxicol 90(5):1117–1128

Shirasaka Y, Mori T, Shichiri M, Nakanishi T, Tamai I (2012) Functional pleiotropy of organic anion transporting polypeptide OATP2B1 due to multiple binding sites. Drug Metab Pharmacokinet 27(3):360–364

Spitsbergen JM, Kleeman JM, Peterson RE (1988) Morphologic lesions and acute toxicity in rainbow trout (*Salmo gairdneri*) treated with 2,3,7,8-tetraclorodibenzo-p-dioxin. J Toxicol Environ Health 23:333–358

Spitsbergen JM, Walker MK, Olson JR, Peterson RE (1991) Pathological alterations in early life stages of lake trout, *Salvelinus namaycush*, exposed to 2,3,7,8-tetrachlorodibenzo-p-dioxin as. Aquat Toxicol 19:41–72

Tilton SC, Omer GA, Benninghoff AD, Carpenter HM, Hendricks JD, Pereira CB, Williams DE (2008) Genomic profiling reveals an alternate mechanism of hepatic tumor promotion by perfluorooctanoic acid in rainbow trout. Environ Health Perspect 116:1047–1055

Tyagi S, Gupta P, Saini AS, Kaushal C, Sharma S (2011) The peroxisome proliferator-activated receptor: a family of nuclear receptors role in various diseases. J Adv Pharm Technol Res 2(4):236–240

U.S. Environmental Protection Agency (USEPA) (1985) Methods for measuring the acute toxicity of effluents and receiving waters to freshwater and marine organisms, 3rd edn. USEPA, Washington, DC, p 216. EPA-600-4-8-5-013

U.S. Environmental Protection Agency (USEPA) (1993) Methods for measuring the acute toxicity of effluents and receiving waters to freshwater and marine organisms, 4th edn. USEPA, Washington, DC, p 273. EPA-600-4-90-027F

U.S. Environmental Protection Agency (USEPA) (1994) Short-term methods for estimating the chronic toxicity of effluents and receiving waters to freshwater organisms, 3rd edn. USEPA, Washington, DC, p 334. EPA-600-4-91-002

U.S. Environmental Protection Agency (USEPA) (1995) Short-term methods for estimating the chronic toxicity of effluents and receiving waters to west coast marine and estuarine organisms, 1st edn. USEPA, Washington, DC, p 15. EPA-600-R-95-136

Ulhaq M, Orn S, Carlsson G, Morrison DA, Norrgren L (2013) Locomotor behavior in zebrafish (*Danio rerio*) larvae exposed to perfluoroalkyl acids. Aquat Toxicol 144-145:332–340

Van Veld PA, Nacci DE (2008) Chap. 13 Toxicity resistance. In: Di Giulio RT, Hinton DE (eds) The toxicology of the fishes. CRC Press, Boca Raton, pp 597–641

Vanden Heuvel JP, Thompson JT, Frame SR, Gillies PJ (2006) Differential activation of nuclear receptors by perfluorinated fatty acid analogs and natural fatty acids: a comparison of human, mouse, and rat peroxisome proliferator-activated receptor-alpha, -beta, and -gamma, liver X receptor-beta, and retinoid X receptor-alpha. Toxicol Sci 92(2):476–489

Walther G, Post E, Convey P, Menzel A, Parmesan C, Beebee TJC, Fromentin J, Guldberg OH, Bairlein F (2002) Ecological responses to recent climate change. Nature 416:389–395
Wan Ibrahim WN, Tofighi R, Onishchenko N, Rebellato P, Bose R, Uhlen P, Ceccatelli S (2013) Perfluorooctane sulfonate induces neuronal and oligodendrocytic differentiation in neural stem cells and alters the expression of PPARγ *in vitro* and *in vivo*. Toxicol Appl Pharmacol 269(1):51–60
Watkins AM, Wood CR, Lin MT, Abbott BD (2015) The effects of perfluorinated chemicals on adipocyte differentiation *in vitro*. Mol Cell Endocrinol 400:90–101
Weaver YM, Ehresman DJ, Butenhoff JL, Hagenbuch B (2010) Roles of rat renal organic anion transporters in transporting perfluorinated carboxylates with different chain lengths. Toxicol Sci 113(2):305–314
Weinberger J II, Klaper R (2014) Environmental concentrations of the selective serotonin reuptake inhibitor fluoxetine impact specific behaviors involved in reproduction, feeding and predator avoidance in the fish *Pimephales promelas* (fathead minnow). Aquat Toxicol 151:77–83
Weis JS, Smith G, Zhou T, Santiago-Bass C, Weis P (2001) Effects of contaminants on behavior: biochemical mechanisms and ecological consequences: killifish from a contaminated site are slow to capture prey and escape predators; altered neurotransmitters and thyroid may be responsible for this behavior, which may produce population changes in the fish and their major prey, the grass shrimp. Bioscience 51(3):209–217
Weis JS, Weis P (1989) Tolerance and stress in a polluted environment. Bioscience 39:89–95
Weis P, Weis JS (1977) Methylmercury teratogenesis in the killifish, *Fundulus heteroclitus*. Experimental. Teratology 16(3):317–325
Wintermyer ML, Cooper KR (2003) Dioxin/furan and polychlorinated biphenyl concentrations in eastern oysters (*Crassostrea virginica* Gmelin) tissues and the effects on egg fertilization and development. J Shellfish Res 22:737–746
Wintermyer ML, Cooper KR (2007) The development of an aquatic bivalve model: evaluating the toxic effects on gametogenesis following 2,3,7,8-tetrachlorodibenzo-p-dioxin (2,3,7,8-TCDD) exposure in the eastern oyster (*Crassostrea virginica*). Aquat Toxicol 81:10–26
Wirgin I, Roy NK, Loftus M, Chambers RC, Franks DG, Hahn ME (2011) Mechanistic basis of resistance to PCBs in Atlantic tomcod from the Hudson River. Science 331(6022):1322–1325
Wisk JD, Cooper KR (1990a) The stage specific toxicity of 2,3,7,8-tetrachlorodibenzo-p-dioxin in embryos of the Japanese medaka (*Oryzias latipes*). Environ Toxicol Chem 9:1159–1169
Wisk JD, Cooper KR (1990b) Comparison of the toxicity of several polychlorinated dibenzo-p-dioxins and 2,3,7,8-tetrachlorodibenzofuran in embryos of the Japanese medaka (*Oryzias latipes*). Chemosphere 20:361–377
Wolf DC, Moore T, Abbott BD, Rosen MB, Das KP, Zehr RD, Lindstrom AB, Strynar MJ, Lau C (2008) Comparative hepatic effects of perfluorooctanoic acid and WY 14,643 in PPAR-alpha knockout and wild-type mice. Toxicol Pathol 36(4):632–639
Worm B, Barbier EB, Beaumont N, Duffy E, Folke C, Halpern BS, Jackson BC, Lotze HK, Micheli F, Palumbi SR, Sala E, Selkoe KA, Stachowicz JJ, Watson R (2006) Impacts of biodiversity loss on ocean ecosystem services. Science 314:787–790
Yang CH, Glover KP, Han X (2010) Characterization of cellular uptake of perfluorooctanoate via organic anion-transporting polypeptide 1A2, organic anion transporter 4, and urate transporter 1 for their potential roles in mediating human renal reabsorption of perfluorocarboxylates. Toxicol Sci 117(2):294–302
Yu N, Wei S, Li M, Yang J, Li K, Jin L, Xie Y, Giesy JP, Zhang X, Yu H (2016) Effects of perfluorooctanoic acid on metabolic profiles in brain and liver of mouse revealed by a high-throughput targeted metabolomics approach. Sci Rep 6:23963
Zhang L, Li YY, Chen T, Xia W, Zhou Y, Wan YJ, Lv ZQ, Li GQ, Xu SQ (2011) Abnormal development of motor neurons in perfluorooctane sulphonate exposed zebrafish embryos. Ecotoxicology 20(4):643–652

The Extent of the Modern Synthesis: The Foundational Framework for Evolutionary Biology

7

Thomas E. Dickins and Benjamin J. A. Dickins

Abstract

In this chapter we present a brief history of the Modern Synthesis and some comments on its logical structure and emphasize its framing role in modern biology. In doing so we clarify what evolution is. We then turn our attention to recent claims that the Modern Synthesis requires extending in order to deal with new findings in biology. At the core of these arguments are two assertions: (1) that other inheritance systems have conceptual parity with genetic inheritance and (2) that development introduces new variation over which selection can operate. In order to make these arguments, models of causation within the Modern Synthesis are challenged. We demonstrate that these challenges arise from analogical rather than formal reasoning and are false.

7.1 What Is Evolution?

The term evolution refers to a kind of change. Darwin (1859) made clear the nature of evolution, stating that it relied upon trait variation, differential success and inheritance. This is sometimes boiled down to the simpler list of variation, competition and inheritance. A system that incorporates these properties can evolve. More precisely, such a system allows selection to occur. Evolutionary biology focuses upon populations as systems, consisting of individuals (Hull 1980) that vary on trait dimensions in competition with one another.

T. E. Dickins (✉)
Faculty of Science and Technology, Middlesex University, London, UK
e-mail: T.Dickins@mdx.ac.uk

B. J. A. Dickins
School of Science and Technology, Nottingham Trent University, Nottingham, UK
e-mail: Benjamin.Dickins@ntu.ac.uk

© Springer International Publishing AG, part of Springer Nature 2018

W. Burggren, B. Dubansky (eds.), *Development and Environment*,
https://doi.org/10.1007/978-3-319-75935-7_7

Competition, or differential success, refers to the likelihood of some traits reproducing more than others because they are better suited to particular demands of the local environment. Selection refers to the statistical outcome of environmental constraints sifting competitor traits. Those traits that win out are passed on across generations and increase in relative frequency. To this extent some discuss the notion of a general theory of evolution (Lewontin 1970; Webb 2011) in which the means of variation, competition and inheritance remain unspecified (Box 7.1). All that is important is that these things happen and particular traits increase in relative number. Over time certain traits will come to dominate as they best solve the competitive problems of the environment. It is worth noting this last point—evolutionary theory is fundamentally an environmental theory.[1]

The general theory of evolution can be opposed to a special theory of evolution (Webb 2011). A special theory will specify the means and nature of variation, competition and inheritance. The Modern Synthesis, the current orthodoxy within evolutionary biology, is an example of a special theory of evolution.

7.2 What Is the Modern Synthesis?

The Modern Synthesis is a term widely used to refer to the paradigm that frames modern evolutionary biology. The term was coined by Julian Huxley (Huxley 1942) as the subtitle to his original and extensive review of developments in biology since the rediscovery of Mendel. For Huxley the synthesis was between Mendelian genetics and Darwinian evolution and central to this was the development of population genetics (Provine 2001; Mayr 1993).

To the philosophically minded the use of the synthesis implies a prior period of thesis and antithesis, and this was indeed the case. The initial thesis was that of Darwin and his account of evolutionary change through natural selection. Natural selection was a statistical outcome, dependent upon heritable trait variation and differential success of those traits under some form of competition. Crucial to his theory was a mechanism of inheritance and a mechanism for generating new variation: this piece of the theory was famously missing from Darwin's work. Nonetheless, the basic idea of Darwinian evolution by natural selection was powerful and was thought by many to explain the patterns of variation in life and the great variety of forms of life.

Darwinians regarded evolution as gradual and incremental leading to continuous variation, and this became the hallmark of Darwinian thinking (Kutschera and Niklas 2004; Provine 2001). But by the early twentieth century, the particulate

[1] Charles Darwin and Alfred Russel Wallace disagreed on the relative importance of environmental causes in evolution. Darwin felt that competition between individuals was more important than environmental changes in driving selection. Wallace felt the reverse. For the purposes of exposition, we will see both as environmental causes as both are extrinsic to the units of inheritance in the Modern Synthesis. Undoubtedly both contribute to competition.

views of inheritance first proposed by Mendel had been rediscovered. This was a theory of inheritance that provided both the mechanisms for passing on traits but also accounted for the emergence of new variants through mutation[2]. Mendelians saw particulate inheritance as incompatible with Darwinian theory for a number of reasons. For example, Thomas Huxley (the great defender of Darwin) and Bateson (who coined the term mutation) felt that continuous variations were too small to generate significant selection pressures. Galton argued that selection on continuous variation would be limited by regression to the mean and De Vries felt that large, discontinuous mutations were the source of variation and essential to evolution. As Provine (2001) documents, this division between Darwinian and Mendelian views of evolution persisted for 25 years and not for entirely scientific reasons. There was much personal animosity between several leading characters on both sides of the debate, which delayed the inevitable synthesis. As with the rest of science, there were social processes in play (Reif et al. 2000; Smocovitis 1992).

The synthesis was due to the mathematical modelling of the population dynamics of gene pools by Fisher, Haldane and Wright, which saw the emergence of population genetics. Evolutionary processes could account for changes in gene frequencies within populations such that mutations were one key source of new variation and natural selection one key way of sorting variants. Mendelians and Darwinians were effectively discussing the same phenomena but at different levels of explanation (Scott-Phillips et al. 2011; Dickins and Rahman 2012) (see Sect. 7.3.). More precisely, Mendelians were only interested in the production of the phenotype, whereas Darwinians were focused upon the phenotypic makeup of the population over time.

The introduction of formal population thinking can be seen as the first stage in the synthesis and has been characterized as one of restriction (Gould 2002), as some associate it with the removal of Lamarckianism, saltationism and orthogenesis. However, these developments are perhaps better associated with adjustments to Darwinian thought and the rise of Neo-Darwinism (Futuyma 2015; Kutschera and Niklas 2004) rather than as specifically rooted in population genetics. What this stage of the synthesis did not do was to enforce a particular focus upon levels of selection nor upon natural selection as the predominant force or process responsible for evolutionary change. In particular, genetic drift (allele frequency change owing to random sampling of gametes between generations) could still be seen as a significant force. It was a pluralistic stage.

Gould claims that the second stage of the synthesis saw the emergence of a much narrower and more dogmatic focus upon natural selection as the predominant evolutionary process. To support this, Gould cites Ernst Mayr's definition of the Modern Synthesis, some 20 years after Julian Huxley:

> The proponents of the synthetic theory maintain that all evolution is due to the accumulation of small genetic changes, guided by natural selection, and that transpecific evolution is nothing but an extrapolation and magnification of the events that take place within populations and species. (Mayr 1963, p. 586)

[2] The association of mutations arose during the development of Mendelism.

Here we see that natural selection operating over genes within populations is regarded as the predominant force of microevolutionary change and that microevolutionary changes also account for macroevolutionary change. However, a further 23 years later, a leading textbook summarizes the situation in a more pluralistic vein than Gould and Mayr:

> The major tenets of the evolutionary synthesis, then, were that populations contain genetic variation that arises by random (i.e. not adaptively directed) mutation and recombination; that populations evolve by changes in gene frequency brought about by random genetic drift, gene flow, and especially natural selection; that most adaptive genetic variants have individually slight phenotypic effects so that phenotypic changes are gradual (although some alleles with discrete effects may be advantageous, as in certain colour polymorphisms); that diversification comes about by speciation, which normally entails the gradual evolution of reproductive isolation among populations; and that these processes, continued for sufficiently long, give rise to changes of such great magnitude as to warrant the designation of higher taxonomic levels (genera, families, and so forth). (Futuyma 1986, p. 12)

Natural selection is not regarded as the sole cause, and we see a commitment to gradualism, microevolutionary changes accounting for all, and, as with Mayr, a focus on the gene as the level of selection (Williams 1996). We shall return to these definitional statements below (Sect. 7.4).

7.3 Proximate and Ultimate Causation

During the development of the Modern Synthesis, Mayr wished to highlight the distinguishing features of biological explanation. He did this in a landmark paper in which he made clear the distinction between proximate and ultimate causation (Mayr 1961)[3].

Mayr discussed the three elements of causality—an explanation of past events, a prediction of future events and an interpretation of goal-directed, or teleological, phenomena. He argued that biological causation is not predictive in the classical sense (it is statistically predictive) and nor is it teleological. Instead he highlighted teleonomic purposiveness, a property of programmed systems, such as those that have evolved[4], as the core concern in biology. In other words, natural selection can build goal-directed organisms, and this leads to two kinds of causal focus—ultimate and proximate:

> (Ultimate causes) are causes that have a history and that have been incorporated into the system through many thousands of generations of natural selection.
>
> …

[3] While Mayr clearly explained and advanced the use of proximate and ultimate causation, the distinction is found in earlier biological work. For example, the introductory comments in a special issue of Ibis from 1950 make mention of ultimate factors (Thomson 1950).

[4] Note: natural selection itself is not purposive. This is not orthogenesis.

Box 7.1 General and Special Theories of Evolution

The general theory of evolution can be opposed to a special theory of evolution (Webb 2011). A special theory will specify the means and nature of variation, competition and inheritance. The Modern Synthesis is an example of a special theory of evolution, as we have seen. For the Modern Synthesis, trait variation is a consequence in part of genetic variation; those variants are differentially successful within environments such that some reproduce more than others and as a consequence the underlying genetic variants that contribute to those traits increase in relative frequency within the gene pool.

There are a number of ways we might try to formally capture evolution. One way would be as follows. Let[5]:

Competition = M
Trait variation = V
Inheritance = I
Selection = S
Evolution = D
We can then say that:

$$(M\&V)\ \pi\ S \qquad (1)$$

$$(S\&I)\ \pi\ D \qquad (2)$$

which means that competition and variation lead to selection and selection and inheritance lead to evolution. We thus describe a two-stage process. But as this is a formal presentation, we are most strongly saying that selection is conditional upon the presence of competition and variation, in conjunction. Also, evolution is conditional upon the conjunction of selection and inheritance. This whole relationship might be expressed as follows:

$$(((M\&V)\ \pi\ S)\ \&\ I)\ \pi\ D \qquad (3)$$

Such a proposition amounts to the following format:

$$(x\&y)\ \pi\ z$$

where ((M&V) π S) becomes x. A structure like this is not a satisfying definition not least because the entire statement can be true both when x and y are false or when just x or just y are false. In fact, the only occasion under which this statement can be nothing other than false is when z is false.

In general one should expect a different logical relationship to prevail for a fundamental definition, as follows:

$$S\ \sqrt{}(M\&V) \qquad (4)$$

$$D\ \sqrt{}(S\&I) \qquad (5)$$

(continued)

[5] M = Malthus; D = Darwin.

Box 7.1 (continued)

These formulae rely upon a biconditional structure. This establishes a contingent relationship, as in the previous expressions, but to be true all of the constants must also be true. It is all or nothing. Thus we only have evolution if and only if there is selection and inheritance, and we have selection if and only if there is competition and variation.

We can readily produce a single statement as before by using a conjunction.

$$D\surd(I \,\&\, (S\surd(M\&V))) \quad (6)$$

This two-stage, biconditional definition is a nominalist definition pace Popper (Popper 1945). It does not capture an essence of something but is rather a pragmatic action. As Popper notes, definitions are to be read from right to left. Where we might say that a mouse is a small generalist rodent with a long tail, we are really answering the question: "what should we call small, generalist rodents with a long tail?" We are not attributing a fundamental mouse essence. As such, scientific definitions are a little loose and act to name things we are interested in. Moreover, this definition is not an axiomatic statement from which specific hypotheses can be derived and tested. Given this we might wish to slightly modify the general theory classification and simply refer to a kind of change captured by this definition. Special theories will deal with special instances of this kind of change, and they will develop their own fundamental axioms from which to derive hypotheses. In the Modern Synthesis, axiomatic statements relate to the concept of fitness and genetic variation, and expressions such as Hamilton's rule and the Price equation capture them (Hamilton 1964; Price 1970).

With our definition in place, we can sketch some entailments. First, evolution is a kind or type of change and can be seen as a set. Within this set are other sets. These will include the set captured by the Modern Synthesis. It is also the case that the component constants of our definition—M, V and I—are sets. So, competition can be established in a number of different ways, as can variation and inheritance. Indeed, it is the nature of these configurations that will be reflected within any special theories of evolution. But even within the special theory of the Modern Synthesis, the set of competition contains the set of competition caused by environmental changes, and within that set there are numerous further sets. This is a property of loose, nominalist definitions—much is captured, but it is not imprecise. Rather an area to work in is referenced.

The breadth of these constant sets—M, V and I—is large and only constrained or focused by the business at hand for a particular theoretical activity. How one chooses to delineate set characteristics will reflect fundamental theoretical assumptions about how the world is organized. Those assumptions will also impact upon how one thinks about the interaction between different M, V and I sets. For example, in the Modern Synthesis variation and inheritance are

Box 7.1 (continued)

delivered through genes. Competition is between genetic variants (alleles) that build traits that outreproduce others such that the relative frequency of the underlying alleles is increased. Modern cultural evolutionists can be concerned with the variation and inheritance of ideas (memes, perhaps) (Mesoudi et al. 2006; Dawkins 1989). Those ideas can increase in relative frequency, or they can decline, as a consequence of how well they fare in a variety of problem spaces. Thus far we have defined two special spheres of evolutionary change—the Modern Synthesis and memetics—each dealing in a specific subset of the overall sets of M, V and I.

Box 7.2 False Dichotomies: Preformation Versus Epigenesis

The tendency to dichotomize problem spaces is both a common trait and a milestone in the history of many disciplines. Accounting for something in terms of an either/or proposition causes a gross level sorting of possibilities that can be useful in order to generate data that challenges either or both theoretical propositions. De novo dichotomies, based on no prior observation or even intuition, are usually false, and they are also rare. More generally, dichotomous problem spaces emerge from a period of inductive observation, the theoretical terms that are coined effectively capturing a long, descriptive list. As such they are not fully scientific categories (Popper 1979), for they do no real predictive and falsifiable work.

The biological sciences are no exception to this. Three common and related dichotomies that are still very much in use are nature/nurture, innate/learnt and robust/plastic (Bateson and Gluckman 2011). All three are category errors if interpreted as oppositional dichotomies. Put simply, nurture is something that is found in nature such that nurturing behaviours are part of the set of naturally occurring things. Learning is the product of learning mechanisms, which facilitate the acquisition of new knowledge but are not themselves the product of learning (if only for the fear of an infinite regress). The hardwired properties of such mechanisms will permit certain kinds of things to be learnt. Plasticity is one way of being robust, for it enables an organism to weather the vicissitudes of life.

Nature/nurture effectively opposes something fixed, innate and fully formed at birth with something flexible and acquired and in need of learning and development. This is often repackaged as nativism versus empiricism and, as they are described above, captures all three of our dichotomies. Within psychological circles, this has been treated as a philosophical problem (Fodor 1983; Pinker 2002) and less an empirical question, in part due to the legacy of Locke and Kant. Nonetheless, some have marshalled evidence in favour of one view or another (Karmiloff-Smith 1996; Samuels 2002).

(continued)

Box 7.2 (continued)

The antecedents of nativism and empiricism are to be found in Pythagoras and the distinction between preformation and epigenesis (Coall et al. 2014). Various preformation theories were in place during this period. Many were homuncular theories, suggesting that a small but fully formed adult was present within the paternal or maternal substrate in some manner. Preformationism, then, is the nativist idea that all traits have a complete form at conception, when male and female contributions meet[6]. For Pythogoras the male provided the essential contribution and the female an environment in which it could thrive and grow. Aristotle promoted epigenesis in which material causes were contributed by both parents; thereafter efficient causes led to the emergence of the formal cause which has final purpose. Aristotelian epigenesis is an empiricist idea—form emerges over time. This ancient discussion introduces another distinction, that between growth and development. Growth is merely an increase and development a change in form and function.

> (P)roximate causes govern the responses of the individual (and his organs) to immediate factors of the environment while ultimate causes are responsible for the evolution of the particular DNA code of information with which every individual of every species is endowed. (Mayr 1961, p. 1503)

This distinction is often referred to as that between how and why (Scott-Phillips et al. 2011). A proximate account will answer the question of how an individual operates within a particular environment. An ultimate explanation will address why that individual operates in that way. Thus, ultimate accounts address issues around the evolved function of a proximate mechanism, and this is invariably made sense of in terms of average lifetime inclusive fitness. To adapt Mayr's own example (1961, pp. 1502–1503), we can say that a bird migrates in order to find insects in other regions as winter arrives in its current location, thereby avoiding death by starvation, or that the bird has a genetic constitution that enables it to track particular elements of its environment such that it can maximize its fitness. Both of these are ultimate accounts that tell us why the bird behaves in this way. An account that discussed the bird's physiological responses to hours of daylight, or temperature, would be a proximate one, telling us how the bird produces this behaviour. The two are complementary and produce a total account.

It is worth noting that the distinction between proximate and ultimate causation captures the distinction between Mendelian (proximate) processes and Darwinian (ultimate) causation. To that end, the Modern Synthesis was in part a realization of these kinds of explanation and their relation to one another (Scott-Phillips et al. 2011).

---

[6] We will only discuss sexually reproducing species here, for ease of exposition. But note that all of these arguments apply to asexual organisms too.

7.4 The Parameters of the Extended Evolutionary Synthesis

7.4.1 The Basic Claim

As the title implies, the extended evolutionary synthesis (EES) is a claim that the Modern Synthesis, as outlined above, requires additional conceptual architecture to deal with the most recent findings in biology. This is therefore a claim about a special theory of evolution, and as such it addresses issues of inheritance, variation and causation. While the title, EES, and derivations thereof, has been used by a number of authors, it is Pigliucci and Muller (Pigliucci and Muller 2010) who have most coherently anthologized this movement in Evolution: The Extended Synthesis an historical nod to Huxley implying a sense of continued work. Their book has six core sections: Variation and Selection, Evolving Genomes, Inheritance and Replication, Evolutionary Developmental Biology, Macroevolution and Evolvability and Philosophical Dimensions, which introduce the reader to a plethora of interesting phenomena and recent work. The rhetorical claim of the book is best summed by the following quote:

> The commonest reaction to our explanations is something along the lines of "But that is already understood as part of the Modern Synthesis anyway." We beg to differ. Many of the empirical findings and ideas discussed in this volume are simply too recent and distinct from the framework of the (Modern Synthesis) to be reasonably attributed to it without falling into blatant anachronism. (Pigliucci and Muller 2010, p. 4)

To this end Pigliucci and Muller are claiming that there is a requirement for new theory. However the book fails to deliver a decisive statement on what the underlying assumptions and predictions might be for an extended evolutionary theory (EET). Most recently Laland and key colleagues from the EES field (including Muller) have delivered on this task and synthesized the claims of EES theorists into a useful review paper (Laland et al. 2015).

Early on in their paper, Laland et al. (2015) list some of the core assumptions of the Modern Synthesis:

> These include: (i) evolutionarily significant phenotypic variation arises from genetic mutations that occur at a low rate independently of the strength and direction of natural selection; (ii) most favourable mutations have small phenotypic effects, which results in gradual phenotypic change; (iii) inheritance is genetic; (iv) natural selection is the sole explanation for adaptation; and, (v) macro-evolution is the result of accumulation of differences that arise through micro-evolutionary processes. (Laland et al. 2015, p. 2)

The citation given to support this list is the 1998 edition of Futuyma's textbook, *Evolutionary Biology*. Here is how Futuyma most recently characterizes the Modern Synthesis:

> The major elements of the (Modern Synthesis), which remain major elements of evolutionary theory today, include (1) the units of evolution are populations of organisms, not types or single organisms ("population thinking"); (2) evolution is based on mutations that are random with respect to the adaptive needs of the organism (but are not necessarily random in other respects), resulting in inherited variation that may be amplified by recombination;

> (3) natural selection (at the level of individual organisms), acting on inherited variation, is the major cause of evolution of adaptive characteristics; (4) changes in the genetic composition of populations can also result from random genetic drift, especially in small populations; (5) new species are formed by divergence between populations of an ancestral species, owing to factors that reduce or prevent gene flow between populations that undergo different evolutionary changes; (6) gradual accumulation of changes by these same factors results in character differences that distinguish higher taxa. i.e. macroevolution. (Futuyma 2015, p. 31)

As can be seen from these two juxtaposed quotations and also those given previously (Sect. 7.2), there is much conceptual overlap. What perhaps does differ is the nature of emphasis. Laland et al. (2015) reference the pre-eminence of natural selection in the Modern Synthesis. For them, the Modern Synthesis claims that the "major directing or creative influence in evolution is natural selection, which alone explains why the properties of organisms match the properties of their environments" (Laland et al. 2015, p. 2). Futuyma indeed claims natural selection as a major cause of evolution but then goes on to discuss genetic drift as another. However, a little further on Laland et al. (2015) note that the Modern Synthesis regards populations as evolving *through changes in gene frequencies brought about through natural selection, drift, mutation and gene flow* (p. 2).

7.4.2 Constructive Development

Laland et al. (2015) argue for another source of evolutionary change, that of constructive development, which refers to "the ability of an organism to shape its own developmental trajectory by constantly responding to, and altering, internal and external states" (p. 6). This is resonant of Mayr's point (Sect. 7.3), about teleonomic purposiveness, in that it references the abilities of organisms. But where Mayr (Mayr 1961) sees such abilities as the consequence of natural selection, Laland and colleagues are claiming that these abilities have conceptual causal parity with selection. This claim makes clear that the prime interest for the EES is the phenotype, and thus natural selection and development see to the presence of specific phenotypes in the population. This is trivially true but loses sight of what evolution is. Evolution is change at the population level, and this is what the Modern Synthesis is addressed to. The developmental processes that build a phenotype only provide antecedent conditions for a bout of selection; there must be something that varies and that can be differentiated through competition. Thus, put crudely, if some environmental condition arises such that individuals that develop an aspect of their phenotype in a particular way thrive relative to others that cannot develop in this way, selection will favour the former. This is all about individual differences in the developmental response, something captured by the concept of reaction norms in biology (Ghalambor et al. 2015).

Developmental processes, like many others, can be characterized as conditional architectures (Nettle and Bateson 2015; Dickins and Dickins 2008), such that $p \rightarrow q$ where p is an antecedent condition, for example, a particular environmental parameter value, and q is a phenotypic output. p and q are effectively kinds, or sets, within which a finite array of variant tokens can be accommodated. This is one way of describing a reaction norm. Individuals may well differ in the amount of *q*-type responsivity they have available for development, making some individuals potentially more plastic with reference to the number of *p*-type environmental conditions they can deal with. Key individual differences, where heritable, are sources of variation for selection. Therefore, we would anticipate the array of possible *q* response to a *p* input to be the product of evolutionary processes.

Given this characterization, one can begin to model plastic response. For example, exclusive-or decisions can be built into development ($[p \rightarrow q) v (p' \rightarrow q']$) such that the organism has the ability to deal with a *p* or *p'* input during its lifetime and deliver an adapted outcome. Nettle and Bateson (2015) add a probability structure to this form of model in order to capture the full subtlety of adapted plasticity (and see Sect. 7.5.2). It is possible that these potential pathways might appear to be variation over which selection is operating: the external environment is in a *p'* state so therefore not-*q*; *q'* is in the ascendant. If many organisms can do this in a population that finds itself under *p'*, then there will be a population shift in the representation of *q'* at a point in time. It is an easy next step to argue that the *p'* environment/ecological niche is inherited (Laland et al. 2000) and to claim a form of a generalized Darwinism.

In addition, if *p*' goes on for a long time, it is possible that a secondary process may switch off the capacity for $p \rightarrow q$; perhaps, this might prove metabolically costly to build. However, this order of secondary process is more like standard Modern Synthetic natural selection as ultimately this will come under some order of genetic control (or absence thereof post-negative selection). Therefore, any population claims that seek to expunge $p \rightarrow q$ from the phenotype proper (including the mechanisms responsible for building it) can only rely on the processes outlined within the Modern Synthesis.

The reason for establishing this argument in this way is to show what the Modern Synthesis framework does—it deals with a particular clause (the expunging clause in this example). When you say this, you say a Modern Synthetic thing. When you say that the population is more *q* at time t_1 and more *q'* at t_2, you are not saying anything specifically evolutionary in the Modern Synthesis sense. If you add a cause—a general shift from *p* to *p'* over the course of time from t_1 to t_2—you capture a key element that can affect phenotypic expression (from *q* to *q'*); but that is all. To be an evolutionary claim, in the Modern Synthetic sense, then the underlying population level variation in reaction norms must be altered. This is the distinction between latent phenotypic expression and absolute phenotypic range. The latter is captured by a list of all *p*-to-*q*-type transitions that a given population representative organism can produce; the former is captured by a list of the specific *p*-to-*q*-tokens currently exhibited by *n* organisms. It is change in the absolute range, available at the population level, that is the business of evolution.

7.4.3 Reciprocal Causation

By implication natural selection builds something fixed, when described at the individual level, while constructive development is plastic. It is made very clear that this new process is not equivalent to the concept of gene-environment interactions because the focus is upon the mechanisms of development. These mechanisms are not to be seen as proximate mechanisms pace Mayr, because the extended evolutionary theory has a different model of causation, referred to as reciprocal causation (Laland et al. 2011).

Laland et al. (2011) introduce reciprocal causation during a critical analysis of the proximate-ultimate distinction in which they argue that development is crucially missing from what they term Mayr's unidirectional model of biological causation. Laland and colleagues appear to regard development as a set of processes that can introduce trait variation during lifespan. As an example, they use intersexual selection in the peacock and peahen:

> The ultimate explanation for the male trait is the prior existence of the female preference, proximately manifest in the peahen mate-choice decisions, shaped by inherited tendencies and modified by experience throughout development. Likewise, the ultimate explanation for the peahen's mating preferences is the prior existence of variation in the trait associated with fitness. (Laland et al. 2011, p. 1512)

Laland et al. (2011) state that in the Mayrian example of migration, the evolutionary process is unidirectional and the organism has been shaped to match the external environment, whereas for intersexual selection, the causation is reciprocal; proximate mechanisms *both shape and respond to selection, allowing developmental processes to feature in proximate and ultimate explanations* (p. 1512). So, the standard Modern Synthetic story is that the heritable proximate mate preferences of an historical peahen led to the greater success of a peacock with a particular heritable tail phenotype. That phenotype went to fixity in subsequent populations, and the underlying genes increased in relative frequency. The reciprocal causation account adds that peahen preferences can be changed by developmental processes and as a consequence of these developmental changes a particular and new peacock phenotype could be favoured. Thus, developmental processes, that permit phenotypic plasticity, can affect evolution by introducing new variation. There are two consequences of reciprocal causation: that phenotypic plasticity can generate selection pressures and lead to evolutionary change and that the origins of evolutionary episodes are ambiguous. Laland and colleagues claim that this ongoing process breaks down the association of ontogeny with proximate causes and phylogeny with ultimate causes.

The argument for reciprocal causation clearly relies upon notions of proximate and ultimate causation to run (Dickins and Barton 2012). Again this enables the trivially true interpretation of the EES position, above, that development can lay antecedent conditions for a bout of selection, which we argue can only really be understood in terms of reaction norms. This does not establish true causal parity between development and selection; it is merely a relationship between the two, and one entirely in accord with the Modern Synthesis because the focus of the Modern Synthesis is upon evolution, which is a population phenomenon.

Put more bluntly: advocates of the EES are not arguing that development and evolution are type-identical categories; they are arguing that they are logically discrete categories that can interact. The claim is that this interaction is reciprocal, which would imply bidirectional equality of influence between development and evolution, and yet the examples they use simply show that developmental processes exist and can be externally influenced and this can lay conditions for selection dynamics.

From this emphasis on development, it would appear that those seeking an EES have effectively returned to the distinction between preformation and epigenesis, nativism and empiricism (Box 7.2). However, at no point is there any claim for infinite plasticity from this group. Effectively their claim is that developmental processes operate within short time frames (lifetimes) and at different spans and rates. By implication, developmental processes have an outcome, which is fixed relative to some point in time prior to the death of the organism. Moreover, the claim is firmly embedded at the individual organism level, with due recognition that changes in individuals impart changes in evolutionary dynamics[7].

7.4.4 Inclusive Inheritance

The constructive development argument also relies upon the concept of inclusive inheritance, where soft and hard forms of inheritance are at least equal partners (Box 7.3). For example:

> (T)he merging of Darwinism with genetics into the modern synthesis has led to a semantic shift, resulting in the tendency to assume that only the DNA sequence is inherited across generations. However, evolution acts on any phenotypic differences that are stable across generations. According to this view, phenotypic variation should be partitioned into its transmitted versus non-transmitted components rather than into its classical genetic and environmental components. (Danchin et al. 2011, p. 475)

The inclusive inheritance claim is that evolution[8] is something that operates on phenotypes and that phenotypes can be inherited by genetic and non-genetic means, such as epigenetic processes (Sect. 7.5.2). As a consequence, advocates say, we must consider moving beyond the Modern Synthesis. This argument is neither formal nor empirical. At best it is a claim that phenotypic change as a consequence of transgenerational epigenetic processes appears to conform to the general theory of evolution, which would be of no consequence to the Modern Synthesis.

To claim conceptual parity between soft and hard inheritance would be to conflate proximate and ultimate causation (Dickins and Rahman 2012). Ultimate causes are to be understood in terms of the interaction between organismic traits and their ecology and the evolutionary history of that relationship as embedded within DNA sequences. The details of how particular DNA sequences operate to control gene expression, or

[7]The precise interaction with population level process is never explicated, not least because the assumption is that all new mechanisms of soft inheritance have conceptual and ontological parity with genetic mechanisms. In our view this is an unresolved empirical matter.

[8]This use of the term evolution is odd. Evolution does not act on anything; "evolution" describes a kind of change, change that can be caused by natural selection, drift, etc.

Box 7.3 Hard and Soft Inheritance

The distinction between hard and soft inheritance is attributed to Ernst Mayr (Mayr 1982). It is the distinction between genetic inheritance, where genetic sequence is seen as impervious to extrinsic effects, and Lamarckian inheritance. In the Lamarckian view, the environment shapes what is inherited from an individual directly (not via its impact on individual survival or reproductive value). This is often characterized as the inheritance of characteristics acquired during an individual organism's life span. Lamarckian ideas were removed from biological thought during the emergence of the Modern Synthesis owing to a developing understanding of reproductive biology. This is particularly associated with the germ plasm theory of August Weismann, in which reproductive potential is limited to developing germ cells (which are usually sequestered), while the majority of cells (somatic cells) comprising the body of a multicellular organism are excluded. As molecular biology emerged, it was also argued that while nucleic acids, in particular DNA, acted as the heritable material and specified the sequence of amino acids in proteins, proteins were not used as templates for the sequence of nucleotides in RNA or DNA, an idea referred to as the central dogma of molecular biology (Crick 1970).

In recent years transgenerational epigenetic effects have been recruited by those wishing to support Lamarckian ideas through an extended evolutionary synthesis. Jablonka, Lamb and Danchin have been notable in their championing of this position (Jablonka and Lamb 2006, 2008; Danchin and Pocheville 2014; Danchin et al. 2011).

Jablonka and Lamb argue that there are four dimensions or sources of evolutionary change—genetic inheritance—and then three sources of soft inheritance: epigenetic, behavioural and symbolic inheritance systems. Soft inheritance provides phenotypic tailoring during the lifetime development of an organism. However, no account in the coordination of these soft inheritance systems is provided and nor are they related to inclusive fitness. The core claim is that soft inheritance plays a Lamarckian role in the overall evolutionary dynamic such that epigenetic processes, for example, allow for the inheritance of acquired characteristics. Thus, on the one hand epigenetic, behavioural and symbolic systems are seen as proximate mechanisms that provide a phenotypic tailoring service, and on the other, this action introduces new variation into the phenotype, changing frequencies at the population level, being transmitted across generations and looking like evolutionary change. This is a claim to the conceptual and causal parity between hard and soft inheritance mechanisms and therefore to an extension of the Modern Synthesis to include soft systems.

to build proteins, are proximate causes. Genetic inheritance is not natural selection. Natural selection is the outcome of trait variation, inheritance and competition. According to the Modern Synthesis, trait variation is a consequence of genetic mutation (Sect. 7.5.2), inheritance is a result of genes being passed on during reproduction, and competition is endemic where resources are finite. Any trait variation that increases the differential in accessing those resources will be selected, and the genes underpinning this will increase in relative frequency. This is the essence of ultimate causation, how genes work is not. Therefore, to outline other inheritance systems and describe how they work are not to provide an account at the ultimate level.

This situation is perhaps not surprising as the EES rejects the ultimate-proximate distinction in favour of reciprocal causation. But if this distinction is denied, we find it hard to understand how any meaningful questions about evolutionary biology could be framed and how there could be any hope of synthesis, let alone an extended synthesis, between proximate and evolutionary biology. This is because the synthesis of Mendelian genetics and Darwinian evolution was precisely a synthesis of proximate genetic concerns with ultimate evolutionary ones—the how of genetic transmission with the why of evolution. As we have seen, this required the invention of population genetics, and to date no one from the EES movement has proposed a similar theoretical structure. Indeed, those working on soft cultural inheritance have directly borrowed population genetic models in order to hypothesize cultural dynamics (Boyd and Richerson 1985), effectively treating the mathematics as substrate neutral. On the surface this may appear to support the EES position, as surely the mathematics is reflecting a deeper truth about nature, but these models rely upon the assumption that cultural variants are inherited in a manner equivalent to genes, and in so doing simply model cultural change as if it were directly tied to genetic variation. This is not a distinct conceptual departure. The only available counterpoint for the EES community would be to argue that they simply see their soft inheritance systems as fitting into the synthesis as already described—theirs are alternative mechanisms to the genetic, introducing variation through developmental induction and being open to the same competition. This requires a nontraditional view of the hierarchical nature of biological organization (Eronen 2015).

7.5 Variation and the Modern Synthesis

7.5.1 Mutation

In the Modern Synthesis, the initial source of variation on which selection, drift and other processes can act is mutation. The role that mutation plays in evolution depends on the effects of mutations that arise, on the mutation rate and on these evolutionary processes. The outcome of evolution given various parameters may then be modelled in terms of equilibria between these processes or simulated if transitional or non-equilibrium states are significant.

With regard to adaptation, it is assumed that mutation (as a process) is undirected in the sense that beneficial mutations do not arise because they are needed (see Futuyma 2015 quote in Sect. 7.4.1). This assumption has been challenged by experiments that suggested that mutations are targeted to particular genes in response to selection (reviewed in Rosenberg and Adams 2001). While it appears that these early experiments do not in fact support such targeted molecular events, evidence does indicate that short-term increases in mutation rate can arise in cells under stress, a phenomenon known as transient hypermutability (Hersh et al. 2004). In addition a phenomenon known as phase variation has been investigated in bacterial cell populations (Moxon et al. 1994; van der Woude 2011) and shows that certain classes of reversible mutation occur at elevated rates in genes responding to rapid environmental shifts. Such observations in bacteria and the existence of mutation rate variation within eukaryotic genomes (Hodgkinson and Eyre-Walker 2011) lead to a variety of interesting ideas about why mutation rates have the values they do more generally. It is important, however, to keep in mind that an adaptive explanation for an observed pattern—however intuitively appealing—is not an appropriate null hypothesis and genetic drift plausibly accounts for the lower limit of the mutation rate in eukaryotes and in error-prone DNA replicating enzymes that are active in stressful conditions (Lynch 2011).

7.5.2 Variation-Generating Mechanisms

The foregoing discussion shows that there is diversity in the canonical variation-generating mechanism (VGM), mutation. The recent emphasis on epigenetic mechanisms however raises the possibility of additional VGMs (Sect. 7.4.4). Epigenetics refers to changes in the DNA context that result in changes in gene expression that can persist between cell (and potentially therefore also organismal) replications. Typical epigenetic modifications in eukaryotes include methylation of cytosine bases in CpG dinucleotides of DNA (meaning a sequence in which a guanine nucleotide follows a cytosine reading in the 5′ to 3′ direction) or methylation, acetylation or a variety of other modifications of the amino-terminal tails of histone proteins around which DNA is wrapped in a structure called the nucleosome. These modifications can change the accessibility of DNA to proteins, thereby altering the behaviour of genes. Such modifications are stable but capable of being reversed by either active or passive mechanisms[9].

Methylation of DNA bases can also occur in prokaryotes, and epigenetic changes are known to underlie some types of phase variation in bacteria (van der Woude 2011) in addition to well-established phenomena in eukaryotes such as genomic

[9]Alternative DNA methyltransferases, for example, can bring about de novo methylation of cytosine, or they may recognize and methylate hemi-methylated sequences resulting from replication of a methylated parent molecule.

imprinting in mammals (in which the parental origin of a gene copy determines its expression status). The widespread phylogenetic distribution of the machinery for processing and employing small regulatory RNAs (Cerutti and Casas-Mollano 2006) also provides opportunities for more elaborate forms of epigenetic control (including interactions between alleles).

From an EES perspective, this cornucopia of actual and potential mechanisms shows the importance of reciprocal causation (Sect. 7.5.3). From a Modern Synthetic perspective, the emphasis is on identifying evolutionary processes. In the cases of transient hypermutability and phase variation, these phenomena are related to variable environments including host immune systems or environmental dangers such as ultraviolet light. Variable environments can influence the outcome of selection on a lineage in several ways:

1. They can weaken the power of selection to discriminate between deleterious and beneficial mutations leading to reduced fitness (Cvijovic et al. 2015).
2. They can favour beneficial alleles that have higher fitness in all environments encountered (increasing the arithmetic mean fitness).
3. They can favour beneficial alleles that decrease the arithmetic mean fitness but increase its geometric mean over the long run.

Scenario 3, referred to as bet-hedging, can be achieved in a given lineage by making individuals that are able to persist in different environments (conservative bet-hedging), by making a diversity of individuals at least some of which can persist in each environment encountered (diversified bet-hedging) or by some mixture of these (Starrfelt and Kokko 2012). Enhanced dispersal across environments is also a related concept (Meyers and Bull 2002; Starrfelt and Kokko 2012). In so far as phase variation allows a bacterial population to survive, it would seem to exemplify diversified bet-hedging. The seemingly simple scenario 2 is often incompletely separated from bet-hedging. Fitness may be decreased overall by environmental variability, but mutations favoured by selection may have higher mean fitness than the ancestral genotype in these circumstances. Critically one way to achieve a higher fitness is to respond to the environment (Meyers and Bull 2002) on which more in the next section.

What determines which of the above scenarios predominates? Part of the answer to this relates to the granularity of environmental change. If it occurs on a fine scale, in which single organisms may expect to encounter different environments within a lifespan, this is expected to favour the evolution of conservatively bet-hedged phenotypes, but a more coarse-grained variation in which organisms may encounter different environments from their parents is more likely to result in diversified bet-hedging responses (Botero et al. 2015). Technically very long timescale changes can simply result in continued adaptive evolution, but the inability to produce sufficient, or the right type of, variation may result in an uncompensated decline in fitness (Cvijovic et al. 2015) or even population extinction (Botero et al. 2015).

7.5.3 From Variation to Regulation

The discussion in the previous section assumed that environmental variability is unpredictable. However, one simple way in which it is predictable is that, if environmental changes occur relatively rarely, the current environmental state may be approximately predictive of the state experienced by offspring. Modelling work exploring these parameters indicates that this logic may give an adaptive explanation for the existence of incomplete resetting of epigenetic marks between generations (Uller et al. 2015; Sect. 7.5.4) or, more generally, patterns of early irreversible plasticity (Botero et al. 2015).

Predictability can operate on shorter timescales. Depending upon the frequency of change relative to organisms' lifespans (Botero et al. 2015), we would expect selection to favour any mechanism that enables an organism to respond to environmental contingencies as they arise. Much gene regulation, from the simple bacterial operon or regulon (e.g. the lac operon), which enables a cell to respond to the availability of carbon sources (e.g. lactose) by producing digestive enzymes (e.g. β-galactosidase), to the more complex regulatory networks in eukaryotes, would appear to be consistent with this. So here it is apparent that we can understand phenotypic plasticity and its conditional architecture (e.g. lactose present → produce β-galactosidase) by invoking natural selection as the ultimate explanation (Sect. 7.5.2). However even when modelling adaptive phenotypic plasticity, it is apparent that, whether the response is to extrinsic factors or to internal states modified by them, adaptive constraints are as significant to their architecture as is adaptation (Nettle and Bateson 2015). More strongly not all the products of a genetic system are expected to be under ongoing selection (Graur et al. 2015), and the extent to which molecular processes are functional in this sense is likely determined by energetic costs considered relative to the power of genetic drift (Lynch and Marinov 2015).

As discussed in Sect. 7.5.3, variation in developmental (metabolic and regulatory) pathways is present and can lay the foundations for further evolution. The origin of novel metabolic pathways plausibly lies in residual enzyme activities associated with existing pathways (Notebaart et al. 2014), and, among other things, the relative flexibility of transcription factor binding permits rewiring of regulatory networks (Li and Johnson 2010).

7.5.4 The Role of Ecology

The ecological emphasis in the Modern Synthesis encourages us to view VGMs in terms of the ecology in which they arise with key parameters being the granularity and predictability of environmental change as described in the previous two sections. This is how we can make evolutionary sense of the architecture of developmental processes. This does not mean that the organism and its development are unimportant. Indeed the granularity and predictability of environmental change are partly functions of organisms' lifespans and levels of sophistication. Nor does it mean that the question of the origins of VGMs is settled. Indeed it is likely that mechanisms recruited for one purpose are exapted in the course of their evolution.

Should we be disappointed that the Modern Synthesis does not yield a universal abstraction, applying to all organisms at all times, that predicts the outcome of evolution? No. Rather than doing this, it provides a framework for understanding which parameters to attend to. From this perspective, the existence of any given VGM can be seen in its ecological context as a plausibly selected mechanism. This shows much of EES's reciprocal causation in a different light. It would be odd to describe gene expression as a fundamental challenge to the Modern Synthesis because it is intuitively easy to grasp that it is an outcome of selection (other evolutionary processes usually being overlooked by those without a background in population genetics because they are harder to intuit). This is true even though gene expression changes the interactions between genes and influences the outcome of selection. It is therefore apparent that standard evolutionary theory, the Modern Synthesis, provides a sensible approach to understanding variation in VGMs considered more generally and one that puts the details in their proper ecological context.

Conclusion

The Modern Synthesis in evolutionary biology is a theory in the sense that it provides a framework for understanding diverse phenomena. Although it emerged during an earlier period of expansion in our understanding of genetics and genetic mechanisms (Sect. 7.2), it provides a way to understand the architecture of development in terms of responses to ecological contingencies (including where these, in concert with demographic factors, limit the power of selection).

Because evolutionary change occurs between generations and in response to the environment, some plausible or actual developmental mechanisms, such as transgenerational epigenetic inheritance, have attracted attention and interpretation from supporters of an extended evolutionary synthesis (Box 7.3). We believe this partly reflects reasoning by analogy (Sect. 7.5.2) and have attempted to show how these may be profitably understood from a Modern Synthetic perspective.

The Modern Synthesis is mathematical and includes a variety of formalisms, from the Price equation to Hamilton's rule (Price 1970; Hamilton 1964), addressed to understanding the fundamental causes of biological variation. Formalisms can be a barrier to entry for investigators looking to develop an evolutionary perspective. Fortunately many of the core arguments, at least in reference to selective processes, can be understood without these. When this effort is undertaken, it is apparent that evolutionary theory offers a fruitful way to understand biological organization. While it does not offer an exhaustive state description, it does explain some of the most interesting features of organismal design in terms of their ecological context.

References

Bateson P, Gluckman P (2011) Plasticity, robustness, development and evolution. Cambridge University Press, Cambridge

Botero CA et al (2015) Evolutionary tipping points in the capacity to adapt to environmental change. Proc Natl Acad Sci 112(1):184–189

Boyd R, Richerson PJ (1985) Cultural and the evolutionary process. University of Chicago Press, Chicago

Cerutti H, Casas-Mollano JA (2006) On the origin and functions of RNA-mediated silencing: from protists to man. Curr Genet 50(2):81–99

Coall DA et al (2014) Evolution and prenatal development: an evolutionary perspective. In: Lamb M (ed) Handbook of child psychology and developmental science. Cambridge University Press, Cambridge, pp 57–105

Crick F (1970) Central dogma of molecular biology. Nature 227(5258):561–563

Cvijovic I et al (2015) The fate of a mutation in a fluctuating environment. bioRxiv: 016709. https://doi.org/10.1101/016709

Danchin É, Pocheville A (2014) Inheritance is where physiology meets evolution. J Physiol 592(11):2307–2317. Available at: http://doi.wiley.com/10.1113/jphysiol.2014.272096

Danchin É et al (2011) Beyond DNA: integrating inclusive inheritance into an extended theory of evolution. Nat Rev Genet 12(7):475–486. Available at: http://www.ncbi.nlm.nih.gov/pubmed/21681209 [Accessed October 26, 2012]

Darwin C (1859) On the origin of species by means of natural selection, or the preservation of favoured races in the struggle for life. John Murray, London

Dawkins R (1989) The extended phenotype: the long reach of the gene. Oxford University Press, Oxford. Available at: https://books.google.co.uk/books?id=U-ZzcgAACAAJ

Dickins T, Dickins B (2008) Mother nature's tolerant ways: why non-genetic inheritance has nothing to do with evolution. New Ideas Psychol 26:41. Available at: http://www.sciencedirect.com/science/article/pii/S0732118X07000281 [Accessed February 10, 2013]

Dickins TE, Barton RA (2012) Reciprocal causation and the proximate–ultimate distinction. Biol Philos 28:747. Available at: http://www.springerlink.com/index/10.1007/s10539-012-9345-z [Accessed February 10, 2013]

Dickins TE, Rahman Q (2012) The extended evolutionary synthesis and the role of soft inheritance in evolution. Proc R Soc B Biol Sci 279(1740):2913–2921. Available at: http://www.pubmedcentral.nih.gov/articlerender.fcgi?artid=3385474&tool=pmcentrez&rendertype=abstract [Accessed November 5, 2012]

Eronen MI (2015) Levels of organization: a deflationary account. Biol Philos 30(1):39–58. Available at: http://link.springer.com/10.1007/s10539-014-9461-z

Fodor JA (1983) The modularity of the mind. MIT Press, Cambridge

Futuyma DJ (1986) Evolutionary biology. Sinauer Associates, Sunderland

Futuyma DJ (2015) Can modern evolutionary theory explain macroevolution? In: Macroevolution: interdisciplinary evolution research. Springer, Cham, pp 29–86. Available at: http://link.springer.com/10.1007/978-3-319-15045-1

Ghalambor CK, Martin LB, Woods HA (2015) Plasticity, complexity, and the individual. In: Integrative organismal biology. Wiley-Blackwell, Hoboken, pp 1–21

Gould SJ (2002) The structure of evolutionary theory. Belknap Press of Harvard University Press, Cambridge

Graur D, Zheng Y, Azevedo RBR (2015) An evolutionary classification of genomic function. Genome Biol Evol 7(3):642–645

Hamilton WD (1964) The genetical evolution of social behaviour I & II. J Theor Biol 7(1):1–32. Available at: http://www.sciencedirect.com/science/article/pii/0022519364900384 [Accessed July 5, 2014]

Hersh MN et al (2004) Adaptive mutation and amplification in Escherichia coli: two pathways of genome adaptation under stress. Res Microbiol 155(5):352–359

Hodgkinson A, Eyre-Walker A (2011) Variation in the mutation rate across mammalian genomes. Nat Rev Genet 12(11):756–766

Hull DL (1980) Individuality and selection. Annu Rev Ecol Syst 11(1980):311–332

Huxley JS (1942) Evolution: the modern synthesis. Allen & Unwin, London

Jablonka E, Lamb MJ (2006) Evolution in four dimensions: genetic, epigenetic, behavioral and symbolic variation in the history of life. MIT Press, Cambridge

Jablonka E, Lamb MJ (2008) Soft inheritance: challenging the modern synthesis. Genet Mol Biol 31(2):389–395
Karmiloff-Smith A (1996) Beyond modularity. MIT Press, Cambridge
Kutschera U, Niklas K (2004) The modern theory of biological evolution: an expanded synthesis. Naturwissenschaften 91(6):255–276. Available at: http://link.springer.com/10.1007/s00114-004-0515-y
Laland KN, Odling-Smee J, Feldman MW (2000) Niche construction, biological evolution, and cultural change. Behav Brain Sci 23(1):131–146; discussion 146–75. Available at: http://www.ncbi.nlm.nih.gov/pubmed/11303338
Laland KN et al (2011) Cause and effect in biology revisited: is Mayr's proximate-ultimate dichotomy still useful? Science 334(6062):1512–1516. Available at: http://www.ncbi.nlm.nih.gov/pubmed/22174243 [Accessed October 28, 2012]
Laland KN et al (2015) The extended evolutionary synthesis: its structure, assumptions and predictions. Proc R Soc B Biol Sci 282:20151019
Lewontin R (1970) The units of selection. Annu Rev Ecol Syst 1(1970):1–18. Available at: http://www.jstor.org/stable/10.2307/2096764 [Accessed April 9, 2013]
Li H, Johnson AD (2010) Evolution of transcription networks—lessons from yeasts. Curr Biol 20(17):R746–R753
Lynch M (2011) The lower bound to the evolution of mutation rates. Genome Biol Evol 3:1107–1118
Lynch M, Marinov GK (2015) The bioenergetic costs of a gene. Proc Natl Acad Sci 112:15690
Mayr E (1961) Cause and effect in biology. Science 134:1501–1506. Available at: https://www.ifor.math.ethz.ch/education/bachelor/lectures/fs2011/math/Biomath/Mayr.pdf [Accessed November 26, 2012]
Mayr E (1963) Animal species and evolution. Harvard Univeristy Press, Cambridge
Mayr E (1982) The growth of biological thought: diversity, evolution and inheritance. Belknap Press, Cambridge
Mayr E (1993) What was the evolutionary synthesis? Trends Ecol Evol 8(1):31–34. Available at: http://www.ncbi.nlm.nih.gov/pubmed/21236096
Mesoudi A, Whiten A, Laland KN (2006) Towards a unified science of cultural evolution. Behav Brain Sci 29(4):329–347; discussion 347–83. Available at: http://www.ncbi.nlm.nih.gov/pubmed/17094820
Meyers LA, Bull JJ (2002) Fighting change with change: adaptive variation in an uncertain world. Trends Ecol Evol 17(12):551–557. Available at: http://linkinghub.elsevier.com/retrieve/pii/S0169534702026332
Moxon ER et al (1994) Adaptive evolution of highly mutable loci in pathogenic bacteria. Curr Biol 4(1):24–33
Nettle D, Bateson M (2015) Adaptive developmental plasticity: what is it, how can we recognize it and when can it evolve? Proc R Soc B 282(1812):20151005
Notebaart RA et al (2014) Network-level architecture and the evolutionary potential of underground metabolism. Proc Natl Acad Sci 111:11762
Pigliucci M, Muller GB (2010) Evolution: the extended synthesis. MIT Press, Cambridge
Pinker S (2002) How the mind works. Penguin, London
Popper KR (1945) The open society and its enemies. Routledge, London
Popper KR (1979) Objective knowledge: an evolutionary approach (revised edition). Oxford University Press, Oxford
Price GR (1970) Selection and covariance. Nature 227:520–521
Provine WB (2001) The origins of theoretical population genetics second. University of Chicago Press, Chicago
Reif WE, Junker T, Hossfeld U (2000) The synthetic theory of evolution: general problems and the German contribution to the synthesis. Theory Biosci 119(1):41–91. Available at: <Go to ISI>://WOS:000087151000004
Rosenberg SM, Adams D (2001) Evolving responsively: adaptive mutation. Nat Rev Genet 2:504–515

Samuels R (2002) Nativism in cognitive science. Mind Lang 17(3):233–265. Available at: http://doi.wiley.com/10.1111/1468-0017.00197

Scott-Phillips T, Dickins TE, West SA (2011) Evolutionary theory and the ultimate–proximate distinction in the human behavioral sciences. Perspect Psychol Sci 6(1):38–47. Available at: http://pps.sagepub.com/content/6/1/38.short [Accessed November 5, 2012]

Smocovitis VB (1992) Unifying biology: the evolutionary synthesis and evolutionary biology. J Hist Biol 25(1):1–65

Starrfelt J, Kokko H (2012) Bet-hedging—a triple trade-off between means, variances and correlations. Biol Rev 87(3):742–755

Thomson A (1950) Factors determining the breeding seasons of birds: an introductory review. Ibis 92(2):173–184. Available at: http://onlinelibrary.wiley.com/doi/10.1111/j.1474-919X.1950.tb01748.x/full [Accessed July 5, 2014]

Uller T, English S, Pen I (2015) When is incomplete epigenetic resetting in germ cells favoured by natural selection? Proc R Soc B 282(1811):20150682. Available at: http://rspb.royalsocietypublishing.org/content/282/1811/20150682.long

Webb RH (2011) If evolution is the answer, what is the question? J Evol Psychol 9(2):91–107. Available at: http://www.akademiai.com/openurl.asp?genre=article&id=doi:10.1556/JEP.9.2011.2.2 [Accessed March 6, 2013]

Williams GC (1996) Adaptation and natural selection: a critique of some current evolutionary thought, 3rd edn. Princeton University Press, New Jesey

van der Woude MW (2011) Phase variation: how to create and coordinate population diversity. Curr Opin Microbiol 14(2):205–211

Part II

Contemporary Experimental Approaches

Physiological and Genomic Mechanisms of Resilience to Multiple Environmental Stressors

8

Fernando Galvez

Abstract

Though much attention has focused on single environmental variables, most notably temperature and acidification, global climate change is most realistically to manifest as co-occurring and sustained variations in multiple environmental variables or in more frequent, but episodic, fluctuations in environmental conditions. Environmental variability is likely to produce physiological stress to organisms and may supersede the organismic capacity to handle stressor(s) when their rate or magnitude of change is high. Unfortunately, multiple stressor experiments predictive of natural systems remain difficult to perform. Multiple stressors may produce additive, synergistic, or antagonistic effects that are not always predictable from the impacts of the stressors in isolation. Furthermore, physiological variation is harbored within species and individuals, and this natural variation for resistance or resilience to one stressor may be attenuated by co-occurrence of additional stressors. As such, the combination of factors that limit physiological resilience in at-risk populations remains elusive.

After a brief description of stress biology, this chapter will describe the sources of ocean acidification and its biological impacts on the biota of marine systems. It will next describe the major drivers of oceanic deoxygenation and temperature warming as well as provide a brief discussion of the effects of these environmental stressors on aquatic animals. Next, it will discuss the environmental conditions that favor co-occurrence of these stressors in nature and how global climate change has exacerbated the magnitude and frequency of these multiple stressor interactions in nature. Finally, it will investigate how chronic exposure to new baselines in baselines in environments will sensitize or buffer organisms from acute fluctuations in environmental parameters and discuss how natural

F. Galvez
Department of Biological Sciences, Louisiana State University, Baton Rouge, LA, USA
e-mail: Galvezf@lsu.edu

© Springer International Publishing AG, part of Springer Nature 2018
W. Burggren, B. Dubansky (eds.), *Development and Environment*,
https://doi.org/10.1007/978-3-319-75935-7_8

evolved variation among populations and species may sensitize or buffer wild animals from altered environments. This will address one of the grand challenges in organismal biology, which is the effective integration of molecular through whole animal responses to natural systems.

8.1 Introduction to Stress Biology

The origins of stress biology trace back to at least the early twentieth century when Walter B. Cannon first described the fight-or-flight response in mammals. Cannon and de la Paz (1911) and Hans Selye (Selye 1936) introduced the concept of stress and its impact on human disease. Building upon Cannon's earlier work, Selye proposed stress as a "non-specific response of the body" to a demand imposed on it by any generic noxious agent. He later referred to the generic agents as "stressors," which consisted of environmental or intrinsic variables that initiate stress and a compensatory "stress response" to cope with impairment of homeostasis. Selye emphasized that a stressor may be physical (e.g., cold and heat), chemical (e.g., environmental toxicant such as a metal or an industrial organic), or psychologic in nature and that stressors produced non-specific responses as part of a general adaptation syndrome with an alarm reaction stage, a resistance stage, and an exhaustion stage.

From these early beginnings, the field of stress biology has expanded tremendously into multiple independent fields; however, stress is now viewed more than simply a non-specific, neuroendocrine response to a stressor. Although the concept of stress is well described in biology, the field of stress biology struggles to provide a unifying concept of the term (Reviewed by (Schulte 2014)). Furthermore, quantitative models of stress that accurately predict the impact of stressors on individuals and populations are still far from well described. This is particularly problematic when trying to predict stress in wild populations of animals that may be exposed to multiple stressors of varying intensity and for varying periods of time and experienced in often unpredictable manners.

8.1.1 Multiple Stressor Interactions in Nature

Variability in abiotic environmental factors such as pH, dissolved oxygen, temperature, salinity, reactive nitrogen and phosphorus, and chemical pollutants, or biotic factors such as altered intraspecies and interspecies interactions, is the norm in nature. However, acidification, elevated temperature, and low oxygen are common environmental factors experienced by marine biota. Although these environmental parameters only represent a small subset of factors that will be impacted by global climate change, they are considered important determinants in the distribution of animals in marine habitats. Furthermore, the magnitude and the frequency of environmental variation in these factors are expected to increase with global climate change. This will likely lead to the co-occurrence of multiple stressor interactions over space and time.

Multiple stressor interactions are characterized as being additive, antagonistic, or synergistic. An additive response in multiple stressor exposures will have a net effect equal to the sum of the quantitative responses of each the stressors if tested individually. Antagonistic interactions occur when one stressor interferes with the response of another stressor in such a way as to produce a total outcome that is less than the sum. In contrast, the terms potentiation and synergism are contrary to antagonism. Synergism occurs when the combined response is greater or less than the sum response of the component parts of the exposure tested in isolation. A related concept is that of potentiation. Although most commonly used in toxicology research, the term potentiation occurs when one or more substance, which individually have a negligible effect, enhances the efficacy of another stressor. In this example, the stressor is often a toxicant. Potentiation may even occur when one or more of the stressors are individually exposed at subthreshold levels but combine to produce a significant biological response. For simplification, potentiation will simply be referred to as synergism in the review.

8.2 Anthropogenic Ocean Acidification and Alteration of the Inorganic Carbon Cycle

Since the start of the industrial revolution, approximately 2040 gigatonnes of anthropogenic CO_2 have been discharged to the atmosphere, primarily due to fossil fuel combustion, cement production, flaring, and alterations in land-use patterns such as forestry (Orr et al. 2005; Doney et al. 2009; IPCC 2014). This enormous quantity of CO_2 discharge has driven the atmospheric carbon dioxide (CO_2) concentration from the preindustrial level of ~280 μAtm to its current level of ~400 μAtm (IPCC 2014). Based on models, approximately 30% of the total anthropogenic CO_2 discharged to the environment has entered the world's oceans due to oceanic-atmospheric exchange and altered the inorganic carbon cycle and pH of the world's oceans. Dissolved CO_2 partitions into carbonic acid (H_2CO_3), bicarbonate (HCO_3^-), and carbonate ions (CO_3^{2-}) in the ocean. This results in the simultaneous increase in the partial pressure of CO_2 (P_{CO_2}), an increase in the concentrations of HCO_3^- and H^+ ions, and a decrease in the concentration of CO_3^{2-} ion.

Calcium carbonate minerals will also tend to dissolve due to the decrease in pH and CO_3^{2-} ions in the ocean. The net flux of ~600 gigatonnes of anthropogenic CO_2 into oceans in the last 250 years has decreased mean oceanic pH by approximately 0.1 pH units, which represents a 26% increase in hydrogen ion concentration. According to the IPCC (IPCC 2014), oceanic P_{CO_2} will reach approximately 1000 μAtm by the year 2100 (RCP 8.5), further reducing the CO_3^{2-} ion concentration and the calcium carbonate saturation state and acidifying the ocean by an additional 0.3 units (IPCC 2014). By the year 2300, it is expected that oceanic dissolved CO_2 values will reach 1900 μAtm, and pH will lower by 0.77 units (Caldeira and

Wickett 2003). While decreased pH is the hallmark of ocean acidification, the entire inorganic carbon system is affected concurrently by the increase in P_{CO_2}. As such, the increase in atmospheric CO_2 can affect oceanic P_{CO_2}, pH, and calcium carbonate saturation state, as well as total inorganic carbon and total alkalinity.

These individual stressors can affect biological systems in potentially additive, synergistic, and antagonistic modes of action (Cole et al. 2016). Ocean acidification is likely to change marine climate and threaten marine organisms and ecosystems around the world (Kroeker et al. 2013). Given that the ocean carbonate cycle varies in space and time, global climate change is expected to impact biota in some regions more than in others. Environments with naturally high carbonate saturation state and pH due to high levels of dissolved inorganic carbon from phytoplankton fixation may be less affected. In contrast, regions subject to upwelling events such as the western coasts of North and South America, and the equatorial Pacific Ocean, already experience bouts of severe depression in pH, which are already at levels projected for the end of the twenty-first century in the open ocean. Furthermore, coastal waters and estuaries experience episodes of acidification due to eutrophication, high organic loads, and riverine inputs of acidic water. These factors need to be considered when assessing the impacts of natural populations of animals to ocean acidification.

8.2.1 The Impacts of Ocean Acidification to Marine Biota

Ocean acidification is expected to influence the fitness, phenotype, distribution, and abundance of many marine taxa. Over the last decade, the focus of ocean acidification research has expanded greatly to include many understudied taxa, inclusion of more biological endpoints in studies, and exposure of biota to conditions representative of real-world conditions. The rapid proliferation in the number of studies has warranted an excellent synthesis of the emerging threats of ocean acidification in several meta-analytical assessments (Przeslawski et al. 2015; Harvey et al. 2013; Kroeker et al. 2013; Kroeker et al. 2010). Kroeker and colleagues (Kroeker et al. 2013) expanded on an earlier analysis conducted in 2010 (Kroeker et al. 2010) by describing acidification effects in marine taxa from 228 studies. Overall, ocean acidification is shown to decrease survival; affect biological processes such as calcification, growth, and development; and reduce species abundance but to have limited effects on metabolism and photosynthesis of marine biota (Kroeker et al. 2013). Survival and calcification are the responses most impacted by acidification at levels found currently in upwelling zones and at the mean oceanic levels predicted before the year 2100. Ocean acidification is particularly deleterious to calcifying organisms, including single-cell organisms such as foraminifers (Uthicke et al. 2013) and coccolithophores (Beaufort et al. 2011) and invertebrate calcifiers like corals (Andersson and Gledhill 2013), pteropods (Seibel et al. 2012), bivalves (Talmage and Gobler 2010), crustaceans (Whiteley 2011), and echinoderms (Dupont et al. 2010). This is likely due to the fact unicellular organisms have less capacity to buffer themselves from direct exposure to their environment as is possible by eukaryotes with the capability of regulating the extracellular environment (Melzner et al.

2009). However, acidification will most impact corals, coccolithophores, and mollusks, compared to echinoderms and crustaceans, due to greater effects on calcification. The impacts of acidification to calcifying biota are consistent with the predicted effect of acidification on the carbonate saturation state of oceans. In fact, by the end of the century, the biocalcification potential of marine calcifiers is expected to decrease by over half (Ries 2010).

A concerted effort has been made to assess biological impacts in a larger number of species and using a greater diversity of biological endpoints to assess impact. In a comprehensive review by Heuer and Grosell (Heuer and Grosell 2014), nine categories of organismic endpoints were identified as impacted by hypercapnia in laboratory and field studies in marine fishes. These endpoints included (1) osmoregulation; (2) cardiorespiratory physiology; (3) mitochondrial physiology; (4) metabolic and fitness effects; (5) growth, development, and survival; (6) neurosensory and behavioral effects; (7) calcification; and (8) acid-base balance. Across the nine categories, the lowest effect levels to CO_2 were observed below 600 μAtm. Although fish had not been considered sensitive to ocean acidification, all nine categories were affected in at least a few fish species at P_{CO_2} levels below 2000 μAtm (Heuer and Grosell 2014).

The literature on the effects of ocean acidification to fish tells two distinct messages. Before 2009, the prevailing view was that fish are extremely tolerant to hypercapnia. This perception was driven by several studies in multiple fish species that found no effects on growth and survival to P_{CO_2} levels of ~17,000 to ~50,000 μAtm. It was felt that contemporary or near-future levels of marine acidification would pose little to no negative effects in fish. The impressive hypercapnia tolerance capacity of fish has been associated with the compensatory mechanisms to acid-base disturbances (Claiborne et al. 2002; Perry and Gilmour 2006; Melzner et al. 2009). Hypercapnia will elicit an acute respiratory acidosis, which can be compensated by metabolic adjustments involving either an increase in net H^+ excretion, an increase in net HCO_3^- reabsorption rates, or both (Evans et al. 2005; Marshall and Grosell 2005; Perry and Gilmour 2006). The net effect of this metabolic compensation is a restoration of extracellular pH despite continued exposure to elevated CO_2 at the expense of an increase in internal HCO_3^- and reduction in plasma Cl^-.

Fish have a remarkable capacity to regulate acid-base balance during hypercapnia; however, recent studies suggest that there are physiological consequences to these compensatory responses to ocean acidification. Heuer and Grosell (2014) described that multiple biological endpoints, including neurosensory and behavioral effects, are adversely affected by low to moderately low levels of CO_2. In 2009, a seminal study by Munday and colleagues (Munday et al. 2009) found that exposing orange clownfish to 1000 μAtm P_{CO_2} for 11 days post hatch altered the behavioral response of fish to chemical cues derived from anemones, plants, conspecifics, or parents. This study led to a rapid reemergence of interest in studies in fish and other taxa, particularly in endpoints unrelated to calcification. P_{CO_2} exposures at ranges between ~500 μAtm and ~2000 μAtm have identified alterations to auditory (Simpson et al. 2011), olfactory (Cripps et al. 2011; Dixson et al. 2010; Devine et al.

2012; Munday et al. 2009), and visual (Forsgren et al. 2013) functions and impairment of anti-predatory defenses and prey detection mechanisms in fish (Briffa et al. 2012; Ferrari et al. 2012). Fish also experience a loss of lateralization and a reduced capacity to learn (Domenici et al. 2012; Ferrari et al. 2012). Behavioral lateralization, which has an important anti-predatory function and regulates shoaling behavior, is significantly impacted in several larval marine fishes including the reef fish, *Neopomacentrus azysron*, at 880 μAtm (Domenici et al. 2012).

Many of the behavioral responses during ocean acidification exposure have been repeatedly reversed with gabazine, an antagonist of the GABA-A receptor (Tresguerres and Hamilton 2017; Nilsson et al. 2012). These studies add to the growing body of evidence that ocean acidification alters central neural processing, eliciting a wide assortment of sensory and behavioral alterations. The prevailing hypothesis is that metabolic compensation to hypercapnia-induced acidosis alters extracellular Cl^- and HCO_3^- levels and reverses the reversal potential of the GABA-A receptor and potentially attenuates the inhibitory effects of GABA in the central nervous system (Regan et al. 2016). CO_2 alters sensory function leading to abnormal behavior in fishes due to a reversible impairment of the GABA-A neurotransmitter receptor (Nilsson et al. 2012; Regan et al. 2016). Overall, these behavioral effects are likely capable of threatening the recruitment success of fish, with the potential to affect fish populations and communities.

8.2.2 The Effects of Ocean Acidification at Early-Life Stages

A recent focus of ocean acidification research has been on early-life stages of fish. During larval development, fishes rely on the skin for gas diffusion and ion exchange until the gill epithelium becomes fully differentiated and functional (Rombough 2007). When P_{CO_2} in water is high, CO_2 can accumulate readily by passive diffusion through the skin, owing in part to the large surface area to volume ratios of larvae (Melzner et al. 2009). Even when net P_{CO_2} flux is directed outward, hypercapnia may impair CO_2 efflux and lead to a transient acidosis. However, this is less likely a concern in multicellular animals with high metabolic activities owing to the relatively higher levels of extracellular P_{CO_2} in these animals (Melzner et al. 2009). Although larval fish are capable of acid-base regulation, regulatory mechanisms may not be fully developed in early stages, particularly in pelagic marine species that must undergo significant ontogenic development following hatch. Similarly, physiological systems, which include transport proteins for the transepithelial movement of acid and base equivalents, may not be fully expressed. Coincidently, an animal may have limited capacity to counteract extracellular and intracellular acid-base disturbances during bouts of hypercapnia. Recent studies suggest that the primordial gills of larval fish play a more critical role in acid-base regulation than they do in gas exchange (Brauner and Rombough 2012).

The effects of elevated P_{CO_2} on early-life stages of fish are diverse and often contradictory. Some studies have described increased mortality in embryos and larvae (Forsgren et al. 2013; Baumann et al. 2012) and decreased size in larvae exposed to moderate levels of CO_2. Embryonic and larval fish have higher mortality rates to ocean acidification (Chambers et al. 2013), and the larvae that do survive experience a higher frequency of developmental abnormalities (Chambers et al. 2013; Baumann et al. 2012; Frommel et al. 2013). A recent long-term exposure of Atlantic silverside to 2300 μAtm from hatch through 135 days of development showed no effects on survivorship, although there was a slight reduction in length, weight, condition factor, and fatty acid composition (Murray et al. 2016). Furthermore, effects were most pronounced when fish were not given excess food. Similarly, inland silverside (*Menidia beryllina*) exposed to a P_{CO_2} level of ~1000 μAtm for 10–12 days from fertilization experienced a large reduction in survival and a marginal effect on growth. However, despite these effects, several studies found that even at high environmental CO_2, no effect, or even slightly positive effects, on fish growth survival was produced even among closely related species.

Given the influence of ocean acidification on calcification, several studies have studied the effects of elevated P_{CO_2} on otolith formation. The otolith is an aragonitic structure in the inner ear of fishes that is used by fish to monitor orientation and control acceleration. Elevated CO_2 has a highly variable effect on otolith accretion in marine fishes. Hypercapnia impairs otolith growth in marine medaka (Mu et al. 2015); however, exposure to elevated CO_2 forms paradoxically large otoliths in larval cobia (Bignami et al. 2013), *Solea senegalensis* (Pimentel et al. 2014), and the three-spined stickleback (Schade et al. 2014). In contrast, otolith size is not affected by ocean acidification in Baltic cod (Frommel et al. 2013) and larval mahi-mahi (Bignami et al. 2014). Despite this longstanding interest in otolith biology, especially in fisheries, the mechanisms behind otolith accretion are still poorly understood (Payan et al. 1997; Payan et al. 1999). A better understanding of the physiological mechanisms of otolith calcification in marine species will provide a more nuanced prediction of the impacts of ocean acidification. However, the important implication of these findings is that a reduced carbonate saturation state of water does not necessarily confer reduced calcification potential in biota, especially when the mineral structure is not in direct contact with the environment, as is the case with otoliths.

8.2.3 Anthropogenic CO_2: Driver of Global Climate Change

Beginning the discussion focusing on the effects of anthropogenic carbon dioxide seems appropriate given that increasing atmospheric carbon dioxide levels is a major driver and a proximate cause of global climate change. As discussed, dissolution of CO_2 in sea water is the primary reason for acidification and the alteration of the inorganic carbon cycle of oceans. On its own, ocean acidification will produce a myriad of confounding issues that serve as stressors to marine organisms. However, atmospheric carbon dioxide will also produce conditions that will affect ocean temperatures and dissolved oxygen levels. Most apparent is the fact atmospheric carbon dioxide serves as a "greenhouse" gas, which will produce warming of the atmosphere and waters (see Sect. 8.2.5 on *Carbon Dioxide Induced Warming*). One of the side effects of oceanic warming is that it will decrease the solubility of oceans to oxygen, enhance thermal stratification of the ocean, and directly and indirectly alter the biological productivity of vulnerable habitats. Collectively, these factors are important causes for the deoxygenation of marine environments due to global climate change. Furthermore, global climate change is expected to exacerbate the magnitude of variation in environmental conditions in marine environments globally. Models predict that extremes in these environmental parameters are expected to increase in severity and in duration in future environments (IPCC 2014). This change has the potential to overwhelm the capacity of organisms to physiologically compensate to these stressors in the short term.

8.2.4 Environmental Hypoxia

Oxygen is a limiting resource to aerobic organisms. Due to its critical role in adenosine triphosphate synthesis in the electron transport system, oxygen fuels metabolically active activities such as locomotion, digestion, reproduction, and osmoregulation. Aquatic hypoxia is often defined as the partial pressure of oxygen (P_{O_2}) below which whole animal metabolism is no longer regulated and physiological functions are impaired. This dissolved oxygen tension is referred to as the critical oxygen tension (termed P_{crit}), representing the P_{O_2} level that an animal transitions from oxyregulation to oxyconformation (Rogers et al. 2016). At oxygen tensions above P_{crit}, oxyregulators can "homeostatically" regulate metabolic rate over a range of decreasing water P_{O_2}. As dissolved oxygen drops below P_{crit}, oxygen becomes limiting, and animals must rely on time-limited anaerobic metabolism. Alternatively, some animals can maintain metabolic scope and survive extreme hypoxia by depressing metabolism or enhancing anaerobic metabolic efficiency (Boutilier 2001). Although P_{crit} is generally considered an indicator of hypoxia tolerance, other physiological processes are more sensitive to oxygen depletion. Physiological effects at other levels of biological organization are often impacted at higher thresholds of low water P_{O_2}. For instance, behavior responses, which involve sensation of environmental P_{O_2}, may occur at oxygen levels higher than P_{crit}. In addition to behavioral responses, various molecular, biochemical, cellular, and tissue responses can compensate against hypoxia and be used to delineate an animal's tolerance to low environmental P_{O_2}. A recently articulated "grand challenge" in organismal biology is effective integration from molecules to whole organisms to offer holistic systems-level insight into the mechanisms that govern physiological processes (Mykles et al. 2010).

Hypoxia is one of the most pervasive environmental stressors affecting aquatic biota. Hypoxia is a natural phenomenon in stagnant and poorly mixed water, particularly at high temperatures; however, its incidence and severity in marine environments have increased in the past three to four decades. A meta-analysis of oxygen levels of coastal sites worldwide found an increase in hypoxia prevalence of 5.54% sites/year, with a notable exponential rise in frequency since the 1980s

(Vaquer-Sunyer and Duarte 2008). Although this increase in frequency is partly associated with improved monitoring efforts, hypoxia is developed by factors associated with global climate change. Approximately 400 sites along the oceans' coasts worldwide are termed hypoxic, and this value is likely to increase considering factors contributing to hypoxia remain unabated (Diaz and Rosenberg 2008). Eutrophication can lead to diel cycles in dissolved oxygen by bacteria and phytoplankton during low light conditions when photosynthetic activity is minimal. As water temperatures warm and cellular respiration increases during summer, dissolved oxygen levels in the coastal waters can drop, producing hypoxic zones that are almost completely devoid of oxygen (Scavia et al. 2017). Riverine inputs to coastal regions deliver agricultural runoff and other anthropogenic waste during periods of high flow. For example, the large inputs of fresh water from the Mississippi River drain approximately half of the continental United States into the northern Gulf of Mexico and produce seasonal "dead zones" totaling approximately 20,000 km^2. The size of the zone is influenced by the flow rate and nitrogen concentration of the Mississippi River and other factors inducing stratification such as temperature and salinity (Scavia et al. 2017). Human activities have increased the organic inputs into coastal waters, causing the exacerbation of severe bouts of seasonal hypoxia. Further expansion of hypoxic zones is predicted under different scenarios of global climate change, particularly with the expected increase in baseline surface water temperatures due to the greenhouse effect.

Aquatic hypoxia is often characterized as dissolved oxygen levels at 2 mg/L and below; however, some studies define it as high as 5 mg/L and above (Vaquer-Sunyer and Duarte 2008). Despite the lower hypoxia threshold, a review of 872 published studies on 206 benthic marine species across a full taxonomic range reported a median sublethal concentration of 2.61 mg/L (Vaquer-Sunyer and Duarte 2008). Over half the species tested are known to exhibit sublethal and even lethal effects at 2 mg/L, and the most sensitive species are affected by P_{O_2} at levels as high as 8.6 mg/L. Crustacea, and fishes are among the most hypoxia-sensitive species with median sublethal limits approximately 2.7-fold above the 2 mg/L oxygen threshold (sublethal median limits of 3.21 and 4.41 mg/L, respectively). Hypoxia-tolerance limits of benthic marine species varied by almost two orders of magnitude. Most of the variance in hypoxia tolerance is associated with differences between taxa and on the ability of species to actively avoid hypoxia in nature. However, hypoxia-tolerant species must have underlying cellular and molecular adaptations to low dissolved oxygen. Aquatic hypoxia strongly influences the distribution of fish and the structure of marine communities in the environment (Chu and Tunnicliffe 2015). Furthermore, these data suggest the risks to biota posed by low dissolved oxygen may occur even at levels not operationally defined as hypoxia. Given the prevalence of aquatic hypoxia in nature, there is a significant need to characterize hypoxia tolerance and the capacity for physiological acclimation or adaptation in fish from hypoxia-prone environments.

Hypoxia can have numerous negative effects on the early-life stages of marine biota including mortality at extreme hypoxia. Extreme hypoxia can be lethal (Hassell et al. 2008); however, numerous sublethal effects are observed at higher threshold levels of dissolved oxygen. Moderate levels of hypoxia will elicit sublethal effects that limit the home range size of species (Rankin et al. 2013) or produce developmental deformities such as small size at hatch and failure to survive long past hatch (Hassell et al. 2008), respiratory stress, and metabolic depression (Boutilier 2001) and attenuate evasive responses to predatory cues and alter schooling behaviors. Sublethal impacts will translate into substantial depression of recruitment success due to the importance of small changes in survival during critical early-life stages (Houde 1989).

8.2.5 Carbon Dioxide Induced Warming

Global climate change models predict that an increase in temperature will be a hallmark of future terrestrial and marine environments. The "greenhouse" effect resulting from the increase in atmospheric CO_2 levels is predicted to increase average global temperatures by 2–5 °C by the end of the twenty-first century (IPCC 2014). Baseline habitat temperatures and acidity are expected to rise in tandem, yet their combined biological effects have only recently been studied together (Munday et al. 2009). This increased temperature is driven by elevated atmospheric CO_2, which also drives increased acidification and altered carbonate chemistry in the oceans.

Elevated temperature is expected to have a large influence on the distribution and abundance of species in the environment due to its large effects on cellular physiology and biochemistry (Hochachka and Somero 2002). Species vary in the range of temperatures across which they perform best, although outside this temperature range, the energetic costs interfere with other biological processes including fecundity and other life history traits. Increased temperatures can drive a species outside its optimal thermal niche or even past its tolerance window by altering the costs of metabolic processes (Farrell 2009; Farrell et al. 2009; Farrell 2016).

Absolute aerobic scope (AAS) for activity has been proposed as a framework to define the influence of environment on the performance capacity of animals under varying temperatures (Portner and Peck 2010; Portner 2010; Kassahn et al. 2009). Aerobic scope, which is calculated from the difference between standard metabolic rate (SMR) and maximal aerobic metabolic rate (MMR), provides an integrative measure of aerobic metabolism including physiological processes such as development and growth, reproduction, behavior, and locomotion and ultimately affecting organism fitness. Temperature exponentially increases SMR until it approaches or reaches that of the MMR, which becomes limiting due to physiological constraints of oxygen delivery (Farrell et al. 2009; Portner et al. 2009; Martins et al. 2011). Furthermore, it proposes that temperature influences blood P_{O_2} and subsequently tissue oxygenation, which collectively affect aerobic scope. At temperatures deviating from optimum, a mismatch exists between oxygen availability and delivery to tissues (Portner and Knust 2007). The paradigm predicts that the peak in aerobic scope will coincide with an animal's optimal temperature but decline at exposure temperatures above and below the optimal temperatures. Furthermore, additional environmental stressors, such as hypoxia and salinity, can attenuate aerobic scope by increasing SMR, decreasing MMR, or both. Reduction in aerobic capacity will manifest in fitness level impairments (Farrell et al. 2009; Portner 2010; Pörtner and Farrell 2008), which are likely to have long-term consequences in conducting routine biological functions essential for survival in a changing environment. Furthermore, it may have the effect of narrowing the thermal range over which aerobic metabolism can fuel biological function (Portner and Peck 2010; Kassahn et al. 2009). Collectively, this oxygen- and capacity-limited thermal tolerance (OCLTT) can help integrate the specific physiological effects of temperature with those of other environmental factors like hypoxia and salinity (Portner and Peck 2010). If one population is more tolerant of an environmental stress than another, then population variation in resilience of aerobic scope upon environmental challenge is one framework within which to capture and characterize evolved differences. Ultimately, the goal would be to gain a better understanding of the physiological costs and mechanisms of tolerance to multi-stressor exposure.

The OCLTT paradigm assumes that heart physiology plays an integral role in defining thermal tolerance limits (Pörtner et al. 2017). Cardiac anatomy and physiology are plastic during embryonic and larval development (Pelster 2002), and heart form and function are highly sensitive to environmental perturbations in temperature (Gamperl and Farrell 2004), which can contribute to long-term performance effects (Farrell et al. 2009; Portner et al. 2009). Furthermore, recent studies suggest that cardiovascular limitations of animal distribution in the environment may be a predictor of the effects of climate change to fish (Farrell 2009; Eliason et al. 2011). For example, comparative studies among migrating sockeye salmon populations show that scope for aerobic metabolic and cardiovascular capacities scale according to the historic river temperatures and migration distance variability among populations (Eliason et al. 2011). By comparing contemporary and predicted future water temperatures to thermal performance characters, the authors could predict which salmon populations were at greatest risk from future climate change (Eliason et al. 2011). Species of porcelain crabs from higher-temperature habitats (lower latitude and higher vertical niche in an intertidal zone) exhibit higher thermal tolerance compared to related species from lower temperature niches, but the thermally tolerant taxa lived closer to their thermal limit and have limited scope for acclimating to higher temperatures (Stillman 2003; Stillman and Somero 2000). *F. heteroclitus* can maintain aerobic scope over a very wide temperature range (Healy and Schulte 2012), though critical thermal maxima are higher in southern than in northern populations (Fangue et al. 2006). In addition to variation in thermal tolerance between northern and southern populations, population variability also exists for stressors that may interact with temperature such as salinity (Scott et al. 2004; Whitehead et al. 2011) and hypoxia (Everett and Crawford 2010).

Despite the attractiveness of the OCLTT hypothesis as a framework for integration of temperature effects in ecotherms, it has also created considerable debate and criticism. Several studies and reviews have added to this ongoing debate with some supporting the OCLTT paradigm and others providing evidence that the oxygen limitation fails to describe thermal sensitivity in many scenarios (Ern et al. 2016;

Clark et al. 2013; Norin et al. 2014). Many of the underlying assumptions of the model are, in practice, difficult to measure particularly at high temperatures nearing thermal maxima, even with the advent new technologies for measuring oxygen consumption with high temporal resolution (Clark et al. 2013). Recent studies have also suggested that the connection between tissue hypoxia and thermal limits is not apparent (Jorgensen et al. 2017; Holmstrup et al. 2010), either because it is difficult to separate time-limiting anaerobic metabolism from aerobic metabolism (Ejbye-Ernst et al. 2016) or because other physiological mechanisms may better predict thermal tolerance. For example, the temperature at which aerobic scope is at its maxima may not necessarily correspond to the temperatures that other biochemical or physiological processes or which growth and reproduction work optimally at. Clearly, there is a need to understand why temperature-induced impairment of aerobic scope predicts thermal tolerance in some species, whereas it continues to increase with temperature almost to point thermal tolerance is exceeded.

8.2.6 Interaction Between Ocean Acidification and Hypoxia

Low dissolved oxygen and acidification can routinely co-occur in coastal waters and in the open ocean at varying spatial and temporal scales. Although dissolved oxygen and pH can vary independently, they most often fluctuate together in marine ecosystems due to their link in cellular metabolism. Coastal regions are prone to diel cycles in dissolved oxygen and pH, which are driven extensively by biotic factors such as metabolic respiration, and affected by physical factors such as temperature and light, which stimulate eutrophication. In these regions, cellular respiration from microbes leads to the consumption of dissolved oxygen and the production of dissolved carbon dioxide in water. A 4-year study monitoring the dissolved oxygen and pH of a tidal salt marsh in coastal United States found that pH and dissolved oxygen were typically lowest when low tides occurred concomitant with pre-dawn times and very high during the mid-summer months when dissolved P_{CO_2} could increase as high as 4000 μatm during a single tidal cycle (Baumann et al. 2015). The study also suggested that outwelling during the ebb of tidal flow had the potential to distribute beyond the tidal marsh region. Similarly, tide pools in the intertidal rocky zone along the western coast of Canada experienced diel cycles of hypoxia and acidification during the light conditions followed by hyperpoxia and alkalification during the night (Richards 2011).

Thermal stratification and poor mixing of waters are also likely to expand the size of oxygen minimum zones that exist globally in the open oceans at depths of approximately 200–1000 m. These hypoxic zones form due to high rates of oxygen consumption from cellular respiration and the low rates of advective supply of oxygen-rich waters. In coastal regions on the eastern continental boundaries of the Pacific Ocean and the Atlantic Ocean, upwelling events deliver acidic and hypoxic waters from the oxygen minimum zones. Although these cold, nutrient-rich waters are critical for the microbial biogeochemistry of oceans and contribute to the high productivity of coastal regions, human activities are adding to the acidity of upwelled waters (Capone and Hutchins 2013). Collectively, these factors are likely to increase the extent and intensity of hypoxia and acidification in coastal upwelling waters. Models estimate that occurrence of hypoxia and acidification in coastal regions and in the open ocean are likely to become exacerbated by an increase in cellular respiration associated with eutrophication. Cai et al. (2011) have established pH and oxygen relationships for present-day and future conditions and compared these to preindustrial levels (Cai et al. 2011).

Considering the likelihood of hypoxia and acidification association in the environment, research on the combined effects of both stressors are critical. Gobler and Baumann (2016) recently reviewed studies describing the effects of continual or fluctuating exposures to a combination of hypoxia and acidification (Gobler and Baumann 2016). To date, the small number of studies on carbon dioxide and oxygen interactions has made meta-analyses unfeasible; however, several studies on fish and bivalves suggest that combined exposures produce effects that are mostly additive (i.e., no interactions between environmental variables). Although low dissolved oxygen and pH each had no effect on juvenile bivalve growth when exposed individually, collectively they produced a synergistic negative effect by reducing growth rate by approximately 35% compared to control animals (Gobler et al. 2014). Bay scallops experienced only an additive effect of high CO_2 and low O_2 on growth using size as a metric of effect.

8.3 Introduction to Multiple Stressor Interactions

Multiple stressor exposures may also alter physiologies or impact populations and ecosystems in ways that are not easily predictable (Christensen et al. 2006; Crain et al. 2008; Pörtner et al. 2005). The dynamicity of the environment requires that animals detect changes in environmental conditions and respond appropriately to the effects of concurrent exposure to multiple stressors. However, the impacts of multiple interacting stressors, especially when superimposed on new baselines of acidification, oxygenation, and temperature, will likely affect physiological compensatory abilities.

There is an urgent need for more studies that investigate multi-stressor interactions in marine biota; however, several factors have impeded progress. One limiting factor in the design of multiple stressor experiments is the multiplicative effect of testing more than stressor individually and in combination to identify interactions between environmental factors. This effect is only confounded if multiple concentrations of each stressor are tested. Statistical analyses such as principal component analysis and multivariate analyses are particular useful for identifying the most critical stressors and levels of stressors to study in experiments (Ivanina and Sokolova 2016). In general, these analyses can associate most of the variance in natural populations to a relatively small suite of environmental factors such as temperature and/or pH. These variables should form the backbone of multiple stressor experiments, particularly when the goal is to draw conclusions on the effects of global climate change to marine biota. Another important consideration is the magnitude and relative timing of the stressor exposures. In nature, environmental stressors fluctuate continuously such that the timing and magnitude of stress events may not always occur synchronously (Gunderson et al. 2016). The fluctuation in the intensity of environmental factors is likely to produce different effects than if the environmental factor was maintained at a constant value throughout an experiment. Furthermore, in multiple stressor experiments, variation in the timing of stressor events may have large consequences on the strength of the biological response. Gunderson et al. (2016) provide an informative review of the effects of concomitant stressor exposures versus exposures in which varying amounts of time elapse between stressor treatments (Gunderson et al. 2016). They propose that consecutive exposure of stressors in close succession is more likely to produce synergistic or antagonistic effects than if ample time exists between stressor treatments. In the latter case, stressors are expected to produce additive effects. Finally, developmental and physiological plasticity is important. At different stages during embryonic and larval development, aquatic animals may have periods of time of high sensitivity to stressors, or exposure to stressors could have latent effects that express themselves well after exposure, even on later reproductive fitness. Epigenetic effects can also potentially extend the window of efficacy of stressor exposures across generational timelines. Based on this complexity, new methods need to be developed to optimize the study of stressor exposures.

8.3.1 Physiological Plasticity and Genetic Variation

Reviews have highlighted the potential importance of acclimation and adaptation (Chevin et al. 2010; Chown et al. 2010; Fuller et al. 2010; Hoffmann and Sgro 2011) and have called for more empirical studies that include consideration of genetic variation on phenotypic plasticity in environmental stress research. Physiological plasticity may manifest as chronic acclimation ability to altered environmental baselines and ability to compensate for acute exposures to new environmental extremes. For instance, estuarine animals must cope with the often-harsh conditions imposed by multiple, co-occurring environmental stressors. In general, intertidal animals must regularly compensate physiologically to perturbations induced by environmental extremes, and this acclimation ability can vary among taxa that occupy different thermal niches. Estuaries experience periodic and stochastic variability in pH, dissolved oxygen, temperature, and salinity. Although each variable poses significant stress on animals when exposed alone, their combination in exposures has the potential for additive and synergistic interactions (Wolanski 2007). Even when animals are adapted to tolerate these extreme environments, the variability of estuaries is likely to be considered stressful to animals inhabiting these conditions (Schulte 2014). Shorter time-scale variations in environmental conditions are associated with tides and weather conditions and, in the case of the former, can lead to the cycling of variables such as those listed above. As an example, populations within species of Fundulid killifish harbor natural evolved variation for tolerance to temperature, salinity, and hypoxia, and have provided a nuanced understanding of relevant physiological mechanisms, and for more broadly exploring the scope of adaptive variation harbored for combinations of stressors (Whitehead et al. 2012; Whitehead et al. 2012; Whitehead et al. 2013).

Much effort has focused on studying the biological diversification of animals under different environmental scenarios. In part, the focus has been on how dispersal abilities might affect species' ranges and available niche space. However, given the rate and magnitude of anthropogenic environmental change, and habitat fragmentation that is already limiting the connectivity of suitable and available niche space, demographic mechanisms may be unable to compensate for contemporary and projected global climate extremes (Schloss et al. 2012). Models are currently being developed as a means to understand these difficult, multifactorial influences on ecological alteration. For instance, bioclimatic envelope modeling, which uses an interdisciplinary approach to predict the shaping of future ecosystems under global climate change projections, may provide future insights into novel climates along the leading edge of the shifting climate envelope (Mahony et al. 2017). In addition, niche-level modeling has used physiological data to provide estimates of distribution probabilities of species under different environmental conditions. Clearly, the dispersal ability of organisms, and their capacity to detect and to avoid adverse environments, must be considered when interpreting the ecological effects of environmental stressors to organisms.

8.3.2 Comparative Physiological Approaches in Multiple Stressor Research

Application of comparative physiological methods is a powerful and appropriate way forward for forecasting the impacts of climate change (Helmuth 2009; Pörtner and Farrell 2008; Somero 2010; Somero 2011). In the spirit of August Krogh's principle, articulated as "For many problems there is an animal on which it can be most conveniently studied" (Krebs 1975), these combined attributes of physiological plasticity and variable tolerance to individual or multiple interacting environmental stressors make intertidal species particularly well-qualified as a comparative physiological model system for addressing our research questions. Contemporary organisms represent the successful products of evolutionary testing, and found within this biodiversity storehouse are evolutionary solutions to various environmental problems (Schwenk et al. 2009).

For instance, species living in habitats often depleted by oxygen, such as estuaries, are often physiologically more resilient to hypoxia than species living in more stable, well-oxygenated environments (Richards 2011; Vaquer-Sunyer and Duarte 2008). Killifish are an emerging estuarine model (Burnett et al. 2007) that are particularly appropriate for environmental stress research because much natural physiological variation for temperature (Fangue et al. 2006), salinity (Whitehead et al. 2012; Whitehead et al. 2012), and hypoxia (Everett and Crawford 2010) tolerance is harbored within species, populations and species have known evolutionary histories (Duvernell et al. 2008; Whitehead 2010; Williams et al. 2008), and support the high-standing genetic variation in killifish populations (Reid et al. 2016). Estuarine habitats are characterized by frequent periods of hypoxia, especially along the northern Gulf of Mexico (Engle et al. 1999) where the *Fundulus grandis* thrives. Killifish as a group are relatively tolerant to low oxygen (Nordlie 2006), including *F. grandis* (Virani and Rees 2000), although populations of *F. grandis* vary in their ability to compensate to low oxygen (Everett and Crawford 2010).

Killifish have also repeatedly diversified between marine and freshwater environments and have experienced exceptional speciation rates among vertebrates (Alfaro et al. 2009). Within *Fundulus*, moderate to extreme euryhalinity appears to be ancestral and retained throughout the genus; however, multiple clades of killifish have lost tolerance to saline waters. Within the genus is repeated diversification into dilute freshwater systems accompanied by repeated loss of osmotic plasticity (Whitehead 2010). Within *F. heteroclitus*, derived freshwater-native populations exhibit slightly diminished resilience to high salinity, consistent with an intermediate stage of freshwater specialization. At the macro-evolutionary scale, osmotic plasticity has been repeatedly lost, anciently and recently and to various degrees, upon diversification into freshwater niches. Since several killifish species exhibit extreme physiological flexibility, this flexibility is shared by many species and has been repeatedly assimilated. As such, killifish provide a particularly well-suited comparative system to study mechanisms conferring differences in environmental stressor tolerance. *Fundulus* species can provide novel insights into the molecular and physiological mechanisms of survival in stable environments and the phenotypic plasticity of life in variable environments.

Populations and species traits will vary, but not all variances will be physiologically or ecologically important. Some traits can diverge by neutral drift, others may be constrained by strong purifying selection, and yet others will be influenced by habitat with divergence driven by natural selection. Many expression patterns are going to vary between species; however, by using the comparative approach and strategically selected species contrasts, it is possible to distinguish those that are most likely of physiological and ecological consequence. A phylogenetic comparative approach can distinguish variances that appear neutral or highly conserved, from those that appear to be affected by habitat. The principle is that if evolutionary distance accounts for trait variation, then neutral evolutionary processes are most likely responsible. Remaining variance (i.e., variance corrected for non-independence due to shared ancestry) can be examined for habitat correlations.

Determining the mechanisms that enable compensatory responses offers greater scope for inference about the physiological consequences of multiple stressor exposures. For example, upon thermal challenge, greater regulation of proteasome-encoding genes in a more thermally sensitive species of mussel was predictive of higher levels of protein ubiquitination in that species relative to its more thermally tolerant congener, indicating that species variation in thermal tolerance was related to different capacities to protect protein structure (Lockwood et al. 2010; Tomanek and Zuzow 2010). An integrative approach may also offer insight into mechanisms of additivity, synergy, or antagonism, among interacting stressors. For example, mussel exposures to temperature and salinity induced expression of an overlapping set of genes but most of them in opposite directions (Lockwood and Somero 2011). The identity of these interacting genes implicated disruption of membranes and ion transport as targets for potential interacting effects of co-occurring temperature and osmotic stress. Such coupling of comparative physiology and genome expression profiling offers integrative systems-level insights into the physiological mechanisms that mediate resistance, resilience, or impairment (Whitehead et al. 2011; Whitehead et al. 2011; Whitehead et al. 2012; Whitehead et al. 2013).

8.4 Concluding Remarks

Dispersal ability, phenotypic plasticity, and evolutionary potential are key factors affecting the vulnerability and persistence of wild populations in stressful environments (Chevin et al. 2010; Hoffmann and Sgro 2011; Somero 2010; Davis et al. 2005; Jump and Penuelas 2005; Whitehead et al. 2011; Whitehead et al. 2012). It is expected that the type of interactive effect between stressors will often vary between traits, individuals, species, and populations. A mechanistic understanding of responses to multiple stressor exposures will help predict the likelihood of resilience and persistence of resident species to persistent climate change. This information will be essential for predicting the population variants with highest fitness for future environments and estimating the capacity of different populations to colonize new environments. Evolutionary history and genomic variation will presumably contribute to the sensitivity and resilience of animals to multi-stressor challenges that involve co-exposure to extremes in pH, dissolved oxygen, temperature, and salinity. There is also a need to develop a conceptual model of stress capable of integrating the biological effects of stress from receptor to whole animal and that can predict the cumulative impacts of multiple stressors on animals and populations of animals.

References

Alfaro ME, Santini F, Brock C, Alamillo H, Dornburg A, Rabosky DL, Carnevale G, Harmon LJ (2009) Nine exceptional radiations plus high turnover explain species diversity in jawed vertebrates. Proc Natl Acad Sci 106:13410–13414

Andersson AJ, Gledhill D (2013) Ocean acidification and coral reefs: effects on breakdown, dissolution, and net ecosystem calcification. Annu Rev Mar Sci 5:321–348

Baumann H, Talmage SC, Gobler CJ (2012) Reduced early life growth and survival in a fish in direct response to increased carbon dioxide. Nat Clim Chang 2:38–41

Baumann H, Wallace RB, Tagliaferri T, Gobler CJ (2015) Large natural pH, CO_2 and O_2 fluctuations in a temperate tidal salt marsh on diel, seasonal, and interannual time scales. Estuar Coasts 38:220–231

Beaufort L, Probert I, De Garidel-Thoron T, Bendif EM, Ruiz-Pino D, Metzl N, Goyet C, Buchet N, Coupel P, Grelaud M, Rost B, Rickaby RE, De Vargas C (2011) Sensitivity of coccolithophores to carbonate chemistry and ocean acidification. Nature 476:80–83

Bignami S, Enochs IC, Manzello DP, Sponaugle S, Cowen RK (2013) Ocean acidification alters the otoliths of a pantropical fish species with implications for sensory function. Proc Natl Acad Sci 110:7366–7370

Bignami S, Sponaugle S, Cowen RK (2014) Effects of ocean acidification on the larvae of a high-value pelagic fisheries species, mahi-mahi *Coryphaena hippurus*. Aquat Biol 21:249–260
Boutilier RG (2001) Mechanisms of cell survival in hypoxia and hypothermia. J Exp Biol 204:3171–3181
Brauner CJ, Rombough PJ (2012) Ontogeny and Paleophysiology of the gill: new insights from larval and air-breathing fish. Respir Physiol Neurobiol 184:293–300
Briffa M, De La Haye K, Munday PL (2012) High CO_2 and marine animal behaviour: potential mechanisms and ecological consequences. Mar Pollut Bull 64:1519–1528
Burnett KG, Bain LJ, Baldwin WS, Callard GV, Cohen S, Di Giulio RT, Evans DH, Gomez-Chiarri M, Hahn ME, Hoover CA, Karchner SI, Katoh F, Maclatchy DL, Marshall WS, Meyer JN, Nacci DE, Oleksiak MF, Rees BB, Singer TD, Stegeman JJ, Towle DW, Van Veld PA, Vogelbein WK, Whitehead A, Winn RN, Crawford DL (2007) Fundulus as the premier teleost model in environmental biology: opportunities for new insights using genomics. Comp Biochem Physiol Part D Genomics Proteomics 2:257–286
Cai W-J, Hu X, Huang W-J, Murrell MC, Lehrter JC, Lohrenz SE, Chou W-C, Zhai W, Hollibaugh JT, Wang Y, Zhao P, Guo X, Gundersen K, Dai M, Gong G-C (2011) Acidification of subsurface coastal waters enhanced by eutrophication. Nat Geosci 4:766–770
Caldeira K, Wickett ME (2003) Oceanography: anthropogenic carbon and ocean pH. Nature 425:365
Cannon WB, de la Paz D (1911) Emotional stimulation of adrenal secretion. Am J Physiol 28:64–70
Capone DG, Hutchins DA (2013) Microbial biogeochemistry of coastal upwelling regimes in a changing ocean. Nat Geosci 6:711–717
Chambers R, Candelmo AC, Habeck EA, Poach EM, Wieczorek D, Cooper K, Greenfield CE, Phelan B (2013) Ocean acidification effects in the early life-stages of summer flounder, *Paralichthys dentatus*. Biogeosci Discuss 10:13897–13929
Chevin LM, Lande R, Mace GM (2010) Adaptation, plasticity, and extinction in a changing environment: towards a predictive theory. PLoS Biol 8:E1000357
Chown SL, Hoffmann AA, Kristensen TN, Angilletta MJ, Stenseth NC, Pertoldi C (2010) Adapting to climate change: a perspective from evolutionary physiology. Clim Res 43:3–15
Christensen MR, Graham MD, Vinebrooke RD, Findlay DL, Paterson MJ, Turner MA (2006) Multiple anthropogenic stressors cause ecological surprises in boreal lakes. Glob Chang Biol 12:2316–2322
Chu JWF, Tunnicliffe V (2015) Oxygen limitations on marine animal distributions and the collapse of epibenthic community structure during shoaling hypoxia. Glob Chang Biol 21:2989–3004
Claiborne JB, Edwards SL, Morrison-Shetlar AI (2002) Acid-Base regulation in fishes: cellular and molecular mechanisms. J Exp Zool 293:302–319
Clark TD, Sandblom E, Jutfelt F (2013) Aerobic scope measurements of fishes in an era of climate change: respirometry, relevance and recommendations. J Exp Biol 216:2771–2782
Cole VJ, Parker LM, O'connor SJ, O'connor WA, Scanes E, Byrne M, Ross PM (2016) Effects of multiple climate change stressors: ocean acidification interacts with warming, hyposalinity, and low food supply on the larvae of the brooding flat oyster *Ostrea angasi*. Mar Biol 163:17
Crain CM, Kroeker K, Halpern BS (2008) Interactive and cumulative effects of multiple human stressors in marine systems. Ecol Lett 11:1304–1315
Cripps IL, Munday PL, Mccormick MI (2011) Ocean acidification affects prey detection by a predatory reef fish. PLoS One 6:E22736
Davis MB, Shaw RG, Etterson JR (2005) Evolutionary responses to changing climate. Ecology 86:1704–1714
Devine BM, Munday PL, Jones GP (2012) Homing ability of adult cardinalfish is affected by elevated carbon dioxide. Oecologia 168:269–276
Diaz RJ, Rosenberg R (2008) Spreading dead zones and consequences for marine ecosystems. Science 321:926–929
Dixson DL, Munday PL, Jones GP (2010) Ocean acidification disrupts the innate ability of fish to detect predator olfactory cues. Ecol Lett 13:68–75

Domenici P, Allan B, Mccormick MI, Munday PL (2012) Elevated carbon dioxide affects behavioural lateralization in a coral reef fish. Biol Lett 8:78–81
Doney SC, Fabry VJ, Feely RA, Kleypas JA (2009) Ocean acidification: the other CO_2 problem. Annu Rev Mar Sci 1:169–192
Dupont S, Ortega-Martinez O, Thorndyke M (2010) Impact of near-future ocean acidification on echinoderms. Ecotoxicology 19:449–462
Duvernell DD, Lindmeier JB, Faust KE, Whitehead A (2008) Relative influences of historical and contemporary forces shaping the distribution of genetic variation in the Atlantic killifish, *Fundulus heteroclitus*. Mol Ecol 17:1344–1360
Ejbye-Ernst R, Michaelsen TY, Tirsgaard B, Wilson JM, Jensen LF, Steffensen JF, Pertoldi C, Aarestrup K, Svendsen JC (2016) Partitioning the metabolic scope: the importance of anaerobic metabolism and implications for the oxygen- and capacity-limited thermal tolerance (OCLTT) hypothesis. Conserv Physiol 4:13
Eliason EJ, Clark TD, Hague MJ, Hanson LM, Gallagher ZS, Jeffries KM, Gale MK, Patterson DA, Hinch SG, Farrell AP (2011) Differences in thermal tolerance among sockeye salmon populations. Science 332:109–112
Engle VD, Summers JK, Macauley JM (1999) Dissolved oxygen conditions in northern Gulf of Mexico estuaries. Environ Monit Assess 57:1–20
Ern R, Norin T, Gamperl AK, Esbaugh AJ (2016) Oxygen dependence of upper thermal limits in fishes. J Exp Biol 219:3376–3383
Evans DH, Piermarini PM, Choe KP (2005) The multifunctional fish gill: dominant site of gas exchange, osmoregulation, acid-base regulation, and excretion of nitrogenous waste. Physiol Rev 85:97–177
Everett MV, Crawford DL (2010) Adaptation versus allometry: population and body mass effects on hypoxic metabolism in *Fundulus grandis*. Physiol Biochem Zool 83:182–190
Fangue NA, Hofmeister M, Schulte PM (2006) Intraspecific variation in thermal tolerance and heat shock protein gene expression in common killifish, *Fundulus heteroclitus*. J Exp Biol 209:2859–2872
Farrell AP (2009) Environment, antecedents and climate change: lessons from the study of temperature physiology and river migration of salmonids. J Exp Biol 212:3771–3780
Farrell AP (2016) Pragmatic perspective on aerobic scope: peaking, plummeting, pejus and apportioning. J Fish Biol 88:322–343
Farrell AP, Eliason EJ, Sandblom E, Clark TD (2009) Fish cardiorespiratory physiology in an era of climate change. Can J Zool 87:835–851
Ferrari MCO, Manassa RP, Dixson DL, Munday PL, Mccormick MI, Meekan MG, Sih A, Chivers DP (2012) Effects of ocean acidification on learning in coral reef fishes. PLoS One 7:10
Forsgren E, Dupont S, Jutfelt F, Amundsen T (2013) Elevated CO_2 affects embryonic development and larval phototaxis in a temperate marine fish. Ecol Evol 3:3637–3646
Frommel AY, Schubert A, Piatkowski U, Clemmesen C (2013) Egg and early larval stages of Baltic cod, *Gadus morhua*, are robust to high levels of ocean acidification. Mar Biol 160:1825–1834
Fuller A, Dawson T, Helmuth B, Hetem RS, Mitchell D, Maloney SK (2010) Physiological mechanisms in coping with climate change. Physiol Biochem Zool 83:713–720
Gamperl AK, Farrell AP (2004) Cardiac plasticity in fishes: environmental influences and intraspecific differences. J Exp Biol 207:2539–2550
Gobler CJ, Baumann H (2016) Hypoxia and acidification in ocean ecosystems: coupled dynamics and effects on marine life. Biol Lett 12:8
Gobler CJ, Depasquale EL, Griffith AW, Baumann H (2014) Hypoxia and acidification have additive and synergistic negative effects on the growth, survival, and metamorphosis of early life stage bivalves. PLoS One 9:E83648
Gunderson AR, Armstrong EJ, Stillman JH (2016) Multiple stressors in a changing world: the need for an improved perspective on physiological responses to the dynamic marine environment. Annu Rev Mar Sci 8:357–378
Harvey BP, Gwynn-Jones D, Moore PJ (2013) Meta-analysis reveals complex marine biological responses to the interactive effects of ocean acidification and warming. Ecol Evol 3:1016–1030

Hassell KL, Coutin PC, Nugegoda D (2008) Hypoxia impairs embryo development and survival in black bream (*Acanthopagrus butcheri*). Mar Pollut Bull 57:302–306
Healy TM, Schulte PM (2012) Thermal acclimation is not necessary to maintain a wide thermal breadth of aerobic scope in the common killifish (*Fundulus heteroclitus*). Physiol Biochem Zool 85:107–119
Helmuth B (2009) From cells to coastlines: how can we use physiology to forecast the impacts of climate change? J Exp Biol 212:753–760
Heuer RM, Grosell M (2014) Physiological impacts of elevated carbon dioxide and ocean acidification on fish. Am J Phys Regul Integr Comp Phys 307:R1061–R1084
Hochachka PW, Somero GN (2002) Biochemical adaptation: mechanism and process in physiological evolution. Oxford University Press, New York
Hoffmann AA, Sgro CM (2011) Climate change and evolutionary adaptation. Nature 470:479–485
Holmstrup M, Bindesbol AM, Oostingh GJ, Duschl A, Scheil V, Kohler HR, Loureiro S, Soares A, Ferreira ALG, Kienle C, Gerhardt A, Laskowski R, Kramarz PE, Bayley M, Svendsen C, Spurgeon DJ (2010) Interactions between effects of environmental chemicals and natural stressors: a review. Sci Total Environ 408:3746–3762
Houde ED (1989) Subtleties and episodes in the early life of fishes. J Fish Biol 35:29–38
IPCC (2014) Climate change 2014: synthesis report. In: Pachauri RK, Meyer LA (eds) Contribution of working groups I, II and III to the Fifth Assessment Report of the Intergovernmental Panel on Climate Change. IPCC, Geneva, Switzerland
Ivanina AV, Sokolova IM (2016) Effects of intermittent hypoxia on oxidative stress and protein degradation in molluscan mitochondria. J Exp Biol 219:3794–3802
Jorgensen LB, Macmillan HA, Overgaard J (2017) Cold mortality is not caused by oxygen limitation or loss of ion homeostasis in the tropical freshwater shrimp *macrobrachium rosenbergii*. Cryobiology 76:146–149
Jump AS, Penuelas J (2005) Running to stand still: adaptation and the response of plants to rapid climate change. Ecol Lett 8:1010–1020
Kassahn KS, Crozier RH, Portner HO, Caley MJ (2009) Animal performance and stress: responses and tolerance limits at different levels of biological organisation. Biol Rev 84:277–292
Krebs HA (1975) The August Krogh principle: "for many problems there is an animal on which it can be most conveniently studied". J Exp Zool 194:221–226
Kroeker KJ, Kordas RL, Crim R, Hendriks IE, Ramajo L, Singh GS, Duarte CM, Gattuso J-P (2013) Impacts of ocean acidification on marine organisms: quantifying sensitivities and interaction with warming. Glob Chang Biol 19:1884–1896
Kroeker KJ, Kordas RL, Crim RN, Singh GG (2010) Meta-analysis reveals negative yet variable effects of ocean acidification on marine organisms. Ecol Lett 13:1419–1434
Lockwood BL, Sanders JG, Somero GN (2010) Transcriptomic responses to heat stress in invasive and native blue mussels (genus *Mytilus*): molecular correlates of invasive success. J Exp Biol 213:3548–3558
Lockwood BL, Somero GN (2011) Transcriptomic responses to salinity stress in invasive and native blue mussels (genus *Mytilus*). Mol Ecol 20:517–529
Mahony CR, Cannon AJ, Wang TL, Aitken SN (2017) A closer look at novel climates: new methods and insights at continental to landscape scales. Glob Chang Biol 23:3934–3955
Marshall WS, Grosell M (2005) Ion transport, osmoregulation, and acid-base balance. In: Evans DH, Claiborne JB (eds) Physiology of fishes. CRC Press, Boca Raton
Martins EG, Hinch SG, Patterson DA, Hague MJ, Cooke SJ, Miller KM, Lapointe MF, English KK, Farrell AP (2011) Effects of river temperature and climate warming on stock-specific survival of adult migrating Fraser River sockeye salmon (*Oncorhynchus nerka*). Glob Chang Biol 17:99–114
Melzner F, Gobel S, Langenbuch M, Gutowska MA, Portner HO, Lucassen M (2009) Swimming performance in Atlantic Cod (*Gadus morhua*) following long-term (4–12 months) acclimation to elevated seawater PCO_2. Aquat Toxicol 92:30–37
Melzner F, Gutowska MA, Langenbuch M, Dupont S, Lucassen M, Thorndyke MC, Bleich M, Pörtner HO (2009) Physiological basis for high CO_2 tolerance in marine ectothermic animals: pre-adaptation through lifestyle and ontogeny? Biogeosciences 6:2313–2331

Mu J, Jin F, Wang JY, Zheng N, Cong Y (2015) Effects of CO_2-driven ocean acidification on early life stages of marine medaka (*Oryzias melastigma*). Biogeosciences 12(12):3861–3868. https://doi.org/10.5194/Bg-12-3861-2015

Munday PL, Crawley NE, Nilsson GE (2009) Interacting effects of elevated temperature and ocean acidification on the aerobic performance of coral reef fishes. Mar Ecol Prog Ser 388:235–242

Munday PL, Dixson DL, Donelson JM, Jones GP, Pratchett MS, Devitsina GV, Doving KB (2009) Ocean acidification impairs olfactory discrimination and homing ability of a marine fish. Proc Natl Acad Sci 106:1848–1852

Murray CS, Fuiman LA, Baumann H (2016) Consequences of elevated CO_2 exposure across multiple life stages in a coastal forage fish. ICES J Mar Sci 74:1051

Mykles DL, Ghalambor CK, Stillman JH, Tomanek L (2010) Grand challenges in comparative physiology: integration across disciplines and across levels of biological organization. Integr Comp Biol 50:6–16

Nilsson GE, Dixson DL, Domenici P, Mccormick MI, Sorensen C, Watson S-A, Munday PL (2012) Near-future carbon dioxide levels alter fish behaviour by interfering with neurotransmitter function. Nat Clim Chang 2:201–204

Nordlie FG (2006) Physicochemical environments and tolerances of cyprinodontoid fishes found in estuaries and salt marshes of eastern North America. Rev Fish Biol Fish 16:51–106

Norin T, Malte H, Clark TD (2014) Aerobic scope does not predict the performance of a tropical eurythermal fish at elevated temperatures. J Exp Biol 217:244–251

Orr JC, Fabry VJ, Aumont O, Bopp L, Doney SC, Feely RA, Gnanadesikan A, Gruber N, Ishida A, Joos F, Key RM, Lindsay K, Maier-Reimer E, Matear R, Monfray P, Mouchet A, Najjar RG, Plattner G-K, Rodgers KB, Sabine CL, Sarmiento JL, Schlitzer R, Slater RD, Totterdell IJ, Weirig M-F, Yamanaka Y, Yool A (2005) Anthropogenic ocean acidification over the twenty-first century and its impact on calcifying organisms. Nature 437:681–686

Payan P, Edeyer A, De Pontual H, Borelli G, Boeuf G, Mayer-Gostan N (1999) Chemical composition of saccular endolymph and otolith in fish inner ear: lack of spatial uniformity. Am J Phys Regul Integr Comp Phys 277:R123–R131

Payan P, Kossmann H, Watrin A, Mayer-Gostan N, Boeuf G (1997) Ionic composition of endolymph in teleosts: origin and importance of endolymph alkalinity. J Exp Biol 200:1905–1912

Pelster B (2002) Developmental plasticity in the cardiovascular system of fish, with special reference to the zebrafish. Comp Biochem Physiol A Mol Integr Physiol 133:547–553

Perry SF, Gilmour KM (2006) Acid-base balance and CO_2 excretion in fish: unanswered questions and emerging models. Respir Physiol Neurobiol 154:199–215

Pimentel MS, Faleiro F, Dionisio G, Repolho T, Pousao-Ferreira P, Machado J, Rosa R (2014) Defective skeletogenesis and oversized otoliths in fish early stages in a changing ocean. J Exp Biol 217:2062–2070

Portner HO (2010) Oxygen- and capacity-limitation of thermal tolerance: a matrix for integrating climate-related stressor effects in marine ecosystems. J Exp Biol 213:881–893

Pörtner H-O, Bock C, Mark FC (2017) Oxygen- and capacity-limited thermal tolerance: bridging ecology and physiology. J Exp Biol 220:2685–2696

Pörtner HO, Farrell AP (2008) Ecology physiology and climate change. Science 322:690–692

Portner HO, Farrell AP, Knust R, Lannig G, Mark FC, Storch D (2009) Adapting to climate change response. Science 323:876–877

Portner HO, Knust R (2007) Climate change affects marine fishes through the oxygen limitation of thermal tolerance. Science 315:95–97

Pörtner HO, Langenbuch M, Michaelidis B (2005) Synergistic effects of temperature extremes, hypoxia, and increases in CO_2 on marine animals: from earth history to global change. J Geophys Res Oceans 110:C09s10

Portner HO, Peck MA (2010) Climate change effects on fishes and fisheries: towards a cause-and-effect understanding. J Fish Biol 77:1745–1779

Przeslawski R, Byrne M, Mellin C (2015) A review and meta-analysis of the effects of multiple abiotic stressors on marine embryos and larvae. Glob Chang Biol 21:2122–2140

Rankin PS, Hannah RW, Blume MTO (2013) Effect of hypoxia on rockfish movements: implications for understanding the roles of temperature, toxins and site fidelity. Mar Ecol Prog Ser 492:223–234

Regan MD, Turko AJ, Heras J, Andersen MK, Lefevre S, Wang T, Bayley M, Brauner CJ, Huong DTT, Phuong NT, Nilsson GE (2016) Ambient CO_2, fish behaviour and altered GABAergic neurotransmission: exploring the mechanism of CO_2-altered behaviour by taking a hypercapnia dweller down to low CO_2 levels. J Exp Biol 219:109–118

Reid NM, Proestou DA, Clark BW, Warren WC, Colbourne JK, Shaw JR, Karchner SI, Hahn ME, Nacci D, Oleksiak MF, Crawford DL, Whitehead A (2016) The genomic landscape of rapid repeated evolutionary adaptation to toxic pollution in wild fish. Science 354:1305–1308

Richards JG (2011) Physiological, behavioral and biochemical adaptations of intertidal fishes to hypoxia. J Exp Biol 214:191–199

Ries JB (2010) Geological and experimental evidence for secular variation in seawater Mg/Ca (calcite-aragonite seas) and its effects on marine biological calcification. Biogeosciences 7:2795–2849

Rogers NJ, Urbina MA, Reardon EE, Mckenzie DJ, Wilson RW (2016) A new analysis of hypoxia tolerance in fishes using a database of critical oxygen level (P_{crit}). Conserv Physiol 4:Cow012

Rombough P (2007) The functional ontogeny of the teleost gill: which comes first, gas or ion exchange? Comp Biochem Physiol A Mol Integr Physiol 148:732–742

Scavia D, Bertani I, Obenour DR, Turner RE, Forrest DR, Katin A (2017) Ensemble modeling informs hypoxia management in the northern Gulf of Mexico. Proc Natl Acad Sci 114:8823–8828

Schade FM, Clemmesen C, Wegner KM (2014) Within- and transgenerational effects of ocean acidification on life history of marine three-spined stickleback (*Gasterosteus aculeatus*). Mar Biol 161:1667–1676

Schloss CA, Nuñez TA, Lawler JJ (2012) Dispersal will limit ability of mammals to track climate change in the western hemisphere. Proc Natl Acad Sci 109:8606–8611

Schulte PM (2014) What is environmental stress? Insights from fish living in a variable environment. J Exp Biol 217:23–34

Schwenk K, Padilla DK, Bakken GS, Full RJ (2009) Grand challenges in organismal biology. Integr Comp Biol 49:7–14

Scott GR, Rogers JT, Richards JG, Wood CA, Schulte PM (2004) Intraspecific divergence of ionoregulatory physiology in the euryhaline teleost *Fundulus heteroclitus*: possible mechanisms of freshwater adaptation. J Exp Biol 207:3399–3410

Seibel BA, Maas AE, Dierssen HM (2012) Energetic plasticity underlies a variable response to ocean acidification in the pteropod, *Limacina helicina antarctica*. PLoS One 7:E30464

Selye H (1936) A syndrome produced by diverse nocuous agents. Nature 138:32–33

Simpson SD, Munday PL, Wittenrich ML, Manassa R, Dixson DL, Gagliano M, Yan HY (2011) Ocean acidification erodes crucial auditory behaviour in a marine fish. Biol Lett 7:917–920

Somero GN (2010) The physiology of climate change: how potentials for acclimatization and genetic adaptation will determine 'winners' and 'losers'. J Exp Biol 213:912–920

Somero GN (2011) Comparative physiology: a "crystal ball" for predicting consequences of global change. Am J Phys Regul Integr Comp Phys 301:R1–R14

Stillman JH (2003) Acclimation capacity underlies susceptibility to climate change. Science 301:65–65

Stillman JH, Somero GN (2000) A comparative analysis of the upper thermal tolerance limits of eastern Pacific porcelain crabs, genus *Petrolisthes*: influences of latitude, vertical zonation, acclimation, and phylogeny. Physiol Biochem Zool 73:200–208

Talmage SC, Gobler CJ (2010) Effects of past, present, and future ocean carbon dioxide concentrations on the growth and survival of larval shellfish. Proc Natl Acad Sci 107:17246–17251

Tomanek L, Zuzow MJ (2010) The proteomic response of the mussel congeners *Mytilus galloprovincialis* and *M. trossulus* to acute heat stress: implications for thermal tolerance limits and metabolic costs of thermal stress. J Exp Biol 213:3559–3574

Tresguerres M, Hamilton TJ (2017) Acid–base physiology, neurobiology and behaviour in relation to CO_2-induced ocean acidification. J Exp Biol 220:2136–2148
Uthicke S, Momigliano P, Fabricius KE (2013) High risk of extinction of benthic foraminifera in this century due to ocean acidification. Sci Rep 3:1769
Vaquer-Sunyer R, Duarte CM (2008) Thresholds of hypoxia for marine biodiversity. Proc Natl Acad Sci 105:15452–15457
Virani NA, Rees BB (2000) Oxygen consumption, blood lactate and inter-individual variation in the gulf killifish, *Fundulus grandis*, during hypoxia and recovery. Comp Biochem Physiol A Physiol 126:397–405
Whitehead A (2010) The evolutionary radiation of diverse osmotolerant physiologies in killifish (*Fundulus* sp.) Evolution 64:2070–2085
Whitehead A, Galvez F, Zhang SJ, Williams LM, Oleksiak MF (2011) Functional genomics of physiological plasticity and local adaptation in killifish. J Hered 102:499–511
Whitehead A, Pilcher W, Champlin D, Nacci D (2012) Common mechanism underlies repeated evolution of extreme pollution tolerance. Proc R Soc Lond B Biol Sci 279:427–433
Whitehead A, Roach JL, Zhang S, Galvez F (2011) Genomic mechanisms of evolved physiological plasticity in killifish distributed along an environmental salinity gradient. Proc Natl Acad Sci 108:6193–6198
Whitehead A, Roach JL, Zhang SJ, Galvez F (2012) Salinity- and population-dependent genome regulatory response during osmotic acclimation in the killifish (*Fundulus heteroclitus*) gill. J Exp Biol 215:1293–1305
Whitehead A, Zhang S, Roach JL, Galvez F (2013) Common functional targets of adaptive micro- and macro-evolutionary divergence in killifish. Mol Ecol 22:3780–3796
Whiteley NM (2011) Physiological and ecological responses of crustaceans to ocean acidification. Mar Ecol Prog Ser 430:257
Williams DA, Brown SD, Crawford DL (2008) Contemporary and historical influences on the genetic structure of the estuarine-dependent gulf killifish *Fundulus grandis*. Mar Ecol Prog Ser 373:111–121
Wolanski E (2007) Estuarine ecohydrology. Elsevier, Amsterdam

Toxicity in Aquatic Environments: The Cocktail Effect

9

Andrew J. Esbaugh, Alexis Khursigara, and Jacob Johansen

Abstract

Anthropogenically sourced chemicals are commonly released into aquatic habitats where they can have a multitude of detrimental effects on organisms and ecosystems. Aquatic toxicants typically exert impacts through perturbation of physiological function; however, these effects can vary greatly depending on the life stage and environment of the organism. Using a case study approach, this chapter will explore the influence of life stage and environment on observed toxicity, with a particular emphasis on the physiological underpinnings that result in toxicity. This will include examining how toxicity can vary across developmental stages and environments and also how sublethal toxicity can impact the ability of organisms to perform and cope with typical environmental stressors, such as hypoxia. We will first examine how aquatic toxicants can have very different effects and sensitivities depending on the life stage of the affected organism using examples of neuroendocrine disruption in fish and amphibians. Such chemicals can have a suite of impacts including altered developmental progression, sexual differentiation, and reproduction, as well as disrupting the basic function of the endocrine stress axis. Through the lens of the biotic ligand model, we will also explore how specific environmental characteristics can impact metal toxicity and how these factors can dramatically alter the fitness consequences to a wide array of aquatic organisms. Finally, we will use polycyclic aromatic hydrocarbon toxicity to demonstrate that even transient exposure to toxicants at various life stages can have persistent physiological consequences that can hamper an organism's performance and ability to cope with environmental challenges. Specifically, we will use embryonic fish models to demonstrate how

A. J. Esbaugh (✉) · A. Khursigara · J. Johansen
Department of Marine Science, University of Texas at Austin, Austin, TX, USA
e-mail: A.Esbaugh@austin.utexas.edu

© Springer International Publishing AG, part of Springer Nature 2018

W. Burggren, B. Dubansky (eds.), *Development and Environment*,
https://doi.org/10.1007/978-3-319-75935-7_9

transient exposure to polycyclic aromatic hydrocarbons can result in cardiorespiratory perturbations that can have long-lasting physiological consequences for individuals that range from mortality to reduced physiological performance.

9.1 Introduction

Anthropogenic-driven contamination of aquatic habitats is a common occurrence worldwide, with impacted environments ranging from isolated streams and lakes to wide swaths of coastal and offshore ocean habitat. The sources of toxicants that enter aquatic ecosystems are varied in scope but among others include industrial by-products, mining and refining effluents, agricultural and urban runoff, as well as point-source events such as oil spills. Regardless of the source, the overarching objective of aquatic toxicology research is to understand the ecological consequences of toxicants on aquatic species with the ultimate goal of providing sound information to guide regulatory decisionmaking. The diversity of aquatic environments, both with respect to the physical characteristics and the biological composition, can make this challenging. A recently proposed mechanistic framework—the adverse outcome pathway (Ankley et al. 2010; Kramer et al. 2011)—provides a useful conceptual tool to connect chemical characteristics of toxicants to ecological risk at the population level. This framework begins with the basic toxicity pathway that describes the interaction of chemicals with cellular macromolecules that ultimately leads to a cellular response. These responses are diverse but typically involve activation or depletion of proteins through gene regulation, signaling cascades, or other posttranslational processes. The second phase of this framework extends cellular responses to organs and ultimately to an organismal response, such as lethality or impaired performance, owing to altered physiology or disrupted homeostasis. In severe cases individual responses can be extended to population or species levels. The role of organismal and cellular physiology in toxicity is clear within the context of this framework. Unique physiological characteristics may make a species particularly sensitive to a given toxicant that targets such pathways. Conversely, some species may have the physiological capacity to detoxify certain chemicals and thereby mitigate adverse outcomes.

The role the environment plays in determining an organism's adverse outcome pathway is not as obvious; however, as this chapter will discuss, the environmental factors that accompany toxicant exposure can have a profound role in determining the ultimate fate of exposed organisms. The first important consideration is that the environment can impact the chemical properties of the toxicant itself and thereby impact bioaccumulation and other macromolecular interactions inherent to toxicity pathways. In this way, some environments can be described as protective, while others can be highlighted as at risk. A second important environmental consideration is the concept of additive or synergistic effects that magnify the outcome of toxicant exposure alone. This influence can be of the steady-state variety, for example, a toxicant that inhibits ion transport in organisms living in an ion-poor environment. Similarly, non-steady-state style stressors, such as hypoxia or thermal stress as well as physiological stresses related to exercise and social interactions, can exacerbate the effects of a toxicant. This is particularly relevant for sublethal impacts as the additive effects of toxicant and environmental stress can greatly enhance the potential for ecological death. This latter concept is the premise that animals unable to successfully compete and reproduce in an environment are effectively dead from a fitness perspective.

On this background, the purpose of this chapter is not to provide a comprehensive review of the field of aquatic toxicology, which would be a monumental task. Instead, in the spirit of this book series, we will attempt to highlight the effect that an organism's environment, physiology, and life stage can have on adverse outcomes stemming from toxicant exposure. We will first briefly explain the concepts of bioconcentration through biomagnification, after which we will explore aspects of three distinct classes of toxicants: (1) metals, (2) oil mixtures, and (3) endocrine disruptors. Each of these cases was selected to highlight particular toxicological occurrences that are related to physiology and the environment. Wherever possible we attempt to highlight subject-specific comprehensive reviews to aid the reader.

9.2 Bioconcentration, Bioaccumulation, and Biomagnification

Bioconcentration refers to the tendency of a chemical to accumulate within the tissues of an organism at a higher concentration than found in the aquatic environment. This is defined using a bioconcentration factor (Chapman et al. 1996), which is simply the relative increase of a chemical within the body compared to that in the environment. Bioaccumulation is similar but refers to concentration relative to both the water and diet (Chapman et al. 1996) and therefore has a stronger ecological basis. Aquatic toxicants have dramatically different profiles with respect to bioaccumulation. Many organic pollutants bioaccumulate independent of the environmental concentrations, while metal bioaccumulation factors are typically inversely related to environmental concentration (McGeer et al. 2003; DeForest et al. 2007). This is due in large part to the lipophilicity of organic pollutants, which results in a relatively consistent accumulation based on their respective octanol-water partitioning coefficients (Di Toro et al. 2000; Kipka and Di Toro 2009). In fact, bioconcentration factors are combined with relative species sensitivities within the target lipid model to predict toxicity of organisms to a complex mixture of narcotic chemicals, with the assumption that mortality results when the total concentration of narcotics exceed a species' tolerance threshold (Di Toro and McGrath 2000; Di Toro et al. 2000, 2007). Bioaccumulation can also be defined based on specific tissues, as chemical concentration is often heterogeneous across organs and body compartments. For instance, many metals will be preferentially stored to specific tissues, with the liver being a relatively common target of bioaccumulation (e.g., Cu, Cd) (McGeer et al. 2011; Grosell 2011). Of course, this is highly variable, especially with respect to essential metals, such as Zn, that show tissue-specific bioaccumulation in the muscle, bone, and skin (Hogstrand 2011).

An important aspect of bioaccumulation is the balance between uptake, storage, and clearance, which strongly relates to the inherent physiological pathways of the respective organisms, as well as the toxicant in question. Upon uptake many metals are detoxified via cellular pathways that effectively limit the intracellular free metal ion concentration. The two most common such pathways are metallothionein (Mao et al. 2012) and glutathione (Hernandez et al. 2015), which represent protein and nonprotein reservoirs for metal chelation. In the case of some organic pollutants, such as PAHs, chemicals can metabolize to enhance excretion; however, the physiological capacity for metabolism is very different across species. Many xenobiotics are metabolized through the AHR-mediated induction of cytochrome p450, or CYP1A, which makes the chemicals more hydrophilic through oxidation and thereby enhances their excretion (Mimura and Fujii-Kuriyama 2003; Denison and Nagy 2003; Zhou et al. 2010). An important trade-off in this system is the production of reactive oxygen species, which are capable of cellular damage on their own and have been suggested as the cause of embryonic cardiac impairment following exposure to 4-ring PAHs. Similarly, oxidation can also activate certain chemicals and thereby enhance toxicity. While the AHR and CYP1A pathways are evolutionarily conserved across many animal groups, the actual ability to metabolize PAHs and other petroleum hydrocarbons is limited to vertebrates, crustaceans, and polychaetes (Zhou et al. 2010). Organisms with no such capacity, such as bivalves (Rebelo et al. 2003), therefore exhibit more prolonged bioaccumulation. With respect to excretion, information is limited and pertains mostly to fishes. The generally high levels of metals and xenobiotic metabolites observed in the bile relative to the plasma infer a hepatobiliary excretion route (Lee and Anderson 2005); however, bile storage does not necessarily equate to significant bile secretion (Grosell et al. 2000). Nonetheless, there is little evidence for urinary and branchial excretion of toxicants in fishes (see review Wood 2011).

Biomagnification refers to the tendency of toxicant concentration to increase in animals of higher trophic levels and has become particularly relevant in terms of human health through dietary exposure. For the most part, biomagnification is most relevant for toxicants that are not effectively broken down by exposed organisms or the natural environment and presumably can be absorbed through dietary exposure. Three of the most well-known examples are the lipophilic persistent organic pollutant polychlorinated biphenyls (PCBs), Hg, and Se (see reviews Hamilton 2004; Kidd and Batchelar 2011; Letcher et al. 2010). All three of these pollutants have been shown to biomagnify across trophic levels in both marine and freshwater food webs. An important aspect of biomagnification is that higher trophic levels may be more sensitive than lower trophic levels to a given toxicant, and therefore any bioaccumulation in lower trophic levels can make seemingly low water-borne concentrations more dangerous. Such is the case for Se whereby fish are much more sensitive than their prey items (see review Janz 2011). This makes the ecological aspects of an organism particularly relevant as discrete feeding patterns may dramatically affect observed toxicity.

9.3 Metals in the Aquatic Environment

Metals are a broadly defined group that consists of the vast majority of naturally occurring elements. Owing to their many beneficial characteristics – malleability when heated, electrical and heat conduction, and ability to blend into alloys – they are prized for many industrial applications. The universal exploitation of metals means that these elements will inevitably be dispersed back into the environment at unnaturally high concentrations. In fact, anthropogenic sources drive metal flux into the environment at rates that greatly exceed natural rates of renewal (Rauch 2009), making metals nearly ubiquitous potential pollutants in aquatic environments and a major focus of regulatory effort. Importantly, scientific advances relating environmental factors and organismal physiology to observed toxicity have greatly improved the ability of regulatory agencies to perform proper risk assessment. These advances are perfectly exemplified by the biotic ligand model framework (Di Toro et al. 2001; Paquin et al. 2002, 2011), which is usurping the roles of traditional regulatory tools for the purposes of risk assessment. The underlying premise of this framework is that toxicity is a combination of the bioavailability of particular metals given the environment and the physiological perturbation of the given organism (Fig. 9.1). While the biotic ligand model is far from the only risk assessment model in use for metals (Paquin et al. 2011), its mechanistic nature and integration of environmental factors and organismal physiology make it an ideal framework to approach metal toxicology for the purposes of this chapter.

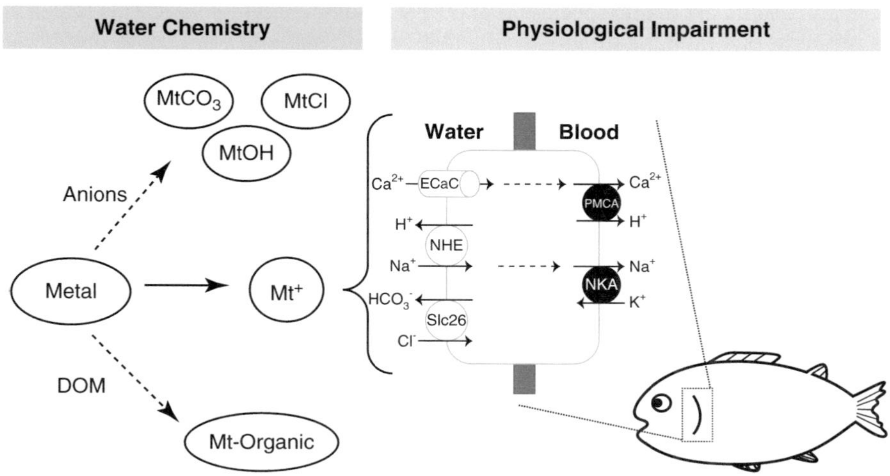

Fig. 9.1 Diagrammatic representation of the biotic ligand model that is used to predict metal toxicity in aquatic organisms. The left portion of the figure represents the influence that water chemistry components in the environment (dissolved organic matter, chlorides, hydroxides, carbonates, and sulfides) can have on metal speciation. In this representation, the bioavailable species is the ionic metal form. The bioavailable species subsequently reacts with the biotic ligand (organism), which manifests in a physiological impairment, represented in this case by the ion transport pathways in the fish gill. Modified from Paquin et al. (2002)

9.3.1 The Role of Environmental Chemistry in Determining Species Sensitivity

One underlying premise of the biotic ligand model is that metals in the environmental exist in a variety of chemical species, and only a specific subset of these chemical species are bioavailable and drive toxicity. The importance of this concept, known as complexation, is illustrated in Table 9.1. In this example, an equal amount of Pb (100 μg l^{-1}) is found in environments with substantially different chemistries: (1) an alkaline hard water environment with low organic loading, (2) a soft water environment with low organic loading, and (3) a soft water environment with high organic loading. We can see that the chemical speciation of Pb is very different across the three environments, and if we assume the ionic Pb^{2+} species drives toxicity, it is clear that the same overall Pb concentration will have very different effects. In scenario 1, only 0.5% of the total Pb load remains as bioavailable Pb^{2+} while 21.7% is bound by carbonate. In contrast, scenario 2 has very little carbonate complexation, owing to the low pH, and therefore has nearly 20x more bioavailable Pb^{2+}. The majority of the remaining Pb in both scenarios is associated with organic material, which is typical of most environments. Also notable is the relative increase in organic Pb speciation in scenario 2, as this illustrates the competitive aspect of complexation. Not only does scenario 2 have less carbonate to bind Pb^{2+}, and thus compete with organic material, but the much lower ionic strength means there will be fewer cations to compete with Pb^{2+} for binding sites on the organic material. A more extreme example of this is shown in scenario 3, where DOC is increased by 100-fold and results in organic speciation of over 99% of the total Pb load. This scenario is meant to highlight the important role that organic material can play in providing protection against metal pollution, as the free Pb^{2+} is reduced by almost 50-fold solely due to a high, yet environmentally realistic, organic load.

Table 9.1 The influence of three hypothetical water chemistry scenarios on metal (Pb) speciation assuming a 100 μg l^{-1} total load

		Scenario #1	Scenario #2	Scenario #3
Water parameters (uM)	Na^+	1250	100	100
	Ca^{2+}	1000	25	25
	Mg^{2+}	1200	25	25
	Cl^-	950	65	65
	SO_4^{2-}	500	20	20
	DIC	2900	70	70
	pH	8.4	6.7	6.7
	DOC	75	75	750
Metal speciation[a] (μg l^{-1})	Pb-organic	73.1	87.8	99.3
	$PbCO_3$	21.7	0.2	0
	Pb^{2+}	0.5	9.7	0.2

[a]Speciation estimates were performed using the freeware program Visual MINTEQ. Dissolved organic matter was assumed to contain 50% carbon by weight and 65% fulvic acid content

This illustrative example holds true for many metals whereby environmental alkalinity, ion concentration, and dissolved organic matter can provide a high level of protection for organisms. Generally, ion-poor waters with limited organic loading are considered the most at-risk habitats for metal pollution as the bioavailable species make up the largest percentage of the total dissolved metal load. However, it is important to not overgeneralize, and metal speciation should be interpreted on a case-by-case basis. Moreover, the concept of complexation makes it imperative that thorough water chemistry is obtained and reported during any study of metal toxicity in aquatic environments. This is especially true in freshwater environments, which can vary substantially in ion content, acidity, and organic loading in different geographic regions. Where possible, it is also beneficial to further characterize dissolved organic matter by defining the relative contribution of fulvic and humic acids, as these have different binding properties. However, in the absence of more specific characterizations, a 65% fulvic acid composition can be assumed based on prior characterization of natural waters (Tipping 2002).

9.3.2 The Interplay Between Metals and Physiology in Aquatic Organisms

The second aspect of the biotic ligand model is the mechanism by which bioavailable metal species interact with an organism, the biotic ligand, to exert a toxicological effect. This aspect of the model relates more directly to the mode of uptake and action and as such can vary depending on the species. Let us again use Pb to illustrate this concept. Pb is a nonessential metal that has no dedicated uptake mechanisms. Instead it has been demonstrated that Pb acts as a calcium mimic and therefore competes with calcium for biotic ligand-binding sites. This is demonstrated by a dose-dependent reduction in calcium uptake rates in the presence of Pb (Rogers and Wood 2004; Grosell and Brix 2009). In fact, the very high sensitivity of snails to chronic Pb exposure, as determined through reduced growth rates, was thought to be driven largely by the high calcium uptake demand in these animals (Grosell and Brix 2009). Similar ionic mimicry with calcium occurs for many other free metal species, including Zn^{2+}, Cd^{2+}, Sr^{2+}, and Co^{2+}. Consequently, higher concentrations of calcium in the environment should provide increased protection against these metals, based on competition for calcium-binding sites. This relationship proves true for Pb toxicity in fathead minnows (*P. promelas*), where the concentration of calcium in the test medium is strongly predictive of the Pb^{2+} concentration required to drive toxicity. Conversely, data from a series of acute and chronic studies on *C. dubia* and *L. stagnalis* show no relationship with observed Pb^{2+} toxicity and environmental calcium. Instead both of these species show strong pH effects. This should serve to illustrate the difficulty of making cross-species comparisons about the mechanisms that drive toxicity and thus the water chemistry components that may influence protection profiles and thus be incorporated into species biotic ligand models.

This basic framework can be more complicated for essential metals, such as Cu and Zn, as these can have dedicated transport pathways. These metals can act both competitively and noncompetitively with water chemistry constituents. Cu toxicity is one such example, whereby Cu accumulation in the gills of rainbow trout has been shown to be competitive with Na^+ for apical entry during the initial 2 h of exposure. However, prolonged exposure results in a mixed competitive and noncompetitive inhibition of Na^+ uptake, with the noncompetitive aspects owing to inhibition of basolateral Na^+ K^+ ATPase activity. In the latter case, excessive Cu^{2+} uptake can occur through the normal apical routes such as the divalent metal transporter (dmt1) or the copper transporter (ctr1). Essential metals can also be more complex owing to the greater complexity of internal homeostatic regulatory mechanisms. While the internal handling pathways for nonessential metals are largely limited to simple depuration, essential metals have regulatory mechanisms that tightly control cellular concentrations. For example, Zn uptake is inversely related to the concentration of Zn in the environment because low levels stimulate increased expression and activity of a variety of Zn transporters. It is therefore important to consider the role of acclimation during studies of toxicity, which in the toxicological literature is typically defined as an increase in lethal resistance as a consequence of sublethal exposure. Similarly, toxicological impairments can also be overcome through the inherent plasticity of the impaired physiological system. It is well established that chronic Cu exposure stimulates Na^+ K^+ ATPase expression, which acts to offset the impaired Na^+ homeostasis mentioned above. Chronic Cu exposure also increased the number of mitochondrial-rich cells found in the gills, which can further serve to correct osmoregulatory disturbances. Similar ionoregulatory compensation mechanisms in fish have been observed in response to Pb, Al, and Ag, among others. An important consideration on this point is the inherent level of plasticity within the given physiological systems. Ionoregulatory compensation in response to environmental stress is common in fishes and other osmoregulating organisms; however, behavioral, reproductive, and neurological impairments are much less likely to exhibit recovery with chronic exposure.

9.3.3 Life Stage and Life Cycle Considerations

To this point we have discussed metal toxicology mostly in the framework of survival, which is obviously an important endpoint owing to its clear ecological relevance. Almost without exception early life stage organisms are the most sensitive from a survival perspective (e.g., Calfee et al. 2014), and this can typically extend to a multitude of other endpoints. This is also clear from the life stage recommendations within the US Environmental Protection Agency methods for acute and chronic toxicity testing of effluents and receiving waters, which range from $\leq$24 h for daphnia to just after yolk absorption (15–60 days) in salmonids (USEPA 2002a, b). The underlying explanation for the sensitivity of early life organisms is likely multifaceted but has much to do with the surface to volume challenges that govern many aspects of early life physiology. The rate constants of metal absorption from water across a variety of aquatic organisms are known to decrease with increasing size (Hendriks and Heikens 2001), which increases bioaccumulation and thus sensitivity in smaller organisms. The physiological aspects of allometric scaling likely also play a role, as the high surface to volume ratios result in higher osmoregulatory pressures (Gibbs and Somero 1990; Zanotto et al. 2004) and mass-specific metabolic rate (e.g., Pan et al. 2016). As discussed above, metals can disrupt both ionoregulatory and respiratory function and as such would exert increased stress upon early life organisms. In fact, differential ionoregulatory disturbances were specifically highlighted as a mechanistic underpinning of increased sensitivity of freshwater daphnia to Cu and Ag relative to freshwater fish (Grosell et al. 2002).

While acute toxicity is typically assessed through survival, or a suitable equivalent, chronic toxicity is often tested on an ecologically relevant sublethal endpoint. In most cases this is growth or reproduction; however, other sublethal sensitive endpoints have also been assessed (e.g., USEPA 2002b). From an ecological standpoint, reproduction is clearly the most pertinent parameter, and the documented disruption of endocrine function by metals can have clear reproductive implications. For example, many metals have been shown to disrupt gametogenesis and embryogenesis; but these effects can be the end result of chronic exposure over a lifetime. As such, it remains best practice to begin chronic exposures at similar life stages as recommended for acute exposures. While full life cycle exposures are easy for invertebrates, such as rotifers or daphnia, it is much more challenging for species slow to reach sexual maturity. In these cases, growth is used as a surrogate. However, a common complaint is that this endpoint overestimates toxicity as there is potential for slow-growing organisms to catch up at later stages or maintain reproductive output despite size differences. This was an especially compelling argument when assessing the relevance of reduced snail growth (*L. stagnalis*), which is the most sensitive endpoint of some metals and as such should drive most water quality criteria determinations (Grosell et al. 2006; Schlekat et al. 2010; De Schamphelaere et al. 2008). Snail growth is typically assessed over a 14–28-day period that occurs immediately after the animals emerge from their egg casing and start feeding on solid food (7–10 days post hatch) (Grosell et al. 2006). Note that 14- and 28-day growth tests showed similar results for Pb exposure (Esbaugh et al. 2012). Importantly, a recent study using Pb demonstrated that while partial growth recovery did occur in snails between 28 and 56 days, the sensitivity of 28-day growth and cumulative reproductive output was similar (Munley et al. 2013). As such, the authors argued that partial life cycle growth tests were reasonably predictive of full life cycle tests, which is analogous to fish where partial life cycle tests are commonly used to assess chronic toxicity.

9.4 The Toxicity of Oil Mixtures

With the ever-increasing global demand for energy, crude oil and related petroleum-based products are an omnipresent environmental toxicant of concern. The chemical constituents of crude oil can enter aquatic ecosystems in a variety of ways, ranging from natural oil seep to anthropogenic sources that include wood treatment, refining, oil spills, and urban runoff. The US Bureau of Energy Management (BOEM) estimates that within the last 40 years, over 850 million liters of crude oil have been released into oceans as a result of over 300 major oil spills associated with US Outer Continental Shelf exploration alone (Anderson et al. 2012); however, the vast majority of this was associated with the 2010 Deepwater Horizon oil spill. Oil by nature is a complex mixture that includes many classes of chemicals, but polycyclic aromatic hydrocarbons (PAHs) have been identified as the major toxicant driving toxicity in aquatic environments (Di Toro and McGrath 2000; Di Toro et al. 2000). These chemicals are divided into specific subclasses based on the number of benzene rings, and the specific composition can provide both a unique identifier that allows oil to be traced to back to its source, as well as a unique toxicological profile. This is further complicated because many constituents in oil are not readily soluble in water and therefore the total PAH (ΣPAH) load can exist in both a dissolved and non-dissolved microdroplet state. While the vast majority of the ΣPAH load is found in microdroplets, the prevailing evidence suggests that dissolved components drive toxicity (Carls et al. 2008).

9.4.1 The Influence of Environmental Factors on Oil Toxicity

As mentioned above, the chemical composition of oil can be highly variable between geographically disparate sites, which can dramatically influence the overall toxicity of oil water mixtures. While environmental factors do not influence oil toxicity in a manner equivalent to metal speciation (see Sect. 9.3.1), there are a number of considerations that pertain specifically to the impact on oil spills. These will only be discussed briefly here, but we refer you to Chap. 10 in this volume for more details related specifically to the Deepwater Horizon oil spill. The influence of the environment on oil composition following a spill is collectively termed weathering and can substantially influence effective toxicity. To generalize, weathering alters the composition of oil-water mixtures by selectively removing lightweight chemical constituents, which effectively concentrates higher weight constituents through evaporation, dissolution, and biodegradation (Liu et al. 2012). The toxicology characteristics of the PAH subclasses, and even within a subclass, can vary (e.g., Incardona et al. 2004; Di Toro et al. 2000), which makes the study of oil toxicology highly complex.

The ultimate influence of weathering processes on oil toxicity is currently of debate and likely depends on the specific organism of study. The target lipid model proposes that oil toxicity is the net result of bioaccumulation of all PAHs until a threshold is exceeded, which results in mortality (Di Toro and McGrath 2000; Di Toro et al. 2000). Under this framework, non-weathered oils should be more toxic because lightweight constituents more easily bioaccumulate. An advancement of this framework, the toxic potential model (Di Toro et al. 2007), combines water-octanol coefficients (a measure of lipid solubility) with overall water solubility to assess the toxic potential of oil mixtures. As above, lower molecular weight constituents are also more soluble in water, which infers greater toxicity of non-weathered oils rich in low molecular weight PAHs. Nonetheless, toxicity testing in fish has routinely demonstrated the opposite using survival assays. In mahi-mahi, Pacific herring, and pink salmon, weathered oil was significantly more toxic to early life stages (Esbaugh et al. 2016; Carls et al. 1999; Heintz et al. 1999). The proposed explanation for the increased toxicity of weathered oil is simply that toxicity in fish is not the result of net bioaccumulation of PAHs but instead is influenced more strongly by 3-ring PAHs. This is supported by single-compound studies on zebrafish that have shown that these compounds—specifically phenanthrene, dibenzothiophene, and fluorene—are sufficient to create a toxic phenotype, while similar concentrations of naphthalene (a 2-ring PAH) have no effect (Incardona et al. 2004). Surprisingly, recent studies on red drum (*Sciaenops ocellatus*) found no difference in toxicity between weathered and non-weathered oils (Khursigara et al. 2017), suggesting there is still more to be resolved regarding the influence of weathering on marine biota.

Another important environmental aspect to consider is the degree to which PAHs are exposed to ultraviolet (UV) light. Many PAHs are known photodynamic chemicals whereby exposure to UV light leads to an increase in toxicity (Newsted and Giesy 1987)—a phenomenon known as photoinduced toxicity. The influence of UV light on PAH toxicity is well documented (Oris and Giesy 1985, 1987; Barron et al. 2003; Alloy et al. 2015) and can be dramatic, with some studies on fish embryos showing an increase in observed toxicity by approximately an order of magnitude (e.g., Alloy et al. 2016; Stieglitz et al. 2016). The ultimate influence of UV light on toxicity is a product of the intensity, duration, specific UV wavelengths, as well as the target PAHs, as only a subset are photoactive. As such, factors such as water turbidity and the depth that an organism inhabits can influence the extra risk associated with UV. With this in mind, a particular concern for risk assessment are newly fertilized fish embryos. In many cases these sensitive early life stages will be buoyant (e.g., pelagic fishes) or deposited in shallow near shore environments, both of which can lead to high-intensity UV exposure. It is also noteworthy that while embryonic survival is the most commonly assessed endpoint with respect to photoinduced toxicity, a number of other endpoints also show photoinduced toxicity in invertebrates including fecundity (Holst and Giesy 1989) and feeding behaviors (Hatch and Burton 1999).

9.4.2 Sensitivity of Early Life Stages

As mentioned previously in our discussion of metal toxicity, early life stages are generally the most sensitive to toxicant exposure, and this can be extended to oil and PAH toxicity. There are a number of examples in the literature demonstrating high survival sensitivity of larval fish using water-accommodated fraction, gravel column, and sediment exposures. A non-exhaustive list includes reduced hatching success in Gulf killifish when exposed to oiled sediments (Dubansky et al. 2013;

Whitehead et al. 2012), dose-dependent mortality in Pacific herring (Carls et al. 1999) and pink salmon (Heintz et al. 1999) beginning at 8 μg l^{-1} and >45 μg l^{-1}, respectively, as well as recent LC_{50}s ranging between 12.3 μg l^{-1} and 14.6 μg l^{-1} mahi-mahi (Esbaugh et al. 2016) and red drum (Khursigara et al. 2017). While it can be difficult to quantify the magnitude of environmental contamination following severe oil spill events, these concentrations are well within environmental relevance for such events.

Oil exposure also results in a characteristic suite of defects in embryonic and larval fish that are collectively termed cardiotoxicity. This is generally defined by pronounced fluid accumulation (edema) in the pericardial cavity (Fig. 9.2). This has been shown to occur at ΣPAH concentrations below 15 μg l^{-1} in bluefin tuna, yellowfin tuna, amberjack (Incardona et al. 2014), red drum (Khursigara et al. 2017), mahi-mahi (Esbaugh et al. 2016), pink salmon (Heintz et al. 1999), and Pacific herring (Carls et al. 1999). This general toxic phenotype can be associated with a

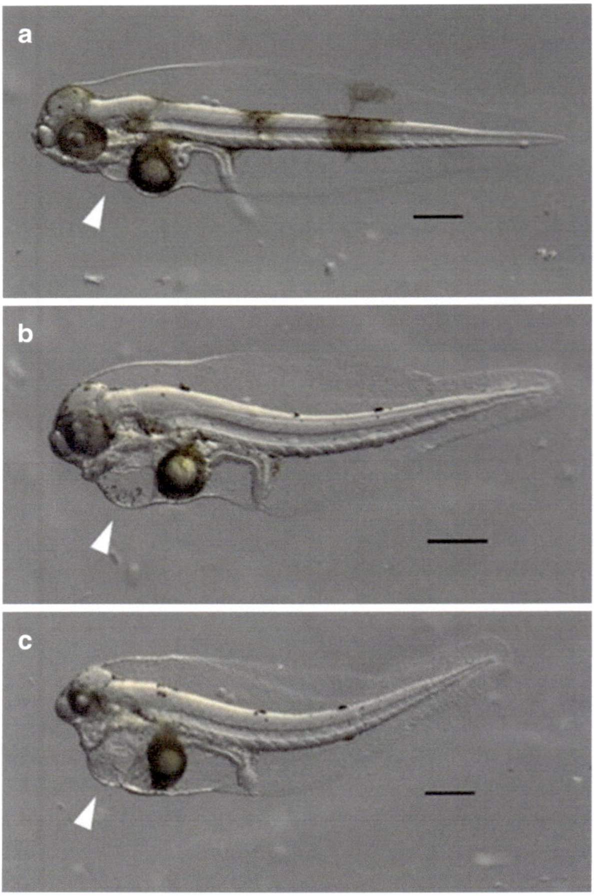

Fig. 9.2 Representative dose response of cardiac malformation in embryonic fish larvae. Images show 48 h postfertilization red drum (*S. ocellatus*) exposed control (**a**), 1.8 (**b**), and 2.6 (**c**) μg l^{-1} ΣPAH50 via a high-energy oil water accommodated fraction protocol. Note the progressive increase in pericardial area (arrowheads) with increasing oil dose, as well as the increased degree of spinal curvature and craniofacial malformation. Scale bars = 250 μm. Images courtesy of Khursigara et al. (2017)

collection of other morphological deformities such as craniofacial malformations, lack of fin ray precursors, spinal curvature, and failed or delayed swim bladder inflation (Incardona et al. 2004; Incardona et al. 2014; Khursigara et al. 2017). Many of the cardiotoxic and craniofacial deformities are likely associated with impaired cranioneural crest development (De Soysa et al. 2012).

Pericardial edema has also been associated in a number of physiologically relevant impairments to cardiac performance, which include bradycardia, arrhythmia, and reduced contractility (Esbaugh et al. 2016; Incardona et al. 2014; Edmunds et al. 2015; Incardona et al. 2009; Incardona et al. 2004). More recently, investigators have begun to explore the impacts on cardiac output, which is a more complete measure of cardiac function that incorporates both heart rate and stroke volume. Studies on the 5-ring PAH benzo[k]fluoranthene (BkF; 40 μM) demonstrated that the observed pericardial edema was associated with a significant bradycardia with no observable impacts on stroke volume (Incardona et al. 2011). While cardiac output was not assessed on an individual basis, and therefore could not be assessed statistically, the mean change in cardiac parameters suggests a decrease in cardiac output of approximately 20%. Conversely, studies on red drum showed a tight correlation between pericardial edema and stroke volume, with only modest reductions in heart rate. These studies calculated an EC_{50} of 2.2 μg l^{-1} for cardiac output, which is approximately 10 times more sensitive than survival at the equivalent life stage (Khursigara et al. 2017).

9.4.3 Ecophysiological Implications of Oil Exposure

One of the more interesting aspects of oil toxicity, and why we have chosen to highlight it in this chapter, is that acute sublethal exposure events can have a number of ecological implication related to impaired physiological performance. As mentioned in the previous section (Sect. 9.4.2), oil exposure in early life causes reduced cardiac performance, but this impairment is not limited to early life. Acute exposure of subadult mahi-mahi to 9.6 μg l^{-1} resulted in 44, 39, and 52% reductions in stroke volume, cardiac output, and stroke work when assessed using an isolated heart in situ preparation (Nelson et al. 2016). Presumably, this results from disrupted excitation-contraction coupling in the heart, which has been demonstrated using in vitro cardiomyocyte preparations from bluefin and yellowfin tuna (Brette et al. 2014).

The physiological implications of reduced cardiac function are significant for all vertebrates, but the particular nature of the aquatic environment and fish circulatory physiology can make these impacts even more pronounced. Water has much less oxygen than an equivalent volume of air, which means that it is a greater challenge for aquatic organisms to obtain oxygen. This basic challenge has shaped many aspects of fish cardiorespiratory physiology, from the oxygen-based ventilatory drive to the basic morphology of the gills. A second important factor for fish is that many do not have dedicated coronary circulation and instead obtain oxygen via the blood in the respective cardiac chambers. The heart is placed last in the circulatory system relative to the gills and as such receives relatively oxygen-poor blood. This can create a negative cascade effect whereby reduced cardiac function limits oxygen supply capacity, which can feed back on cardiac function. This basic principle is the underlying mechanism of the widely applied oxygen supply capacity of thermal tolerance theory (Portner and Knust 2007), which is used to predict and define ecological outcomes stemming from thermal sensitivity of cardiac function (Portner and Peck 2010). While the ubiquitous application of this theory for climate change research is currently debated (Ern et al. 2016; Grans et al. 2014; Norin et al. 2014; Verberk et al. 2016), the basic framework that extends physiological impairment to ecological risk is of tremendous value and, more importantly, can be directly applied to oil toxicity.

The most obvious implication for cardiac impairment is maximum metabolic rate and, by corollary, maximum swim performance. These are ecologically relevant measures that encompass the full energetic capacity for individuals to perform activities related to resource acquisition and predator avoidance. There is now ample evidence that acute exposure as embryos or juveniles results in impairment in one or both of these parameters. For example, juvenile exposure to oil has been shown to decrease maximum metabolic rate in mahi-mahi (Mager et al. 2014), both maximum and standard metabolic rate in sole (Claireaux et al. 2004), and routine metabolic rate in chub mackerel (Klinger et al. 2015). While the changes in maximum metabolic rate are likely related to cardiac impairment, it has been hypothesized that the changes in routine metabolic rate stem from baseline physiological stress, as indicated by elevated plasma cortisol (Klinger et al. 2015). The observed reductions in maximum metabolic rate are also matched by reduced swim performance, measured either as critical swim speed (U_{crit}) (Mager et al. 2014), time to exhaustion (Mauduit et al. 2016), or swim velocity (Gravato and Guilhermino 2009). Interestingly, the cardiac impairment observed in embryonic fish not only occurs at much lower exposure concentrations but also appears to result in long-lasting damage that continues well past when the animal has been returned to control environments. This is best exemplified by a study on mahi-mahi, where embryos exposed to 1.2 μg l^{-1} ΣPAH50 exhibited a significantly decreased U_{crit} after 25 days of growth and recovery. Acutely exposed individuals of the same size and age didn't exhibit significantly impaired U_{crit} until 30 μg l^{-1} (Mager et al. 2014). These results are consistent with embryonic exposure and recovery studies in other species (Incardona et al. 2015; Hicken et al. 2011) and highlight an important sublethal ecological adverse outcome surely to be associated with ecological death.

While maximum swim performance and general activity are obvious endpoints that would be impacted by reduced cardiorespiratory performance, the consequences of reduced aerobic scope may also extend to growth and reproduction. A series of long-term growth studies on a variety of species have returned mixed outcomes. Studies on common sole and European sea bass using both oil and oil-dispersant mixtures have shown that short-term exposure (5 days and 48 h, respectively) of juvenile life stages resulted in reduced growth and survival rates 6 months after exposure (Claireaux et al. 2004; Claireaux et al. 2013). Conversely, a more recent study on European sea bass showed no such impacts after only a month, despite similar exposure concentrations (Mauduit et al. 2016). There are two important considerations when assessing long-term growth studies as those highlighted above. The first is that feeding rate can dramatically impact outcomes. Many laboratories prescribe to ad libitum feeding regimens that may allow unfit individuals to grow simply because of an overabundance of resources. This experimental design could mask increases in standard metabolic rate by allowing individuals to effectively eat their way out of trouble, while reduced physical performance can be masked by the lack of conspecific competition for food. The three studies highlighted above all utilized a seminatural mesocosm design that provides natural fauna; however, in all cases, the feeding design is unlikely to constrain growth. A second important factor to consider is the life stage, as smaller animals exhibit much higher growth rates than older individuals.

It is also important to consider the potential additive or synergistic effects that result from exposure to other environmental stressors. Drawing again on the mechanisms set forth in the OCLTT framework, the two most concerning environmental stressors are hypoxia and temperature (Fig. 9.3). Hypoxia is a particularly common stress in aquatic environments and is relevant in the northern Gulf of Mexico region impacted by the Deepwater Horizon oil spill owing to the prevalent occurrence of oxygen minimum zones. To our knowledge, there has been little attempt to quantify the impacts of oil exposure on hypoxia tolerance through analysis of P_{crit}; however, there is evidence that oil exposure impacts LOE endpoints (Mauduit et al. 2016). Similarly, reduced reproductive output and larval survival in *C. variegatus* were observed only when hypoxia was included as a secondary stressor, but these results

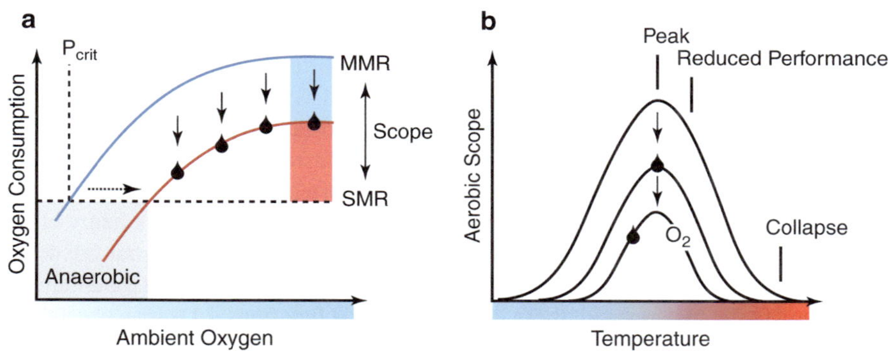

Fig. 9.3 Theoretical representation of the additive and synergistic impacts of oil exposure and common aquatic environmental stressors. (**a**) Environmental hypoxia constrains maximum metabolic rate (MMR) and progressively decreases aerobic scope until the critical oxygen partial pressure (P_{crit}) is reached. Oil exposure is known to impact MMR with no effect on standard metabolic rate (SMR), which would increase P_{crit} and exacerbate the effects of environmental hypoxia. (**b**) Many organisms exhibit oxygen-dependent thermal tolerance whereby increasing temperature progressively reduced aerobic scope until the point of collapse, which represents the upper thermal limitations. Under this framework, oil-exposed organisms will exhibit a reduced thermal tolerance, which will be further exacerbated by environmental hypoxia

occurred regardless of the oil dose and as such may simply represent a hypoxia effect (Hedgpeth and Griffitt 2016). Hypoxia has also been shown to interfere with detoxification mechanisms in *C. variegatus*—a maladaptive response that leads to an increase in DNA damage relative to exposures performed in normoxia (Dasgupta et al. 2016). There is less support for potential additive effects with temperature. In theory, any reduction in aerobic scope should result in reduced critical thermal tolerance (T_{crit}); however, the limited data suggest that thermal tolerance is not impaired during loss of equilibrium tests at several postexposure time points (Claireaux et al. 2013; Mauduit et al. 2016). In fact, the only significant effects were found in European sea bass, which actually exhibited increased time to loss of equilibrium at 1 week post-exposure.

Of course, cardiac impairment is not the only sublethal effect of PAH oil exposure (see reviews Collier et al. 2013; Buskey et al. 2016). Disruption in the endocrine system alters steroid production and metabolism through the aryl hydrocarbon receptor (AHR), other receptors, and responsive elements that play a role in regulation of endocrine processes. Oil is known to effect growth condition and energy metabolism through disruption in thyroid function altering plasma concentrations of hormones. It also impairs corticosteroid secretion, immune function, growth, and reproduction (Dubansky et al. 2013; Klinger et al. 2015). Another consequence of oil exposure is increased occurrence of histopathologic lesions. Studies on southern flounder and alligator gar have reported histopathology in the liver and gills (Buskey et al. 2016) and in the gill, trunk kidney, epidermis, and stomach of menhaden (Bentivegna et al. 2015).

9.5 Endocrine Disruption

One of the most sensitive pathways for chemicals to affect wildlife is via endocrine disruption. In complex organisms (invertebrates and vertebrates alike), hormones regulate critical developmental, growth, and homeostatic mechanisms, including sex differentiation during development, immune and reproductive functions, and maintenance of body metabolism. In higher-order vertebrates, hormones also regulate levels of glucose and ions in the blood, blood pressure, as well as many other glandular, muscle, and nervous system functions (e.g., Ogawa and Parhar 2014; do Rego and Vaudry 2016). An environmental endocrine disruptor (ED) is essentially a compound that interferes with these normal hormone functions. EDs may exert their influence by (1) mimicking the effects of endogenous hormones, such as estrogens and androgens, (2) agonizing or antagonizing the actions of endogenous hormones, (3) altering the synthesis and metabolism of normal hormones, and (4) modifying hormone receptor levels (Crisp et al. 1998; Pait et al. 2002). A crucial aspect of hormone action is that it takes place at extremely low concentrations. Typical physiological levels of hormones are in the range of parts per trillion, which is facilitated by a strong specificity of hormone binding to its receptor within different tissues (Lange et al. 2002). Specificity and extreme sensitivity is biologically necessary because it allows for a large number of signaling molecules to coexist in circulation while limiting erroneous interactions (Welshons et al. 2003). However, extreme sensitivity also means that even minute traces of EDs in nature have the potential for large-scale impacts.

There are three primary pathways for ED impacts. In vertebrates the hypothalamus controls the synthesis and release of endocrine messengers that stimulate the synthesis and release of pituitary hormones, including steroid-stimulating hormones and thyroid-stimulating hormones (see review Arcand-Hoy and Benson 1998). Steroid hormones, such as gonadotropins (GtHs), are those principally involved in reproductive processes and are particularly important during early development. Most notably, GtHs stimulate the synthesis of sex steroids (i.e., androgens, estrogens, and progestins) that regulate sex determination, sexual differentiation, and sexual development in vertebrates (Arcand-Hoy and Benson 1998; Tyler et al. 1998). Both androgen and estrogen receptors are found in many tissues, including the brain, pituitary, gonads, and accessory sex organs (Tyler et al. 1998; Vinas and Watson 2013; Coumailleau et al. 2015), and estrogen receptors are even found in the bone (Manolagas et al. 2013). In later development, thyroid hormones become increasingly important as they play a major role in growth and maturation, as well as a variety of physiological responses to environmental cues. For instance, thyroid hormones such as thyroxine (T4) and triiodothyronine (T3) are believed to aid fish in adapting to changes in temperature and osmotic stress (Pait et al. 2002). A number of hormones are also known as immunomodulators, which act to suppress or stimulate immune responses. Among these is the major stress hormone cortisol, which is thought to mediate immune suppression and increase disease susceptibility in chronically stressed fish (Baker et al. 2013).

9.5.1 Endocrine Disruptors in the Aquatic Environment

The estimated number of chemical compounds with potential ED impacts is growing each year. Although some suspected ED compounds are natural, such as phytoestrogens in plants (Sirotkin and Harrath 2014), it is estimated that more than 100,000 man-made chemicals are in routine use worldwide (Crisp et al. 1998; Pait et al. 2002; Vandenberg et al. 2009; Schug et al. 2013; Orton and Tyler 2015). These chemicals enter the aquatic environment through sewage and industrial effluent, as well as animal waste and agricultural runoff (Vos et al. 2000; Jobling and Tyler 2003; Rahman et al. 2009; Vajda et al. 2011). Thus, chemical traces are found in almost all aquatic ecosystems worldwide, and aquatic organisms, such as amphibians and fish, are most exposed (Kloas et al. 2009). Some ED chemicals have been created explicitly to alter hormone signaling. These include progestogen contraceptives (Kloas et al. 2009), as well as many pesticides, such as dichlorodiphenyltrichloroethane (DDT) (Rahman et al. 2009). Although pesticides are typically synthesized to disrupt hormone pathways of target species, adverse responses in nontarget species have been observed (see review Crisp et al. 1998). These pesticides include insect juvenile hormone mimics and analogs (e.g., methoprene, precocene) and molting disruptors (e.g., fenoxycarb) (Crisp et al. 1998). Other chemicals have sufficient resemblance to vertebrate hormones that they can bind to hormone receptors (Schug et al. 2013) with deleterious endocrine outcomes. These include dioxins, epoxy resins, some heavy metals (e.g., Cd), bisphenol A (BPA, a primary constituent of polycarbonate plastics), and petroleum hydrocarbons (e.g., Crisp et al. 1998; Jobling et al. 1998; Tyler et al. 1998; Henson and Chedrese 2004; Rochman et al. 2014). Many of these chemicals are of particular concern due to their persistence in nature, whereby they are resistant to natural degradation and thus can accumulate in aquatic environment. For example, BPA, DDT, and atrazine are all high-efficacy EDs with classification as persistent organic pollutants (vom Saal and Hughes 2005; Hayes et al. 2006; Rochman et al. 2014). Although many known EDs have lower binding affinity than the respective natural endogenous analog, an increasing number of hormone mimics are now known to provoke biological responses at picomolar levels (Alonso-Magdalena et al. 2012; Schug et al. 2013). One hypothesis to explain the heightened sensitivity is that EDs may be additive or synergistic when present as mixtures (Alonso-Magdalena et al. 2012; Kortenkamp 2007). This has been shown for some steroid estrogens, alkylphenolic chemicals, and pesticides (methoxychlor) (Alonso-Magdalena et al. 2012; Hayes et al. 2006). As such, the impacts of EDs on organisms should be viewed holistically, as opposed to an isolated assessment of each individual chemical of interest.

9.5.2 Assessing the Impact of Endocrine Disruption

Assessing the impacts of EDs can be highly complex. The fact that a compound can bind to a hormone receptor does not necessarily mean that it will stimulate hormone activity. Cause-effect relationships for EDs are also complicated by the fact that the mechanisms of impact are much broader than simple hormone agonist and antagonist relationships. There are a number of non-receptor-binding pathways through which endocrine modulators can exert their effects, including altering the number and affinity of receptors (Tyler et al. 1998; Henson and Chedrese 2004; Alonso-Magdalena et al. 2012) with ultimate impacts on reproduction; metabolism; the liver, bone, and immune system; learning; and behavior. In principal, there is virtually no developmental pathway or organ/tissue function that is immune to disruption (Schug et al. 2013).

Multiple studies have found biphasic dose responses at different endpoints and levels of organization (see reviews Vandenberg et al. 2012; Welshons et al. 2003; Weltje et al. 2005). These non-monotonic dose responses may be inverted U-shaped or U-shaped (Fig. 9.4), with the greatest impact on endpoints like cell proliferation or organ size at the lowest doses (Vandenberg et al. 2012). These responses can be generated by the synergistic effect of two or more monotonic dose response curves that occur through different pathways affecting a common endpoint with opposing effects (Christiansen et al. 2009). Alternatively, this pattern can be explained by the downregulation of receptors at higher hormone levels. For instance, low doses of androgens can increase cell proliferation in androgen-target cells, whereas at high doses inhibit cell proliferation (Vandenberg et al. 2012).

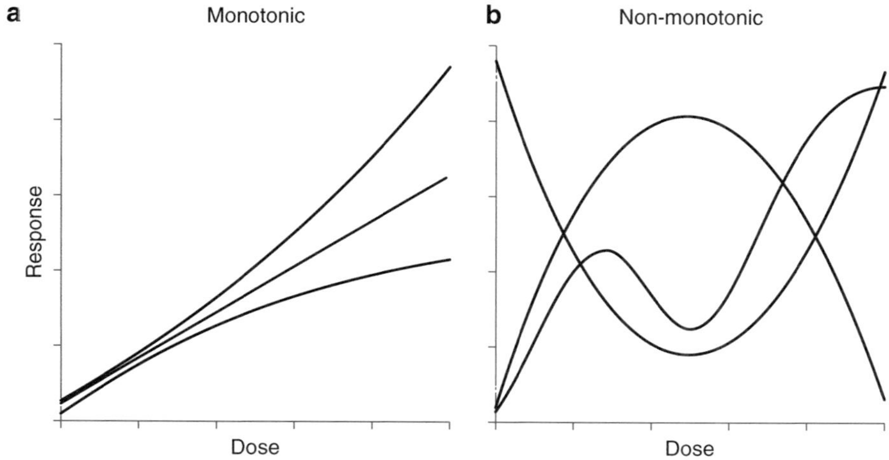

Fig. 9.4 Diagram representing monotonic (**a**) and non-monotonic (**b**) dose-response curves. Monotonic curves represent a range of relationships whereby increasing toxicant dose results in an increased magnitude of response, which can include linear, exponential, and logarithmic style relationships. Non-monotonic relationships are more complex and often involve synergistic relationships of two or more monotonic dose response curves that occur through different pathways while affecting a common endpoint

As described in our early discussion of metals, the best practice for ED studies is to test the effect across multiple concentrations and across the complete life cycle of model organisms. However, the complex biology of endocrine disruption means that no single endpoint can be used to identify and assess chemicals with ED characteristics. Vitellogenin, an egg yolk precursor, is a widely accepted biomarker of estrogenic substances (Sumpter and Jobling 1995), and assessments of ED can include measures of plasma hormone levels (e.g., gonadotropin, sex steroid hormones, and thyroid hormones) in treated animals as compared to controls. Developmental and reproductive effects can also be assessed by means of histopathology analysis. High-throughput genomic and transcriptomic approaches have also been used to reveal novel effects of EDs on sensory systems, brain function, and osmoregulatory pathways, among others (see review Brander 2013). Additionally, many ED compounds can bioaccumulate through the diet and biomagnify within the tissues of higher trophic groups due to their lipophility and persistence in the environment (Geyer et al. 2000; Pait et al. 2002; Kloas et al. 2009). Consequently, there is a possible time lag between the time of exposure and the biological response.

9.5.3 Amphibians and Fish as Model Study Systems

Adverse effects of EDs in nature have been reported for a number of organisms spanning gastropods, reptiles, fish, birds, and mammals, including abnormal thyroid function and development, decreased fertility, decreased hatching success, feminization or masculinization, decreased offspring survival, and alteration of immune and behavioral functions (Crisp et al. 1998; Vos et al. 2000). In most cases, however, a causal link between a specific chemical and a physiological effect has not been established, particularly because of the complexity associated with the vast number of chemicals that enter the environment, the high number of target tissues and feedback mechanisms, and potential synergistic effects of chemical mixtures (Kortenkamp 2007; Brander 2013). Model systems among lower vertebrates have therefore been suggested, particularly within amphibians and fish (Arcand-Hoy and Benson 1998; Hayes et al. 2006), which may also function as bioindicator species for environmental contamination.

Amphibians and fish carry many of the characteristics of higher vertebrates. Their reproductive biology is generally divided into two parts: the phase of sexual differentiation of the developing individual and the maintenance of reproductive functions in adults (Kloas et al. 2009). In principle, these developmental patterns are regulated by comparable hypothalamic-pituitary-gonadal (HPG) and hypothalamic-pituitary-thyroid axes (Villeneuve et al. 2007; Carr and Patino 2011). This includes the natural sex steroid hormone 17ß-estradiol (E2), which has remarkably conserved receptor structures and specificities across all vertebrate groups (Lange et al. 2002). These conserved pathways allow assessment of EDs that may also affect higher vertebrates, including humans (Kloas et al. 2009). It is important to note, however, that although the same hormones may exist in different groups, the specific function may differ among groups. For example, the primary role of prolactin in mammals is stimulation of the mammary gland and lactation, whereas in amphibians' prolactin helps coordinate thyroid-dependent metamorphosis (Wright et al. 1994), and is involved in osmoregulation in fish (Guh and Hwang 2016).

While congruency exists with higher vertebrates, amphibians and fish each have unique attributes that are beneficial for assessing the effects of ED and which make them ideal model species. Embryos develop rapidly, which makes it more logistically feasible to explore dose-dependent interactions between potential EDs and specific endocrine pathways with defined developmental outcomes. They are also known to absorb EDs via the gills and skin (Kloas et al. 2009), which makes them highly susceptible to exogenous exposure. More importantly, amphibians undergo unique changes during metamorphosis, and fish during sexual differentiation, which are tightly endocrine regulated (Kloas et al. 2009; Hayes et al. 2006; Arcand-Hoy and Benson 1998), and therefore allow detailed examination of the specific pathways of impact by suspected ED compounds. Below we will discuss two case studies that have used these models to great effect.

9.5.4 Amphibian Decline: A Case Study

During metamorphosis, amphibian larvae undergo extremely complex, thyroid-regulated morphological changes, such as emergence of limbs, resorption of tail, and reorganization of the gastrointestinal system (Kloas et al. 2009). These changes are all regulated by thyroid hormones, and although amphibians possess all components of the thyroid system found in higher vertebrates, the importance of thyroid hormones for metamorphosis is unique to this group (Tata 2006). Amphibian metamorphosis has therefore been introduced as the most sensitive model to detect EDs affecting the thyroid system of lower and higher vertebrates, including humans (Kloas et al. 2009; Tata 2006). The effect of EDs on amphibians is also highly ecologically relevant. Populations of frogs, toads, and salamanders are declining globally at an alarming rate (Wake and Vredenburg 2008; Rohr and Raffel 2010). As non-amniotes (lacking egg shells) with permeable skin, amphibians have no barriers to chemical contaminants and are very susceptible to ED exposure (Hayes et al. 2006; Schug et al. 2013). Pesticide EDs are therefore among a number of proposed causes for amphibian declines (Kloas et al. 2009; Wake and Vredenburg 2008; Hayes et al. 2006).

A number of detailed study descriptions are available (see reviews Hayes et al. 2006; Wake and Vredenburg 2008; Kloas et al. 2009), but certain studies are worth mentioning here. Atrazine is a pesticide used to retard broadleaf weeds in crops (Ackerman 2007). The pesticide was banned in Europe in 2004 due to its persistence in groundwater contamination but continues to be one of the most heavily applied pesticides in the USA and Australia, where it is one of the most common contaminants of drinking water and surface water (Ackerman 2007). This contaminant is not readily removed by water treatment (Westerhoff et al. 2005), and as a surface spray, it can easily enter waterways and spread many hundreds of miles from the point of application (Hayes et al. 2006). Although initially thought to be safe for humans and wildlife (Ackerman 2007), it is now documented as a potent ED with a number of adverse effects on estrogen and thyroid-dependent processes that are fundamental for amphibian metamorphosis (Rohr and McCoy 2010; Brodeur et al. 2013; Orton and Tyler 2015). Specifically, atrazine has a non-monotonic effect on the timing and size of individuals at metamorphosis, with the largest effects found at intermediate doses (Brodeur et al. 2013). These effects are thought to be caused by an interaction with the estrogen or glucocorticoid receptors, highlighting direct impacts on the thyroid pathway (Brodeur et al. 2013; Rohr and McCoy 2010). It is also shown to cause chemical castration and feminization of male larvae by reducing testosterone and increasing estrogenic activity (Hayes et al. 2010). Finally, it disrupts cardiac function (Lenkowski et al. 2008) and immune function (Brodkin et al. 2007) and alters multiple behavioral traits related to swimming patterns and activity (Ehrsam et al. 2016), although the exact hormonal pathways for these latter effects are less understood. The effects of atrazine also appear to be additive as the impacts are aggravated in chemical mixtures (e.g., S-metolachlor) (Hayes et al. 2006). Pesticide mixtures can negate or reverse the correlation between time to metamorphosis and size, increase stress (via corticosterone), induce damage to the thymus, and cause immunosuppression (Hayes et al. 2006). It is presently unknown whether atrazine simply works as an effector, or whether it also enhances impacts of other chemicals, but the implication is clear. Estimates of pesticide ED potency using single compounds may be grossly underestimating actual impacts in the wild and hence their role in amphibian declines.

9.5.5 Fish Intersex: A Case Study

Fish have been proposed as the most ecologically relevant model for the study of sex steroid disruption in vertebrates (Arcand-Hoy and Benson 1998). As the most speciose group of vertebrates, fish are found in nearly all aquatic ecosystems worldwide. They are particularly sensitive to ED exposure during early life stages and during the period of active reproduction in adult fish (Hutchinson and Pickford 2002; Schug et al. 2013), because the many processes involved with sex determination and gonadal differentiation require tight, coordinated endocrine regulation (Hutchinson and Pickford 2002). For instance, exposure to exogenous estrogen just 10 days either side of hatching can cause feminization of the subsequent fry (see review Tyler et al. 1998), whereas impacts on the hypothalamo-pituitary-gonadal axis are likely to occur in adults (Schug et al. 2013). Additionally, the major constituents of the developing fish egg (i.e., yolk and chorion proteins) are synthesized in the liver and maternally transported to the oocyte for uptake during development (Arukwe and Goksoyr 2003). As a consequence, ED contaminants that bioaccumulate in adult individuals can reach much higher concentrations in lipid reserves within the egg. These high concentrations of contaminants are then exposed to early life stages when the egg lipid reserves are metabolized to fuel embryo development (Arukwe and Goksoyr 2003). As early life stages of fish have a limited capacity to metabolize and excrete xenobiotics and other contaminants, the end result is aggravated, but relatively easily detectable impacts on offspring.

The most studied pathway for ED impact in fish is via estrogenic compound influences on gonadal differentiation and sex determination. Specifically, endocrine disruption has been reported in freshwater and marine fish populations around the world, ranging from subtle changes in the physiology and sexual behavior of fish to permanently altered sexual differentiation and impairment of fertility (Crisp et al. 1998; Pait et al. 2002; Jobling and Tyler 2003). Although numerous examples exist, several case studies are particularly informative. Bisphenol-A is a monomer of polycarbonate plastics which is known to have the efficacy and effects of estradiol (vom Saal and Hughes 2005; Westerhoff et al. 2005). BPA is one of the highest volume chemicals produced worldwide, with over 6 billion pounds produced each year, much of which eventually ends up in landfills and waterways due to leaching and slow rate of decomposition (vom Saal and Hughes 2005). BPA and similar highly potent estrogenic ED compounds in fish are frequently found in waters near sewage treatment plants and heavy industry like pulp and paper plants (Westerhoff et al. 2005). In North America and the UK, wild fish populations exposed to industrial effluents have high incidences of reduced estrogens and androgens, masculinization, feminization, altered sexual development, and decreased fertility due to smaller gonads, and fewer and smaller eggs (vom Saal and Hughes 2005; Jobling and Tyler 2003). These developmental defects are known to be associated with gonadal differentiation in fish following exposure to natural and/or synthetic estrogens. More critical changes include intersex conditions, whereby individuals may have both testicular and ovarian tissue, as well as fully developed sexual phenotypes that differ from the genetic composition (Arcand-Hoy and Benson 1998; Piferrer

2001; Tyler and Jobling 2008). Intersex species often have reduced fecundity. For example, the absence of a sperm duct in a male fish would prevent the release of viable sperm, which has been demonstrated for roach in some UK rivers (Nolan et al. 2001). Additionally, the offspring of fecund intersex individuals may all become genetically male or female (Piferrer 2001). The aquaculture industry has exploited these intersex conditions for several decades to produce all female or male offspring (Piferrer 2001), but the practice could have severe implications for inadvertently affected wild populations.

Conclusion

Like many areas of biology, high-throughput sequencing techniques are greatly impacting the field of aquatic toxicology. These approaches not only allow the simultaneous assessment of tens of thousands of transcripts as a consequence of exposure but when combined with biological pathway and process analysis can provide an excellent overview of the impacted physiological pathways. As such, these transcriptome-wide approaches are especially useful when attempting to ascertain cellular and subcellular aspects of toxicology pathways or explore the effects of emerging pollutants. In fact, the efficacy of high-throughput approaches should make them the starting point for any study of emerging pollutants on demonstrated at risk organisms. However, these approaches are not limited to new pollutants, as these approaches also hold real promise in expanding our understanding of known pollutants. An excellent example of this was recently shown for oil toxicity. As described above, the paradigm for oil toxicity in fish is that PAHs cause mortality and reduced performance through cardiac impairment, and hypothesis-driven gene expression studies corroborate this view (Edmunds et al. 2015). However, using a high-throughput approach, Xu et al. (2016) recently demonstrated a suite of other impacted pathways at low-dose concentrations, including visual perception and neural degeneration. These can both substantially increase the ecological risks associated with oil exposure. We would also like to stress that these high-throughput techniques should not be viewed as an endpoint for toxicology studies but instead as a way to generate well-reasoned mechanistic hypotheses. Researchers must remember that gene expression can be influenced in many ways, and not all of these result in phenotypic consequences. For example, gene expression may increase simply because of increased protein turnover or as a secondary compensation response to a related impacted pathway. As such, these high-throughput studies should be viewed as a complement, not replacement, for hypothesis-driven studies that focus on phenotypic level toxicity responses.

The concept of transgenerational and epigenetic effects is also of great interest to aquatic toxicologists. This concept is simply that exposure to harmful toxicants may illicit effects that can be passed on to later generations. The simplest type of transgenerational effect is when toxicants that bioaccumulate within breeding individuals are deposited in eggs and ultimately impact fertilization, development or other measures of early life performance. A recent example demonstrated such effects in zebrafish following exposure to tris (1,3-dichloro-2-

propyl) phosphate that reduced survival and heart rate in the F_1 generation (Yu et al. 2016). More complex transgenerational effects can work through epigenetic mechanisms whereby toxicant exposure alters F_1 gene expression through non genomic mechanisms, such as DNA methylation. This may cause underexpression or overexpression of genes that alter physiological function in early development or later life. Importantly, these effects can conceivably be passed on for generations or take generations to show any effects. Such impacts have recently been documented in *C. elegans* following arsenite exposure. In this example, dietary exposure of the F_0 generation resulted in reduced brood sizes in F_1–F_5 generations, presumably owing to a combination of demethylation and downregulation (Yu and Liao 2016). These examples suggest that single life cycle studies may be inadequate to fully evaluate the impacts of toxicant exposure. Government agencies and other regulatory bodies routinely use monotonic, linear, and single life cycle threshold models for risk assessments, but clearly this approach should be augmented with transgenerational approaches.

References

Ackerman F (2007) The economics of atrazine. Int J Occup Environ Health 13(4):437–445. https://doi.org/10.1179/oeh.2007.13.4.437

Alloy M, Baxter D, Stieglitz J, Mager E, Hoenig R, Benetti D, Grosell M, Oris J, Roberts A (2016) Ultraviolet radiation enhances the toxicity of Deepwater Horizon oil to mahi-mahi (Coryphaena hippurus) embryos. Environ Sci Technol 50(4):2011–2017. https://doi.org/10.1021/acs.est.5b05356

Alloy MM, Boube I, Griffitt RJ, Oris JT, Roberts AP (2015) Photo-induced toxicity of Deepwater Horizon slick oil to blue crab (Callinectes sapidus) larvae. Environ Toxicol Chem 34(9):2061–2066. https://doi.org/10.1002/etc.3026

Alonso-Magdalena P, Ropero AB, Soriano S, Garcia-Arevalo M, Ripoll C, Fuentes E, Quesada I, Nadal A (2012) Bisphenol-A acts as a potent estrogen via non-classical estrogen triggered pathways. Mol Cell Endocrinol 355(2):201–207. https://doi.org/10.1016/j.mce.2011.12.012

Anderson CJ, Mayes M, LaBelle R (2012) Update of occurrence rates for offshore oil spills. OCS Report BOEM 69:2012

Ankley GT, Bennett RS, Erickson RJ, Hoff DJ, Hornung MW, Johnson RD, Mount DR, Nichols JW, Russom CL, Schmieder PK, Serrrano JA, Tietge JE, Villeneuve DL (2010) Adverse outcome pathways: a conceptual framework to support ecotoxicology research and risk assessment. Environ Toxicol Chem 29(3):730–741. https://doi.org/10.1002/etc.34

Arcand-Hoy LD, Benson WH (1998) Fish reproduction: an ecologically relevant indicator of endocrine disruption. Environ Toxicol Chem 17(1):49–57. https://doi.org/10.1002/etc.5620170108

Arukwe A, Goksoyr A (2003) Eggshell and egg yolk proteins in fish: hepatic proteins for the next generation: oogenetic, population, and evolutionary implications of endocrine disruption. Comp Hepatol 2(1):4

Baker MR, Gobush KS, Vynne CH (2013) Review of factors influencing stress hormones in fish and wildlife. J Nat Conserv 21(5):309–318. https://doi.org/10.1016/j.jnc.2013.03.003

Barron MG, Carls MG, Short JW, Rice SD (2003) Photoenhanced toxicity of aqueous phase and chemically dispersed weathered Alaska North Slope crude oil to Pacific herring eggs and larvae. Environ Toxicol Chem 22(3):650–660

Bentivegna CS, Cooper KR, Olson G, Pena EA, Millemann DR, Portier RJ (2015) Chemical and histological comparisons between Brevoortia sp. (menhaden) collected in fall 2010 from Barataria Bay, LA and Delaware Bay, NJ following the DeepWater Horizon (DWH) oil spill. Mar Environ Res 112:21–34

Brander SM (2013) Chap. 5 Thinking outside the box: assessing endocrine disruption in aquatic life A2. In: Ahuja S (ed) Monitoring water quality. Elsevier, Amsterdam, pp 103–147. https://doi.org/10.1016/B978-0-444-59395-5.00005-4

Brette F, Machado B, Cros C, Incardona JP, Scholz NL, Block BA (2014) Crude oil impairs cardiac excitation-contraction coupling in fish. Science 343(6172):772–776. https://doi.org/10.1126/science.1242747

Brodeur JC, Sassone A, Hermida GN, Codugnello N (2013) Environmentally-relevant concentrations of atrazine induce non-monotonic acceleration of developmental rate and increased size at metamorphosis in Rhinella arenarum tadpoles. Ecotoxicol Environ Saf 92:10–17. https://doi.org/10.1016/j.ecoenv.2013.01.019

Brodkin MA, Madhoun H, Rameswaran M, Vatnick I (2007) Atrazine is an immune disruptor in adult northern leopard frogs (Rana pipiens). Environ Toxicol Chem 26(1):80–84

Buskey EJ, White HK, Esbaugh AJ (2016) Impact of oil spills on marine life in the Gulf of Mexico: effects on plankton, nekton, and deep-sea benthos. Oceanography 29(3):174–181

Calfee RD, Little EE, Puglis HJ, Scott E, Brumbaugh WG, Mebane CA (2014) Acute sensitivity of white sturgeon (Acipenser transmontanus) and rainbow trout (Oncorhynchus mykiss) to copper, cadmium, or zinc in water-only laboratory exposures. Environ Toxicol Chem 33(10):2259–2272. https://doi.org/10.1002/etc.2684

Carls MG, Holland L, Larsen M, Collier TK, Scholz NL, Incardona JP (2008) Fish embryos are damaged by dissolved PAHs, not oil particles. Aquat Toxicol 88(2):121–127. https://doi.org/10.1016/j.aquatox.2008.03.014

Carls MG, Rice SD, Hose JE (1999) Sensitivity of fish embryos to weathered crude oil: part I. Low-level exposure during incubation causes malformations, genetic damage, and mortality in larval pacific herring (Clupea pallasii). Environ Toxicol Chem 18(3):481–493. https://doi.org/10.1002/etc.5620180317

Carr JA, Patino R (2011) The hypothalamus-pituitary-thyroid axis in teleosts and amphibians: endocrine disruption and its consequences to natural populations. Gen Comp Endocrinol 170(2):299–312. https://doi.org/10.1016/j.ygcen.2010.06.001

Chapman PM, Allen HE, Godtfredsen K, Z'Graggen MN (1996) Policy analysis, peer reviewed: evaluation of bioaccumulation factors in regulating metals. Environ Sci Technol 30(10):448A–452A. https://doi.org/10.1021/es962436d

Christiansen S, Scholze M, Dalgaard M, Vinggaard AM, Axelstad M, Kortenkamp A, Hass U (2009) Synergistic disruption of external male sex organ development by a mixture of four antiandrogens. Environ Health Perspect 117(12):1839–1846. https://doi.org/10.1289/ehp.0900689

Claireaux G, Désaunay Y, Akcha F, Aupérin B, Bocquené G, Budzinski H, Cravedi J-P, Davoodi F, Galois R, Gilliers C, Goanvec C, Guérault D, Imbert N, Mazéas O, Nonnotte G, Nonnotte L, Prunet P, Sébert P, Vettier A (2004) Influence of oil exposure on the physiology and ecology of the common sole Solea solea: experimental and field approaches. Aquat Living Resour 17(3):335–351

Claireaux G, Theron M, Prineau M, Dussauze M, Merlin FX, Le Floch S (2013) Effects of oil exposure and dispersant use upon environmental adaptation performance and fitness in the European sea bass, Dicentrarchus labrax. Aquat Toxicol 130-131:160–170. https://doi.org/10.1016/j.aquatox.2013.01.004

Collier TK, Anulacion BF, Arkoosh MR, Dietrich JP, Incardona JP, Johnson LL, Ylitalo GM, Myers MS (2013) Chap. 4 Effects on fish of polycyclic aromatic hydrocarbons (PAHS) and naphthenic acid exposures. In: Keith B, Tierney APF, Colin JB (eds) Fish physiology, vol 33. Academic Press, New York, pp 195–255. https://doi.org/10.1016/B978-0-12-398254-4.00004-2

Coumailleau P, Pellegrini E, Adrio F, Diotel N, Cano-Nicolau J, Nasri A, Vaillant C, Kah O (2015) Aromatase, estrogen receptors and brain development in fish and amphibians. Biochim Biophys Acta 1849(2):152–162. https://doi.org/10.1016/j.bbagrm.2014.07.002

Crisp TM, Clegg ED, Cooper RL, Wood WP, Anderson DG, Baetcke KP, Hoffmann JL, Morrow MS, Rodier DJ, Schaeffer JE, Touart LW, Zeeman MG, Patel YM (1998) Environmental endocrine disruption: an effects assessment and analysis. Environ Health Perspect 106(Suppl 1):11–56

Dasgupta S, DiGiulio RT, Drollette BD, Plata DL, Brownawell BJ, McElroy AE (2016) Hypoxia depresses CYP1A induction and enhances DNA damage, but has minimal effects on antioxidant responses in sheepshead minnow (Cyprinodon variegatus) larvae exposed to dispersed crude oil. Aquat Toxicol 177:250–260. https://doi.org/10.1016/j.aquatox.2016.05.022

De Schamphelaere KA, Koene JM, Heijerick DG, Janssen CR (2008) Reduction of growth and haemolymph Ca levels in the freshwater snail Lymnaea stagnalis chronically exposed to cobalt. Ecotoxicol Environ Saf 71(1):65–70. https://doi.org/10.1016/j.ecoenv.2007.07.004

De Soysa TY, Ulrich A, Friedrich T, Pite D, Compton SL, Ok D, Bernardos RL, Downes GB, Hsieh S, Stein R, Lagdameo MC, Halvorsen K, Kesich LR, Barresi MJ (2012) Macondo crude oil from the Deepwater Horizon oil spill disrupts specific developmental processes during zebrafish embryogenesis. BMC Biol 10:40. https://doi.org/10.1186/1741-7007-10-40

DeForest DK, Brix KV, Adams WJ (2007) Assessing metal bioaccumulation in aquatic environments: the inverse relationship between bioaccumulation factors, trophic transfer factors and exposure concentration. Aquat Toxicol 84(2):236–246. https://doi.org/10.1016/j.aquatox.2007.02.022

Denison MS, Nagy SR (2003) Activation of the aryl hydrocarbon receptor by structurally diverse exogenous and endogenous chemicals. Annu Rev Pharmacol Toxicol 43:309–334. https://doi.org/10.1146/annurev.pharmtox.43.100901.135828

Di Toro DM, Allen HE, Bergman HL, Meyer JS, Paquin PR, Santore RC (2001) Biotic ligand model of the acute toxicity of metals. 1. Technical basis. Environ Toxicol Chem 20(10):2383–2396

Di Toro DM, McGrath JA (2000) Technical basis for narcotic chemicals and polycyclic aromatic hydrocarbon criteria. II. Mixtures and sediments. Environ Toxicol Chem 19(8):1971–1982. https://doi.org/10.1002/etc.5620190804

Di Toro DM, McGrath JA, Hansen DJ (2000) Technical basis for narcotic chemicals and polycyclic aromatic hydrocarbon criteria. I. Water and tissue. Environ Toxicol Chem 19(8):1951–1970. https://doi.org/10.1002/etc.5620190803

Di Toro DM, McGrath JA, Stubblefield WA (2007) Predicting the toxicity of neat and weathered crude oil: toxic potential and the toxicity of saturated mixtures. Environ Toxicol Chem 26(1):24–36. https://doi.org/10.1897/06174R.1

Dubansky B, Whitehead A, Miller JT, Rice CD, Galvez F (2013) Multitissue molecular, genomic, and developmental effects of the Deepwater Horizon oil spill on resident Gulf killifish (*Fundulus grandis*). Environ Sci Technol 47:5074–5082

Edmunds RC, Gill JA, Baldwin DH, Linbo TL, French BL, Brown TL, Esbaugh AJ, Mager EM, Stieglitz J, Hoenig R, Benetti D, Grosell M, Scholz NL, Incardona JP (2015) Corresponding morphological and molecular indicators of crude oil toxicity to the developing hearts of mahi mahi. Sci Rep 5:17326. https://doi.org/10.1038/srep17326

Ehrsam M, Knutie SA, Rohr JR (2016) The herbicide atrazine induces hyperactivity and compromises tadpole detection of predator chemical cues. Environ Toxicol Chem 35(9):2239–2244. https://doi.org/10.1002/etc.3377

Ern R, Norin T, Gamperl AK, Esbaugh AJ (2016) Oxygen-dependence of upper thermal limits in fishes. J Exp Biol 219:3376. https://doi.org/10.1242/jeb.143495

Esbaugh AJ, Brix KV, Mager EM, De Schamphelaere K, Grosell M (2012) Multi-linear regression analysis, preliminary biotic ligand modeling, and cross species comparison of the effects of water chemistry on chronic lead toxicity in invertebrates. Comp Biochem Physiol Part C Toxicol Pharmacol 155(2):423–431. https://doi.org/10.1016/j.cbpc.2011.11.005

Esbaugh AJ, Mager EM, Stieglitz JD, Hoenig R, Brown TL, French BL, Linbo TL, Lay C, Forth H, Scholz NL, Incardona JP, Morris JM, Benetti DD, Grosell M (2016) The effects of weathering and chemical dispersion on Deepwater Horizon crude oil toxicity to mahi-mahi (Coryphaena hippurus) early life stages. Sci Total Environ 543(Pt A):644–651. https://doi.org/10.1016/j.scitotenv.2015.11.068

Geyer HJ, Rimkus GG, Scheunert I, Kaune A, Schramm K-W, Kettrup A, Zeeman M, Muir DCG, Hansen LG, Mackay D (2000) Bioaccumulation and occurrence of endocrine-disrupting chemicals (EDCs), persistent organic pollutants (POPs), and other organic compounds in fish and other organisms including humans. In: Beek B (ed) Bioaccumulation—new aspects and developments. Springer, Berlin, pp 1–166. https://doi.org/10.1007/10503050_1

Gibbs A, Somero GN (1990) Na+-K+-adenosine triphosphatase activities in gills of marine teleost fishes: changes with depth, size and locomotory activity level. Mar Biol 106(3):315–321. https://doi.org/10.1007/bf01344307

Grans A, Jutfelt F, Sandblom E, Jonsson E, Wiklander K, Seth H, Olsson C, Dupont S, Ortega-Martinez O, Einarsdottir I, Bjornsson BT, Sundell K, Axelsson M (2014) Aerobic scope fails to explain the detrimental effects on growth resulting from warming and elevated CO2 in Atlantic halibut. J Exp Biol 217(Pt 5):711–717. https://doi.org/10.1242/jeb.096743

Gravato C, Guilhermino L (2009) Effects of Benzo(a)pyrene on Seabass (Dicentrarchus labrax L.): biomarkers, growth and behavior. Hum Ecol Risk Assess Int J 15(1):121–137. https://doi.org/10.1080/10807030802615659

Grosell M (2011) Chap. 2 Copper. In: Chris M, APF W, Colin JB (eds) Fish physiology, vol 31, Part A. Academic Press, New York, pp 53–133. https://doi.org/10.1016/S1546-5098(11)31002-3

Grosell M, Brix KV (2009) High net calcium uptake explains the hypersensitivity of the freshwater pulmonate snail, Lymnaea stagnalis, to chronic lead exposure. Aquat Toxicol 91(4):302–311. https://doi.org/10.1016/j.aquatox.2008.10.012

Grosell M, Gerdes RM, Brix KV (2006) Chronic toxicity of lead to three freshwater invertebrates--Brachionus calyciflorus, Chironomus tentans, and Lymnaea stagnalis. Environ Toxicol Chem 25(1):97–104

Grosell M, Nielsen C, Bianchini A (2002) Sodium turnover rate determines sensitivity to acute copper and silver exposure in freshwater animals. Comp Biochem Physiol C 133(1-2):287–303

Grosell M, O'Donnell MJ, Wood CM (2000) Hepatic versus gallbladder bile composition: in vivo transport physiology of the gallbladder in rainbow trout. Am J Physiol Regul Integr Comp Physiol 278(6):R1674–R1684

Guh YJ, Hwang PP (2016) Insights into molecular and cellular mechanisms of hormonal actions on fish ion regulation derived from the zebrafish model. Gen Comp Endocrinol 251:12. https://doi.org/10.1016/j.ygcen.2016.08.009

Hamilton SJ (2004) Review of selenium toxicity in the aquatic food chain. Sci Total Environ 326(1-3):1–31. https://doi.org/10.1016/j.scitotenv.2004.01.019

Hatch AC, Burton GA Jr (1999) Photo-induced toxicity of PAHs to Hyalella azteca and Chironomus tentans: effects of mixtures and behavior. Environ Pollut 106(2):157–167

Hayes TB, Case P, Chui S, Chung D, Haeffele C, Haston K, Lee M, Mai VP, Marjuoa Y, Parker J, Tsui M (2006) Pesticide mixtures, endocrine disruption, and amphibian declines: are we underestimating the impact? Environ Health Perspect 114(Suppl 1):40–50. https://doi.org/10.1289/ehp.8051

Hayes TB, Khoury V, Narayan A, Nazir M, Park A, Brown T, Adame L, Chan E, Buchholz D, Stueve T, Gallipeau S (2010) Atrazine induces complete feminization and chemical castration in male African clawed frogs (Xenopus laevis). Proc Natl Acad Sci 107(10):4612–4617. https://doi.org/10.1073/pnas.0909519107

Hedgpeth BM, Griffitt RJ (2016) Simultaneous exposure to chronic hypoxia and dissolved polycyclic aromatic hydrocarbons results in reduced egg production and larval survival in the sheepshead minnow (Cyprinodon variegatus). Environ Toxicol Chem 35(3):645–651. https://doi.org/10.1002/etc.3207

Heintz RA, Short JW, Rice SD (1999) Sensitivity of fish embryos to weathered crude oil: part II. Increased mortality of pink salmon (Oncorhynchus gorbuscha) embryos incubating downstream from weathered Exxon Valdez crude oil. Environ Toxicol Chem 18(3):494–503. https://doi.org/10.1002/etc.5620180318

Hendriks AJ, Heikens A (2001) The power of size. 2. Rate constants and equilibrium ratios for accumulation of inorganic substances related to species weight. Environ Toxicol Chem 20(7):1421–1437

Henson MC, Chedrese PJ (2004) Endocrine disruption by cadmium, a common environmental toxicant with paradoxical effects on reproduction. Exp Biol Med (Maywood) 229(5):383–392

Hernandez LE, Sobrino-Plata J, Montero-Palmero MB, Carrasco-Gil S, Flores-Caceres ML, Ortega-Villasante C, Escobar C (2015) Contribution of glutathione to the control of cellular redox homeostasis under toxic metal and metalloid stress. J Exp Bot 66(10):2901–2911. https://doi.org/10.1093/jxb/erv063

Hicken CE, Linbo TL, Baldwin DH, Willis ML, Myers MS, Holland L, Larsen M, Stekoll MS, Rice SD, Collier TK, Scholz NL, Incardona JP (2011) Sublethal exposure to crude oil during embryonic development alters cardiac morphology and reduces aerobic capacity in adult fish. Proc Natl Acad Sci U S A 108(17):7086–7090. https://doi.org/10.1073/pnas.1019031108

Hogstrand C (2011) Chap. 3 Zinc. In: Chris M, APF W, Colin JB (eds) Fish physiology, vol 31, Part A. Academic Press, New York, pp 135–200. https://doi.org/10.1016/S1546-5098(11)31003-5

Holst LL, Giesy JP (1989) Chronic effects of the photoenhanced toxicity of anthracene on Daphnia magna reproduction. Environ Toxicol Chem 8(10):933–942. https://doi.org/10.1002/etc.5620081012

Hutchinson TH, Pickford DB (2002) Ecological risk assessment and testing for endocrine disruption in the aquatic environment. Toxicology 181-182:383–387. https://doi.org/10.1016/S0300-483X(02)00471-7

Incardona JP, Carls MG, Day HL, Sloan CA, Bolton JL, Collier TK, Scholz NL (2009) Cardiac arrhythmia is the primary response of embryonic Pacific herring (Clupea pallasii) exposed to crude oil during weathering. Environ Sci Technol 43(1):201–207

Incardona JP, Carls MG, Holland L, Linbo TL, Baldwin DH, Myers MS, Peck KA, Tagal M, Rice SD, Scholz NL (2015) Very low embryonic crude oil exposures cause lasting cardiac defects in salmon and herring. Sci Rep 5:13499. https://doi.org/10.1038/srep13499. http://www.nature.com/articles/srep13499#supplementary-information

Incardona JP, Collier TK, Scholz NL (2004) Defects in cardiac function precede morphological abnormalities in fish embryos exposed to polycyclic aromatic hydrocarbons. Toxicol Appl Pharmacol 196(2):191–205. https://doi.org/10.1016/j.taap.2003.11.026

Incardona JP, Gardner LD, Linbo TL, Brown TL, Esbaugh AJ, Mager EM, Stieglitz JD, French BL, Labenia JS, Laetz CA, Tagal M, Sloan CA, Elizur A, Benetti DD, Grosell M, Block BA, Scholz NL (2014) Deepwater Horizon crude oil impacts the developing hearts of large predatory pelagic fish. Proc Natl Acad Sci U S A 111(15):E1510–E1518. https://doi.org/10.1073/pnas.1320950111

Incardona JP, Linbo TL, Scholz NL (2011) Cardiac toxicity of 5-ring polycyclic aromatic hydrocarbons is differentially dependent on the aryl hydrocarbon receptor 2 isoform during zebrafish development. Toxicol Appl Pharmacol 257(2):242–249. https://doi.org/10.1016/j.taap.2011.09.010

Janz DM (2011) Chap. 7 Selenium. In: Chris M, Wood APF, Colin JB (eds) Fish physiology, vol 31, Part A. Academic Press, New York, pp 327–374. https://doi.org/10.1016/S1546-5098(11)31007-2

Jobling S, Nolan M, Tyler CR, Brighty G, Sumpter JP (1998) Widespread sexual disruption in wild fish. Environ Sci Technol 32(17):2498–2506. https://doi.org/10.1021/es9710870

Jobling S, Tyler CR (2003) Endocrine disruption, parasites and pollutants in wild freshwater fish. Parasitology 126(7):S103–S108

Khursigara AK, Perrichon P, Martinez Bautista N, Burggren WW, Esbaugh AJ (2017) Cardiac function and survival are affected by crude oil in larval red drum (Sciaenops ocellatus). Sci Total Environ 579:797

Kidd K, Batchelar K (2011) Chap. 5 Mercury. In: Chris M, Wood APF, Colin JB (eds) Fish physiology, vol 31, Part B. Academic Press, New York, pp 237–295. https://doi.org/10.1016/S1546-5098(11)31027-8

Kipka U, Di Toro DM (2009) Technical basis for polar and nonpolar narcotic chemicals and polycyclic aromatic hydrocarbon criteria. III. A polyparameter model for target lipid partitioning. Environ Toxicol Chem 28(7):1429–1438. https://doi.org/10.1897/08-364.1

Klinger DH, Dale JJ, Machado BE, Incardona JP, Farwell CJ, Block BA (2015) Exposure to Deepwater Horizon weathered crude oil increases routine metabolic demand in chub mackerel, Scomber japonicus. Mar Pollut Bull 98(1-2):259–266. https://doi.org/10.1016/j.marpolbul.2015.06.039

Kloas W, Urbatzka R, Opitz R, Wurtz S, Behrends T, Hermelink B, Hofmann F, Jagnytsch O, Kroupova H, Lorenz C, Neumann N, Pietsch C, Trubiroha A, Van Ballegooy C, Wiedemann C, Lutz I (2009) Endocrine disruption in aquatic vertebrates. Ann N Y Acad Sci 1163:187–200. https://doi.org/10.1111/j.1749-6632.2009.04453.x

Kortenkamp A (2007) Ten years of mixing cocktails: a review of combination effects of endocrine-disrupting chemicals. Environ Health Perspect 115(Suppl 1):98–105. https://doi.org/10.1289/ehp.9357

Kramer VJ, Etterson MA, Hecker M, Murphy CA, Roesijadi G, Spade DJ, Spromberg JA, Wang M, Ankley GT (2011) Adverse outcome pathways and ecological risk assessment: bridging to population-level effects. Environ Toxicol Chem 30(1):64–76. https://doi.org/10.1002/etc.375

Lange IG, Hartel A, Meyer HH (2002) Evolution of oestrogen functions in vertebrates. J Steroid Biochem Mol Biol 83(1-5):219–226

Lee RF, Anderson JW (2005) Significance of cytochrome P450 system responses and levels of bile fluorescent aromatic compounds in marine wildlife following oil spills. Mar Pollut Bull 50(7):705–723. https://doi.org/10.1016/j.marpolbul.2005.04.036

Lenkowski JR, Reed JM, Deininger L, McLaughlin KA (2008) Perturbation of organogenesis by the herbicide atrazine in the amphibian Xenopus laevis. Environ Health Perspect 116(2):223–230. https://doi.org/10.1289/ehp.10742

Letcher RJ, Bustnes JO, Dietz R, Jenssen BM, Jorgensen EH, Sonne C, Verreault J, Vijayan MM, Gabrielsen GW (2010) Exposure and effects assessment of persistent organohalogen contaminants in arctic wildlife and fish. Sci Total Environ 408(15):2995–3043. https://doi.org/10.1016/j.scitotenv.2009.10.038

Liu Z, Liu J, Zhu Q, Wu W (2012) The weathering of oil after the Deepwater Horizon oil spill: insights from the chemical composition of the oil from the sea surface, salt marshes and sediments. Environ Res Lett 7(3):035302

Mager EM, Esbaugh AJ, Stieglitz JD, Hoenig R, Bodinier C, Incardona JP, Scholz NL, Benetti DD, Grosell M (2014) Acute embryonic or juvenile exposure to Deepwater Horizon crude oil impairs the swimming performance of mahi-mahi (Coryphaena hippurus). Environ Sci Technol 48(12):7053–7061. https://doi.org/10.1021/es501628k

Manolagas SC, O'Brien CA, Almeida M (2013) The role of estrogen and androgen receptors in bone health and disease. Nat Rev Endocrinol 9(12):699–712. https://doi.org/10.1038/nrendo.2013.179

Mao H, Wang DH, Yang WX (2012) The involvement of metallothionein in the development of aquatic invertebrate. Aquat Toxicol 110-111:208–213. https://doi.org/10.1016/j.aquatox.2012.01.018

Mauduit F, Domenici P, Farrell AP, Lacroix C, Le Floch S, Lemaire P, Nicolas-Kopec A, Whittington M, Zambonino-Infante JL, Claireaux G (2016) Assessing chronic fish health: an application to a case of an acute exposure to chemically treated crude oil. Aquat Toxicol 178:197–208. https://doi.org/10.1016/j.aquatox.2016.07.019

McGeer JC, Brix KV, Skeaff JM, DeForest DK, Brigham SI, Adams WJ, Green A (2003) Inverse relationship between bioconcentration factor and exposure concentration for metals: implications for hazard assessment of metals in the aquatic environment. Environ Toxicol Chem 22(5):1017–1037

McGeer JC, Niyogi S, Scott Smith D (2011) Chap. 3 Cadmium. In: Chris M, Wood APF, Colin JB (eds) Fish physiology, vol 31, Part B. Academic Press, New York, pp 125–184. https://doi.org/10.1016/S1546-5098(11)31025-4

Mimura J, Fujii-Kuriyama Y (2003) Functional role of AhR in the expression of toxic effects by TCDD. Biochim Biophys Acta 1619(3):263–268

Munley KM, Brix KV, Panlilio J, Deforest DK, Grosell M (2013) Growth inhibition in early life-stage tests predicts full life-cycle toxicity effects of lead in the freshwater pulmonate snail, Lymnaea stagnalis. Aquat Toxicol 128:60–66. https://doi.org/10.1016/j.aquatox.2012.11.020

Nelson D, Heuer R, Cox GK, Stieglitz JD, Hoenig R, Mager EM, Benetti DD, Grosell M, Crossley D (2016) Effects of crude oil on in situ cardiac function in young adult mahi-mahi (Coryphaena hippurus). Aquat Toxicol 180:274

Newsted JL, Giesy JP (1987) Predictive models for photoinduced acute toxicity of polycyclic aromatic hydrocarbons to Daphnia magna, strauss (cladocera, crustacea). Environ Toxicol Chem 6(6):445–461. https://doi.org/10.1002/etc.5620060605

Nolan M, Jobling S, Brighty G, Sumpter JP, Tyler CR (2001) A histological description of intersexuality in the roach. J Fish Biol 58(1):160–176. https://doi.org/10.1111/j.1095-8649.2001.tb00505.x

Norin T, Malte H, Clark TD (2014) Aerobic scope does not predict the performance of a tropical eurythermal fish at elevated temperatures. J Exp Biol 217(Pt 2):244–251. https://doi.org/10.1242/jeb.089755

Ogawa S, Parhar IS (2014) Structural and functional divergence of gonadotropin-inhibitory hormone from jawless fish to mammals. Front Endocrinol 5:177. https://doi.org/10.3389/fendo.2014.00177

Oris JT, Giesy JP (1985) The photoenhanced toxicity of anthracene to juvenile sunfish (Lepomis spp.) Aquat Toxicol 6(2):133–146. https://doi.org/10.1016/0166-445X(85)90012-8

Oris JT, Giesy JP (1987) The photo-induced toxicity of polycyclic aromatic hydrocarbons to larvae of the fathead minnow (Pimephales promelas). Chemosphere 16(7):1395–1404. https://doi.org/10.1016/0045-6535(87)90079-8

Orton F, Tyler CR (2015) Do hormone-modulating chemicals impact on reproduction and development of wild amphibians? Biol Rev Camb Philos Soc 90(4):1100–1117. https://doi.org/10.1111/brv.12147

Pait AS, Nelson JO, Center for Coastal Monitoring and Assessment (2002) Endocrine disruption in fish: an assessment of recent research and results. U.S. Dept. of Commerce, National Oceanic and Atmospheric Administration, National Ocean Service, National Centers for Coastal Ocean Science, Center for Coastal Monitoring and Assessment, Silver Spring

Pan YK, Ern R, Esbaugh AJ (2016) Hypoxia tolerance decreases with body size in red drum Sciaenops ocellatus. J Fish Biol 89(2):1488–1493. https://doi.org/10.1111/jfb.13035

Paquin P, Redman A, Ryan A, Santore R (2011) Chap. 9 Modeling the physiology and toxicology of metals. In: Chris M, Wood APF, Colin JB (eds) Fish physiology, vol 31, Part B. Academic Press, New York, pp 429–484. https://doi.org/10.1016/S1546-5098(11)31031-X

Paquin PR, Gorsuch JW, Apte S, Batley GE, Bowles KC, Campbell PG, Delos CG, Di Toro DM, Dwyer RL, Galvez F, Gensemer RW, Goss GG, Hostrand C, Janssen CR, McGeer JC, Naddy RB, Playle RC, Santore RC, Schneider U, Stubblefield WA, Wood CM, Wu KB (2002) The biotic ligand model: a historical overview. Comp Biochem Physiol C 133(1-2):3–35

Piferrer F (2001) Endocrine sex control strategies for the feminization of teleost fish. Aquaculture 197(1-4):229–281. https://doi.org/10.1016/S0044-8486(01)00589-0

Portner HO, Knust R (2007) Climate change affects marine fishes through the oxygen limitation of thermal tolerance. Science 315(5808):95–97. https://doi.org/10.1126/science.1135471

Portner HO, Peck MA (2010) Climate change effects on fishes and fisheries: towards a cause-and-effect understanding. J Fish Biol 77(8):1745–1779. https://doi.org/10.1111/j.1095-8649.2010.02783.x

Rahman MF, Yanful EK, Jasim SY (2009) Endocrine disrupting compounds (EDCs) and pharmaceuticals and personal care products (PPCPs) in the aquatic environment: implications for the drinking water industry and global environmental health. J Water Health 7(2):224–243. https://doi.org/10.2166/wh.2009.021

Rauch JN (2009) Global mapping of Al, Cu, Fe, and Zn in-use stocks and in-ground resources. Proc Natl Acad Sci U S A 106(45):18920–18925. https://doi.org/10.1073/pnas.0900658106

Rebelo M d F, Moreira JC, Moraes MO, Coutinho D, Torres JPM (2003) Is there a CYP1A1 in marine bivalves? Expressed sequences related to cytochrome P450 in Crassostrea rhizophorae. Comment Toxicol 9(5-6):279–287. https://doi.org/10.1080/08865140390450377

do Rego JL, Vaudry H (2016) Comparative aspects of neurosteroidogenesis: from fish to mammals. Gen Comp Endocrinol 227:120–129. https://doi.org/10.1016/j.ygcen.2015.05.014

Rochman CM, Kurobe T, Flores I, Teh SJ (2014) Early warning signs of endocrine disruption in adult fish from the ingestion of polyethylene with and without sorbed chemical pollutants from the marine environment. Sci Total Environ 493:656–661. https://doi.org/10.1016/j.scitotenv.2014.06.051

Rogers JT, Wood CM (2004) Characterization of branchial lead-calcium interaction in the freshwater rainbow trout Oncorhynchus mykiss. J Exp Biol 207(5):813–825. https://doi.org/10.1242/Jeb.00826

Rohr JR, McCoy KA (2010) A qualitative meta-analysis reveals consistent effects of atrazine on freshwater fish and amphibians. Environ Health Perspect 118(1):20–32. https://doi.org/10.1289/ehp.0901164
Rohr JR, Raffel TR (2010) Linking global climate and temperature variability to widespread amphibian declines putatively caused by disease. Proc Natl Acad Sci 107(18):8269–8274. https://doi.org/10.1073/pnas.0912883107
vom Saal FS, Hughes C (2005) An extensive new literature concerning low-dose effects of bisphenol A shows the need for a new risk assessment. Environ Health Perspect 113(8):926–933. https://doi.org/10.1289/ehp.7713
Schlekat CE, Van Genderen E, De Schamphelaere KA, Antunes PM, Rogevich EC, Stubblefield WA (2010) Cross-species extrapolation of chronic nickel biotic ligand models. Sci Total Environ 408:6148–6157
Schug TT, Abagyan R, Blumberg B, Collins TJ, Crews D, DeFur PL, Dickerson SM, Edwards TM, Gore AC, Guillette LJ, Hayes T, Heindel JJ, Moores A, Patisaul HB, Tal TL, Thayer KA, Vandenberg LN, Warner JC, Watson CS, vom Saal FS, Zoeller RT, O'Brien KP, Myers JP (2013) Designing endocrine disruption out of the next generation of chemicals. Green Chem 15(1):181–198. https://doi.org/10.1039/C2GC35055F
Sirotkin AV, Harrath AH (2014) Phytoestrogens and their effects. Eur J Pharmacol 741:230–236. https://doi.org/10.1016/j.ejphar.2014.07.057
Stieglitz JD, Mager EM, Hoenig RH, Alloy M, Esbaugh AJ, Bodinier C, Benetti DD, Roberts AP, Grosell M (2016) A novel system for embryo-larval toxicity testing of pelagic fish: applications for impact assessment of Deepwater Horizon crude oil. Chemosphere 162:261–268. https://doi.org/10.1016/j.chemosphere.2016.07.069
Sumpter JP, Jobling S (1995) Vitellogenesis as a biomarker for estrogenic contamination of the aquatic environment. Environ Health Perspect 103(Suppl 7):173–178
Tata JR (2006) Amphibian metamorphosis as a model for the developmental actions of thyroid hormone. Mol Cell Endocrinol 246(1-2):10–20. https://doi.org/10.1016/j.mce.2005.11.024
Tipping E (2002) Cation binding by humic substances. In: Campbell PGC, Harrison RM, de Mora SJ (eds) Cambridge environmental chemistry series. Cambridge University Press, New York
Tyler CR, Jobling S (2008) Roach, sex, and gender-bending chemicals: the feminization of wild fish in English rivers. Bioscience 58(11):1051–1059. https://doi.org/10.1641/b581108
Tyler CR, Jobling S, Sumpter JP (1998) Endocrine disruption in wildlife: a critical review of the evidence. Crit Rev Toxicol 28(4):319–361. https://doi.org/10.1080/10408449891344236
USEPA (2002a) Methods for measuring the acute toxicity of effluents and receiving waters to freshwater and marine organisms, 5th edn. United States Environmental Protection Agency, Washington, DC. EPA-821-R-02-012
USEPA (2002b) Short-term methods for estimating the chronic toxicity of effluents and receiving waters to freshwater organisms, 4th edn. United States Environmental Protection Agency, Washington, DC. EPA-821-R-02-013
Vajda AM, Barber LB, Gray JL, Lopez EM, Bolden AM, Schoenfuss HL, Norris DO (2011) Demasculinization of male fish by wastewater treatment plant effluent. Aquat Toxicol 103(3-4):213–221. https://doi.org/10.1016/j.aquatox.2011.02.007
Vandenberg LN, Colborn T, Hayes TB, Heindel JJ, Jacobs DR Jr, Lee DH, Shioda T, Soto AM, vom Saal FS, Welshons WV, Zoeller RT, Myers JP (2012) Hormones and endocrine-disrupting chemicals: low-dose effects and nonmonotonic dose responses. Endocr Rev 33(3):378–455. https://doi.org/10.1210/er.2011-1050
Vandenberg LN, Maffini MV, Sonnenschein C, Rubin BS, Soto AM (2009) Bisphenol-A and the great divide: a review of controversies in the field of endocrine disruption. Endocr Rev 30(1):75–95. https://doi.org/10.1210/er.2008-0021
Verberk WC, Overgaard J, Ern R, Bayley M, Wang T, Boardman L, Terblanche JS (2016) Does oxygen limit thermal tolerance in arthropods? A critical review of current evidence. Comp Biochem Physiol A Mol Integr Physiol 192:64–78. https://doi.org/10.1016/j.cbpa.2015.10.020
Villeneuve DL, Larkin P, Knoebl I, Miracle AL, Kahl MD, Jensen KM, Makynen EA, Durhan EJ, Carter BJ, Denslow ND, Ankley GT (2007) A graphical systems model to facilitate hypothesis-

driven ecotoxicogenomics research on the teleost brain–pituitary–gonadal axis. Environ Sci Technol 41(1):321–330. https://doi.org/10.1021/es061739x

Vinas R, Watson CS (2013) Bisphenol S disrupts estradiol-induced nongenomic signaling in a rat pituitary cell line: effects on cell functions. Environ Health Perspect 121(3):352–358. https://doi.org/10.1289/ehp.1205826

Vos JG, Dybing E, Greim HA, Ladefoged O, Lambre C, Tarazona JV, Brandt I, Vethaak AD (2000) Health effects of endocrine-disrupting chemicals on wildlife, with special reference to the European situation. Crit Rev Toxicol 30(1):71–133. https://doi.org/10.1080/10408440091159176

Wake DB, Vredenburg VT (2008) Are we in the midst of the sixth mass extinction? A view from the world of amphibians. Proc Natl Acad Sci 105(Suppl 1):11466–11473. https://doi.org/10.1073/pnas.0801921105

Welshons WV, Thayer KA, Judy BM, Taylor JA, Curran EM, vom Saal FS (2003) Large effects from small exposures. I. Mechanisms for endocrine-disrupting chemicals with estrogenic activity. Environ Health Perspect 111(8):994–1006

Weltje L, vom Saal FS, Oehlmann J (2005) Reproductive stimulation by low doses of xenoestrogens contrasts with the view of hormesis as an adaptive response. Hum Exp Toxicol 24(9):431–437

Westerhoff P, Yoon Y, Snyder S, Wert E (2005) Fate of endocrine-disruptor, pharmaceutical, and personal care product chemicals during simulated drinking water treatment processes. Environ Sci Technol 39(17):6649–6663

Whitehead A, Dubansky B, Bodinier C, Garcia TI, Miles S, Pilley C, Raghunathan V, Roach JL, Walker N, Walter RB, Rice CD, Galvez F (2012) Genomic and physiological footprint of the Deepwater Horizon oil spill on resident marsh fishes. Proc Natl Acad Sci U S A 109:20298–20302

Wood CM (2011) Chap. 1 An introduction to metals in fish physiology and toxicology: basic principles. In: Chris M, Wood APF, Colin JB (eds) Fish physiology, vol 31, Part A. Academic Press, New York, pp 1–51. https://doi.org/10.1016/S1546-5098(11)31001-1

Wright ML, Cykowski LJ, Lundrigan L, Hemond KL, Kochan DM, Faszewski EE, Anuszewski CM (1994) Anterior pituitary and adrenal cortical hormones accelerate or inhibit tadpole hind limb growth and development depending on stage of spontaneous development or thyroxine concentration in induced metamorphosis. J Exp Zool 270(2):175–188. https://doi.org/10.1002/jez.1402700207

Xu EG, Mager EM, Grosell M, Pasparakis C, Schlenker LS, Stieglitz JD, Benetti D, Hazard ES, Courtney SM, Diamante G, Freitas J, Hardiman G, Schlenk D (2016) Time- and oil-dependent transcriptomic and physiological responses to Deepwater Horizon oil in mahi-mahi (Coryphaena hippurus) embryos and larvae. Environ Sci Technol 50(14):7842–7851. https://doi.org/10.1021/acs.est.6b02205

Yu CW, Liao VH (2016) Transgenerational reproductive effects of arsenite are associated with H3K4 dimethylation and SPR-5 downregulation in Caenorhabditis elegans. Environ Sci Technol 50:10673. https://doi.org/10.1021/acs.est.6b02173

Yu L, Jia Y, Su G, Sun Y, Letcher RJ, Giesy JP, Yu H, Han Z, Liu C (2016) Parental transfer of tris (1,3-dichloro-2-propyl) phosphate and transgenerational inhibition of growth of zebrafish exposed to environmentally relevant concentrations. Environ Pollut 220:196. https://doi.org/10.1016/j.envpol.2016.09.039

Zanotto FP, Wheatly MG, Reiber CL, Gannon AT, Jalles-Filho E (2004) Allometric relationship of postmolt net ion uptake, ventilation, and circulation in the freshwater crayfish Procambarus clarkii: intraspecific scaling. Physiol Biochem Zool 77(2):275–284. https://doi.org/10.1086/378924

Zhou H, Wu H, Liao C, Diao X, Zhen J, Chen L, Xue Q (2010) Toxicology mechanism of the persistent organic pollutants (POPs) in fish through AhR pathway. Toxicol Mech Methods 20(6):279–286. https://doi.org/10.3109/15376516.2010.485227

Case Study: The 2010 Deepwater Horizon Oil Spill and Its Environmental Developmental Impacts

10

John P. Incardona and Nathaniel L. Scholz

Abstract

Field and laboratory studies following the 1989 Exxon Valdez oil spill in Prince William Sound, Alaska, revealed new insights into the interactions between aquatic pollution and heart development in fish. The observation of cardiac-related defects in wild herring and salmon spawned near oiled shorelines led to more than two decades of research to characterize the developmental toxicity of crude oil and crude oil-derived compounds, especially polycyclic aromatic hydrocarbons (PAHs). By the time of the 2010 Deepwater Horizon incident, this mechanistic understanding had advanced to the point where potential impacts to the developing heart were a key focus for natural resource injury assessment activities focused on mahi mahi, tunas, and other species that spawn in the northern Gulf of Mexico. For both types of crude oil, subclasses of PAHs containing three aromatic rings, as well as complex mixtures, were found to have disruptive effects on cardiomyocyte physiology, blocking key ion channels involved in excitation-contraction (E-C) coupling. Disruption of E-C coupling by crude oil leads to rhythm and contractility defects at the whole-heart level, which concomitantly leads to abnormal development if exposure occurs during cardiac morphogenesis. Studies spanning a range of oil exposure concentrations have now identified a spectrum of developmental defects, from acute embryo-larval heart failure and related secondary malformations to more subtle anatomical changes in outflow tract and ventricular structure, the latter corresponding to a permanent reduction of cardiorespiratory performance later in life. The state of the science now indicates that virtually all teleosts show a cardiotoxic response to crude oils. Moreover, nuanced differences in the nature of the cardiotoxicity

J. P. Incardona (✉) · N. L. Scholz
National Oceanic and Atmospheric Administration, National Marine Fisheries Service, Northwest Fisheries Science Center, Environmental and Fisheries Sciences Division, Seattle, WA, USA
e-mail: John.Incardona@noaa.gov; Nathaniel.Scholz@noaa.gov

© Springer International Publishing AG, part of Springer Nature 2018

W. Burggren, B. Dubansky (eds.), *Development and Environment*,
https://doi.org/10.1007/978-3-319-75935-7_10

across taxa are likely to be more influenced by species life history than by the geological origins of different oils. This is attributable to variations in the roles of specific macromolecular targets for PAHs in the hearts of fish with different swimming physiologies. Specifically, these targets include ion channels that control cardiomyocyte repolarization (e.g., rectifying potassium channels) and contraction (e.g., sarcoplasmic reticulum calcium channels). Overall, research on oil spills has shown how the early development and ecophysiology of fishes shape their vulnerability to a global environmental threat.

10.1 Introduction

10.1.1 General Considerations

On April 20, 2010, an uncontrollable blowout occurred at a wellhead on the seafloor in the Mississippi Canyon Block 252 (MC252) of the Macondo Prospect in the northern Gulf of Mexico. Soon thereafter a catastrophic explosion set fire to the Deepwater Horizon drilling rig, which sank 2 days later. The wellhead's blowout preventer and an attempt to plug the well with a blind shear ram both failed. In the ensuing weeks, an estimated 4.9 million barrels of crude oil was released into the northern Gulf until the wellhead was successfully capped on July 15. The Deepwater Horizon disaster was the largest spill in history for US waters, and it oiled fish spawning habitats from the seafloor to the ocean surface, from the pelagic zone to shoreline beach and marsh habitats (Fig. 10.1). The spill posed a major threat to commercial and recreational fisheries, including yellowfin tuna, red snapper, amberjack, groupers, mahi mahi, menhaden, and other species.

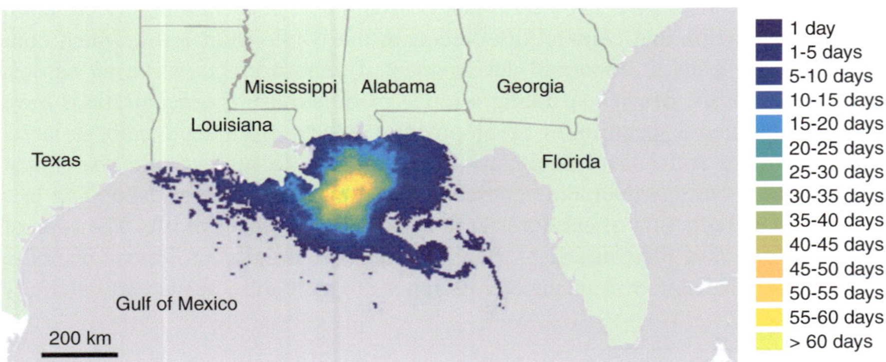

Fig. 10.1 Footprint of the Deepwater Horizon spill in the northern Gulf of Mexico. Indicated color gradient represents the modeled cumulative oiling in surface waters by number of days. These data represent the toxic exposure risk for pelagic-spawning fish species, most of which develop rapidly, e.g., 24–48 h. Map was generated with data from ERMA (National Oceanic and Atmospheric Administration 2017)

This chapter describes the impacts of crude oil on early development in fish, using assessment science that was implemented in the aftermath of Deepwater Horizon as a case study. However, oil toxicity to fish embryos and larvae has been an important area of research for decades, particularly in the years following the 1989 Exxon Valdez tanker spill in Prince William Sound, Alaska. At the time, Exxon Valdez was the largest spill in US history, and it spurred the congressional enactment of the Oil Pollution Act the next year. Fish early life stages are exceptionally vulnerable to crude oil-derived contaminants, and this chapter explores how our understanding of developmental toxicity evolved and expanded in the years leading up to Deepwater Horizon. As common threads through the literature, research to date has (for the most part) focused on four priority areas. The first is the identification of the most important target organs for crude oil toxicity in developing fish. Second, crude oils are highly complex mixtures, and there has been a considerable push to identify the primary chemical drivers of toxicity. The third aim has been an elucidation of mechanisms underlying developmental abnormalities at morphological, functional, and increasingly molecular scales. Finally, there is an ongoing need to better understand nuanced but important effects on physiology that are delayed in time, as contributors to delayed mortality. The long-term goal is a more holistic foundation for assessing individual and population-scale impacts of future oil spills.

10.1.2 Oil Spill Science Before Deepwater Horizon

10.1.2.1 1989 Exxon Valdez Oil Spill

Among aquatic species and life stages, it was known long before the Deepwater Horizon disaster that developing fish are among the most sensitive to low levels of crude oil. Prior to the 1989 Exxon Valdez spill in Prince William Sound, Alaska, toxicity studies with petroleum focused on traditional lethality endpoints (i.e., LC_{50} assays), usually with adult or juvenile organisms. For example, the opening of the trans-Alaska pipeline prompted an assessment of the sensitivity of a broad range of native Alaskan marine organisms in 96-h LC_{50} tests (Rice et al. 1979). These tests represented pelagic, benthic, and intertidal habitats and included fish, crustaceans, mollusks, echinoderms, and worms. While oil-derived hydrocarbon exposures were quantified using less detailed and less sensitive techniques relative to conventional modern methods, calculated LC_{50} values for these species were all in the mg/L range or parts per million (ppm). Pelagic fish and shrimp were identified as the most sensitive, and yet mortality projections for these species were in the range of 1–3 mg/L aromatic hydrocarbons. These and other studies (e.g., Anderson et al. 1974) generally showed that it takes a fairly high concentration of oil to kill even relatively small juvenile fish and invertebrates. However, other studies demonstrated that morphological abnormalities could arise in developing fish exposed to lower levels of oil, in some cases leading to the death of the test species (Linden 1978; Smith and Cameron 1979).

In the early stages of the Exxon Valdez spill, PAHs (Fig. 10.2b) were measured in surface waters of the spill zone as part of an extensive environmental surveillance

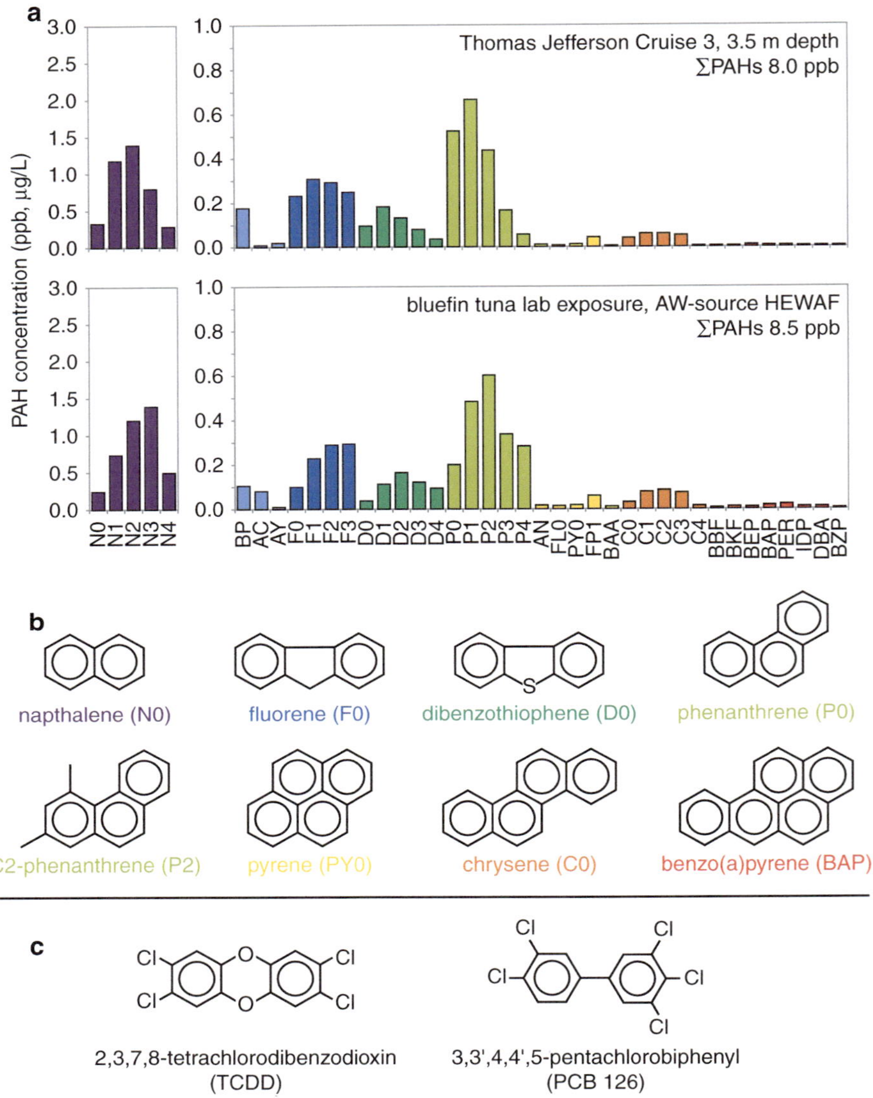

Fig. 10.2 Polycyclic aromatic hydrocarbons in crude oil and some organochlorine compounds. (**a**) Levels of PAHs measured in (top) a surface water sample from the Deepwater Horizon spill zone in the Gulf of Mexico taken June 2010 and (bottom) a high-energy WAF (HEWAF) preparation with Deepwater Horizon-MC252 crude oil used in laboratory exposure studies with bluefin tuna embryos (Incardona et al. 2014). (**b**) Structures of representative individual PAHs, color-coded to match the measured families of compounds in (**a**). Lower molecular weight 2–4 rings typically occur in homologous series of compounds with increasing alkylation (C1-, C2-, represented as N0–N4, F0–F3, etc.) that includes methyl and larger side groups. Focal compounds are naphthalenes (N0–N4, purple), fluorenes (F0–F4, blue), dibenzothiophenes (D0–D4, green), phenanthrenes (P0–P4, olive), pyrene/fluoranthenes (e.g., PY0, yellow), chrysenes (C0–C4, orange), and benzo(a)pyrene (BAP, red). (**c**) Structures of representative organochlorine compounds, 2,3,7,8-tetrachlorodibenzodioxin (TCDD) and 3,3′,4,4′,5-pentachlorobiphenyl (PCB126)

program. Even in water directly beneath surface oil slicks, maximum detections were in the μg/L or parts per billion (ppb) range and thus orders of magnitude below laboratory-derived LC_{50}s (Short and Harris 1996). However, the timing of the spill coincided with the spawning of Pacific herring (*Clupea pallasi*) and pink salmon (*Oncorhynchus gorbuscha*) in Prince William Sound. Both species deposit eggs in nearshore or intertidal zones. Their developing embryos were therefore particularly susceptible to PAH exposure from oil stranded on the shoreline. Based on the prior studies suggesting higher sensitivity of fish early life history stages, sampling efforts focused on collecting herring larvae along oiled shorelines and pink salmon alevins from redds (nests) in the gravel beds of stream deltas where they crossed oiled shorelines. These field studies demonstrated an increase in morphological abnormalities in herring larvae and increased mortality in pink salmon alevins associated with oiled shorelines (Bue et al. 1996; Hose et al. 1996; Marty et al. 1997a and 1997b; Norcross et al. 1996).

Controversy surrounding the potential for developmental injury in response to oil exposures that, at the time, were considered to be very low prompted a series of laboratory studies led, in part, by researchers at the NOAA Auke Bay research laboratory. These investigations established a connection between oil exposure and a specific syndrome of morphological defects in embryos and larvae. The groundbreaking work further implicated PAHs as the causative components in crude oil. Oiled gravel columns were developed to emulate the transport of water-soluble components of oil from contaminated beach substrates into adjacent water (Short and Heintz 1997). Incubation of embryos in effluent from these oiled gravel columns demonstrated that direct contact with oiled substrates was unnecessary for toxicity. Both Pacific herring and pink salmon embryos exposed this way developed a suite of visible defects that mirrored those documented in field-collected samples. These included, most evidently, the accumulation of fluid (edema or ascites) in the pericardial and yolk sac spaces (Marty et al. 1997a and 1997b) (Fig. 10.3). Importantly, direct measures of petroleum-derived chemicals accumulated by embryos pointed to PAHs and not other common chemicals such as alkanes or monoaromatic compounds such as benzene or toluene (Marty et al. 1997b). Additional dose-response studies demonstrated that very low levels of dissolved PAHs from crude oil were highly concentrated by herring and salmon embryos, consistently leading to the same suite of abnormalities (Carls et al. 1999; Heintz et al. 1999; Marty et al. 1997b). Effects concentrations in the smaller herring embryos were in the range of 1–10 μg/L (ppb) total ($\sum$)PAHs, with fluid accumulation around the heart or in the yolk sac being the most sensitive indicator of toxicity (Carls et al. 1999).

The Alaskan studies also yielded the first evidence of delayed mortality, specifically in pink salmon. Conventional hatchery operations made it possible to design experiments wherein very large numbers of embryos were exposed to a range of oil concentrations. Embryos at the low end of the dose-response curve were subsequently reared in clean water to outwardly normal-appearing juveniles, tagged, and then released into the ocean. Upon returning to fresh water 2 years later, pink salmon in the oil exposure treatments showed consistently lower marine survival (Heintz 2007; Heintz et al. 2000). These findings suggested that crude oil impacts to fisheries resources could extend well beyond grossly abnormal embryos and larvae.

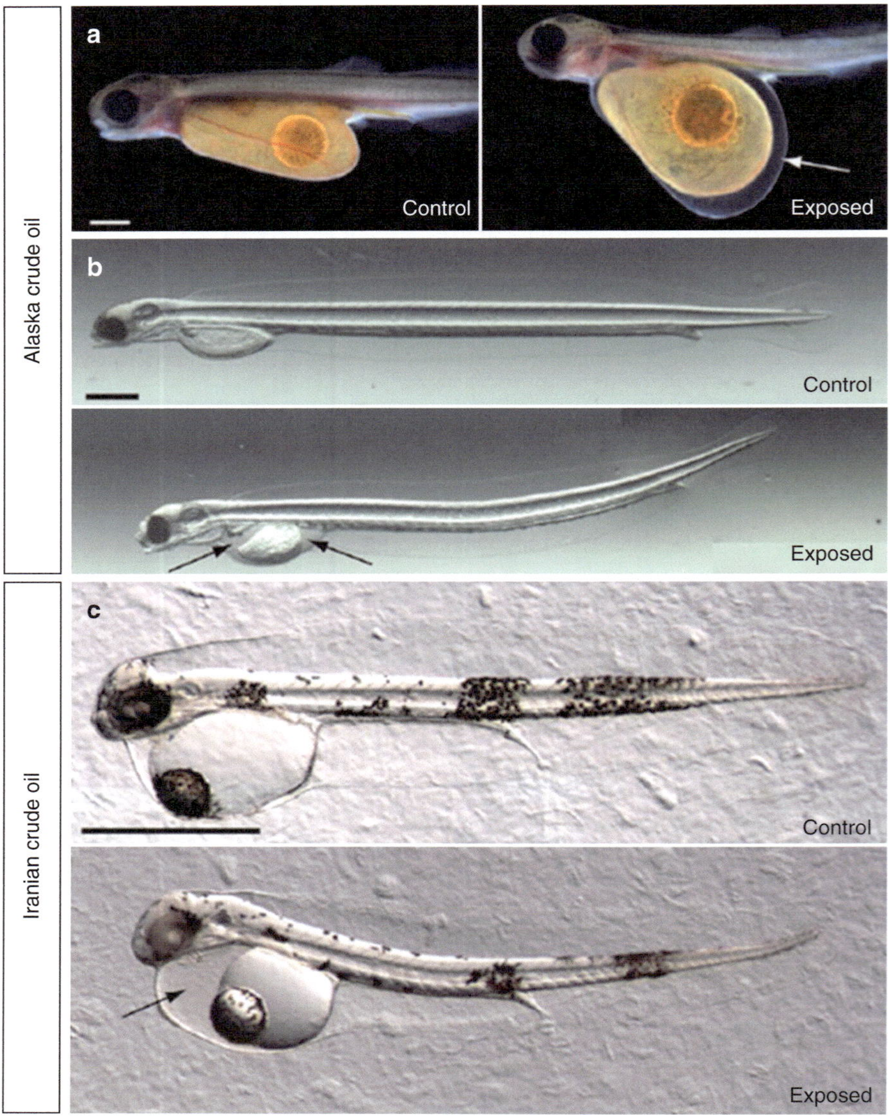

Fig. 10.3 Gross phenotypes associated with exposure to crude oil in fish species studied prior to the Deepwater Horizon spill. Each pair represents a control and exposed animal, as indicated in the lower right corner. (**a**, **b**) Crude oil exposure syndrome resulting from Alaska North Slope crude oil in (**a**) Pink salmon and (**b**) Pacific herring. (**c**) Japanese sea perch exposed to Iranian heavy crude oil from the Hebei Spirit spill in South Korea. Arrows indicate yolk sac and/or pericardial edema. Scale bars are 2 mm (**a**) and 1 mm (**b**, **c**)

10.1.2.2 Cosco Busan and Hebei Spirit Spills

As detailed below, in the years leading up to the Deepwater Horizon spill, laboratory studies involving both wild marine species and the zebrafish model identified the developing heart as the primary target of PAH developmental toxicity. However, the lessons learned from Exxon Valdez were also applied to two other significant oil spills in the intervening period. In late 2007 the container ship Cosco Busan collided with the San Francisco-Oakland Bay Bridge, spilling residual fuel (bunker) oil into San Francisco Bay. A few weeks later, off the west coast of South Korea, a barge collided with the oil tanker Hebei Spirit, causing a very large spill of Middle Eastern crude oil. Both spills threatened regional fisheries resources. The Cosco Busan spill occurred several weeks before the annual spawning of Pacific herring in San Francisco Bay, while the Hebei Spirit spill contaminated shorelines along some of Korea's most productive aquaculture and fisheries areas on the Taean Peninsula.

Given that edema is a very sensitive indicator of oil exposure in herring embryos, field assessments of the Cosco Busan spill relied on deployment of caged Pacific herring embryos along oiled and unoiled (reference) shorelines. Embryos were allowed to incubate to just before hatching, whereupon they were retrieved and returned to a laboratory for live assessment of cardiac function and measurement of edema by video microscopy. Naturally spawned herring embryos were also collected from oil and unoiled sites and analyzed. These field studies identified an unexpectedly lethal phototoxic effect associated with bunker oil (Incardona et al. 2012a; Incardona et al., 2012b). However, this study also showed that pericardial edema was an extremely sensitive indicator of oil exposure in the field, with increased incidence at oiled sites up to 2 years following the spill (Incardona et al. 2012a). The release of Iranian heavy crude oil from the Hebei Spirit spill prompted comparative and detailed studies of crude oils from distinct geological origins. Direct comparisons of the Iranian oil to the Alaska North Slope crude oil studied after Exxon Valdez demonstrated nearly identical cardiotoxic responses in zebrafish embryos (Jung et al. 2013). The Korean assessment effort also included a broader array of species, focusing on pelagic-spawning perciforms such as olive flounder (*Paralichthys olivaceus*) and Japanese sea bass (*Lateolabrax japonicus*) (Jung et al. 2015). Both species (Fig. 10.3c and not shown) displayed a developmental toxicity syndrome virtually indistinguishable from that described for Pacific herring (Incardona et al. 2009; Jung et al. 2015). To date, the canonical crude oil developmental toxicity syndrome has been demonstrated in at least 19 fish species (Table 10.1).

10.1.3 A Common Oil Toxicity Syndrome with Species-Specific Differences

10.1.3.1 Heart Anatomy and Development in Fish

Exposure to crude oil or other sources of complex PAH mixtures generally leads to a common syndrome of edema accumulation in fish embryos, often in association with craniofacial defects. At a more detailed level, there are often distinct

Table 10.1 Fish species tested for crude oil developmental toxicity

Species	Reference
Atlantic herring (*Clupea harengus*)	McIntosh et al. (2010)
Pacific herring (*Clupea pallasi*)	Carls et al. (1999)
Mahi mahi (*Coryphaena hippurus*)	Edmunds et al. (2015)
Zebrafish (*Danio rerio*)	Incardona et al. (2005)
Mummichog (*Fundulus heteroclitus*)	Couillard (2002)
Gulf killifish (*Fundulus grandis*)	Dubansky et al. (2013)
Atlantic cod (*Gadus morhua*)	Sørensen et al. (2017)
Japanese sea perch (*Lateolabrax japonicus*)	Jung et al. (2015)
Atlantic haddock (*Melanogrammus aeglefinus*)	Sørhus et al. (2016)
Crimson-spotted rainbowfish (*Melanotaenia fluviatilis*)	Pollino and Holdway (2002)
Silversides (*Menidia beryllina*)	Adeyemo et al. (2015)
Pink salmon (*Oncorhynchus gorbuscha*)	Marty et al. (1997)
Medaka (*Oryzias latipes*)	González-Doncel et al. (2008)
Marine medaka (*Oryzias melastigma*)	Mu et al. (2014)
Olive flounder (*Paralichthys olivaceus*)	Jung et al. (2015)
Red drum (*Sciaenops ocellatus*)	Khursigara et al. (2017)
Yellowtail amberjack (*Seriola lalandi*)	Incardona et al. (2014)
Yellowfin tuna (*Thunnus albacares*)	Incardona et al. (2014)
Southern bluefin tuna (*Thunnus maccoyii*)	Incardona et al. (2014)

species-specific differences in toxicity phenotypes, involving both cardiac and extracardiac features. Our understanding of the syndrome has been expanded considerably in the aftermath of the 2010 Deepwater Horizon wellhead blowout, which spurred additional studies on a diversity of Gulf of Mexico fish species. The current state of the science suggests that the precise impacts of crude oil on early development stem from a number of differences between species. These include variation in basic heart forms among fish families as well as ecophysiological traits such as temperature range, demersal or pelagic larval form, and egg size (Incardona 2017; Incardona and Scholz 2016).

The fish heart consists of two main chambers, a single atrium and ventricle, with additional "chambers" preceding the atrium (sinus venosus) and forming the outflow tract from the ventricle (bulbus arteriosus). Superimposed on this common structure, are three basic ventricular forms that appear to be related to swimming biology (Santer et al. 1983). The most common form is the saccular or purse-like ventricle found in species that are primarily burst swimmers or cryptic predators, such as most gadids (cod family), sculpins, and flatfish. Fish that are high-speed predators (such as tunas), continuous swimmers (pelagic planktivores such as herring), or highly migratory (salmonids) have a pyramidal ventricle that is optimized for high cardiac output (Claireaux et al. 2005). The third and least common form is a tubular heart, found in some Gadiformes such as European hake (*Merluccius merluccius*). Another structural distinction in heart morphology among species is the form of the ventricular myocardium. All fish have a highly trabeculated or spongy myocardium, lined by the endocardium. Fish with higher cardiorespiratory performance and metabolic demand have a compact myocardial layer surrounding the spongy layer and infused with a coronary blood supply. In general, a compact myocardium is present in species with pyramidal ventricles and absent in those with saccular ventricles (Agnisola and Tota 1994; Farrell 1987; Pieperhoff et al. 2009; Sanchez-Quintana et al. 1995; Santer and Greer Walker 1980; Tiitu and Vornanen 2002), although there are exceptions (Simoes et al. 2002).

Heart development in fish has been characterized most extensively at both the morphological and molecular level for zebrafish, a fast-swimming species with a pyramidal ventricle and compact myocardium (Burggren and Bagatto 2008; Burggren et al. 2017; Hu et al. 2000, 2001). While later stages of heart development are likely to differ among the three basic fish heart forms, early heart development appears to be very similar across species (Fig. 10.4) (Glickman and Yelon 2002; Incardona et al. 2014; Sørhus et al. 2017). Briefly, cardiac progenitor cells are determined in bilateral fields of mesoderm during early somitogenesis stages, segregating into pools of atrial and ventricular precursors. These fields fuse at the midline to form the cardiac cone, in which atrial and ventricular cells form concentric rings (viewed dorsoventrally), with the ventricle at the cone apex and oriented dorsally. The cone then rotates laterally, with the atrium to the left, followed by the chambers bending or "looping" to bring the atrium and ventricle into a side-by-side orientation.

Typically near the hatching stage for most species, ventricular cardiomyocytes begin to proliferate to increase chamber wall thickness. In the early larval stage, ventricular cardiomyocyte proliferation accelerates as trabeculation commences. Formation of the compact myocardium also begins at this stage and has been described in detail in zebrafish and other species (Burggren et al. 2017; Gupta and Poss 2012). A sophisticated clonal analysis identified three adult ventricular cardiomyocyte lineages created in sequence from each other during embryogenesis and early larval development. Cardiomyocytes in the early single-cell thick ventricular wall give rise to a "primordial" monolayer of cells that persist in the juvenile compact myocardium. During trabeculation, a new cardiomyocyte lineage is derived from this primordial layer by delamination and subsequent migration, intermixing, and proliferation. During juvenile to adult growth, the primordial layer retains its single-cell thickness but becomes covered by an outer "cortical" layer of compact myocardium. The origin of these cardiomyocytes in the cortical compact myocardium appears to be clones of cells that emerge from the trabeculae and migrate through the primordial layer. While the primordial layer descends from roughly several dozen embryonic ventricular cardiomyocytes, the cortical layer arises from fewer than ten clonally dominant cells derived from the embryonic ventricle. Therefore, a very small number of embryonic cells ultimately contribute to the adult compact myocardium.

For fish with a compact myocardium, a perturbation of the normal ontogeny of these cells during early development as a consequence of oil exposure is likely to cause lasting changes in heart form and function later in life. The morphogenetic processes that produce a compact myocardium are also a basis for species-specific differences in oil injury phenotypes.

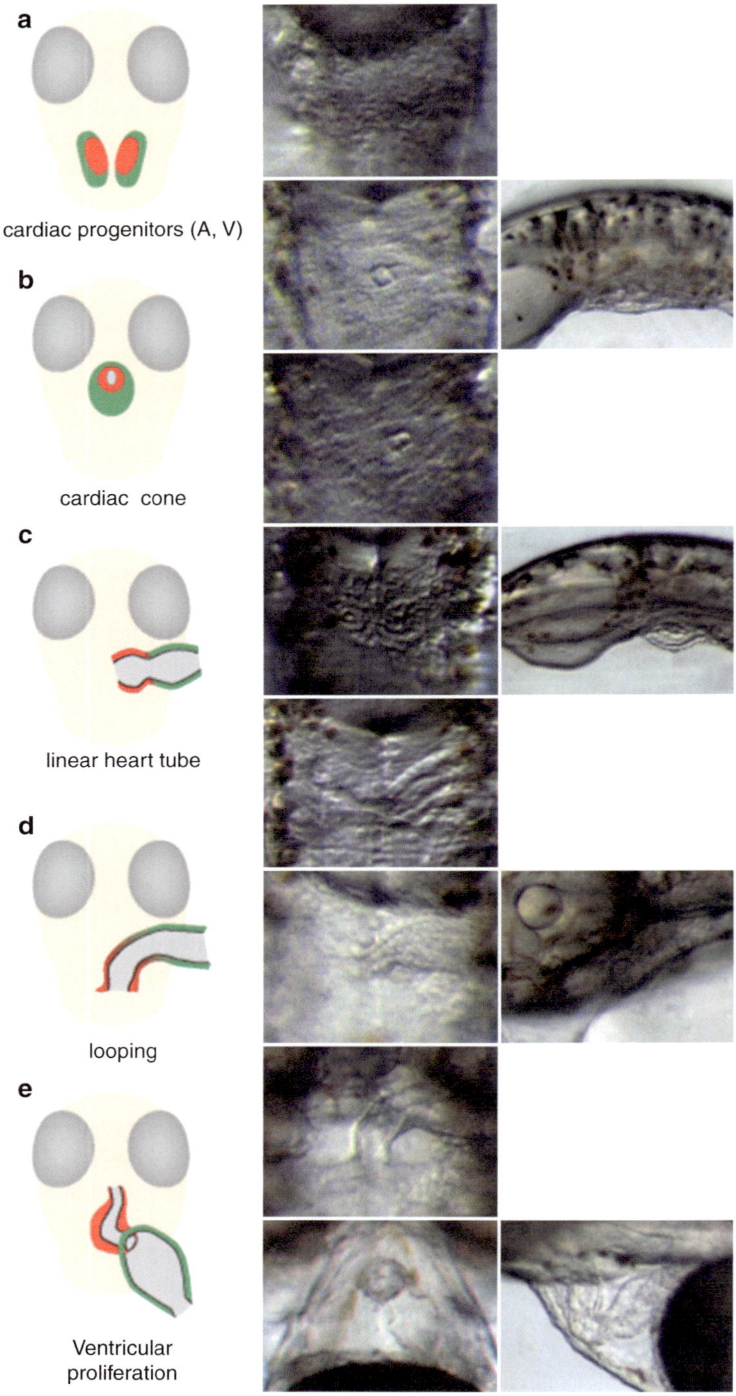
a
cardiac progenitors (A, V)
b
cardiac cone
c
linear heart tube
d
looping
e
Ventricular
proliferation

10.1.3.2 Larval Fish Forms in Relation to Physiological Constraints of Habitat

At the larval stage, the nature and severity of both cardiac and extracardiac abnormalities in fish exposed embryonically to oil also appear to depend on general larval shape, which in turn is related to larval habitat. Larvae occupying demersal vs. pelagic habitats have distinct body forms, as first described by Shelbourne in the mid-twentieth century (Fig. 10.5). Shelbourne was also the first to make a connection between these differences and osmoregulation and buoyancy control (Shelbourne 1955, 1956). Fish fins develop within epidermal finfolds, with the median or unpaired fins (dorsal, anal, and caudal) evolutionarily older than the paired fins (pectoral and pelvic). All fish larvae have an initially continuous median finfold that extends along the dorsal and ventral midline. The rayed fins develop at the appropriate position within the finfold from condensations derived from somatic mesoderm (Mabee et al. 2002; Freitas et al. 2006). However, the structure of the median finfold differs dramatically between pelagic and demersal larvae (Fig. 10.5a). Pelagic larvae have extremely large median finfolds that enclose an extensive subdermal space, whereas demersal larvae generally have much smaller finfolds with little subdermal space—i.e., an epidermis closely opposed to the underlying mesoderm. In pelagic larvae, moreover, this median subdermal space is also relatively enlarged over the cephalic region. While the adult form of fin rays begin to appear soon after hatching in demersal freshwater larvae such as zebrafish, the pelagic median finfold remains rayless for a much longer interval (Shelbourne 1956).

The differences in marginal finfold structure and associated subdermal space between pelagic and demersal fish larvae are also linked to differences in vascularization of the yolk sac. Demersal larvae are characterized by early development of yolk sac blood vessels, in many cases having highly vascularized yolk sacs as epitomized by the salmonids (Fig. 10.5b). In contrast, pelagic larvae have avascular yolk sacs, with the yolk mass enclosed in a yolk sac sinus that is continuous with the marginal finfold subdermal space and topologically equivalent to the intravascular space (Fig. 10.5c). In a series of papers, Shelbourne related both the presence of the

Fig. 10.4 Heart development in fish. Left column shows a cartoon schematic of normal heart development based largely on zebrafish. The embryonic head region is shown as a ventral view (anterior at top) and atrial and ventricular cell types indicated by green and red, respectively. Photomicrographs show comparable stages in red drum embryos, with ventral views in the middle column and lateral views in the right column. For ventral view micrographs anterior is at top; for lateral views anterior is to the left. (**a**) Cardiac progenitors are determined in bilateral fields of mesoderm during early somitogenesis stages. After midline fusion, the cardiac cone is formed (**b**), with the ventricle at the apex oriented dorsally and appearing as a ring in dorsal/ventral views. Coordinated contractions typically begin by the linear heart tube stage (**c**), after the cone rotates laterally. Looping (**d**) involves the bending of the ventricle to bring it adjacent to the atrium. By the early hatching stage, ventricular cardiomyocytes proliferate (**e**) to increase the thickness of the chamber's myocardial wall. After hatching, ventricular proliferation accelerates with trabeculation of the spongy myocardium (not shown), but studies described here did not extend to this stage

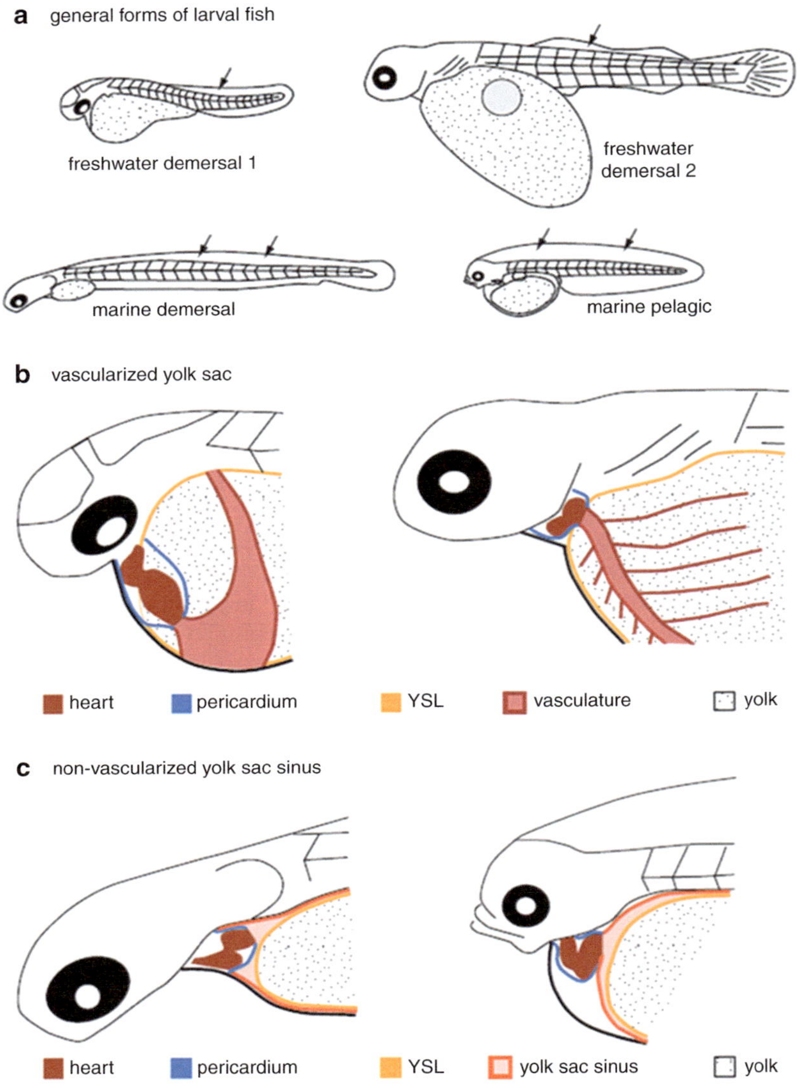

Fig. 10.5 Anatomy of larval fish in relation to habitat. (**a**) Representative general larval forms associated with the three major spawning types, freshwater demersal (top), marine demersal (bottom left), and marine pelagic (bottom right). Two general forms are associated with freshwater demersal spawners, type 1, characteristic of species with small eggs and rapid development (e.g., zebrafish), and type 2, characteristic of species with larger eggs and yolk mass and more protracted development (e.g., salmonids, *Fundulus*). Arrows indicate marginal finfolds. After Shelbourne (Shelbourne 1956). (**b**) Schematic representation of cardiovascular and yolk sac anatomy in freshwater demersal species with vascularized yolk sacs (left, type1; right, type 2). Type 2 species tend to have much more highly vascularized yolk sacs. (**c**) Anatomy of the cardiovascular system and yolk sac in marine demersal (left) and marine pelagic species (right) with non-vascularized yolk sac sinuses. In all fish larvae, the yolk mass is separated from the vascular space by the yolk syncytial layer (YSL), a transient multicellular membrane that first plays important roles in early embryonic patterning then a later metabolic role during yolk absorption

extensive subdermal space and the avascular yolk sac sinus to the osmoregulatory and buoyancy control demands of the pelagic marine environment (Shelbourne 1955, 1956, 1957). The dorsal subdermal spaces of pelagic larvae are connected to the yolk sac sinus via channels in the vicinity of the pectoral finfold. Shelbourne (1955) demonstrated that all of the peripheral subdermal spaces act as a vascular space continuous with the yolk sac, which in many species has the highest density of cells involved in osmoregulation. Thus, in early pelagic larvae and prior to the development of the gills, the yolk sac functions as the primary osmoregulatory and respiratory organ. At the same time, extracellular fluid circulation through these subdermal spaces is the likely delivery pathway for yolk-derived nutrients to developing tissues. Therefore, the finfold and other subdermal compartments in pelagic larvae are topologically "intravascular," while the corresponding compartments in demersal larvae are extravascular. This leads to different patterns of edema formation and accumulation in oil-exposed pelagic larvae and hence different outcomes in terms of extracardiac deformities that are secondary to cardiotoxicity. These differences in anatomy are subsequently intertwined with other species-specific differences relating to the uptake and elimination of oil-derived chemicals (toxicokinetics), and precise molecular mechanisms (toxicodynamics), contributing to variability in responses to oil exposure.

10.1.3.3 The Roles of Toxicokinetics vs. Toxicodynamics

The specific responses of organisms to toxic chemical exposure are influenced by contaminant uptake, distribution, and elimination, or toxicokinetics (Nikinmaa 2014a, b, c, d). At the same time, individual species may differ physiologically, by expressing different forms or amounts of a cellular target for PAHs and other toxicants (e.g., a specific protein). In this case, differences in toxicodynamics can influence relative susceptibility. Species differences in toxicokinetics and/or toxicodynamics present an ongoing challenge to understanding the mechanisms underlying the crude oil toxicity syndrome, as well as the development of quantitative tools to establish resource injury following spills in fish spawning habitats. In developing fish, toxicokinetic differences can arise in many ways, from very simple differences in structure, such as size, to more complicated variations, such as the time at which key metabolic organs such as the liver are formed. In addition, there may be genetic differences between species, such as variants of detoxifying enzymes, leading to differences in toxicokinetics.

In general, organisms have evolved mechanisms of defense against toxic chemicals, and these are active in developing fish embryos. These include pumps and transporters that use ATP to transfer foreign chemicals (xenobiotics) across membranes (e.g., ATP-binding cassette proteins (Ferreira et al. 2014)) and enzyme systems that metabolize or degrade xenobiotics (Nikinmaa 2014c). For crude oil and PAHs, the latter is exemplified by the aryl hydrocarbon (AH)—cytochrome P4501A (CYP1A) pathway (Fig. 10.6). A fish's ability to metabolize and eliminate PAHs (and other compounds) is a critical determinant of crude oil toxicity. Many (but not all) PAHs are strong ligands for the AH receptor (AHR), a ligand-activated transcription factor that controls expression of a gene battery dedicated to metabolism and elimination of primarily planar polyaromatic compounds. This includes Phase I

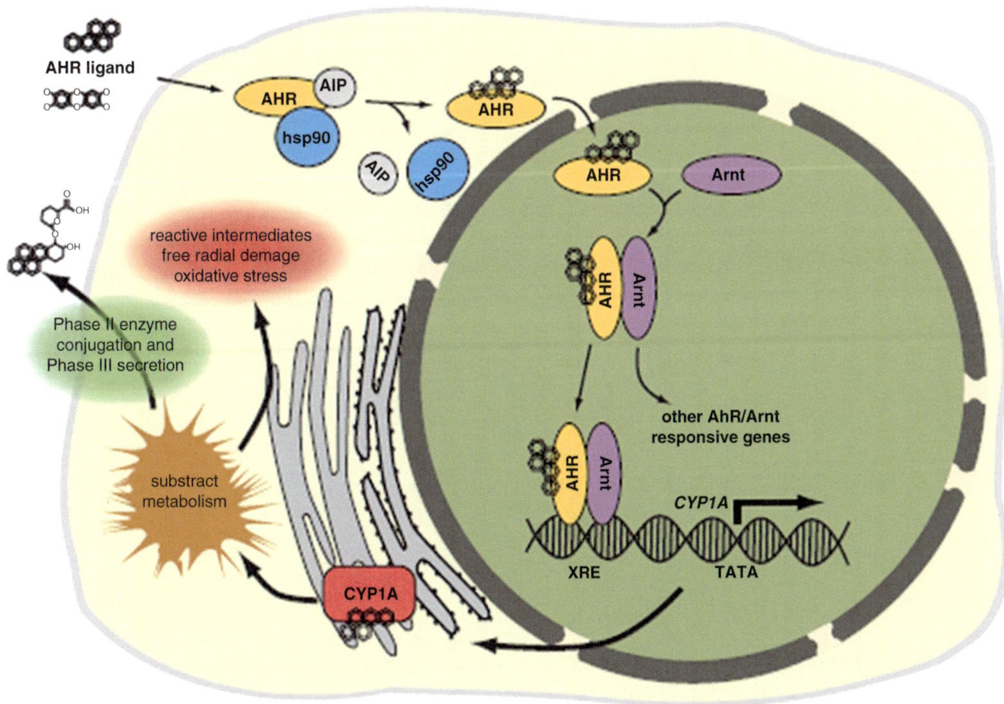

Fig. 10.6 Schematic of the basic xenobiotic metabolism pathway controlled by the aryl hydrocarbon receptor. The AHR resides in a cytoplasmic complex with hsp90 (heat-shock protein 90) and AIP (AHR-interacting protein), which dissociates upon ligand binding, allowing the ligand-bound AHR to translocate into the nucleus. After dimerizing with ARNT, the AHR-ARNT dimer binds to xenobiotic response elements (XRE) in the promoter region of genes involved in detoxification, with *CYP1A* being the most prominent. CYP1A enzyme accumulates in the endoplasmic reticulum and oxidizes planar aromatic compounds to produce reactive intermediates that are modified by Phase II enzymes for secretion. Uncoupling of CYP1A metabolism from Phase II enzymes can lead to accumulation of these reactive intermediates, which can cause cellular damage by creating DNA adducts, for example. This is the basis of carcinogenesis by PAHs such as benzo(a)pyrene

enzymes such as CYP1A, which oxidize PAHs to reactive intermediates, and Phase II enzymes such as glutathione-S-transferases and glucuronosyltransferases that conjugate the reactive intermediates to hydrophilic moieties allowing secretion from cells and ultimately excretion from the animal (Kohle and Bock 2007; Nikinmaa 2014c). The induction of CYP1A expression in fish tissues is a rapid and robust indicator of PAH exposure and for decades has been used as an extremely sensitive biomarker of exposure in oiled habitats (Payne 1976).

In the disciplines of both toxicology and pharmacology, the distribution of chemicals within organisms has been studied most intensively in adult life stages, a time at which key metabolic organs such as the liver and kidney are fully formed and functional. The volume of distribution concept dictates that, all else being equal, animals of different sizes will require more or less of a chemical exposures (dose) to achieve the same tissue concentration. This is primarily why juvenile or adult fish are much less sensitive to crude oil toxicity than developing embryos and larvae. PAHs are rapidly eliminated from fish largely by hepatic metabolism via the AHR/CYP1A pathway and excretion through bile (Kammann et al. 2017; Pampanin et al. 2014; Sette et al. 2013; extensively reviewed by Varanasi 1989). Post-metamorphic fish with scales take up crude oil-derived chemicals either through ingestion or across the gills. In each case, PAHs that reach the bloodstream will pass through the liver, which provides a high level of "first-pass" metabolic detoxification. By contrast, fish embryos and larvae may have only a rudimentary liver or no liver at all, and uptake of crude oil compounds occurs directly across the unscaled epidermis. Consistent with this, fish embryos exposed to crude oil or potent AHR ligands show the strongest induction of CYP1A in the epidermis (Incardona et al. 2009, 2005, 2006, 2011; Jung et al. 2013; Sørhus et al. 2016). The very small size of fish embryos coupled with this low metabolic capacity for detoxification results in a proportionally high bioconcentration of PAHs from even very low water concentrations of crude oil.

Among embryos of different species, toxicokinetic differences will result from size alone. Given that embryos within the chorion are spherical, higher surface-to-volume ratios will lead to more rapid uptake and higher bioconcentration in species with smaller eggs. For hydrophobic compounds such as PAHs, bioconcentration will then also be influenced by lipid content of both yolk and embryonic tissues. This phenomenon is most evident from direct comparisons between Pacific herring embryos (~1.3 mm diameter, low lipid) and pink salmon embryos (~6 mm, high lipid). For these two species, a 50-fold difference in waterborne PAH exposures yielded an equivalent tissue PAH concentration (Incardona et al. 2015). Metabolic capacity and bioconcentration will also be influenced by the degree of gastrointestinal tract and liver organogenesis that occurs during embryogenesis relative to larval development. For example, most marine species developing from small eggs hatch after minimal organogenesis, while species with larger lipid-rich eggs such as salmonids or cyprinodontiformes (killifishes; *Fundulus* spp) hatch at stages with more advanced liver development. As discussed in more detail below, this variation in the timing of organogenesis may also underlie some of the differences in toxic response observed among species. The toxicokinetic attributes of small fish embryos also have implications for elucidating cardiotoxic mechanisms. Early cardiac function in fish embryos is entirely driven by intrinsic activity of cardiomyocytes, as external influences such as the autonomic nervous system develop later (Burggren and Bagatto 2008; Burggren et al. 2017; Schwerte et al. 2006). Coupled with the minimal metabolic protection provided by extracardiac tissues, the embryonic fish heart behaves in many ways like isolated cardiomyocytes in vitro, providing an assay for direct effects of chemicals on cardiomyocytes (e.g., Langheinrich et al. 2003; MacRae 2013; Milan et al. 2003).

10.1.3.4 Species Ecophysiology as a Basis for Differences in Toxicodynamics

Fish have evolved to occupy an exceptionally diverse range of habitats, with some species showing remarkable plasticity. In particular, fish are found across a wide range of temperature regimes. Polar and boreal species tolerate extreme cold, but within a narrow range, whereas some intertidal species or deep-diving predators such as tunas can tolerate rapid, wide swings in temperature from warm to very cold and back. While multiple mechanisms underlie this diversity, such as enzymatic adaptations in intertidal species (Somero 2002) or endothermy in tunas (Block et al. 1993), differences in cardiac physiology are consistently an important factor.

The heart is a working pump driven by physiologically complex excitable cells. The cyclical contraction and relaxation of cardiac muscle represents a rhythmic sequence of mechanical and electrical events at the cellular level. Contraction, or systole, is initiated by depolarization of cardiomyocytes by action potentials, while return of the cardiomyocyte membrane potential to the resting state, or repolarization, leads to relaxation (diastole) in preparation for the next cycle of contraction. The process of excitation-contraction (EC) coupling (Fig. 10.7a) links

→

Fig. 10.7 Excitation-contraction and excitation-transcription coupling in cardiomyocytes and effects of PAHs. (**a**) A simplified schematic of normal E-C and E-T coupling shows the interconnectedness of SR and nuclear Ca^{2+} pools. Generally, depolarization by action potentials causes entry of extracellular Ca^{2+} through L-type calcium channels (LTCC) located in T-tubules, in turn triggering calcium-induced calcium release from the SR via the RyR. Elevated cytoplasmic Ca^{2+} (red arrows) binds actin-myosin myofilaments leading to contraction. Resting Ca^{2+} levels are restored by SERCA2-mediated pumping back into the SR and extracellular transport by the sodium-calcium exchanger NCX1 (green arrows). SERCA2 also restores nuclear envelope Ca^{2+} levels. Resting membrane potentials are restored for the next contraction cycle in part by the activity of rectifying potassium channels such as ERG. In E-T coupling, elevated cytoplasmic and nuclear Ca^{2+} (also derived from release by inositol triphosphate receptors; IP_3R) leads to activation of calmodulin (CaM) and calcineurin (CaN), the latter a phosphatase that modifies and activates calcium-sensitive transcription factors (CaTF) such as NFATc and myocardin. (**b**) Model for the impacts of PAHs on action potential duration, E-C and E-T coupling (detailed in Sects. 10.4 and 10.5.2.3). Crude oil WAFs and phenanthrene have been shown to block the I_{Kr} current/ERG channel, leading to action potential prolongation and arrhythmia. In addition, cardiomyocyte exposure leads to SR Ca^{2+}depletion, either through effects on SERCA2 or RyR. Phenanthrene may also block Ca^{2+} entry through the LTCC. Net effects are overall reduced Ca^{2+} available for E-C and E-T coupling (dashed arrows), with potential cytoplasmic Ca^{2+} overload contributing to reduced myofiber relaxation. The potential action of PAHs on SERCA2 or other proteins controlling nuclear Ca^{2+} suggest a mechanism for direct effects on cardiomyocyte gene expression through altered activation of Ca^{2+}-dependent transcription factors. Model derived from Brette et al. (2014, 2017)

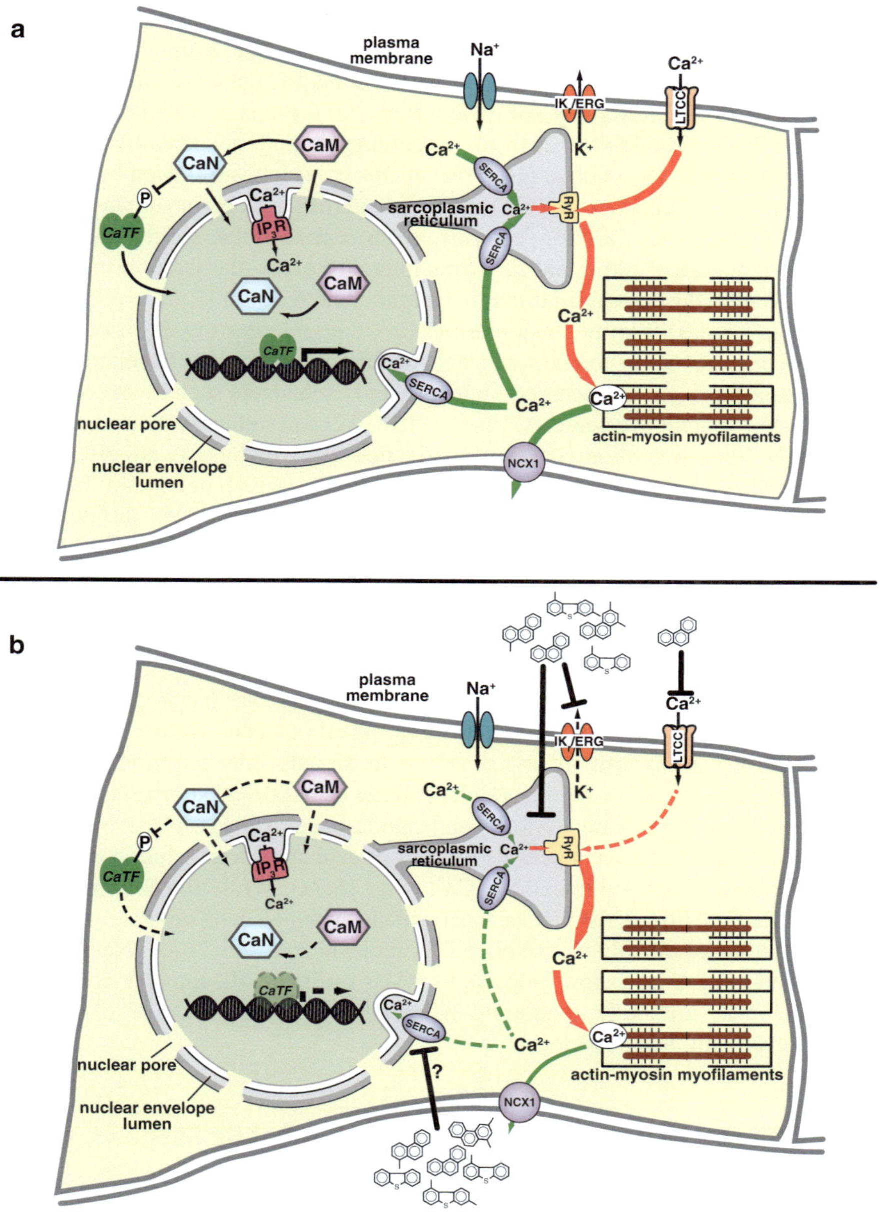
a
plasma membrane
Na+
Ca2+
IK/ERG
LTCC
K+
CaM
CaN
P
CaTF
Ca2+
IP3R
SERCA
sarcoplasmic reticulum
RyR
actin-myosin myofilaments
nuclear pore
nuclear envelope lumen
NCX1
b
?

depolarization (an electrical event) to the contractile activity of actin-myosin myofibrils (a mechanical event) within cardiomyocytes. The action potential and EC-coupling processes are controlled by fluxes of Na^+, K^+, and Ca^{2+} ions, described at a basic level as follows (see for review Bers 2002; Priest and McDermott 2015; Klabunde 2017; Schmitt et al. 2014). First, action potentials are generated by inward Na^+ currents carried by voltage-gated Na^+ channels, which is followed by the opening of L-type Ca^{2+} channels (LTCC) that reside in the cardiomyocyte plasma membrane. This leads to Ca^{2+} ion entry from the extracellular space, triggering a subsequent release of additional Ca^{2+} from internal stores—the sarcoplasmic reticulum (SR)—into the cytoplasm through the ryanodine receptor (RyR; RyR2 isoform in cardiomyocytes). Elevated cytoplasmic Ca^{2+} then initiates myofibril contraction by regulating the interaction between actin and myosin filaments. Baseline levels of cytoplasmic Ca^{2+} ions are restored by pumps and exchangers such as sarcoplasmic-endoplasmic reticulum calcium ATPase 2 (SERCA2) and sodium-calcium exchanger 1 (NCX1). Finally, cardiomyocyte repolarization is driven by outward efflux of K^+ through voltage-gated channels such as *erg* (ether-á-go-go related gene). In actuality, E-C coupling is more complex, involving subtle but important differences in cardiomyocyte cell type (e.g., atrial chamber vs. ventricular chamber) and the life history-based cardiac physiology of a particular species, such as basal heart rate. Moreover, both the passive- and energy-dependent movements of ions are highly temperature-dependent. Overall, there are different and complex suites of ion channel subtypes expressed in the hearts of different species.

In addition to playing a central role in E-C coupling, Ca^{2+} has long been known as a second messenger linking extracellular signals to gene expression. Nuclear signaling by Ca^{2+} is usually associated with signals that act through inositol 1,4,5-triphosphate (IP3) receptors (Hohendanner et al. 2014; Patterson et al. 2004), but RyRs are also implicated (Alonso and Garcia-Sancho 2011). The evolving view is that excitable cells have a parallel process of excitation-transcription (E-T) coupling also mediated by intracellular Ca^{2+} (Fig. 10.7a). While the SR and nuclear pools of Ca^{2+} are linked, the precise mechanisms by which cells separately integrate these signals are not fully characterized (Ljubojevic and Bers 2015). Nevertheless, as detailed in sections below, impacts of PAHs on both E-C and E-T coupling in cardiomyocytes are likely to underlie the most pronounced effects of oil exposure on the developing fish heart.

10.2 Cardiotoxic Severity Determines Early Life Stage vs. Delayed Mortality in Fish

As with most forms of chemical-induced developmental toxicity, the nature and severity of the crude oil injury phenotype changes along a gradient of exposure concentrations. In the related field of developmental genetics, a primary goal is often to isolate one or more total loss-of-function (null) mutants that produce a robust, easily characterized phenotype. The desired outcome is an unequivocal determination of gene function. Typically, such mutations are either lethal or they produce severe malformations in embryos. For the purposes of environmental health research, a central aim of toxicological studies is to relate real-world contaminant exposures to adverse health outcomes in individual organisms, in ways that can ultimately be scaled to the abundance, productivity, and conservation of wild populations (Baldwin et al. 2009; Garcia-Reyero 2015; Scholz and Incardona 2015; Villeneuve et al. 2014).

As discussed in Sect. 10.1 of this chapter, controlled laboratory and field studies on pink salmon following the Exxon Valdez spill provided the first direct evidence of delayed mortality (reduced marine survival) in fish that were transiently exposed to very low levels of oil during early development but appeared outwardly normal when they grew to the ocean migrant (smolt) stage (Heintz 2007; Heintz et al. 2000). Hence, concentration-response studies are essential as well as studies investigating critical windows for exposure for understanding the full spectrum of survival outcomes that may be caused by a developmental exposure to crude oil. Given the complexity of both cardiac function and morphogenesis, it is not surprising that previously observed phenotypes have been variable across a range of exposures. Additionally, given the central and species-specific role of circulation in multiple developmental and homeostatic processes, phenotypic differences can also arise from variation in extracardiac toxicity.

10.2.1 Cardiac Function and Circulation Defects in Response to Relatively High Levels of Oil

Most of the studies described in this chapter used various methods to produce oil-derived compounds at environmentally relevant concentrations—i.e., exposure conditions that would be anticipated in spawning habitats following an oil spill. In terms of aggregate PAHs, this is generally below 100 μg/L ∑PAHs (or ~0.5 μM for compounds with average molecular weight of 200 g/mol). Therefore, our uses of the terms "high" or "low" are in the context of this exposure gradient (0–100 μg/L ∑PAHs), with the caveat that for very small embryos, effective high doses may be in the ~10 μg/L range. For a given fish species, exposures at the high end of concentration-response curves typically produce severe cardiotoxicity characterized by a complete (or near-complete) loss of cardiac function during embryo-larval development, together with extracardiac defects in the jaw, eyes, body axis, and other tissues as a consequence of a disrupted peripheral circulation. For example, in freshwater species such as zebrafish, kidney development normally requires positive vascular pressure during organogenesis (Serluca et al. 2002) and consequently fails in embryos with circulatory defects due to PAH exposure (Incardona et al. 2004). Post-Exxon Valdez studies on Pacific herring and pink salmon in Alaska were unable to differentiate between abnormalities that were direct forms of oil toxicity versus those that were secondary to a loss of cardiac function. The genetic tools available in zebrafish made it possible to experimentally address this important distinction directly. At the upper end of the concentration-response range, zebrafish embryos exposed to either single PAHs or complex mixtures appeared very similar to cardiac loss-of-function mutants (Incardona et al. 2005, 2004; Jung et al. 2013). This was explored further by comparing the familiar and severe PAH-induced syndrome to zebrafish with a morpholino knockdown of the *silent heart* (*sih*) gene (Incardona et al. 2004), which encodes the sarcomeric protein cardiac troponin T (Sehnert et al. 2002). As shown in Fig. 10.8, the gross phenotypes of phenanthrene-exposed and *sih* morphant embryos are virtually indistinguishable. In both cases, extracardiac defects include a dorsally curved body axis (Fig. 10.8a), small eyes, reduced jaw structures, and a reduced brain size (Fig. 10.8b). Importantly, detailed characterization of the cartilaginous elements of the developing jaw showed that PAH exposure reduced the sizes of individual elements, without altering their patterning (Fig. 10.8c). This canonical suite of visible abnormalities is also evident in phylogenetically divergent marine species such as yellowfin tuna exposed to relative higher concentrations of Deepwater Horizon-MC252 oil (Fig. 10.8d). Notably, in the context of abnormal head structures (e.g., the jaw), recent transcriptomic evidence suggests that crude oil may act directly on gene networks involved in craniofacial development, at least in gadid species (Sørhus et al. 2017). Therefore, more work is needed to differentiate between direct and indirect toxicity in these structures.

10.2.2 For Low-Level Sublethal Exposures, Effects on Heart Development Persist

Most fish embryos and newly hatched larvae, like many other lower vertebrates, are small enough to allow tissue oxygenation by diffusion; circulation is not required for oxygen transport until late larval development (Burggren 2005, 2013). Therefore, the disruption of cardiac and circulatory function is not immediately lethal to most fish embryos (Stockard 1915; Warren and Fishman 1998), thereby allowing embryonic and larval development to proceed in fish with significant heart defects. Moreover, fish with milder forms of embryonic cardiotoxicity can survive to later life stages (juveniles and subadults) when reared in a husbandry facility (Kopp et al. 2014). When transferred to clean water, these fish recover at least some degree of circulatory function (Incardona et al. 2004). However, the tight relationship between cardiac function and structural formation during development suggests that fish surviving to later life stages may have lasting modifications to heart shape and, as a consequence, cardiac output. The relationships between ventricular form, cardiac output, and swimming performance have been particularly well characterized (Claireaux et al. 2005; Santer et al. 1983), providing a context for studying delayed reductions in the swimming capability of fish that survive embryonic oil exposures at the low end of the concentration-response range. This delayed-in-time effect was first demonstrated using the zebrafish model, wherein oil-exposed embryos with pericardial edema subsequently developed in clean water into externally normal larvae and adults. However, these adult fish showed reduced critical swimming speed and, despite overtly normal cardiac structure, subtle alterations in ventricular shape consistent with reduced cardiac output (Hicken et al. 2011). These central findings were recently extended to

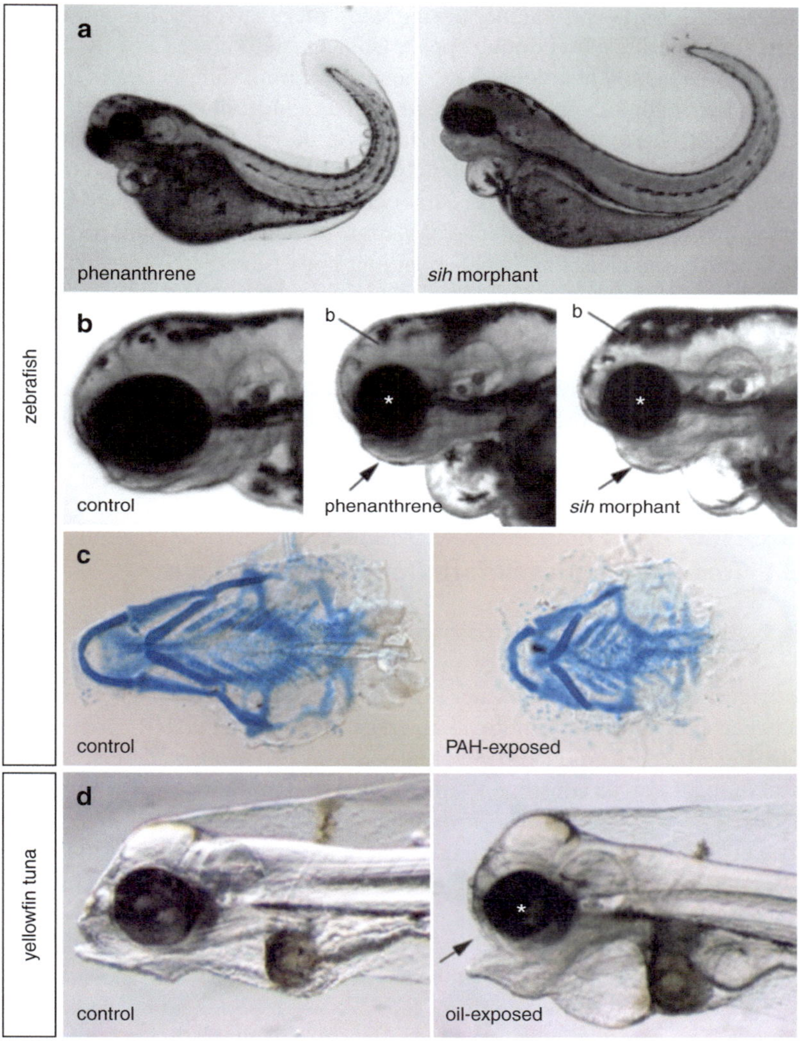

Fig. 10.8 Loss of circulation is sufficient to produce multiple extracardiac defects associated with PAH exposure. As described in Incardona et al. (2004), zebrafish embryos were exposed to either single PAHs such as phenanthrene (**a**, **b**) or simple mixtures of parent PAHs representative of the main families measured in crude oil (**c**). Phenotypes of these animals were compared to *silent heart* (*sih*) morphants, which have no cardiac contraction due to knockdown of the sarcomeric component cardiac troponin T. (**a**) Overview of whole hatching stage larvae exposed to phenanthrene (left) or injected with *sih* morpholino oligonucleotide (right). (**b**) High magnification views of control (left), phenanthrene-exposed (middle), and sih morphant (right) larvae. (**c**) Alcian blue-stained jaw cartilages from control larva (left) and larva exposed to a PAH mixture with an abnormal gross phenotype similar to those in (**b**). Jaw patterning is not changed by PAH exposure, but all elements are reduced in size. (**d**) Late hatching stage yellowfin tuna larvae exposed to MC252 crude oil during embryogenesis, with a gross phenotype very similar to phenanthrene-exposed and *sih* morphant zebrafish. Asterisks indicate reduced eye size, *b* indicates the brain, and arrows indicate abnormal jaws

juvenile pink salmon and Pacific herring, the latter of which also consumed less oxygen at their critical swimming speed. Both species showed lasting changes to ventricular shape, and histological examination of pink salmon hearts revealed an abnormal hypertrophic response coupled with a reduced thickness of the compact myocardium (Incardona et al. 2015). Based on the clonal derivation of compact myocardial cells from early embryonic precursors (Gupta and Poss 2012) (Sect. 10.1.3.1), further study of these precursor cells could reveal the mechanisms linking early but sublethal cardiac dysregulation to latent alterations in ventricular form. Overall, it is clear from this expanding body of research that the heart is a central target organ for crude oil toxicity, yielding a wide range of concentration-dependent cardiac and extracardiac phenotypes as a consequence of transient embryonic exposures. At the relatively higher end of the exposure gradient, the severe crude oil toxicity syndrome has now been extensively characterized in many marine and freshwater fish species. Ongoing research is focused on pathways of toxicity at the very low end of the exposure curve (i.e., at or below 1–5 μg/L ∑PAHs), including losses to individual fitness that are delayed in time.

10.3 Toxicodynamics Before Deepwater Horizon

10.3.1 AHR-Dependent Cardiotoxicity

Investigations prior to the Deepwater Horizon disaster noted the phenotypic similarity between exposure of fish embryos to crude oil and potent AHR agonists such as TCDD (Fig. 10.2c) (Carls et al. 1999; Marty et al. 1997; Walker and Peterson 1991). As a mixture, dissolved constituents of crude oil are strong inducers of the AHR, but individual PAHs in crude oil have potencies ranging from very weak to strong (Barron et al. 2004). Thus, a central question in the decade leading up to Deepwater Horizon was whether crude oil developmental toxicity in fish was mechanistically related to dioxin toxicity.

Due to both whole-genome and individual gene duplications, fish typically have at least two AHR isoforms, AHR1 and AHR2 (Hahn 2002). The tissue-specific activation of AHRs can be observed using CYP1A as a reporter, and gene knockdown techniques (i.e., morpholino oligonucleotides, Bedell et al. 2011; Housden et al. 2017) have been used as a tool in zebrafish and *Fundulus* to explore the role of AHRs in PAH developmental toxicity. First we consider the mechanism demonstrated for the canonical dioxin-family AHR ligand, TCDD. Early studies showed that exposure of embryos to TCDD leads to widespread, strong induction of CYP1A throughout the vasculature, including the heart, and organs associated with first-pass metabolism such as the liver and intestine (Andreasen et al. 2002; Guiney et al. 1997; Toomey et al. 2001). The correlation of vascular endothelial CYP1A induction with appearance of the toxicity phenotype initially suggested that vascular dysfunction or increased permeability were the underlying cause of TCDD-associated malformations (Guiney et al. 1997, 2000), ostensibly through CYP1A-mediated oxidative stress (Dong et al. 2002). Using morpholino knockdown, TCDD developmental toxicity was shown to require function of the AHR2 isoform (Carney et al. 2004). Importantly, the phenotype was identical when *CYP1A* was knocked down, demonstrating that dioxin toxicity was AHR-dependent, but not involving vascular induction of CYP1A or, by inference, oxidative stress. Further studies in zebrafish also demonstrated that TCDD exposure led first to reduced proliferation of cardiomyocytes during initial expansion of the ventricular chamber, followed by a failure in looping and, later still, reductions in cardiac contractility and heart rate (Antkiewicz et al. 2005). These specific cardiac defects were also dependent on AHR2 function (Antkiewicz et al. 2006). Using an elegant transgenic approach in zebrafish (Lanham et al. 2014), researchers recently expressed a constitutively active AHR2 solely in cardiomyocytes, which would be expected to phenocopy TCDD toxicity in the absence of ligand (TCDD). The transgenic zebrafish displayed virtually all aspects of the familiar dioxin toxicity, including cardiac malformation and extracardiac effects (e.g., jaw malformation). This most recent finding unequivocally links the dioxin developmental syndrome to inappropriate activation of AHR2 in the developing heart.

Organochlorine developmental toxicity in fish (TCDD but also PCBs; Fig. 10.2; Clark et al. 2010; Garner et al. 2013) is therefore relatively simple, involving a singular initiating mechanism, i.e., the activation of AHR in cardiomyocytes. The situation is more complex with crude oil and individual PAHs. Single-compound studies showed that some PAHs indeed cause a dioxin-like toxicity that involves cardiomyocyte AHR activation. These include several higher molecular weight compounds that are known to be relatively strong AHR agonists among PAHs, such as the four-ring compound benzo(a)anthracene, the five-ring compound benzo(a) pyrene, and retene, a C4-alkyl phenanthrene (7-isopropyl-1-methylphenanthrene). Each of these compounds produced a similar cardiac phenotype to TCDD, specifically a small ventricle and poor looping that occurred prior to a serious decline in cardiac function (Incardona et al. 2006, 2011; Scott et al. 2011). Also, in each case, CYP1A was induced in cardiomyocytes in addition to the endothelial cells lining the heart (endocardium). AHR2 knockdown led to a rescue from toxicity, normalizing heart development, with blocked CYP1A induction indicative of the requirement for AHR2 activation. While the more potent AHR ligands with ≥4 rings tend to be abundant in combustion-derived PAH sources, they are typically absent from water-soluble mixtures derived from crude oil. However, retene, although primarily associated with wood pulp mill effluent, could be representative of other alkylated phenanthrenes that are very abundant in crude oil. Nevertheless, as described in detail below, embryos exposed to crude oil more often do not show detectable CYP1A induction in cardiomyocytes.

Zebrafish embryos were used to study both the cellular distribution of CYP1A protein following oil exposure and the role of the AHR pathway in toxicity using morpholino knockdown. Initial studies using Alaska North Slope crude oil showed that in embryos with cardiac defects, CYP1A was induced strongly in the outer epidermis and throughout the vascular endothelium but only in the endothelial lining of the heart (endocardium) and not in the myocardium (Incardona et al. 2005). Similar results were obtained in studies with other crude oil sources and other species. Iranian heavy crude oil and oil from the Deepwater Horizon-MC252 well produced nearly identical cardiotoxic effects in zebrafish embryos, along with a pattern of CYP1A induction similar to Alaska North Slope crude oil (Incardona et al. 2013; Jung et al. 2013). Atlantic haddock embryos exposed to Norwegian Sea crude oil displayed severe cardiac malformation in the absence of cardiac CYP1A induction, which was also observed primarily in the outer epidermis (Sørhus et al. 2016). However, after much more prolonged weathering—thereby shifting the PAH composition toward higher levels of alkylation—Alaska North Slope crude oil exposure caused myocardial CYP1A induction in zebrafish embryos (Jung et al. 2013) and Pacific herring embryos (Incardona et al. 2009). In contrast to dioxin, these cumulative findings showed that activation of the AHR pathway in myocardial cells is not necessarily required for PAH cardiotoxicity. As described below, the experimental knockdown of AHR pathway genes has formally demonstrated this.

10.3.2 AHR-Independent Cardiotoxicity

Similar to previous mechanistic studies of organochlorine toxicity, the zebrafish model provided an excellent platform for the investigation of toxic mechanisms related to crude oil and the more abundant PAH classes in petrogenic mixtures. First, studies using single compounds as well as complex crude oil-derived mixtures identified a key phenotypic distinction from dioxin toxicity. Experiments with zebrafish embryos exposed to single-parent (i.e., non-alklyated) compounds representative of the most abundant two-, three-, and four-ring compounds present in water-dissolved mixtures from crude oil demonstrated that single three-ring compounds produced a syndrome generally indistinguishable from crude oil toxicity (Incardona et al. 2005, 2004). However, in contrast to dioxin, embryos exposed to these tricyclic PAHs (fluorene, dibenzothiophene, or phenanthrene) showed defects in cardiac function that preceded heart malformation. Specifically, these compounds reduced the pumping ability of the heart (contractility). In addition, they had a concentration-dependent effect on heart rate and rhythm, including a slowed heart (bradycardia) at low concentrations and serious rhythm abnormalities (conduction blockade) at higher concentrations. These defects were detected at an earlier time point than the cardiomyocyte proliferation defect observed in TCDD-exposed embryos. In zebrafish the heart begins to beat weakly just prior to 24 h post-fertilization (hpf); by 36 hpf, the heartbeat is rapid and regular; and between about 36 and 42 hpf, the specialized cells that coordinate atrial-ventricular conduction begin to form (Burggren et al. 2017; Chi et al. 2008). Embryos exposed to three-ring PAHs or crude oil from fertilization did not show functional defects until the developmental window for the inter-chamber conduction system, which begins at ~36 hpf (Chi et al. 2008). These effects on function were reversible, as embryos moved to clean water at ~36 hpf had normal rhythms by 48 hpf, but they subsequently showed subtle alterations in heart morphology (Incardona et al. 2004). Embryos exposed continuously up to the hatching period showed serious cardiac malformation, a final phenotype very similar to TCDD-exposed embryos.

Similar to the knockdown approach used to study the cardiotoxicity of dioxin (TCDD), morpholinos were also the basis for investigating the role of AHR in PAH toxicity in zebrafish. For individual tricyclic PAHs (phenanthrene and dibenzothiophene), cardiac arrhythmia was more severe in AHR2 knockdown embryos (morphants) (Incardona et al. 2005). In contrast to single PAHs, zebrafish embryos exposed to effluent from gravel coated with Alaska North Slope crude oil did not show arrhythmia but reduced contractility and poor chamber looping, at $\sum$PAH concentrations in the range of 50–75 μg/L. However, this phenotype was also exacerbated by knockdown of *ahr* and *CYP1A* (Incardona et al. 2005). Finally, control zebrafish embryos exposed to even lower levels of highly weathered Alaska crude ($\sum$PAHs 9 μg/L) showed only a very low incidence of cardiotoxicity (3%), while *CYP1A* knockdown led to edema in 92% of embryos exposed to the same concentration (Hicken et al. 2011). Therefore, rather than mediating PAH toxicity, the AHR/CYP1A pathway primarily plays a protective role by detoxifying oil-derived compounds. In addition, AHR activation in the heart is not required for crude oil cardiotoxicity. Although genetic tests of AHR dependency are not readily performed in species such as Pacific herring, the cardiac phenotype in this species is also consistent with AHR-independent cardiotoxicity. As observed in zebrafish embryos exposed to single PAHs, the earliest observed embryonic effects were dose-dependent reduction of heart rate and an increasingly irregular arrhythmia (Incardona et al. 2009).

In summary, in the time leading up to Deepwater Horizon, zebrafish were used to dissect some of the mechanisms underlying the whole-embryo effects caused by exposure to Alaska North Slope crude oil in pink salmon and Pacific herring. Some higher molecular weight PAHs (containing four or more rings) were found to cause a form of developmental cardiotoxicity that was similar to dioxins and PCBs in that heart defects arise from inappropriate activation of the AHR in cardiomyocytes. This form of developmental toxicity leads first to morphological defects (e.g., reduced ventricular cardiomyocyte proliferation), followed by functional defects. However, these particular PAHs are generally not associated with the mixtures derived from crude oil. In contrast, several of the tricyclic PAHs abundant in crude oil caused a developmental cardiotoxicity which was AHR-independent, wherein functional defects (e.g., arrhythmia) preceded morphological defects. While exposure to whole oil in some cases was observed to activate AHR in developing cardiomyocytes, knockdown of AHR indicated that this pathway was not required for the functional defects arising from crude oil or three-ring PAH exposure.

10.4 Toxicodynamics After Deepwater Horizon: K^+ and Ca^{2+} Fluxes and the Myocardial Basis of Whole-Heart Functional Defects

Unlike the Exxon Valdez spill, which primarily impacted nearshore areas, the Deepwater Horizon disaster released oil into open-ocean pelagic and bathypelagic habitats. The spatial and temporal extent of the spill overlapped the spawning ranges of many economically and ecologically important species in the open waters of the northern Gulf of Mexico. These included large and highly migratory top predators, such as bluefin and yellowfin tunas, and abundant species important to commercial and recreational fisheries, such as snappers, groupers, billfish, mahi mahi, and cobia. Compared to nearshore demersal spawners such as herring, embryos of these species are not readily amenable to live collection in the field, as they are very small and fragile components of the complex pelagic planktonic community. Also, most of the potentially affected predators are not routinely maintained as captive brood stocks in land-based husbandry facilities. By contrast, subadult fish are readily captured and held live, providing a source of freshly isolated cardiomyocytes for mechanistic toxicity studies. As discussed above, the developing heart of small fish embryos is similar in many ways to isolated (in vitro) cardiomyocytes. As discussed in detail below, in vitro studies on individual heart muscle cells from subadult tunas and mackerel provided a basis for comparing the cardiotoxicity of oil from the Macondo 252 well in the Gulf of Mexico relative to the much more extensively studied Alaska North Slope crude oil. In addition, these studies also yielded a greater mechanistic understanding of AHR-independent mechanisms of cardiotoxicity.

As described in Sect. 10.1.3.4 above, the rhythmic beating of the heart is dependent on coordinated excitation of cardiomyocytes by regular action potentials and excitation-contraction coupling. Action potentials involve the cyclical depolarization and repolarization of cardiomyocyte plasma membranes, primarily mediated by changes in the electrochemical gradients of Na^+ and K^+ ions, while Ca^{2+} ions act as the primary mediator of E-C coupling (see reviews cited above). The whole-heart phenotypes previously observed in oil-exposed embryos pointed to potential targets that play a role in the cyclical generation of action potentials and intracellular calcium fluxes. Action potential duration, and thus heart rate, is controlled in part by the repolarization phase that is attributable to K^+ efflux through ERG channels (Sanguinetti 2010). In zebrafish embryos, drugs that block ERG channels caused bradycardia and the same types of arrhythmias induced by phenanthrene (Langheinrich et al. 2003; Milan et al. 2003), suggesting this potassium channel might be a direct target of oil-derived chemicals. Similarly, the reduced ventricular contractility observed in zebrafish embryos exposed to several different crude oils (Incardona et al. 2005; Jung et al. 2013), including the MC252 oil (Incardona et al. 2013), suggested impacts on intracellular Ca^{2+} cycling.

To address these questions directly, Brette and colleagues isolated cardiomyocytes from the hearts of scombrid species that were closely related to those in the Gulf of Mexico (subadult bluefin and yellowfin tunas, mackerel) and exposed them to individual PAHs and water-accommodated fractions (WAFs) of Deepwater Horizon-MC252 crude oil (Brette et al. 2014, 2017). The goal was to better understand the cellular basis for the well-known, whole-heart phenotypes previously documented in oil-exposed embryos. These included three categories of functional defects: bradycardia, arrhythmia, and reduced contractility. Repolarizing potassium currents (i.e., I_{Kr}) were measured with the whole-cell patch-clamp technique (Kornreich 2007), and intracellular calcium cycling was assessed with fluorescent calcium-sensing dyes and quantitative confocal microscopy (Bers 2003).

Water-accommodated fractions of oil with total PAHs in the low μg/L (~nM) range prolonged the action potential duration in a dose-dependent manner with the blockade of I_{Kr}. Similar to drugs that block ERG channels, the WAF-induced inhibition of I_{Kr} was rapid, and it was related to the concentrations of measured tricyclic PAHs but not ∑PAHs, the latter including two- and four-ring compounds. The MC252 WAF also reduced intracellular Ca^{2+} levels, depleting the SR of Ca^{2+} by blocking either SERCA2 or promoting Ca^{2+} leakage through the RyR (Brette et al. 2014). Several single PAHs were also tested using similar patch-clamp and calcium-imaging methods. These included naphthalene (two rings); the tricyclics fluorene (with only two aromatic rings) and phenanthrene; the tricyclic nitrogen- and sulfur-containing heterocycles carbazole and dibenzothiophene, respectively; and pyrene (four rings) (Brette et al. 2017). Of these parent (non-alkylated) PAHs, only phenanthrene rapidly blocked I_{Kr}. Phenanthrene also reduced intracellular Ca^{2+} levels but in this case by blocking the LT calcium conductance (I_{CaL} current). While these findings suggest complex interactions of both single tricyclic PAHs and mixtures with several different targets involved in cardiomyocyte electrophysiology (summarized in Fig. 10.7b), effects on K^+ and Ca^{2+} fluxes readily explain the AHR-independent, whole-heart responses of arrhythmia and reduced contractility observed in exposed fish embryos. It is also likely that these same mechanisms operate in subadult fish, leading to impaired cardiorespiratory performance (but not morphological defects) in animals acutely exposed to crude oil (Claireaux and Davoodi 2010; Johansen and Esbaugh 2017; Nelson et al. 2016).

10.5 Species Representing Pelagic and Nearshore Habitats Impacted by Deepwater Horizon

10.5.1 The Crude Oil Exposure Syndrome Across Diverse Species

Assessing the harmful effects of Deepwater Horizon-MC252 oil on developing fish (i.e., injury) was a central aim of the Natural Resource Damage Assessment (NRDA) that followed the spill. In contrast to the Exxon Valdez and Cosco Busan spills (Sect. 10.1), there were few opportunities to capture live animals from fish spawning sites in the weeks and months after the seafloor blowout in the northern Gulf. With the exception of field studies on resident Gulf killifish in oiled marsh habitats, the vast majority of studies were restricted to laboratory determinations of toxicity thresholds, for comparison to PAH concentrations measured in affected marine habitats. A goal of the NRDA was the collection of data that would be representative of a wide range of species and habitats. This, in turn, yielded information that allowed a direct sensitivity comparison for fish species with different egg sizes and different ecophysiological traits. Additional toxicity studies, focusing on other species, were sponsored by the Gulf of Mexico Research Initiative. Overall, the range included fish with a very high cardiorespiratory demand and a wide thermal tolerance (e.g., deep-diving tunas), more sedentary estuarine species (e.g., red drum), and euryhaline species that deposit large demersal eggs in nearshore marshes (e.g., Gulf killifish). Potential impacts on Atlantic bluefin tuna spawning grounds in the northern Gulf were a major concern given critically low adult population numbers (Taylor et al. 2011).

Overall, access to land-based facilities capable of producing synchronized cohorts of newly fertilized eggs was a significant limiting factor in terms of species selection. This was particularly true of the large pelagics such as bluefin tuna. Captive brood stocks in the Gulf and western Panama provided sources of mahi mahi (Stieglitz et al. 2017) and yellowfin tuna (Margulies et al. 2007) embryos, respectively, while Southern bluefin tuna (*Thunnus maccoyii*) and yellowtail amberjack (*Seriola lalandi*) embryos were obtained from a husbandry facility in southeast Australia (Bubner et al. 2012; Bar et al. 2016), as surrogates for the closely related Gulf species. For nearshore species, red drum are routinely cultured as part of stock enhancement programs in several Gulf states, including Texas (Karlsson et al. 2008; Holt 1992), while the Gulf killifish *Fundulus grandis* and its close relative *Fundulus heteroclitus* have been used in laboratory studies (Hsiao and Meier 1989; Oziolor et al. 2016). In this respect, the latter was a useful bridge for laboratory and field studies.

Exposure studies with pelagic embryos used a high-energy method to disperse small droplets of oil into seawater, creating a WAF mixture of particulate oil and dissolved PAHs to mimic the release of oil under high pressure from the sea floor into the water column (Edmunds et al. 2015; Esbaugh et al. 2016; Incardona et al. 2014). As shown above (Fig. 10.2), the resulting PAH compositional profiles and concentrations closely mirrored water samples collected in the pelagic zone during the active spill phase. Test species included Southern bluefin tuna, yellowfin tuna, yellowtail amberjack, and mahi mahi, with egg sizes ranging from just under 1 mm for the tunas to 1.2 mm for amberjack and 1.4 mm for mahi mahi. Corresponding surface-to-volume ratios ranged from 6.7 to 4.3. In general, the responses of the tunas and amberjack were similar (Incardona et al. 2014). Gross visible defects were evident at the high end of the concentration-response curve (i.e., 10–15 μg/L particulate plus dissolved $\sum$PAHs). These included severe edema involving the pericardial space and yolk sac, reduction in the marginal finfolds, and body axis curvature (Fig. 10.9a–c). Mahi mahi embryos responded primarily with edema accumulation, without severe effects on extracardiac structures such as the finfold (Fig. 10.9d) (Edmunds et al. 2015).

Embryos and larvae of nearshore species were more likely exposed to surface slicks or oil stranded in shoreline substrates or on vegetation, so different exposure methods were used for these assessment studies. This included low-energy WAF preparations in which water was mixed beneath an undisturbed surface slick and direct or indirect contact with oiled sediments either collected from the field or produced in the lab. Red drum embryos exposed to both high-energy and low-energy WAFs showed indistinguishable injury phenotypes, with severe yolk sac edema coupled with reduction of the marginal finfolds at higher concentrations (Fig. 10.9e and not shown) (Khursigara et al. 2017; Morris et al. 2017, Crude oil cardiotoxicity to red drum embryos is independent of oil dispersion energy, unpublished). Killifish embryos exposed to oiled sediments from Louisiana marshes also had severe edema and reduced craniofacial structures (Fig. 10.9f) (Dubansky et al. 2013). Similar

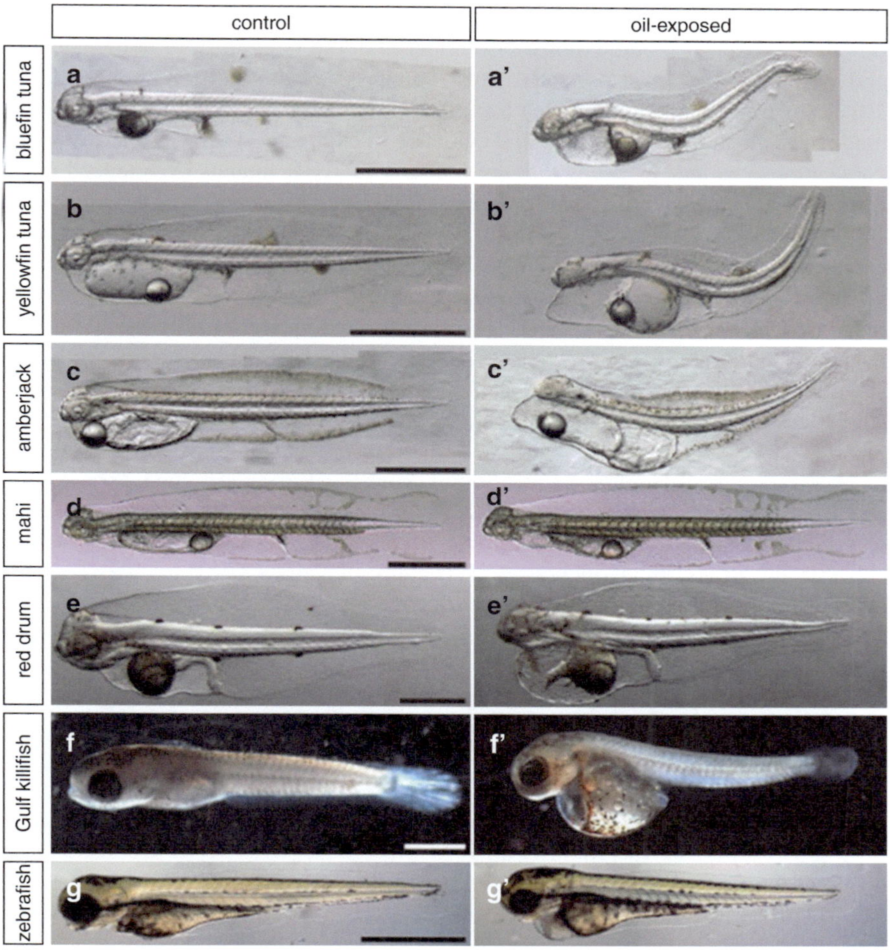

Fig. 10.9 Larval phenotypes of different species exposed to Deepwater Horizon-MC252 crude oil during embryogenesis. Five marine pelagic and two demersal species are represented, with control larvae on the left and exposed larvae on the right. (**a**) Bluefin tuna, (**b**) yellowfin tuna, (**c**) yellowtail amberjack, (**d**) mahi mahi, (**e**) red drum, (**f**) Gulf killifish, and (**g**) zebrafish. All were exposed to high-energy WAFs, except for killifish (**f**), which was exposed to oil-contaminated marsh sediment collected from the Louisiana shoreline (Grand Terre Island). Scale bars are 1 mm

results were obtained with zebrafish embryos exposed to either high-energy WAF preparations (Fig. 10.9g) or sediments spiked with a Louisiana crude with properties similar to MC252 oil (Raimondo et al. 2014).

Highly weathered surface oil from the Deepwater Horizon spill came ashore at numerous locations along the extensive shoreline of the northern Gulf, particularly in marsh habitats between Barataria Bay, Louisiana, and Mobile Bay, Alabama. Field assessments of fish health focused primarily on Gulf killifish, given its wide distribution and abundance. These exceptional studies involved both in situ sampling of resident fish and exposure of embryos in the laboratory to both sediment and water samples from impacted sites (Dubansky et al. 2013; Whitehead et al. 2012). As expected, adult killifish captured in minnow traps from areas where oil made landfall showed transcriptional evidence of exposure to PAHs, in terms of the conventional battery of genes that are induced by AHR ligands (Whitehead et al. 2012). Moreover, sediment collected from oiled marsh locations a year after the spill, after the dissipation of visible oil, was toxic to killifish embryos. In addition to elevated CYP1A levels, this included embryonic bradycardia and the presence of visible edema at the larval stage (Dubansky et al. 2013). Comparison of the genomic responses in field-collected killifish to those in laboratory exposures provided strong evidence that biologically effective PAH concentrations in the lab studies were representative of the conditions in impacted marshes (Pilcher et al. 2014). These latter studies indicate the environmental realism of laboratory exposure scenarios for the Deepwater Horizon spill, despite their limitations. These findings are also consistent with the concordant laboratory and field studies described for the Exxon Valdez and Cosco Busan spills in Sect. 10.1.2. As discussed in the next section, a close examination of functional and morphological defects involving the heart, as well as extracardiac structures, is increasingly revealing novel relationships between fish early development and species ecophysiology.

10.5.2 Phenotypic Differences Among Species with Differing Ecophysiology

10.5.2.1 Effects on Cardiac Function and Morphology

The effects of MC252 crude oil were initially compared to other geologically distinct oils in zebrafish (Incardona et al. 2013; Jung et al. 2013). Despite considerable differences in some of the physical and chemical characteristics of the whole oils, MC252 oil produced the same cardiac defects in zebrafish embryos as Alaska North Slope and Iranian heavy crude oils. Following exposure to high-energy WAFs, all three crudes specifically reduced ventricular but not atrial contractility, as a basis for poor chamber looping. Importantly, this outcome was the same irrespective of whether embryos were exposed shortly after fertilization, or just after the heartbeat was established at 24 h, prior to the development of the inter-chamber conduction pathway. In contrast, both bluefin and yellowfin tuna as well as amberjack embryos showed dose-dependent reduction in heart rate (bradycardia) with irregular arrhythmias appearing at higher concentrations (Incardona et al. 2014). Dose-dependent bradycardia was also observed before hatch in sheepshead minnow embryos exposed to MC252 crude oil (Bosker et al. 2017). However, similar to zebrafish, the response of both mahi mahi and red drum was primarily a reduced contractility, with minimal dose-dependent effect on heart rate and rhythm (Edmunds et al. 2015; Khursigara et al. 2017; Morris et al. 2017, Crude oil cardiotoxicity to red drum embryos is independent of oil dispersion energy, unpublished). The thresholds for these responses across all species were relatively low (≤20 μg/L ∑PAHs). For the species with the smallest eggs, these thresholds were very low—e.g., below 1 μg/L for arrhythmia in yellowfin tuna. These findings are consistent with the aforementioned in vitro cardiomyocyte patch-clamp studies that demonstrated channel-blocking activities of oil-derived compounds in the pharmacologic range.

As expected from prior studies, these functional defects led to changes in cardiac morphology. At higher concentrations, exposed tuna (not shown), amberjack, and red drum embryos (Fig. 10.10a–d) showed completely unlooped or "string" hearts by hatching stage. Failure of looping was dose-dependent, as shown by measurements of the angle between the atrium and ventricle in species with larger larvae such as mahi mahi and red drum. Generally, impacts on cardiac morphology were less severe in mahi mahi, and even the embryos with the most severe edema maintained some circulation of red blood cells. In addition, a reduction in cardiac jelly was apparent in red drum larvae (Fig. 10.10g, h). Cardiac jelly is a transient layer of extracellular matrix in the developing heart between the endocardium and myocardium and contributes to the inherent elasticity at the early tubular stage (Barry 1948). In general, the atrial chamber of the early fish heart has a thicker layer of cardiac jelly. An oil-induced reduction in cardiac jelly and elasticity likely contributes to reduced filling of the chamber (Forouhar et al. 2006; Männer et al. 2008), thereby exacerbating reduced Ca^{2+}-dependent contractility. Each of these effects, alone or in combination, reduce cardiac output (Khursigara et al. 2017).

The window of sensitivity for oil-induced cardiac developmental toxicity parallels that for individual tricyclic PAHs described above in Sect. 10.3.2. The timing of heart development relative to hatching is very similar in both zebrafish and mahi mahi. In both species, cardiogenesis begins in segmentation stage embryos, with heartbeats established by 24 hpf. The hatching stage for mahi mahi is in the 36–48 hpf range, depending on temperature, while zebrafish hatching stage is about 48 hpf. Both species undergo further morphogenesis such as valve development after the third day post-fertilization. Although heart development has not been characterized in detail at the molecular level for any of the Gulf species, it is likely that the temporal expression pattern of genes encoding the targets of crude oil in E-C coupling and action potential repolarization in mahi mahi mirrors that of zebrafish. Transcriptome studies in cold-water species (Atlantic haddock, *Melanogrammus aeglefinus*) support this generally, showing that mRNA levels for the key targets of PAHs peaked after the onset of the regular heartbeat (Sørhus et al. 2016). Consistent with this and the earlier work on single PAHs, both the functional (contractility) and morphological defects in zebrafish embryos were virtually identical following MC252 crude oil exposure starting either shortly after fertilization (4–48 hpf) or after 24 hpf (24–48 hpf) (Incardona et al. 2013). Although cardiac endpoints were not measured, larval lethality tests showed a similar window of sensitivity for mahi mahi, with the most pronounced effects (i.e., lowest concentration causing 50% mortality, LC_{50}) occurring with exposures at 24–48 hpf, with the second most sensitive period being exposure from shortly after fertilization up to 24 hpf (Mager et al. 2017).

These collective findings in zebrafish and mahi mahi bring up a key point about toxicokinetics and highlight the dissimilarity to dioxin/PCB toxicity. Because organochlorine dixoins and PCBs cannot be eliminated from tissues by metabolism, AHR-dependent dioxin cardiotoxicity persists into the hatching stage at a similar

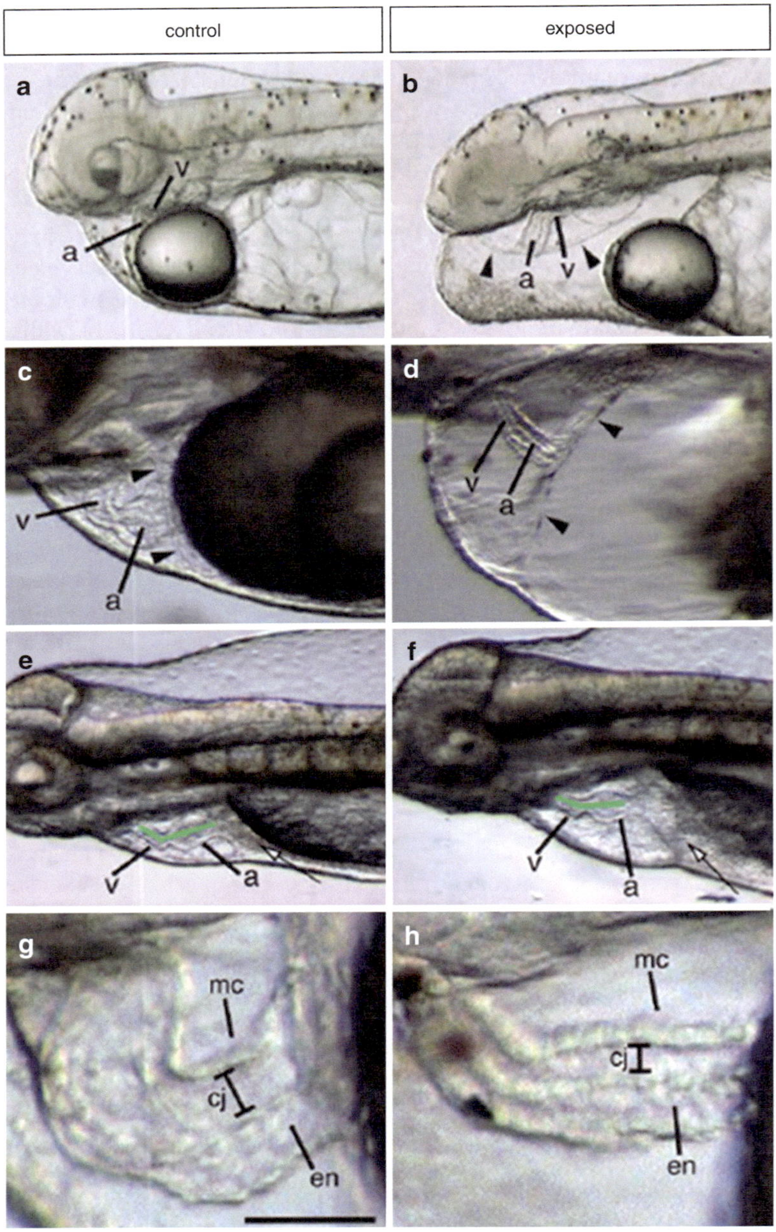

Fig. 10.10 Cardiac abnormalities arising from exposure to MC252 crude oil. (**a**, **b**) amberjack, (**c**, **d**) red drum, (**e**, **f**) mahi mahi, (**g**, **h**) red drum. Left panel of each pair is control; right panel is exposed to MC252 oil. In (**a–d**) atrial (*a*) and ventricular (*v*) chambers are indicated. Arrowheads indicate pericardial membranes. Changes in the atrioventricular angle in mahi mahi (**e**, **f**) are indicated by green lines. White arrows indicate erythrocytes moving through the yolk sac sinus. (**g**, **h**) Reduced cardiac jelly in MC252-exposed red drum. *cj* cardiac jelly, *en* endocardium, *mc* myocardium. Scale bar for (**g**, **h**) is 50 μm

level of sensitivity (Carney et al. 2006). At the same time, the primary mechanism for these AHR agonists is a reduction of cardiomyocyte proliferation, which is the main driver of larval heart growth after primary morphogenesis. For crude oil exposures, there are two reasons the late embryonic stages of cardiac morphogenesis are so sensitive. First is the tight coupling of function (e.g., intracellular Ca^{2+} handling, contractility, rhythm) to form during the dramatic morphogenetic steps associated with transitioning from a linear heart tube to adjacent chambers (looping) and ballooning of the ventricle (Sect. 10.1.3.1 and Fig. 10.4). Second, as fish mature their capacity for metabolic detoxification increases, hence post-hatch larvae with a gastrointestinal system primed for feeding (and metabolism) are more resistant. Embryos are thus at their most vulnerable point for crude oil exposure from a detoxification standpoint during late organogenesis, right when their hearts are most sensitive.

10.5.2.2 Abnormal Development of Extracardiac Structures

As discussed above (Sect. 10.2), morphological defects from oil exposure are concentration-dependent at the level of the whole embryo. At the high end of the concentration-response relationship, severely affected embryos (or hatched larvae) have multiple defects. Despite the inherent challenge of relating effects of a complex chemical mixture on multiple tissues, many of the gross defects appear to be secondary to the loss of embryonic circulation as described above. New toxicity data for pelagic-spawning species has yielded interesting insights into the unique developmental anatomy of pelagic fish larvae. The perturbation of these structures by oil-derived chemicals is shedding new light on fundamental life history attributes that determine the early life stage physiology of these geographically widespread but relatively understudied organisms.

The link between reduced circulation and accumulation of edema is obvious in embryos of freshwater species. First, hydrodynamic forces generated by the beating heart are required for initial formation of the glomerulus in the embryonic (pronephric) kidney, as shown by an analysis of cardiac function mutants in zebrafish (Serluca et al. 2002). Second, as freshwater fish are hyperosmotic relative to their environment, they must continuously generate large volumes of dilute urine to offset water gains. Thus in oil-exposed freshwater embryos, edema accumulation results from reduced kidney perfusion. This becomes especially severe (and ultimately lethal) if cardiac function is so poor that pronephros development fails (Incardona et al. 2004). By contrast, marine fish are hyposmolar to the surrounding water and have a tendency to lose water. They therefore must absorb water through the gastrointestinal tract, and they generally have kidneys in which the glomerulus is either absent or markedly reduced in size and function (Marshall and Smith 1930; Beyenbach 2004; Hickman and Trump 1969). As demonstrated initially by Shelbourne (Shelbourne 1957), the epidermis—and specifically the large surface area yolk sac epidermis—is the primary organ regulating salt and water balance in pelagic larvae. However, the accumulation of edema in marine fish larvae seems counterintuitive. The answer appears evident in the visible changes to the marginal finfold subdermal spaces of oil-exposed embryos. These findings suggest that edema in marine fish is not a consequence of external water uptake but rather the movement of fluid between different compartments in the developing embryo.

Most of the pelagic species exposed to MC252 crude oil showed reductions in the marginal finfolds (Figs. 10.11 and 10.12) (Incardona et al. 2014; Morris et al. 2017, Crude oil cardiotoxicity to red drum embryos is independent of oil dispersion energy, unpublished). Typically, this effect was most evident antero-dorsally, in the vicinity of the head. For example, red drum larvae with severe yolk sac edema also showed a marked depression of the dorsal marginal finfold over the head region (Fig. 10.11a, b). However, there was also an overall reduction in the peripheral subdermal spaces, also visibly affecting the lateral surfaces (Fig. 10.11c, d). In yellowfin tuna, the marginal finfold was reduced circumferentially in a concentration-dependent manner. This occurred in both dorsal and ventral caudal regions (Fig. 10.12a–c), with a similar dorsal depression in the head region of severely impacted larvae (Fig. 10.12c, f). In these larvae, the normal expansion of the hindbrain (or fourth) ventricle by cerebrospinal fluid was reduced or even failed completely (Fig. 10.11a, b; Fig. 10.12d–f). In mahi mahi, these effects on the marginal finfold were less severe (Fig. 10.11e–g), wherein the caudal regions were generally intact in embryos with the most severe yolk sac edema. That said, even mildly affected mahi mahi larvae showed some reduction in both the hindbrain ventricle and the overlying subdermal space (arrows, Fig. 10.11e–g). These findings suggest that when circulation fails in marine pelagic embryos, water shifts from the peripheral subdermal spaces to the yolk sac. As the yolk sac is the largest site for osmoregulation, water is most likely lost locally within the yolk sac sinus. The free connection of the yolk sac sinus to the peripheral subdermal space provides a pathway for fluid shifts. This is consistent with pronounced defects in the dorsal anterior region, closest to the dorsal subdermal space and the yolk sac connection channels near the pectoral fins. While expansion of the hindbrain ventricle was shown to be circulation-dependent in zebrafish (Lowery and Sive 2005), it is possible that osmotic fluid shifts also contribute to this phenotype in oil-exposed pelagic larvae. The severe impacts observed at higher exposure concentrations likely contribute to larval mortality, but smaller fluid shifts represented by milder reductions in the subdermal spaces or hindbrain ventricles could lead to developmental outcomes with latent adverse effects. Although exposures to relatively lower levels of oil do not produce obvious structural defects in the developing central or peripheral nervous systems (Edmunds et al. 2015; Incardona et al. 2013), more detailed studies will be required to discern relevant behavioral or physiological phenotypes. A potential

→

Fig. 10.11 Crude oil cardiotoxicity leads to a shift of fluid from the peripheral subdermal spaces to the yolk sac and pericardium in pelagic marine larvae. Control (**a**, **c**) and MC252 oil-exposed (**b**, **d**) red drum larvae are shown in lateral (**a**, **b**) and dorsal (**c**, **d**) views. Arrowheads indicate either the dorsal margin of the medial finfold (**a**, **b**) or lateral epidermis (**c**, **d**). Arrow in (**a**) indicates the hindbrain ventricle, filled with cerebrospinal fluid in the control but unfilled in the oil-exposed larva. Control (**e**) and oil-exposed (**f**, **g**) mahi mahi larvae shown in lateral views. Pericardial and yolk sac edema are outlined in red. Arrows indicate hindbrain ventricle. For both red drum and mahi mahi, ∑PAH concentrations are indicated for the exposed larvae

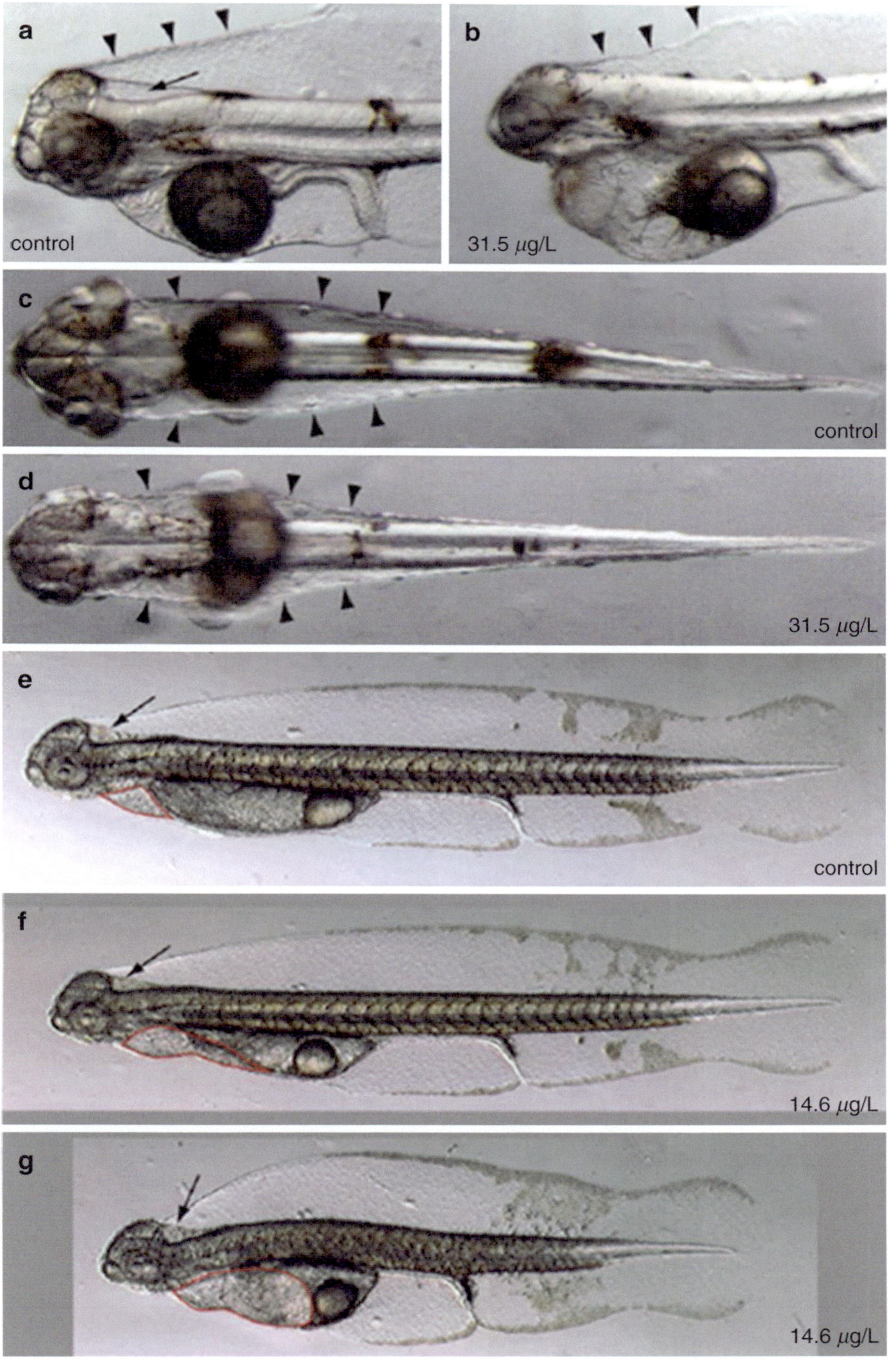
a
b
control
31.5 µg/L
c
control
d
31.5 µg/L
e
control
f
14.6 µg/L
g
14.6 µg/L

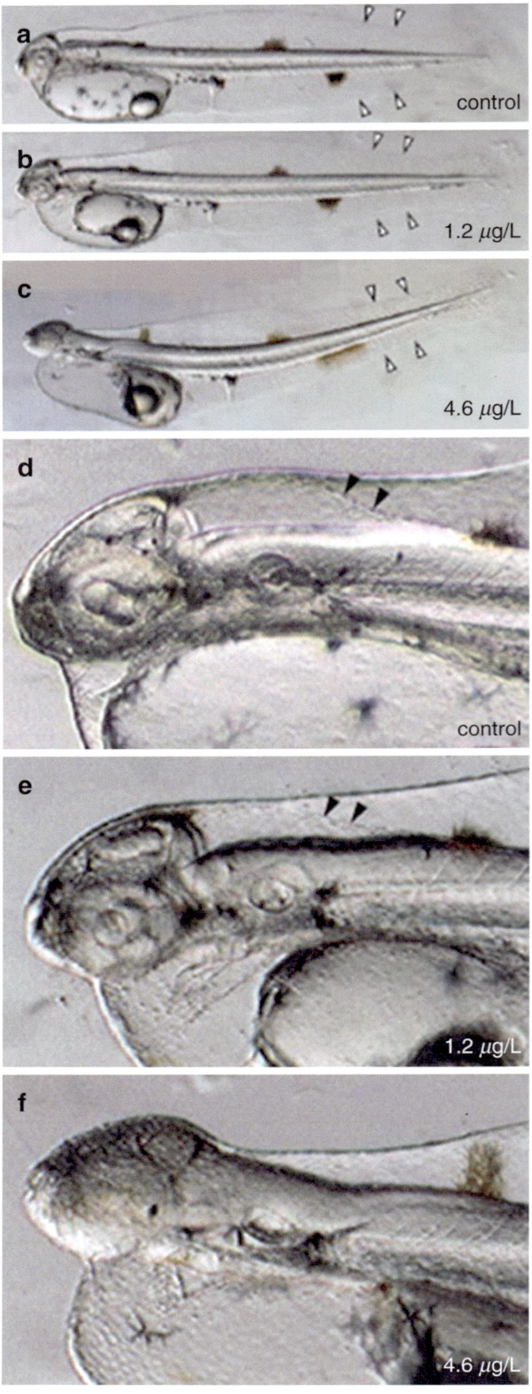

Fig. 10.12 Dose-dependent extracardiac phenotypes in MC252 oil-exposed yellowfin tuna larvae. Yellowfin tuna embryos were exposed to high-energy WAFs made with a weathered oil sample collected from an offshore surface slick. (**a**–**c**) Whole larvae from control treatment (**a**) and MC252 WAF dilutions with total PAHs of 1.2 μg/L (**b**) and 4.6 μg/L (**c**). White arrowheads indicate extent of marginal v in the caudal region. (**d**–**f**) High magnification views of the head regions of the same larvae, control (**d**) and exposed to ∑PAHs of 1.2 μg/L (**e**) and 4.6 μg/L (**f**). Arrowheads indicate dorsal-posterior margin of the hindbrain ventricle, which lacks fluid in the larva exposed to 4.6 μg/L ∑PAHs

reduction in swimming efficiency measured in juvenile mahi mahi that were exposed to MC252 oil during embryogenesis (Mager et al. 2014) could be an indication of a lasting neurodevelopmental defect (Arrenberg et al. 2009; Dunn et al. 2016).

10.5.2.3 Effects on Cardiac Gene Expression

As a premier genetic model for vertebrate cardiac development, zebrafish have produced a wealth of data on changes in gene expression both leading to and resulting from defects in cardiac function (Pott et al. 2014; Staudt and Stainier 2012). Similarly, the high degree of evolutionary conservation between human and fish hearts at the cellular and molecular levels applies to heart failure pathways as well (Bloomekatz et al. 2016). Using a hypothesis-driven approach based on the zebrafish and human health literature, a set of ten genes linked to cardiac function, morphogenesis, and pathological hypertrophy was examined in mahi mahi exposed to MC252 crude oil (Edmunds et al. 2015). MC252 exposure downregulated myofibrillar genes, including atrial and ventricular myosin heavy chain isoforms (*amhc* and *vhmc*, respectively) and the regulatory cardiac myosin light chain 2 gene (*cmlc2*). Also downregulated was the gene *four-and-a-half* LIM domain family protein 2 (fhl2), encoding a protein expressed primarily in the developing heart but playing complex roles through myriad protein-protein interactions. However, fhl2 functions are clearly related to abnormal phenotypes observed in oil-exposed embryos, such as contractility (maintenance of myofiber elasticity) and regulation of repolarizing K^+ currents (reviewed by Tran et al. 2016). Two major transcription factors involved in cardiomyocyte differentiation, NK2 homeobox 5 (nkx2.*5*) and T-box 5 (tbx5), were also downregulated, along with the *nkx2.5* target gene encoding atrial natriuretic peptide (ANP), a homeostatic regulator of contractility. Importantly, these changes in gene expression did not occur until after hatching, indicating an altered developmental trajectory for the heart, downstream of the disruption of normal physiological function during transient embryonic exposures.

Advances in high-throughput DNA sequencing and bioinformatics have markedly improved the ability to carry out functional genomics in non-model fish. Transcriptome sequencing applied to embryonically exposed mahi mahi larvae provided further insight into molecular responses to crude oil exposure and produced some findings consistent with the hypothesis-driven approach (Xu et al. 2016). In this study, gene expression was analyzed at two developmental time points that flanked the previous study (Edmunds et al. 2015), namely, in early hatching stage larvae and 2 days later when the larvae were feeding. Although first discovered as regulators of bone development (Wozney et al. 1988), bone morphogenetic (BMP) family members are potent modulators of cell fate and growth during the formation of multiple organs, including the heart (van Wijk et al. 2007). As members of the transforming growth factor beta superfamily, they are ligands for specific membrane serine/threonine kinase receptors that signal to the nucleus to alter gene expression (Miyazono et al. 2005). Bone morphogenetic protein 4 (bmp4) shows a dynamic expression pattern during zebrafish heart development, initially expressed in both atrial and ventricular chambers and later becoming restricted to the atrial-ventricular boundary in a nkx2.5-dependent manner (Tu et al. 2009). MC252 crude oil exposure altered the dynamic temporal pattern of *bmp4* expression in mahi mahi. As a concentration-dependent morphogen, either excess or reduced BMP4 activity leads to looping defects in zebrafish (Chen et al. 1997) and would be expected to do the same in other species. Consistent with the effects of MC252 crude oil on intracellular calcium handling, exposure also altered expression of several key genes involved in Ca^{2+} homeostasis and E-C coupling. This includes the genes encoding the Ca^{2+}-buffering proteins calsequestrin 1 (*casq1*) and 2 (*casq2*). Importantly, casq2 interacts with and regulates the activity of the RyR in addition to playing a key role in buffering SR Ca^{2+} (Faggioni and Knollmann 2012). It remains to be determined whether this is a compensatory or pathophysiological response. Finally, both the L-type calcium channel 1C subunit gene (*cacna1c*) and *ryr2* (encoding the RyR) were down-regulated at the later time point.

Gene expression studies in Gulf species are challenged by their extremely rapid rate of development (typically hatching at ~24 hpf), which conflicts with the time required to carefully collect replicate samples for high-fidelity preservation of RNA. A limitation of these studies has been identification of very early changes in gene expression that precede alterations in cardiac morphology. For mahi mahi embryos, morphological indicators were consistently more sensitive than changes in gene expression (Edmunds et al. 2015). However, the changes in gene expression observed in these rapidly developing Gulf species are indicative of alterations in developmental trajectory initiated by oil-induced cardiac dysfunction. Transcriptome analysis of red drum embryos exposed to MC252 crude oil confirmed this as well, where activation of cardiac hypertrophy pathways was consistently detected increasingly over time after hatching (Xu et al. 2017). The greater temporal resolution available in slowly developing cold-water species may provide some insight into earlier changes in gene expression that might be applicable in particular to Gulf species that inhabit a broader temperature range, such as the deep-diving tunas. For example, recent studies in Atlantic haddock found that oil exposure caused bradycardia and atrioventricular block phenotypes, linked to reduced expression of the haddock ERG ortholog *kcnh2*, as well as other E-C coupling genes (Sørhus et al. 2016, 2017). In addition, there was evidence of altered BMP signaling (upregulation of *bmp10*) that preceded visible cardiac looping defects. Coupled with the findings in mahi mahi, the weight of evidence suggests that changes in gene expression are likely to closely match specific cardiac phenotypes related to altered Ca^{2+} and K^+ fluxes. Some of these effects are hypothetically mediated through parallel disruption of E-T coupling (Fig. 10.7b). Although a simple set of gene expression markers that are diagnostic of crude oil developmental toxicity remain elusive, this remains an important future objective, to be pursued using hypothesis-driven and robust functional genomic approaches.

10.6 Summary

Oil spills such as the Deepwater Horizon disaster are high-profile, public examples of the impacts of fossil fuel-derived pollution on aquatic systems. However, PAHs have many other sources in the environment, including emissions from vessels, motor vehicles, and industrial facilities. They are major components of nonpoint source urban stormwater runoff, for example, and are correspondingly toxic to fish and other aquatic species (McIntyre and Scholz 2015). The assessment needs of the Deepwater Horizon NRDA and parallel sources of funding for new science have significantly advanced our understanding of PAH toxicity to fish early life stages, in a manner historically similar to increased scientific efforts in the aftermath of the Exxon Valdez spill. The insights gained will have implications beyond future oil spills injury assessments. This includes, for example, the success of cleanup efforts at industrial sites contaminated with legacy PAHs, as well as the implementation of efforts to reduce PAH pollution using green stormwater infrastructure and similar mitigation methods. The new science in the Deepwater era is yielding several new hypotheses and approaches to improving diagnostic tools for the impacts of PAHs. These will be broadly applicable to fish populations worldwide.

The identification of dual impacts on cardiomyocyte function—disruption of K^+ and Ca^{2+} fluxes—provides a key starting point for understanding species-specific cardiotoxicity. Whether a species displays heart rate and rhythm defects in addition to cardiac contractility defects is most likely determined by life history and eco-physiology. Species with profound bradycardia and arrhythmia include the migratory and deep-diving tunas (Incardona et al. 2014) as well as cold-water species such as Pacific herring (Incardona et al. 2009) and Atlantic haddock (Sørhus et al. 2016). Fishes showing primarily reduced contractility with minimal effect on heart rate include zebrafish, red drum, and mahi mahi. Although toxicokinetics could underlie some of this inter-species variation, the much smaller size of red drum eggs compared to mahi mahi suggests otherwise. Rather, the common factor may be a narrower thermal tolerance. Red drum are a nearshore/shelf species that reside in relatively shallow water (<100 m) (Powers et al. 2012) and are broadly adaptable to gradually changing temperatures but intolerant of rapid temperature drops (Gunter 1941; Gunter and Hildebrand 1951; Saillant et al. 2008). Similarly, the diving behavior of mahi mahi is limited to shallower and warmer waters (relative to tunas), despite the fast swimming predatory nature of this species (Merten et al. 2014; Walli et al. 2009; Dagorn et al. 2006). Although both Gulf killifish and zebrafish can be considered eurythermal (Cortemeglia and Beitinger 2005; Umminger 1971), their optimal ranges are very different, and killifish are more resistant to extreme cold than red drum. The density of the delayed rectifying K^+ channels is higher in tunas and other species that are tolerant of rapid temperature fluctuations or extreme cold (Galli et al. 2009; Haverinen and Vornanen 2009). A greater susceptibility to crude oil-induced impairments of heart rate and rhythm is consistent with this, suggesting more generally that the severity of functional defects in heart development may follow a gradient of ERG expression levels. Thus, characterizing the relative expression of *kcnh2* and *kcnh6* genes encoding these channels may provide a broader ability to predict species sensitivity to PAH pollution. At the same time, the diversity of specific cardiotoxic phenotypes also suggests that there may be no single molecular indicator for quantifying PAH cardiotoxicity across all fish species. Ideally, functional genomics targeting relatively small groups of related species will identify suites of genes that could be used more broadly to assess pollution impacts in different habitats.

Finally, observed extracardiac impacts of crude oil, particularly those stemming from disruption of osmoregulation and the movement of fluid between internal compartments in marine fish, suggest additional hypotheses for determining injury. This is a direct result of the intensive focus on pelagic species in support of the Deepwater Horizon impact assessment. Although not studied in detail in pelagic species, genetic and pharmacological studies in demersal euryhaline medaka embryos have shown that even modest reductions in circulation impaired osmoregulation in seawater, increasing whole-embryo osmolality (Miyanishi et al. 2013). This suggests that the accumulation of edema is not a result of net water gain within the embryo but rather the movement of free water into the pericardium and yolk sac from other parts of the body. This is also consistent with the marginal finfold phenotypes of oil-exposed pelagic larvae, which collapse as fluid moves from this most proximal source to the yolk sac sinus. Translocation of water from the peripheral subdermal spaces may lead to increases in osmolality in the adjacent tissues. Such osmotic stress could be particularly detrimental to cells relying on ionic gradients, including excitable cells in the developing nervous system. The apparent defects in expansion of the hindbrain ventricle in oil-exposed fish are a morphological indication of this type of pathophysiology (Gato and Desmond 2009). Altered ion homeostasis in the developing nervous system is likely to produce functional defects, but activity-dependent changes in development may not yield overtly visible anatomical abnormalities. Rather, any impacts on neural development from altered osmoregulation are likely to be subtle. A closer examination of functional and structural defects related to osmoregulation and ion homeostasis should further define potential quantitative tools for measuring oil spill impacts. Given that there may be less plasticity among fish species for osmoregulatory mechanisms compared to cardiac function, molecular indicators relating to this phenotype may be more broadly applicable.

References

Adeyemo OK, Kroll KJ, Denslow ND (2015) Developmental abnormalities and differential expression of genes induced in oil and dispersant exposed Menidia beryllina embryos. Aquat Toxicol 168:60–71. https://doi.org/10.1016/j.aquatox.2015.09.012

Agnisola C, Tota B (1994) Structure and function of the fish cardiac ventricle - flexibility and limitations. Cardioscience 5(3):145–153

Alonso MT, Garcia-Sancho J (2011) Nuclear Ca(2+) signalling. Cell Calcium 49(5):280–289. https://doi.org/10.1016/j.ceca.2010.11.004

Anderson JW, Neff JM, Cox BA, Tatem HE, Hightower GM (1974) Characteristics of dispersions and water-soluble extracts of crude and refined oils and their toxicity to estuarine crustaceans and fish. Mar Biol 27(1):75–88

Andreasen EA, Spitsbergen JM, Tanguay RL, Stegeman JJ, Heideman W, Peterson RE (2002) Tissue-specific expression of AHR2, ARNT2, and CYP1A in zebrafish embryos and larvae: effects of developmental stage and 2,3,7,8-tetrachlorodibenzo-p-dioxin exposure. Toxicol Sci 68(2):403–419

Antkiewicz DS, Burns CG, Carney SA, Peterson RE, Heideman W (2005) Heart malformation is an early response to TCDD in embryonic zebrafish. Toxicol Sci 84(2):368–377

Antkiewicz DS, Peterson RE, Heideman W (2006) Blocking expression of AHR2 and ARNT1 in zebrafish larvae protects against cardiac toxicity of 2,3,7,8-tetrachlorodibenzo-p-dioxin. Toxicol Sci 94(1):175–182. https://doi.org/10.1093/toxsci/kfl093
Arrenberg AB, Del Bene F, Baier H (2009) Optical control of zebrafish behavior with halorhodopsin. Proc Natl Acad Sci U S A 106(42):17968–17973. https://doi.org/10.1073/pnas.0906252106
Baldwin DH, Spromberg JA, Collier TK, Scholz NL (2009) A fish of many scales: extrapolating sublethal pesticide exposures to the productivity of wild salmon populations. Ecol Appl 19(8):2004–2015. https://doi.org/10.1890/08-1891.1
Bar I, Smith AD, Bubner E, Yoshizaki G, Takeuchi Y, Yazawa R, Chen BN, Cummins S, Elizur A (2016) Assessment of yellowtail kingfish (Seriola lalandi) as a surrogate host for the production of southern bluefin tuna (*Thunnus maccoyii*) seed via spermatogonial germ cell transplantation. Reprod Fertil Dev 28(12):2051–2064. https://doi.org/10.1071/RD15136
Barron MG, Heintz RA, Rice SD (2004) Relative potency of PAHs and heterocycles as aryl hydrocarbon receptor agonists in fish. Mar Environ Res 58(2-5):95–100
Barry A (1948) The functional significance of the cardiac jelly in the tubular heart of the chick embryo. Anat Rec 102(3):289–298
Bedell VM, Westcot SE, Ekker SC (2011) Lessons from morpholino-based screening in zebrafish. Brief Funct Genomics 10(4):181–188. https://doi.org/10.1093/bfgp/elr021
Bers DM (2002) Cardiac excitation-contraction coupling. Nature 415(6868):198–205
Bers DM (2003) Dynamic imaging in living cells: windows into local signaling. Sci Signal 2003(177):Pe13. https://doi.org/10.1126/stke.2003.177.pe13
Beyenbach KW (2004) Kidneys sans glomeruli. Am J Physiol Renal Physiol 286(5):F811–F827. https://doi.org/10.1152/ajprenal.00351.2003
Block BA, Finnerty JR, Stewart AF, Kidd J (1993) Evolution of endothermy in fish: mapping physiological traits on a molecular phylogeny. Science 260(5105):210–214
Bloomekatz J, Galvez-Santisteban M, Chi NC (2016) Myocardial plasticity: cardiac development, regeneration and disease. Curr Opin Genet Dev 40:120–130. https://doi.org/10.1016/j.gde.2016.05.029
Bosker T, van Balen L, Walsh B, Sepulveda MS, DeGuise S, Perkins C, Griffitt RJ (2017) The combined effect of Macondo oil and corexit on sheepshead minnow (Cyprinodon variegatus) during early development. J Toxicol Environ Health A 80(9):477–484. https://doi.org/10.1080/15287394.2017.1340208
Brette F, Machado B, Cros C, Incardona JP, Scholz NL, Block BA (2014) Crude oil impairs cardiac excitation-contraction coupling in fish. Science 343:772–776
Brette F, Shiels HA, Galli GLJ, Cros C, Incardona JP, Scholz NL, Block BA (2017) A novel cardiotoxic mechanism for a pervasive global pollutant. Sci Rep 7:41476
Bubner E, Farley J, Thomas P, Bolton T, Elizur A (2012) Assessment of reproductive maturation of southern bluefin tuna (Thunnus maccoyii) in captivity. Aquaculture 364-365:82–95. https://doi.org/10.1016/j.aquaculture.2012.08.005
Bue BG, Sharr S, Moffitt SD, Craig AK (1996) Effects of the Exxon Valdez oil spill on pink salmon embryos and preemergent fry. In: Rice SD, Spies RB, Wolfe DA, Wright BA (eds) Proceedings of the Exxon Valdez Oil Spill Symposium, Anchorage, AK, USA, 25 Feb 1993, American Fisheries Society, pp 619–627
Burggren WW (2005) Developing animals flout prominent assumptions of ecological physiology. Comp Biochem Physiol A Mol Integr Physiol 141(4):430–439. https://doi.org/10.1016/j.cbpb.2005.03.010
Burggren WW (2013) Cardiovascular development and angiogenesis in the early vertebrate embryo. Cardiovasc Eng Technol 4(3):234–245. https://doi.org/10.1007/s13239-013-0118-x
Burggren WW, Bagatto B (2008) Cardiovascular anatomy and physiology of larval fishes. In: Finn N, Kapoor BG (eds) Fish larval physiology. Oxford & IBH, New Delhi
Burggren WW, Dubansky B, Martinez Bautista N (2017) Cardiovascular development of embryonic and larval fishes. In: Gamperl K, Gillis T (eds) Fish physiology. Vol. 35. The cardiovascular system: design, control and function. Elsevier, London

Carls MG, Rice SD, Hose JE (1999) Sensitivity of fish embryos to weathered crude oil: part I. Low-level exposure during incubation causes malformations, genetic damage, and mortality in larval Pacific herring (*Clupea pallasi*). Environ Toxicol Chem 18(3):481–493

Carney SA, Chen J, Burns CG, Xiong KM, Peterson RE, Heideman W (2006) Aryl hydrocarbon receptor activation produces heart-specific transcriptional and toxic responses in developing zebrafish. Mol Pharmacol 70(2):549–561

Carney SA, Peterson RE, Heideman W (2004) 2,3,7,8-Tetrachlorodibenzo-p-dioxin activation of the aryl hydrocarbon receptor/aryl hydrocarbon receptor nuclear translocator pathway causes developmental toxicity through a CYP1A-independent mechanism in zebrafish. Mol Pharmacol 66(3):512–521

Chen JN, van Eeden FJ, Warren KS, Chin A, Nusslein-Volhard C, Haffter P, Fishman MC (1997) Left-right pattern of cardiac BMP4 may drive asymmetry of the heart in zebrafish. Development 124(21):4373–4382

Chi NC, Shaw RM, Jungblut B, Huisken J, Ferrer T, Arnaout R, Scott IC, Beis D, Xiao T, Baier H, Jan LY, Tristani-Firouzi M, Stainier DYR (2008) Genetic and physiologic dissection of the vertebrate cardiac conduction system. PLoS Biol 6(5):e109

Claireaux G, Davoodi F (2010) Effect of exposure to petroleum hydrocarbons upon cardio-respiratory function in the common sole (Solea solea). Aquat Toxicol 98(2):113–119. https://doi.org/10.1016/j.aquatox.2010.02.006

Claireaux G, McKenzie DJ, Genge AG, Chatelier A, Aubin J, Farrell AP (2005) Linking swimming performance, cardiac pumping ability and cardiac anatomy in rainbow trout. J Exp Biol 208(10):1775–1784

Clark BW, Matson CW, Jung D, Di Giulio RT (2010) AHR2 mediates cardiac teratogenesis of polycyclic aromatic hydrocarbons and PCB-126 in Atlantic killifish (Fundulus heteroclitus). Aquat Toxicol 99(2):232–240

Cortemeglia C, Beitinger TL (2005) Temperature tolerances of wild-type and red transgenic zebra danios. Trans Am Fish Soc 134(6):1431–1437. https://doi.org/10.1577/t04-197.1

Couillard CM (2002) A microscale test to measure petroleum oil toxicity to mummichog embryos. Environ Toxicol 17(3):195–202

Dagorn L, Holland KN, Hallier J-P, Taquet M, Moreno G, Sancho G, Itano DG, Aumeeruddy R, Girard C, Million J, Fonteneau A (2006) Deep diving behavior observed in yellowfin tuna (Thunnus albacares). Aquat Living Resour 19(1):85–88

Dong W, Teraoka H, Yamazaki K, Tsukiyama S, Imani S, Imagawa T, Stegeman JJ, Peterson RE, Hiraga T (2002) 2,3,7,8-tetrachlorodibenzo-p-dioxin toxicity in the zebrafish embryo: local circulation failure in the dorsal midbrain is associated with increased apoptosis. Toxicol Sci 69(1):191–201

Dubansky B, Whitehead A, Miller JT, Rice CD, Galvez F (2013) Multitissue molecular, genomic, and developmental effects of the Deepwater Horizon oil spill on resident gulf killifish (*Fundulus grandis*). Environ Sci Technol 47(10):5074–5082. https://doi.org/10.1021/es400458p

Dunn TW, Gebhardt C, Naumann EA, Riegler C, Ahrens MB, Engert F, Del Bene F (2016) Neural circuits underlying visually evoked escapes in larval zebrafish. Neuron 89(3):613–628. https://doi.org/10.1016/j.neuron.2015.12.021

Edmunds RC, Gill JA, Baldwin DH, Linbo TL, French BL, Brown TL, Esbaugh AJ, Mager EM, Stieglitz JD, Hoenig R, Benetti DD, Grosell M, Scholz NL, Incardona JP (2015) Corresponding morphological and molecular indicators of crude oil toxicity to the developing hearts of mahi mahi. Sci Rep 5:17326

Esbaugh AJ, Mager EM, Stieglitz JD, Hoenig R, Linbo TL, Lay C, Forth H, Brown TL, French BL, Scholz NL, Incardona JP, Morris JM, Benetti DD, Grosell M (2016) The effects of weathering and chemical dispersion on Deepwater Horizon crude oil toxicity to mahi-mahi (*Coryphaena hippurus*) early life stages. Sci Total Environ 543:644–651

Faggioni M, Knollmann BC (2012) Calsequestrin 2 and arrhythmias. Am J Physiol Heart Circ Physiol 302(6):H1250–H1260. https://doi.org/10.1152/ajpheart.00779.2011

Farrell AP (1987) Coronary flow in a perfused rainbow trout heart. J Exp Biol 129:107–123

Ferreira M, Costa J, Reis-Henriques MA (2014) ABC transporters in fish species: a review. Front Physiol 5:266. https://doi.org/10.3389/fphys.2014.00266

Forouhar AS, Liebling M, Hickerson A, Nasiraei-Moghaddam A, Tsai H-J, Hove JR, Fraser SE, Dickinson ME, Gharib M (2006) The embryonic vertebrate heart tube is a dynamic suction pump. Science 312(5774):751–753. https://doi.org/10.1126/science.1123775

Freitas R, Zhang G, Cohn MJ (2006) Evidence that mechanisms of fin development evolved in the midline of early vertebrates. Nature 442(7106):1033–1037. http://www.nature.com/nature/journal/v442/n7106/suppinfo/nature04984_S1.html

Galli GLJ, Lipnick MS, Block BA (2009) Effect of thermal acclimation on action potentials and sarcolemmal K+ channels from Pacific bluefin tuna cardiomyocytes. Am J Physiol Regul Integr Comp Physiol 297(2):R502–R509. https://doi.org/10.1152/ajpregu.90810.2008

Garcia-Reyero N (2015) Are adverse outcome pathways here to stay? Environ Sci Technol 49(1):3–9. https://doi.org/10.1021/es504976d

Garner LV, Brown DR, Di Giulio RT (2013) Knockdown of AHR1A but not AHR1B exacerbates PAH and PCB-126 toxicity in zebrafish (Danio rerio) embryos. Aquat Toxicol 142-143:336–346. https://doi.org/10.1016/j.aquatox.2013.09.007

Gato A, Desmond ME (2009) Why the embryo still matters: CSF and the neuroepithelium as interdependent regulators of embryonic brain growth, morphogenesis and histiogenesis. Dev Biol 327(2):263–272. https://doi.org/10.1016/j.ydbio.2008.12.029

Glickman NS, Yelon D (2002) Cardiac development in zebrafish: coordination of form and function. Semin Cell Dev Biol 13(6):507–513

González-Doncel M, Gonzalez L, Fernández-Torija C, Navas JM, Tarazona JV (2008) Toxic effects of an oil spill on fish early life stages may not be exclusively associated to PAHs: studies with Prestige oil and medaka (Oryzias latipes). Aquat Toxicol 87(4):280–268. https://doi.org/10.1016/j.aquatox.2008.02.013

Guiney PD, Smolowitz RM, Peterson RE, Stegeman JJ (1997) Correlation of 2,3,7,8-tetrachlorodi benzo-p-dioxin induction of cytochrome P4501A in vascular endothelium with toxicity in early life stages of lake trout. Toxicol Appl Pharmacol 143(2):256–273

Guiney PD, Walker MK, Spitsbergen JM, Peterson RE (2000) Hemodynamic dysfunction and cytochrome P4501A mRNA expression induced by 2,3,7,8-tetrachlorodibenzo-p-dioxin during embryonic stages of lake trout development. Toxicol Appl Pharmacol 168(1):1–14

Gunter G (1941) Death of fishes due to cold on the Texas coast, January, 1940. Ecology 22(2):203–208. https://doi.org/10.2307/1932218

Gunter G, Hildebrand HH (1951) Destruction of fishes and other organisms on the South Texas coast by the cold wave of January 28-February 3, 1951. Ecology 32(4):731–736. https://doi.org/10.2307/1932740

Gupta V, Poss KD (2012) Clonally dominant cardiomyocytes direct heart morphogenesis. Nature 484(7395):479–484. http://www.nature.com/nature/journal/v484/n7395/abs/nature11045.html-supplementary-information

Hahn ME (2002) Aryl hydrocarbon receptors: diversity and evolution. Chem Biol Interact 141(1-2):131–160

Haverinen J, Vornanen M (2009) Responses of action potential and K+ currents to temperature acclimation in fish hearts: phylogeny or thermal preferences? Physiol Biochem Zool 82(5):468–482. https://doi.org/10.1086/590223

Heintz RA (2007) Chronic exposure to polynuclear aromatic hydrocarbons in natal habitats leads to decreased equilibrium size, growth, and stability of pink salmon populations. Integr Environ Assess Manag 3(3):351–363. https://doi.org/10.1897/1551-3793(2007)3[351:cetpah]2.0.co;2

Heintz RA, Rice SD, Wertheimer AC, Bradshaw RF, Thrower FP, Joyce JE, Short JW (2000) Delayed effects on growth and marine survival of pink salmon *Oncorhynchus gorbuscha* after exposure to crude oil during embryonic development. Mar Ecol Prog Ser 208:205–216

Heintz RA, Short JW, Rice SD (1999) Sensitivity of fish embryos to weathered crude oil: part II. Increased mortality of pink salmon (*Oncorhynchus gorbuscha*) embryos incubating downstream from weathered Exxon Valdez crude oil. Environ Toxicol Chem 18(3):494–503

Hicken CE, Linbo TL, Baldwin DH, Willis ML, Myers MS, Holland L, Larsen M, Stekoll MS, Rice GS, Collier TK, Scholz NL, Incardona JP (2011) Sub-lethal exposure to crude oil during

embryonic development alters cardiac morphology and reduces aerobic capacity in adult fish. Proc Natl Acad Sci U S A 108:7086–7090
Hickman CPJ, Trump BF (1969) The kidney. In: Hoar WS, Randall DJ (eds) Fish physiology. Academic Press, New York, pp 91–239
Hohendanner F, McCulloch AD, Blatter LA, Michailova AP (2014) Calcium and IP3 dynamics in cardiac myocytes: experimental and computational perspectives and approaches. Front Pharmacol 5:35. https://doi.org/10.3389/fphar.2014.00035
Holt GJ (1992) Experimental studies of feeding in larval red drum. J World Aquacult Soc 23(4):265–270. https://doi.org/10.1111/j.1749-7345.1992.tb00789.x
Hose JE, McGurk MD, Marty GD, Hinton DE, Brown ED, Baker TT (1996) Sublethal effects of the *Exxon Valdez* oil spill on herring embryos and larvae: morphological, cytogenetic, and histopathological assessments, 1989–1991. Can J Fish Aquat Sci 53:2355–2365
Housden BE, Muhar M, Gemberling M, Gersbach CA, Stainier DY, Seydoux G, Mohr SE, Zuber J, Perrimon N (2017) Loss-of-function genetic tools for animal models: cross-species and cross-platform differences. Nat Rev Genet 18(1):24–40. https://doi.org/10.1038/nrg.2016.118
Hsiao S-M, Meier AH (1989) Comparison of semilunar cycles of spawning activity in Fundulus grandis and F. heteroclitus held under constant laboratory conditions. J Exp Zool 252(3):213–218. https://doi.org/10.1002/jez.1402520302
Hu N, Sedmera D, Yost HJ, Clark EB (2000) Structure and function of the developing zebrafish heart. Anat Rec 260(2):148–157
Hu N, Yost HJ, Clark EB (2001) Cardiac morphology and blood pressure in the adult zebrafish. Anat Rec 264(1):1–12
Incardona JP (2017) Molecular mechanisms of crude oil developmental toxicity in fish. Arch Environ Contam Toxicol 73:19–32
Incardona JP, Carls MG, Day HL, Sloan CA, Bolton JL, Collier TK, Scholz NL (2009) Cardiac arrhythmia is the primary response of embryonic Pacific herring (Clupea pallasi) exposed to crude oil during weathering. Environ Sci Technol 43(1):201–207
Incardona JP, Carls MG, Holland L, Linbo TL, Baldwin DH, Myers MS, Peck KA, Rice SD, Scholz NL (2015) Very low embryonic crude oil exposures cause lasting cardiac defects in salmon and herring. Sci Rep 5:13499
Incardona JP, Carls MG, Teraoka H, Sloan CA, Collier TK, Scholz NL (2005) Aryl hydrocarbon receptor-independent toxicity of weathered crude oil during fish development. Environ Health Perspect 113:1755–1762
Incardona JP, Collier TK, Scholz NL (2004) Defects in cardiac function precede morphological abnormalities in fish embryos exposed to polycyclic aromatic hydrocarbons. Toxicol Appl Pharmacol 196(2):191–205
Incardona JP, Day HL, Collier TK, Scholz NL (2006) Developmental toxicity of 4-ring polycyclic aromatic hydrocarbons in zebrafish is differentially dependent on AH receptor isoforms and hepatic cytochrome P450 1A metabolism. Toxicol Appl Pharmacol 217:308–321
Incardona JP, Gardner LD, Linbo TL, Brown TL, Esbaugh AJ, Mager EM, Stieglitz JD, French BL, Labenia JS, Laetz CA, Tagal M, Sloan CA, Elizur A, Benetti DD, Grosell M, Block BA, Scholz NL (2014) Deepwater horizon crude oil impacts the developing hearts of large predatory pelagic fish. Proc Natl Acad Sci U S A 111:E1510–E1518
Incardona JP, Linbo TL, Scholz NL (2011) Cardiac toxicity of 5-ring polycyclic aromatic hydrocarbons is differentially dependent on the aryl hydrocarbon receptor 2 isoform during zebrafish development. Toxicol Appl Pharmacol 257(2):242–249. https://doi.org/10.1016/j.taap.2011.09.010
Incardona JP, Scholz NL (2016) The influence of heart developmental anatomy on cardiotoxicity-based adverse outcome pathways in fish. Aquat Toxicol 177:515–525. https://doi.org/10.1016/j.aquatox.2016.06.016
Incardona JP, Swarts TL, Edmunds RC, Linbo TL, Edmunds RC, Aquilina-Beck A, Sloan CA, Gardner LD, Block BA, Scholz NL (2013) Exxon Valdez to Deepwater Horizon: comparable toxicity of both crude oils to fish early life stages. Aquat Toxicol 142-143:303–316

Incardona JP, Vines CA, Anulacion BF, Baldwin DH, Day HL, French BL, Labenia JS, Linbo TL, Myers MS, Olson OP, Sloan CA, Sol S, Griffin FJ, Menard K, Morgan SG, West JE, Collier TK, Ylitalo GM, Cherr GN, Scholz NL (2012a) Unexpectedly high mortality in Pacific herring embryos exposed to the 2007 Cosco Busan oil spill in San Francisco Bay. Proc Natl Acad Sci U S A 109:E51–E58

Incardona JP, Vines CA, Linbo TL, Myers MS, Sloan CA, Anulacion BF, Boyd D, Collier TK, Morgan S, Cherr GN, Scholz NL (2012b) Potent phototoxicity of marine bunker oil to translucent herring embryos after prolonged weathering. PLoS One 7(2):e30116

Johansen JL, Esbaugh AJ (2017) Sustained impairment of respiratory function and swim performance following acute oil exposure in a coastal marine fish. Aquat Toxicol 187:82–89. https://doi.org/10.1016/j.aquatox.2017.04.002

Jung J-H, Hicken CE, Boyd D, Anulacion BF, Carls MG, Shim WJ, Incardona JP (2013) Geologically distinct crude oils cause a common cardiotoxicity syndrome in developing zebrafish. Chemosphere 91:1146–1155

Jung JH, Kim M, Yim UH, Ha SY, Shim WJ, Chae YS, Kim H, Incardona JP, Linbo TL, Kwon JH (2015) Differential toxicokinetics determines the sensitivity of two marine embryonic fish exposed to Iranian heavy crude oil. Environ Sci Technol 49(22):13639–13648. https://doi.org/10.1021/acs.est.5b03729

Kammann U, Akcha F, Budzinski H, Burgeot T, Gubbins MJ, Lang T, Le Menach K, Vethaak AD, Hylland K (2017) PAH metabolites in fish bile: from the Seine estuary to Iceland. Mar Environ Res 124:41–45. https://doi.org/10.1016/j.marenvres.2016.02.014

Karlsson S, Saillant E, Bumguardner BW, Vega RR, Gold JR (2008) Genetic identification of hatchery-released red drum in Texas bays and estuaries. N Am J Fish Manag 28(4):1294–1304. https://doi.org/10.1577/M07-181.1

Khursigara AJ, Perrichon P, Martinez Bautista N, Burggren WW, Esbaugh AJ (2017) Cardiac function and survival are affected by crude oil in larval red drum, *Sciaenops ocellatus*. Sci Total Environ 579:797–804. https://doi.org/10.1016/j.scitotenv.2016.11.026

Klabunde RE (2017) Cardiac electrophysiology: normal and ischemic ionic currents and the ECG. Adv Physiol Educ 41(1):29–37. https://doi.org/10.1152/advan.00105.2016

Kohle C, Bock KW (2007) Coordinate regulation of phase I and II xenobiotic metabolisms by the Ah receptor and Nrf2. Biochem Pharmacol 73(12):1853–1562. https://doi.org/10.1016/j.bcp.2007.01.009

Kopp R, Bauer I, Ramalingam A, Egg M, Schwerte T (2014) Prolonged hypoxia increases survival even in Zebrafish (*Danio rerio*) showing cardiac arrhythmia. PLoS One 9(2):e89099. https://doi.org/10.1371/journal.pone.0089099

Kornreich BG (2007) The patch clamp technique: principles and technical considerations. J Vet Cardiol 9(1):25–37. https://doi.org/10.1016/j.jvc.2007.02.001

Langheinrich U, Vacun G, Wagner T (2003) Zebrafish embryos express an orthologue of Herg and are sensitive toward a range of QT-prolonging drugs inducing severe arrhythmia. Toxicol Appl Pharmacol 193(3):370–382

Lanham KA, Plavicki J, Peterson RE, Heideman W (2014) Cardiac myocyte-specific AHR activation phenocopies TCDD-induced toxicity in zebrafish. Toxicol Sci 141(1):141–154. https://doi.org/10.1093/toxsci/kfu111

Linden O (1978) Biological effects of oil on early development of the Baltic herring *Clupea harengus membras*. Mar Biol 45(3):273–283

Ljubojevic S, Bers DM (2015) Nuclear calcium in cardiac myocytes. J Cardiovasc Pharmacol 65(3):211–217. https://doi.org/10.1097/fjc.0000000000000174

Lowery LA, Sive H (2005) Initial formation of zebrafish brain ventricles occurs independently of circulation and requires the *nagie oko* and *snakehead/atp1a1a.1* gene products. Development 132(9):2057–2067

Mabee PM, Crotwell PL, Bird NC, Burke AC (2002) Evolution of median fin modules in the axial skeleton of fishes. J Exp Zool 294(2):77–90. https://doi.org/10.1002/jez.10076

MacRae CA (2013) Recent advances in in vivo screening for antiarrhythmic drugs. Expert Opin Drug Discovery 8(2):131–141. https://doi.org/10.1517/17460441.2013.743990

Mager EM, Esbaugh AJ, Stieglitz JD, Hoenig R, Bodinier C, Incardona JP, Scholz NL, Benetti DD, Grosell M (2014) Acute embryonic or juvenile exposure to Deepwater Horizon crude oil impairs the swimming performance of mahi-mahi (*Coryphaena hippurus*). Environ Sci Technol 48(12):7053–7061. https://doi.org/10.1021/es501628k

Mager EM, Pasparakis C, Schlenker LS, Yao Z, Bodinier C, Stieglitz JD, Hoenig R, Morris JM, Benetti DD, Grosell M (2017) Assessment of early life stage mahi-mahi windows of sensitivity during acute exposures to Deepwater Horizon crude oil. Environ Toxicol Chem 36(7):1887–1895. https://doi.org/10.1002/etc.3713

Männer J, Thrane L, Norozi K, Yelbuz TM (2008) High-resolution in vivo imaging of the cross-sectional deformations of contracting embryonic heart loops using optical coherence tomography. Dev Dyn 237(4):953–961

Margulies D, Suter JM, Hunt SL, Olson RJ, Scholey VP, Wexler JB, Nakazawa A (2007) Spawning and early development of captive yellowfin tuna (Thunnus albacares). Fish Bull 105(2):249–265

Marshall EKJ, Smith HW (1930) The glomerular development of the vertebrate kidney in relation to habitat. Biol Bull 59(2):135–153. https://doi.org/10.2307/1536983

Marty GD, Hose JE, McGurk MD, Brown ED, Hinton DE (1997a) Histopathology and cytogenetic evaluation of Pacific herring larvae exposed to petroleum hydrocarbons in the laboratory or in Prince William Sound, Alaska, after the Exxon Valdez oil spill. Can J Fish Aquat Sci 54(8):1846–1857

Marty GD, Short JW, Dambach DM, Willits NH, Heintz RA, Rice SD, Stegeman JJ, Hinton DE (1997b) Ascites, premature emergence, increased gonadal cell apoptosis, and cytochrome P4501A induction in pink salmon larvae continuously exposed to oil-contaminated gravel during development. Can J Zool 75(6):989–1007

McIntosh S, King T, Wu D, Hodson PV (2010) Toxicity of dispersed weathered crude oil to early life stages of Atlantic herring (Clupea harengus). Environ Toxicol Chem 29(5):1160–1167. https://doi.org/10.1002/etc.134

McIntyre JK, Scholz NL (2015) Chemical pollution. In: Closs GP, Krkosek M, Olden JD (eds) Conservation of freshwater fishes. Cambridge University Press, Cambridge, pp 149–178

Merten W, Appeldoorn R, Rivera R, Hammond D (2014) Diel vertical movements of adult male dolphinfish (Coryphaena hippurus) in the western central Atlantic as determined by use of pop-up satellite archival transmitters. Mar Biol 161(8):1823–1834. https://doi.org/10.1007/s00227-014-2464-0

Milan DJ, Peterson TA, Ruskin JN, Peterson RT, MacRae CA (2003) Drugs that induce repolarization abnormalities cause bradycardia in zebrafish. Circulation 107:1355–1358

Miyanishi, H, Okubo, K, Kaneko, T, Takei, Y (2013) Role of cardiac natriuretic peptides in seawater adaptation of medaka embryos as revealed by loss-of-function analysis. Am. J. Physiol. Regul. Integr. Comp. Physiol. 304(6):R423-434. https://doi.org/10.1152/ajpregu.00384.2012

Miyazono K, Maeda S, Imamura T (2005) BMP receptor signaling: transcriptional targets, regulation of signals, and signaling cross-talk. Cytokine Growth Factor Rev 16(3):251–263. https://doi.org/10.1016/j.cytogfr.2005.01.009

Mu J, Jin F, Ma X, Lin Z, Wang J (2014) Comparative effects of biological and chemical dispersants on the bioavailability and toxicity of crude oil to early life stages of marine medaka (*Oryzias melastigma*). Environ Toxicol Chem 33(11):2576–2583. https://doi.org/10.1002/etc.2721

Nelson D, Heuer RM, Cox GK, Stieglitz JD, Hoenig R, Mager EM, Benetti DD, Grosell M, Crossley DA 2nd (2016) Effects of crude oil on in situ cardiac function in young adult mahi-mahi (Coryphaena hippurus). Aquat Toxicol 180:274–281. https://doi.org/10.1016/j.aquatox.2016.10.012

Nikinmaa M (2014a) Chap. 7 Chemical uptake by organisms. In: An introduction to aquatic toxicology. Academic Press, Oxford, pp 73–80. https://doi.org/10.1016/B978-0-12-411574-3.00007-4

Nikinmaa M (2014b) Chap. 8 Chemical distribution in organisms. In: An introduction to aquatic toxicology. Academic Press, Oxford, pp 81–85. https://doi.org/10.1016/B978-0-12-411574-3.00008-6

Nikinmaa M (2014c) Chap. 9 Detoxification. In: An introduction to aquatic toxicology. Academic Press, Oxford, pp 87–98. https://doi.org/10.1016/B978-0-12-411574-3.00009-8

Nikinmaa M (2014d) Chap. 10 Excretion of compounds from organisms. In: An introduction to aquatic toxicology. Academic Press, Oxford, pp 99–110. https://doi.org/10.1016/B978-0-12-411574-3.00010-4

NOAA, National Oceanic and Atmospheric Administration (2017) Environmental response management application. http://erma.noaa.gov/gulfofmexico. Accessed 27Aug 2017

Norcross BL, Hose JE, Frandsen M, Brown ED (1996) Distribution, abundance, morphological condition, and cytogenetic abnormalities of larval herring in Prince William Sound, Alaska, following the *Exxon Valdez* oil spill. Can J Fish Aquat Sci 53:2376–2387

Oziolor EM, Dubansky B, Burggren WW, Matson CW (2016) Cross-resistance in gulf killifish (Fundulus grandis) populations resistant to dioxin-like compounds. Aquat Toxicol 175:222–231. https://doi.org/10.1016/j.aquatox.2016.03.019

Pampanin DM, Larssen E, Oysaed KB, Sundt RC, Sydnes MO (2014) Study of the bile proteome of Atlantic cod (Gadus morhua): multi-biological markers of exposure to polycyclic aromatic hydrocarbons. Mar Environ Res 101:161–168. https://doi.org/10.1016/j.marenvres.2014.10.002

Patterson RL, Boehning D, Snyder SH (2004) Inositol 1,4,5-trisphosphate receptors as signal integrators. Annu Rev Biochem 73:437–465. https://doi.org/10.1146/annurev.biochem.73.071403.161303

Payne JF (1976) Field evaluation of benzopyrene hydroxylase induction as a monitor for marine petroleum pollution. Science 191(4230):945–946

Pieperhoff S, Bennett W, Farrell AP (2009) The intercellular organization of the two muscular systems in the adult salmonid heart, the compact and the spongy myocardium. J Anat 215(5):536–547. https://doi.org/10.1111/j.1469-7580.2009.01129.x

Pilcher W, Miles S, Tang S, Mayer G, Whitehead A (2014) Genomic and genotoxic responses to controlled weathered-oil exposures confirm and extend field studies on impacts of the Deepwater Horizon oil spill on native killifish. PLoS One 9(9):e106351. https://doi.org/10.1371/journal.pone.0106351

Pollino CA, Holdway DA (2002) Toxicity testing of crude oil and related compounds using early life stages of the crimson-spotted rainbowfish (*Melanotaenia fluviatilis*). Ecotoxicol Environ Saf 52(3):180–189

Pott A, Rottbauer W, Just S (2014) Functional genomics in zebrafish as a tool to identify novel antiarrhythmic targets. Curr Med Chem 21(11):1320–1329

Powers SP, Hightower CL, Drymon JM, Johnson MW (2012) Age composition and distribution of red drum (Sciaenops ocellatus) in offshore waters of the north Central Gulf of Mexico: an evaluation of a stock under a federal harvest moratorium. Fish Bull 110(3):283–292

Priest BT, McDermott JS (2015) Cardiac ion channels. Channels (Austin) 9(6):352–359. https://doi.org/10.1080/19336950.2015.1076597

Raimondo S, Jackson CR, Krzykwa J, Hemmer BL, Awkerman JA, Barron MG (2014) Developmental toxicity of Louisiana crude oil-spiked sediment to zebrafish. Ecotoxicol Environ Saf 108:265–272. https://doi.org/10.1016/j.ecoenv.2014.07.020

Rice SD, Moles A, Taylor TL, Karinen JF (1979) Sensitivity of 39 Alaskan marine species to Cook Inlet crude oil and no. 2 fuel oil. In: 1979 Oil Spill Conference (Prevention, Behavior, Control, Cleanup), Los Angeles, 1979, American Petroleum Institute, Washington, DC, pp 549–554

Saillant E, Wang XX, Ma L, Gatlin DM, Vega RR, Gold JR (2008) Genetic effects on tolerance to acute cold stress in red drum, Sciaenops ocellatus L. Aquac Res 39(13):1393–1398. https://doi.org/10.1111/j.1365-2109.2008.02008.x

Sanchez-Quintana D, Garcia-Martinez V, Climent V, Hurle JM (1995) Morphological analysis of the fish heart ventricle: myocardial and connective tissue architecture in teleost species. Ann Anat 177(3):267–274

Sanguinetti MC (2010) HERG1 channelopathies. Pflugers Arch 460(2):265–276. https://doi.org/10.1007/s00424-009-0758-8

Santer RM, Greer Walker M (1980) Morphological studies on the ventricle of teleost and elasmobranch hearts. J Zool 190(2):259–272. https://doi.org/10.1111/j.1469-7998.1980.tb07771.x

Santer RM, Greer Walker M, Emerson L, Witthames PR (1983) On the morphology of the heart ventricle in marine teleost fish (Teleostei). Comp Biochem Physiol A Physiol 76(3):453–457

Schmitt N, Grunnet M, Olesen S-P (2014) Cardiac potassium channel subtypes: new roles in repolarization and arrhythmia. Physiol Rev 94(2):609–653. https://doi.org/10.1152/physrev.00022.2013

Scholz NL, Incardona JP (2015) In response: scaling polycyclic aromatic hydrocarbon toxicity to fish early life stages: a governmental perspective. Environ Toxicol Chem 34(3):459–461. https://doi.org/10.1002/etc.2830

Schwerte T, Prem C, Mairosl A, Pelster B (2006) Development of the sympatho-vagal balance in the cardiovascular system in zebrafish (*Danio rerio*) characterized by power spectrum and classical signal analysis. J Exp Biol 209(Pt 6):1093–1100

Scott JA, Incardona JP, Pelkki K, Shepardson S, Hodson PV (2011) AhR2-mediated; CYP1A-independent cardiovascular toxicity in zebrafish (Danio rerio) embryos exposed to retene. Aquat Toxicol 101:165–174

Sehnert AJ, Huq A, Weinstein BM, Walker C, Fishman M, Stainier DY (2002) Cardiac troponin T is essential in sarcomere assembly and cardiac contractility. Nat Genet 31(1):106–110

Serluca FC, Drummond IA, Fishman MC (2002) Endothelial signaling in kidney morphogenesis: a role for hemodynamic forces. Curr Biol 12(6):492–497

Sette CB, Pedrete Tde A, Felizzola J, Nudi AH, Scofield Ade L, Wagener Ade L (2013) Formation and identification of PAHs metabolites in marine organisms. Mar Environ Res 91:2–13. https://doi.org/10.1016/j.marenvres.2013.02.004

Shelbourne JE (1955) Significance of the subdermal space in pelagic fish embryos and larvae. Nature 176:743–744

Shelbourne JE (1956) The effect of water conservation on the structure of marine fish embryos and larvae. J Mar Biol Assoc UK 35(01):275–286. https://doi.org/10.1017/S0025315400009103

Shelbourne JE (1957) Site of chloride regulation in marine fish larvae. Nature 180(4592):920–922

Short JW, Harris PM (1996) Chemical sampling and analysis of petroleum hydrocarbons in near-surface seawater of Prince William Sound after the Exxon Valdez oil spill. In: Rice SD, Spies RB, Wolfe DA, Wright BA (eds) Proceedings of the Exxon Valdez Oil Spill Symposium, vol 18, American Fisheries Society Symposium Series, pp 17–28

Short JW, Heintz RA (1997) Identification of *Exxon Valdez* oil in sediments and tissues from Prince William Sound and the Northwestern Gulf of Alaska based on a PAH weathering model. Environ Sci Technol 31(8):2375–2384

Simoes K, Vicentini CA, Orsi AM, Cruz C (2002) Myoarchitecture and vasculature of the heart ventricle in some freshwater teleosts. J Anat 200(5):467–475. https://doi.org/10.1046/j.1469-7580.2002.00023.x

Smith RL, Cameron JA (1979) Effect of water soluble fraction of Prudhoe Bay crude oil on embryonic development of Pacific herring. Trans Am Fish Soc 108(1):70–75. https://doi.org/10.1577/1548-8659(1979)108<70:EOWSFO>2.0.CO;2

Somero GN (2002) Thermal physiology and vertical zonation of intertidal animals: optima, limits, and costs of living. Integr Comp Biol 42(4):780–789. https://doi.org/10.1093/icb/42.4.780

Sørensen L, Sørhus E, Nordtug T, Incardona JP, Linbo TL, Giovanetti L, Karlsen O, Meier S (2017) Oil droplet binding and differential PAH toxicokinetics in embryos of Atlantic haddock and cod. PLoS One 12:e0180048

Sørhus E, Incardona JP, Furmanek T, Jentoft S, Meier S, Edvardsen RB (2016) Developmental transcriptomics in Atlantic haddock: illuminating pattern formation and organogenesis in non-model vertebrates. Dev Biol 411(2):301–313. https://doi.org/10.1016/j.ydbio.2016.02.012

Sørhus E, Incardona JP, Furmanek T, Scholz NL, Meier S, Edvardsen RB, Jentoft S (2017) Novel adverse outcome pathways revealed by chemical genetics in a developing marine fish. elife 6:e20707

Sørhus E, Incardona JP, Karlsen Ø, Linbo TL, Sørensen L, Nordtug T, van der Meeren T, Thorsen A, Thorbjørnsen M, Jentoft S, Edvardsen RB, Meier S (2016) Effects of crude oil on haddock reveal roles for intracellular calcium in craniofacial and cardiac development. Sci Rep 6:31058

Staudt D, Stainier D (2012) Uncovering the molecular and cellular mechanisms of heart development using the zebrafish. Annu Rev Genet 46:397–418. https://doi.org/10.1146/annurev-genet-110711-155646

Stieglitz JD, Hoenig RH, Kloeblen S, Tudela CE, Grosell M, Benetti DD (2017) Capture, transport, prophylaxis, acclimation, and continuous spawning of mahi-mahi (Coryphaena hippurus) in captivity. Aquaculture 479:1–6. https://doi.org/10.1016/j.aquaculture.2017.05.006

Stockard CR (1915) The origin of blood and vascular endothelium in embryos without a circulation of the blood and in the normal embryo. Am J Anat 18:227–327

Taylor NG, McAllister MK, Lawson GL, Carruthers T, Block BA (2011) Atlantic bluefin tuna: a novel multistock spatial model for assessing population biomass. PLoS One 6(12):e27693. https://doi.org/10.1371/journal.pone.0027693

Tiitu V, Vornanen M (2002) Morphology and fine structure of the heart of the burbot, a cold stenothermal fish. J Fish Biol 61(1):106–121. https://doi.org/10.1006/jfbi.2002.2021

Toomey BH, Bello S, Hahn ME, Cantrell SM, Wright P, Tillitt DE, Di Giulio RT (2001) 2,3,7,8-Tetrachlorodibenzo-p-dioxin induces apoptotic cell death and cytochrome P4501A expression in developing *Fundulus heteroclitus* embryos. Aquat Toxicol 53(2):127–138

Tran MK, Kurakula K, Koenis DS, de Vries CJ (2016) Protein-protein interactions of the LIM-only protein FHL2 and functional implication of the interactions relevant in cardiovascular disease. Biochim Biophys Acta 1863(2):219–228. https://doi.org/10.1016/j.bbamcr.2015.11.002

Tu CT, Yang TC, Tsai HJ (2009) Nkx2.7 and Nkx2.5 function redundantly and are required for cardiac morphogenesis of zebrafish embryos. PLoS One 4(1):e4249. https://doi.org/10.1371/journal.pone.0004249

Umminger BL (1971) Chemical studies of cold death in the Gulf killifish, *Fundulus grandis*. Comp Biochem Physiol A Physiol 39(4):625–632. https://doi.org/10.1016/0300-9629(71)90184-8

Varanasi U (ed) (1989) Metabolism of polycyclic aromatic hydrocarbons in the aquatic environment. CRC Press, Boca Raton

Villeneuve DL, Crump D, Garcia-Reyero N, Hecker M, Hutchinson TH, LaLone CA, Landesmann B, Lettieri T, Munn S, Nepelska M, Ottinger MA, Vergauwen L, Whelan M (2014) Adverse outcome pathway (AOP) development I: strategies and principles. Toxicol Sci 142(2):312–320. https://doi.org/10.1093/toxsci/kfu199

Walker MK, Peterson RE (1991) Potencies of polychlorinated dibenzo-p-dioxin, dibenzofuran, and biphenyl congeners, relative to 2,3,7,8-tetrachlorodibenzo-p-dioxin, for producing early life stage mortality in rainbow trout (Oncorhynchus mykiss). Aquat Toxicol 21(3):219–237. https://doi.org/10.1016/0166-445X(91)90074-J

Walli A, Teo SLH, Boustany A, Farwell CJ, Williams TD, Dewar H, Prince E, Block BA (2009) Seasonal movements, aggregations and diving behavior of Atlantic bluefin tuna (*Thunnus thynnus*) revealed with archival tags. PLoS One 4(7):e6151. https://doi.org/10.1371/journal.pone.0006151

Warren KS, Fishman MC (1998) "Physiological genomics": mutant screens in zebrafish. Am J Physiol Heart Circ Physiol 275(1):1–7

Whitehead A, Dubansky B, Bodinier C, Garcia TI, Miles S, Pilley C, Raghunathan V, Roach JL, Walker N, Walter RB, Rice CD, Galvez F (2012) Genomic and physiological footprint of the Deepwater Horizon oil spill on resident marsh fishes. Proc Natl Acad Sci U S A 109(50):20298–20302. https://doi.org/10.1073/pnas.1109545108

van Wijk B, Moorman AF, van den Hoff MJ (2007) Role of bone morphogenetic proteins in cardiac differentiation. Cardiovasc Res 74(2):244–255. https://doi.org/10.1016/j.cardiores.2006.11.022

Wozney JM, Rosen V, Celeste AJ, Mitsock LM, Whitters MJ, Kriz RW, Hewick RM, Wang EA (1988) Novel regulators of bone formation: molecular clones and activities. Science 242(4885):1528–1534

Xu EG, Khursigara AJ, Magnuson J, Hazard ES, Hardiman G, Esbaugh AJ, Roberts AP, Schlenk D (2017) Larval red drum (*Sciaenops ocellatus*) sublethal exposure to weathered Deepwater Horizon crude oil: developmental and transcriptomic consequences. Environ Sci Technol 51(17):10162–10172. https://doi.org/10.1021/acs.est.7b02037

Xu EG, Mager EM, Grosell M, Pasparakis C, Schlenker LS, Stieglitz JD, Benetti D, Hazard ES, Courtney SM, Diamante G, Freitas J, Hardiman G, Schlenk D (2016) Time- and oil-dependent transcriptomic and physiological responses to Deepwater Horizon oil in mahi-mahi (Coryphaena hippurus) embryos and larvae. Environ Sci Technol 50(14):7842–7851. https://doi.org/10.1021/acs.est.6b02205

Analytical Methods and Trends in Environmental Forensics

11

Phillip M. Mach and Guido F. Verbeck

Abstract

Environmental forensics utilizes analytical scientific approaches to address the release and sources of contamination within the environment. These methods often seek to reconstruct the history of deleterious environmental events, their sources, and amounts of chemicals released into the environment. Forensic methods couple well-regarded scientific approaches within the legal framework. This due diligence provides tangible science-based results that are important in a regulatory context, including chemical identification, transport of contaminants, and determinable operational histories. Problems environmental forensics address include identifying sources of contamination, defining time frames of emission, and coupling observed conditions to potential sources of contamination.

Environmental forensics is at the forefront of innovation, as the present analytical trend concentrates on fieldable and portable instrumentation. Efforts to bring the lab into the field result in numerous improvements to workflow and evidence collection. New technologies being implemented allow for wide breadth of analysis at the source of contamination. Fieldable laboratories overcome the inability to bring the environment of interest, as evidence, back to the lab in sufficient quantity for a thorough analysis. This results in overall cost and time saving for the performing agency. Samples are collected and be subjected to immediate preliminary screening. This new methodology has a twofold improvement, being that only samples of concern are collected and an entire site can be thoroughly and rapidly screened. Subsequent in-laboratory workload is significantly reduced as further analysis on samples has already been determined to be of evidentiary value in the field.

P. M. Mach (✉) · G. F. Verbeck
Department of Chemistry, University of North Texas, Denton, TX, USA
e-mail: Guido.Verbeck@unt.edu

© Springer International Publishing AG, part of Springer Nature 2018

W. Burggren, B. Dubansky (eds.), *Development and Environment*,
https://doi.org/10.1007/978-3-319-75935-7_11

Portable devices have been developed to include a wide range of analytical techniques. This provides a fieldable option for a myriad of chemical analysis techniques. Spectroscopic methods include Raman, infrared, ultraviolet-visible, and cavity ringdown methods that test for a myriad of chemistries, including organic and metal contaminants. Most prevalent in the forensic community for chemical analysis are mass spectrometry (MS) methods. Portable mass spectrometers provide the precise mass analysis in the field as confirmatory tests in the laboratory. MS provides a multitude of sampling options including surface analysis with direct analysis in real time (DART) techniques, liquid introduction, and common chromatographic methods. Environmental forensics is at the forefront of implementing new technologies. Portable and fieldable devices are poised to improve analysis and provide robust analytical methods in the field. Continued efforts and targeted applications will improve forensic methods that rapidly identify and quantify contamination, determine time frames of release, and determine anthropogenic contribution.

11.1 Introduction

A constantly evolving and rapidly growing field is the application of analytical chemistry techniques as they pertain to the environment. Environmental forensics utilizes these chemical practices and the application of the scientific method over the course of an investigation to prove if laws or regulations have been violated (Gallego et al. 2016; Yim et al. 2012; White 2011, 2012; Tudorachi 2014). This often pertains to the release of contaminants into the environment, either into the atmosphere, ground soils, or water sources, that are harmful to the biosphere. There are innumerable potential pollutants, each with an inherent danger, and also a potential forensic method of detection. While there are numerous techniques recorded in the scientific literature, here we will address the current trends, innovative techniques, and a case study exemplifying these practices. Some of these techniques are mainstream, while others are new and novel methods in the early phases of adoption. These new techniques will usher a new paradigm and uplift current practices, by streamlining approaches, increasing speed of analysis, and provide a proper evaluation of the detriment to the environment.

As an environmental forensic practitioner, one must ensure the proper application of the scientific and traditional forensic methods (Gauch 2012; Bell 2009). Applying the scientific method in an environmental perspective requires a practitioner to consider (1) what the source of contamination is, (2) what the nature of the contaminant is, and (3) how long has it been present. The preliminary investigation and initial tour often provides visible cues to the nature and extent of contamination. However, forensics requires evidence and proper analytical techniques to be conclusive in the scope of an investigation. Evidence procurement and proper chain of custody provide the pieces to complete a proper investigation. Further questions will arise regarding the extent of environmental damage and applicable scientific techniques needed to be utilized to offer conclusive proof. Often, results of the forensic science performed will have monetary fines and remediation costs; thus good scientific practices are required (Ram et al. 2005). There are a myriad of analytical techniques, including those quantitative and qualitative in nature, and choosing proper techniques will further direct an environmental investigation. The choice of one experiment to perform may not yield immediate conclusive results, as techniques are usually validated for a specific analyte. Repeated experiment and properly chosen methods will ultimately yield a valid dataset. These datasets are the decisive element in which the final step of the scientific method is applied—the conclusion.

There are two kinds of datasets ultimately obtained by a forensic practitioner. Qualitative data often determines the presence of a compound or a given class of compounds in a given area. These are often physical descriptors, such as an "oily sheen"; they are indicative of the contaminant presence but do not enumerate the extent or cause of the damage. Quantitative datasets are obtained via proper application of scientific techniques, including standard analysis and standard operating procedures (SOPs). These methods have been highly validated by the scientific community and provide results that are exacting in nature. This is highly important as forensic practices are within the scope of law and regulations, such as those from the United States Environmental Protection Agency (EPA). Enforcement of these laws requires collaborative efforts for various agencies (White 2011). These regulations are itemized and provide values as to when contaminations are deemed neglectful and harmful to the environment.

Initial forensic applications further divide methods of testing into two unique realms, related to the type of data obtained. Presumptive testing provides a generalized result to a given sample. These are often inexpensive and can be rapidly performed in the field, for example, with colorimetric chemical tests. Such tests can indicate the presence of organic compounds but not necessarily from a petroleum-related accident. While this often provides a simple way to screen whether to collect and preserve a piece of evidence, it in no way provides conclusive results. Confirmatory testing is often performed in a laboratory, not in the field, and is precise and exacting in its results. These testing procedures follow SOPs, which are "recipes" scientifically validated and accepted by the community as a method for analysis and detection which often provide quantification of the target analyte.

New and innovative techniques in environmental forensics are part of the ever-evolving field, and continued improvement will provide for more rapid and accurate results. These new techniques not only hone the ability to detect but also reduce the rate of false positive events. Current trends see the deployment of fieldable devices, and ruggedized instrumentations are being continually developed. Hurdles to field these new devices include stability of the equipment to deliver accurate and repeatable results. Furthermore, the scientific community must validate these results, eventually offering an SOP, the method for subsequent analysis. Portable devices allow the laboratory to be brought out to the field. Their implementation often centers on small compact devices of similar laboratory instrumentation. Other advances revolve around sample preparation, in which direct analysis methods or minimal preparation procedures are utilized. This reduces time-consuming evidence procurement procedures and chain-of-custody practices. Possible evidence can be immediately screened for meaningful results. Instrumental systems can provide for a "red light green light" approach; while not necessary quantitative in nature, it is more exacting than colorimetric presumptive tests. Portable analytical systems will give exacting results, with quantitation and identification of a given contaminant. Ultimately, these methods reduce the physical quantity of evidence gathered and offer immediate results that can affect the course of an investigation, directing a forensic scientist to more evidence or a conclusive result.

The following summarizes new and innovative applications to environmental forensics. While not all methods elucidated will be wholly innovative, interesting application of previous techniques will be examined. Trends regarding portable and fieldable devices will be further mentioned. Additional techniques will be cost- and time-saving to the practitioner. Some techniques are newly produced and are being vetted by the scientific community. Hopefully, a new technique or application mentioned here will become a future methodology in the realm of environmental forensics.

11.2 Spectroscopy

Spectroscopy utilizes incident radiation, often from a laser, to irradiate a sample of interest and observe the resulting emission or absorption. Raman spectroscopy and infrared spectroscopy fundamentals and practices can be further examined elsewhere (Hollas 2004). Briefly, these techniques incite vibrational and rotational modes in chemistries and observe the scattering of the incident light. Compounds have unique resultant spectra, similar to fingerprints, which are identifying of a given chemistry. These results are nearly qualitative in nature, providing evidence of the presence. Research has begun to provide methods that now quantify the concentration present.

Water samples are a frequent environmental media that face a myriad of contaminants. Industrial runoff and other sources not only contaminate surface water but also subterranean sources, such as aquifers. Immediate contamination is not always manifested, as contaminants must drain into underground sources. In-line infrared instrumentation (Fig. 11.1) has been developed to detect hydrocarbon pollutants from these subterranean sources (Lay-Ekuakille et al. 2013). This method overcomes certain water chemistry issues that can occur, including solubility of analytes and presence of solid particulate matter. Optical spectroscopy requires precise metering of the transmitted light, as light transfer is affected by the suspended particulate and dissolved material (Lay-Ekuakille et al. 2013). This instrument overcomes sample flow problems and provides resultant infrared spectra for detection of hydrocarbons. Uniquely this instrumentation can be operated as a continual monitoring device, by which subtle changes can be tracked.

Infrared-based sensors have also been applied to detected mercury ion in water. A unique experimental setup utilizes DNA aptamers mounted on a gold substrate (Hoang et al. 2013). These aptamers are composed of thymine straight chains,

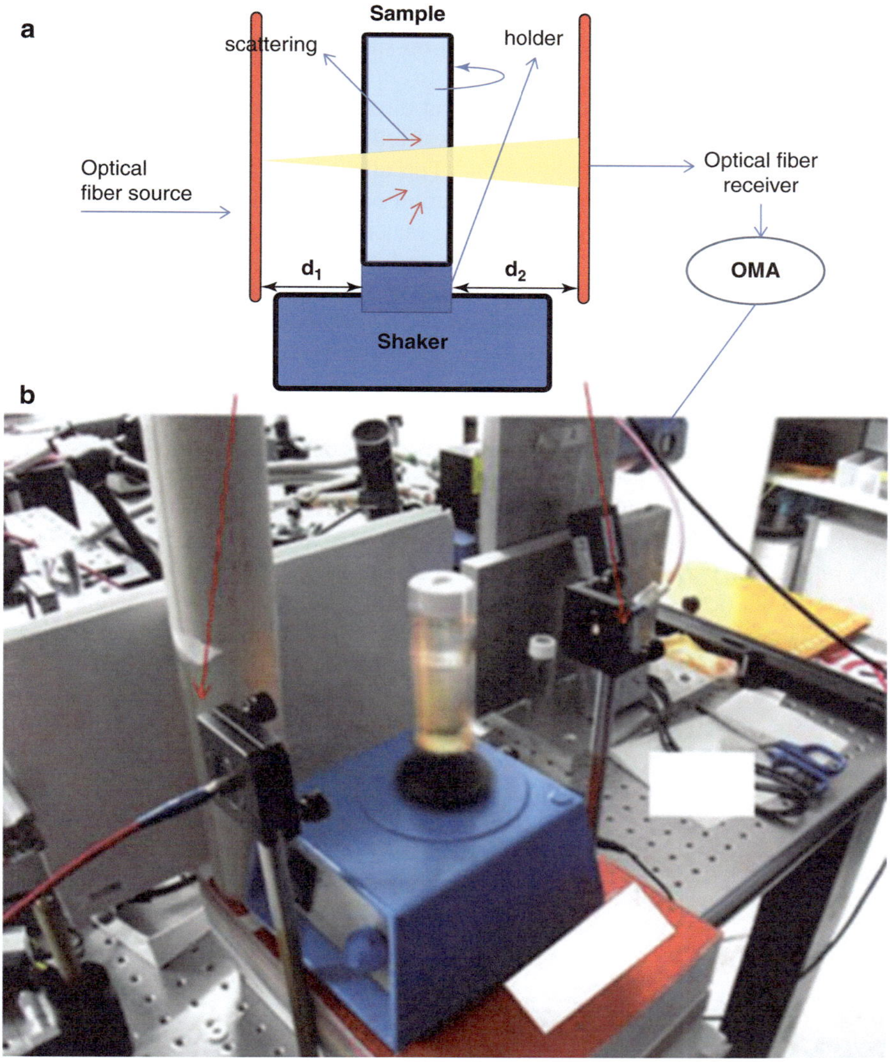

Fig. 11.1 (**a**) Schematic illustration of the experimental setup. The light source region is defined by the light cone. (**b**) Photo of the used experimental setup. Reproduced from Lay-Ekuakille, A., et al. (2013). "Experimental infrared measurements for hydrocarbon pollutant determination in subterranean waters." Review of Scientific Instruments 84(1): 015103. With the full permission of AIP Publishing

which when exposed to mercury cations cause a bend in the conformation, creating a hairpin-like structure, changing the resultant infrared spectrum. The mercury ion creates a bridge between two of the thymine bases, which in turn elicits an observable peak in an infrared spectra. These substrates allow for standard infrared instrumentation to be utilized and require no further sample preparation, with low parts-per-trillion (ppt) detection (Hoang et al. 2013). Further application of this method will augment portable systems for on-site measurements, with quantitative results.

Fiber optics allow for concentration and direction of the incident radiation utilized in infrared spectroscopic techniques. Furthermore, this incident light can be direct into samples via a probing method. This allows for direct analysis of aqueous environments, such as wells and landfill runoff pools. Development of these probes considered the resistance of the fiber-optic material to the analyte and its contaminants and must offer proper transmission and reception of the radiation from the source to the detector. A commercially available infrared spectrometer was adapted to utilize this fiber-optic probe, which allows for "dipping" into various sources of water to determine contaminant levels (Michel et al. 2004). Results of this sampling method showed detection of chlorinated hydrocarbons down to 1 part-per-million (ppm). In the future, combined with portable infrared devices, this sampling probe may offer the ability to rapidly analyze numerous samples in a given region.

Raman spectroscopy offers similar aqueous results. While numerous chemistries can be detected, including organic and metal ion pollutants (Durickovic and Marchetti 2014; Ben-Jaber et al. 2016; Li et al. 2014), physical material should also be considered. Plastic pollution results in microplastics that are common marine environments (Lenz et al. 2015; Claessens et al. 2013; Van Cauwenberghe et al. 2013). These breakdown products of various plastics are increasingly small in size and are often detrimentally ingested by organisms. Visual inspection is often inadequate for identification. Samples were filtered from seawater and observed via Raman spectroscopy. Identification of various plastics, from rubber tires to ABS plastics, was found in oceanic environments (Fig. 11.2) (Lenz et al. 2015). This

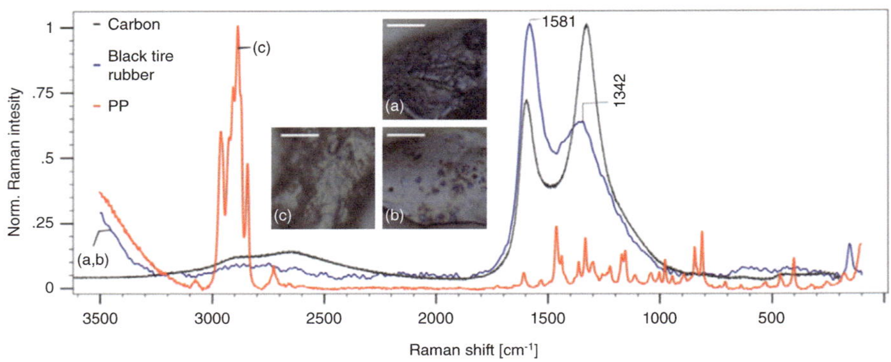

Fig. 11.2 Two spectra (blue and black) showing the vibrational band of amorphous carbon as found for many black particles of diverse morphologies (**a**, **b**). Image (**c**) which is of very similar morphology as (**a**) displays a PP spectrum (red) instead. Scale bars 20 μm. From Lenz, R., et al. (2015). Reprinted from "A critical assessment of visual identification of marine microplastic using Raman spectroscopy for analysis improvement." Marine pollution bulletin 100(1): 82–91. With permission from Elsevier

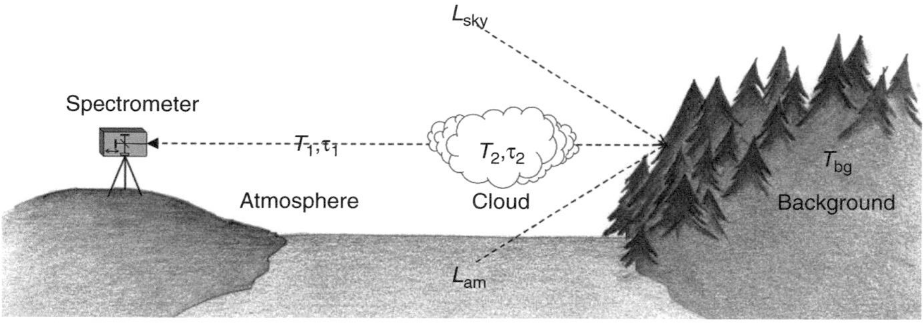

Fig. 11.3 Measurement setup of passive remote sensing of airborne pollutants. Reproduced from Harig, R., et al. "Scanning infrared remote sensing system for identification, visualization, and quantification of airborne pollutants," Proc. SPIE 4574, Instrumentation for Air Pollution and Global Atmospheric Monitoring, 83 (February 12, 2002); doi:https://doi.org/10.1117/12.455146

technique proved useful in various states of degradation and organic matter coating. Libraried spectra provided reference confirmation of plastics and included UV degradation products. Application of this technique includes tracing pollutants carried in ocean currents and their effects on organisms present.

Fugitive atmospheric emissions are more difficult to monitor and detect. As contaminants are rapidly dispersed into the atmosphere. This dispersal can also effect wide areas, effecting many regions. Passive scanning systems observe the composition, size, and direction of movement of a chemical plume, Fig. 11.3 (Harig et al. 2002). These systems capture infrared data regarding the background, the plume, and atmospheric compositions. These systems utilize an array of detectors similar to a digital camera and allow for images of detection to be created. Using algorithms that subtract normal background signals, atmospheric spectra are separated from the plume spectra data. A spectra library of reference data is used to identify the resultant spectra. Datasets allow for conclusive identification, visualization, and quantification of detected plumes. Vehicular and fixed emplacement of these systems are possible (Harig et al. 2002), and additionally aerial (Lega et al. 2012, 2014), allowing for targeted analysis of suspected release sites. Standoff detection via Raman can similarly be performed; gas hydrates and hydrocarbons have been detected out to 120 m (Sharma et al. 2005).

Identification of compounds via infrared spectroscopy relies on databases and libraries of reference spectra. The National Institute of Standards and Technology (NIST) and the Pacific Northwest National Laboratory (PNNL) have created an extensive database of pure chemical compound IR spectra (Sharpe et al. 2004). These databases have been created with consideration to traditional laboratory measurements and passive imaging systems, with differing path length and concentration considerations (Sharpe et al. 2004). Similarly, libraries of Raman spectra have been created for hazardous compounds (Harvey et al. 2002). Mixtures of chemistries are of particular difficulty for spectroscopic techniques, as resultant signals often convolute or mask other identifying characteristics. Statistical analysis, including multivariate techniques, allows for the extraction, evaluation, and identification of the convoluted-obtained spectra (Banas et al. 2010). Pattern recognition techniques have been established using chemometrics for petroleum spectra (Khanmohammadi et al. 2012).

Soils and other surface contaminants are slightly more difficult if there is a substrate present that could interfere with results, as sampling methods are constantly refined (Hadley and Petrisor 2013). A discriminatory pyrolysis method has been developed to help differentiate similar appearing samples; by determining the level of organic breakdown during the pyrolysis event, similar soils can be identified via infrared spectroscopy (Cox et al. 2000). A specific application of infrared identified ratios of heavy metals in soil samples in the Nile Rile delta (Mohamed et al. 2016). Chromium metals have been detected using laser-induced breakdown in soil samples, observing the effect on the technique that different soils have (Bousquet et al. 2007). Raman has been used in various geochemical applications identifying organic and inorganic elements (Vandenabeele et al. 2014; Vítek et al. 2012).

11.3 Mass Spectrometry and Chromatography

Mass spectrometry identifies introduced ions based upon the chemical's mass to charge ratio (m/z). There are numerous methods of ionization, which can be classified into hard and soft methods. Hard ionization methods involve high-energy ionization events which often induce fragmentation of the target molecule, such as from electron ionization. Soft ionization methods ionize a target molecule via less violent processes, first realized by electrospray ionization (ESI). Innovations concentrate on portability and new methods of ionization. For example, ambient ionization methods concentrate on minimal sample preparation in which ionization occurs outside the vacuum of the mass spectrometer, including desorption electrospray ionization (DESI) (Takats et al. 2004, 2005) and DART (Cody et al. 2005). These methods allow for rapid analysis and high throughput. Portable instrumentation brings these chemical analyzers into the field, providing for rapid forensic analysis.

Gas chromatography (GC) as a stand-alone method of analysis in environmental work is presently seeing significant use (Santos and Galceran 2002). These systems are robust, simple to operate, and inexpensive but rely on retention times of the compounds within the chromatographic column for identification (Pies et al. 2008). The advent of inexpensive mass analyzers has ushered in combined gas chromatography-mass spectrometry (GCMS) as an accurate method of identification for forensics applications (Santos and Galceran 2003). The method has offered environmental forensics practitioners, a power combination of chromatographic separations with the identification provided via mass spectrometry (Eckenrode 2001). Further development of portable GCMS systems has seen specialized systems for chemical warfare agents (Smith et al. 2004, 2010; Syage et al. 2001), ignitable liquids (Visotin and Lennard 2016), and volatile organic detection (Beckley et al. 2014; Contreras et al. 2008).

Membrane inlet mass spectrometry (MIMS) utilizes a semipermeable membrane that is selective for certain chemistries. These membranes are permeable to gases and are utilized in both air and water analyses. A typical membrane material is polydimethylsiloxane, but other materials exist that are selective for other types of chemistries (e.g., polar vs nonpolar). Membrane inlet mass spectrometers offer a unique ability to continuously sample a given analyte. Gas-phase chlorinated hydrocarbons have been analyzed with a portable MIMS system (Giannoukos et al. 2016). Prevalent in emissions and burning of hydrocarbons, benzene, toluene, ethylbenzene, and xylenes (BTEX) and polycyclic aromatic hydrocarbons (PAHs) have been measured with MIMS (Mach et al. 2015). Mounted in vehicular systems (Fig. 11.4), these systems have been deployed to measure atmospheric volatile organic concentrations (VOCs) (Mach et al. 2015; Bell et al. 2015; Davey et al. 2014; Hildenbrand et al. 2016). Mobile platforms take repeated measurements while the sample air flowing through the system, each scan is then GPS tagged for so that each moment of analysis has a spatial target. With the advent of geographic information systems, these points are easily plotted, and concentration-related heatmaps can be created for the large areas these systems can cover. More specific and target analysis can then be pinpointed. These vehicular systems provide spatially relevant datasets in real time, over large areas for regional quantitation of environmental species present. The MIMS technique has been further applied to the maritime environment,

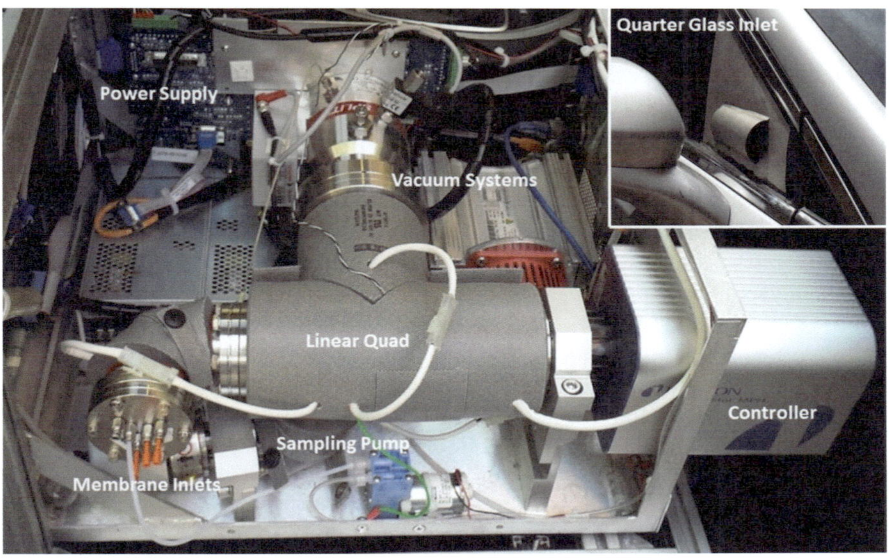

Fig. 11.4 The mobile system replacing the front passenger seat. The unit is placed in a rack mountable case. The inset shows the quarter glass replacement inlet. Reprinted with permission from Mach, P. M., et al. (2015). "Vehicle-mounted portable mass spectrometry system for the covert detection via spatial analysis of clandestine methamphetamine laboratories." Analytical Chemistry 87(22): 11501–11508. Copyright 2015 American Chemical Society

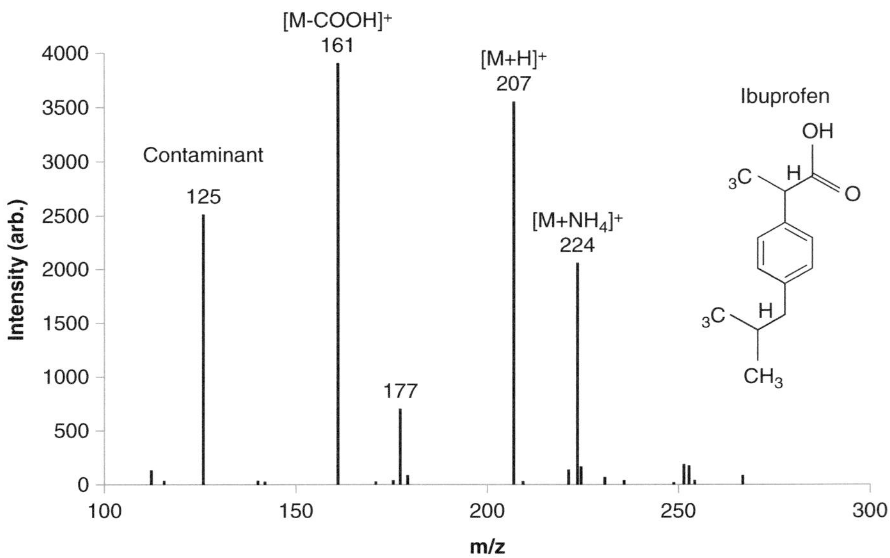

Fig. 11.5 DART MS detection of ibuprofen from the surface of an Advil tablet illustrating the potential for pharmaceutical QA/QC. Reprinted from Wells, J. M., et al. (2008). "Implementation of DART and DESI ionization on a fieldable mass spectrometer." Journal of the American Society of Mass Spectrometry. 19(10),1419–1424. With kind permission from [Springer Science + Business Media]

whereby dissolved gases, including methane, have been detected with dedicated systems (Miranda et al. 2013; Mächler et al. 2012; Bell et al. 2012). Maritime sampling requires and extra dimension to the data, depth. Methane leaks from drill sites in the Gulf of Mexico have been targeted and would prove useful for disasters, such as the Deepwater Horizon accident and remediation efforts (Fig. 11.5).

As previously mentioned, ambient ionization methods are becoming a method of choice for fieldable instrumentation. Developed instrumentation allows for one mass analyzer to have numerous "plug and play" methods of ionization (Wells et al. 2008; O'Leary et al. 2015). Having multiple methods to introduce samples into the system offers a practitioner an opportunity to obtain the best quality dataset within the field. DESI allows for samples on a substrate to be sprayed with an electrically charged mist. Ions of interest are subsequently released from the surface into the vacuum of the mass spectrometer (Takats et al. 2004, 2005). Similarly DART offers an ambient ionization method that requires no ionization liquid and offers direct analysis from surfaces such as plant matter and detecting toxic industrial waste, among a myriad of other substrates and analytes (Cody et al. 2005; Chernetsova et al. 2011; Na et al. 2007). Organic contaminants, such as PAHs and alkylbenzene chemistry, which pose a significant threat to wildlife and humans, due to their carcinogenic nature, have been analyzed with ambient methods (Jjunju et al. 2015). Ambient ionization methods are becoming similar to optical spectroscopy offering "point and shoot analysis" (Wiley et al. 2013; Hendricks et al. 2014).

More traditional methods are still being refined, offering improvements to existing mass spectrometers, via refined method development and novel new extractive techniques. Solid-phase extraction (SPE) is a developed polymer with a sorbent material that is selective for volatile chemistries. This sorbent material allows for enrichment and concentration of present analytes, easing detection from very minute quantities. This sorbent is exposed to the sample liquid and then introduced into a liquid chromatography (LC) instrument coupled to a mass spectrometer. Incredibly low levels of detection were obtained in determining veterinary versus human sewage inputs to surface water (Fenech et al. 2013). Furthermore, SPE has been useful in developing methods for new compounds previously undetected, such as pharmaceuticals (Grabic et al. 2012). Differences between human and veterinary samples included detecting various animal antibiotics to caffeine and acetaminophen attributable to human consumption. Rapid LC techniques utilizing similar statistical methods as mentioned previously, such as principle component analysis, have resulted in samples containing petroleum products being analyzed in 1 min time scales. This process identifies via fingerprinting, utilizing pattern recognition processes (Eide and Zahlsen 2005).

A chromatographic technique emerging as a new tool in environmental forensics is two-dimensional gas chromatography (Frysinger et al. 2002). Initially a sample is injected into a traditional GC column, but the eluted separated analytes are subsequently directed to a second chromatographic column. Working in tandem, the primary column offers one separation mechanism, while the second column provides another separation to ensure isolation of complex mixtures. Normal and reverse-phase chromatographic methods are often paired to ensure polar and nonpolar separations. These separations have isolated BTEX, PAHs, and chlorinated organics. This technique has been applied to oil source identification and tracking of fugitive spilled oil (Nelson et al. 2006; Gaines et al. 1999). Extensive research has been performed to identify ideal methods and column selection, such as in Fig. 11.6, for rapid analysis organohalogenated pollutants, including pesticides (Bordajandi et al. 2008; de Vos et al. 2011).

Unique to mass spectrometry is the use of isotopes for tracing environmental contaminants. Due to differing isotopic distributions, naturally occurring or non-biological samples can be discriminated against as analytes of similar origin may possess similar isotopic distributions (Slater 2003). As many elements may be simultaneously evaluated, a given isotopic fingerprint emerges, unique to a given source. This method of analysis has been practiced in determining the source of methane, be it from bacterial sources, geological, or commercially refined (Benson et al. 2006). Isotope analysis via inductively coupled plasma mass spectrometry has been able to determine groundwater contaminants, including lead and its use in fuels (Aggarwal et al. 2008). Nitrate sources, not normally part of the nitrate cycle, have been differentiated from anthropogenic sources, such as fertilizers or fecal contamination (Fenech et al. 2012). As the fugitive carbon sources are becoming prevalent and subject to regulatory efforts, carbon-13 from methane and carbon dioxide sources can be isotopically identified (Meier-Augenstein 1999).

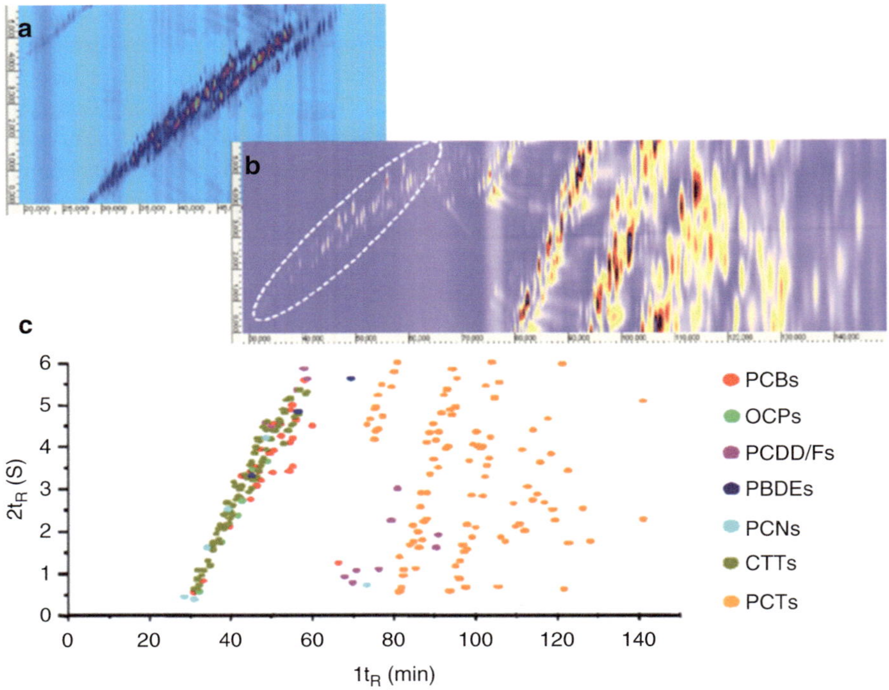

Fig. 11.6 Reconstructed chromatograms obtained from (**a**) CTT (as technical mixture) and (**b**) PCTs (Aroclor 5460), where the PCB position is indicated with a white circle. (**c**) Overlaid GC × GC–microECD chromatogram of the eight POP families investigated on ZB-5 × HT-8 column combination. Reprinted from Bordajandi, L. R., et al. (2008). "Comprehensive two-dimensional gas chromatography in the screening of persistent organohalogenated pollutants in environmental samples." Journal of Chromatography A 1186(1-2): 312–324. With permission from Elsevier

11.4 Conclusion

Many of these methods are still evolving, with new uses for existing methods being validated and novel new techniques highlighted. Optical and radiation-based analysis and mass analysis both have applications in unique scenarios. Utilization of only one technique hamstrings a potential investigation; there is unfortunately no ultimate method of analysis. Each method of analysis overcomes a potential analytical problem, such as matrix interferences. Only a portion of the scientific research regarding environmental forensics was discussed, each with innumerable results and perhaps unrealized uses. Case studies and further application will continue to bring new ideas and analytical practices to this ever growing field. Ultimately, forensics seeks to protect the current state of the environment and offer remediation and improvement when investable disasters and contamination occurs. Adapting to changing regulations and subsequent development of new methods to detect analytes ensure proper application of the law. Preserving the environment will continue to motivate environmental forensics practitioners to hone new methods and techniques.

11.4.1 Environmental Forensics and Persistent Organic Pollutants

The use of persistent organic pollutants (POPs) was widespread and had applicable uses in industrial processes, flame retardants, and pesticides. While POPs had desirable chemical properties such as stability and chemical inertness, this also meant for bioaccumulation and expansive pollution within the environment. This pervasiveness meant that releases into the environment were not localized in their effects as wind and water easily dispersed these chemicals into subsequent environments. Ultimately, POPs produce and use was limited or eliminated via the Stockholm Convention in 2001, which sought to curtail the environmental damage and health effects, while commencing the remediation efforts that have followed (Godduhn and Duffy 2003).

Originally, the Stockholm convention identified 26 target compounds, including polychlorinated biphenyls, polychlorinated dibenzo-p-dioxins, and polychlorinated dibenzofurans. In addition to these chemicals of significance, further research suggested that over 100,000 are in use that could possess similar toxic effects in humans or be detrimental to the environment (Howard and Muir 2010). In a forensic investigation, a single analytical method cannot be targeted toward one class of compound but rather must be a combining tool capable of detecting the myriad of pollutant chemistries. Furthermore forensic investigations should involve, in part, determining the source of the contamination event. An investigation of POPs will often involve determining the biodegradation, biological uptake, and the derivatives that may arise.

There are a variety of analytical techniques available to the investigator when targeting POPs. The EPA has elucidated and validated multiple methods for analysis, such as EPA Method 608 or 8082A, and utilized GC techniques discussed in this chapter. Few methods, however, target an expansive listing of multiple compounds. More recently, EPA Method 1668C has been developed to target over 200 of these compounds in a single sample aliquot. As previously mentioned, extractive techniques, like SPE, are often used to remove the target analyte from the sample matrix prior to analysis.

Remediation efforts for POPs are an ongoing process. New sampling methods are needed to fully realize and support the efforts to remove these toxic pollutants from the environment. It is ultimately up to the practitioner to choose a well validated method of analysis. The most important process is to fully understand and justify the analytical method being used to ensure the best results of an investigation.

References

Aggarwal J, Habicht-Mauche J, Juarez C (2008) Application of heavy stable isotopes in forensic isotope geochemistry: a review. Appl Geochem 23(9):2658–2666

Banas K et al (2010) Multivariate analysis techniques in the forensics investigation of the postblast residues by means of fourier transform-infrared spectroscopy. Anal Chem 82(7):3038–3044

Beckley L et al (2014) On-site gas chromatography/mass spectrometry (GC/MS) analysis to streamline vapor intrusion investigations. Environ Forensic 15(3):234–243

Bell RJ et al (2012) In situ determination of porewater gases by underwater flow-through membrane inlet mass spectrometry. Limnol Oceanogr Methods 10(3):117–128

Bell RJ et al (2015) A field-portable membrane introduction mass spectrometer for real-time quantitation and spatial mapping of atmospheric and aqueous contaminants. J Am Soc Mass Spectrom 26(2):212–223

Bell S (2009) Forensic chemistry. Annu Rev Anal Chem 2:297–319

Ben-Jaber S et al (2016) Photo-induced enhanced Raman spectroscopy for universal ultra-trace detection of explosives, pollutants and biomolecules. Nat Commun 7:12189

Benson S et al (2006) Forensic applications of isotope ratio mass spectrometry—a review. Forensic Sci Int 157(1):1–22

Bordajandi LR et al (2008) Comprehensive two-dimensional gas chromatography in the screening of persistent organohalogenated pollutants in environmental samples. J Chromatogr A 1186(1-2):312–324

Bousquet B, Sirven JB, Canioni L (2007) Towards quantitative laser-induced breakdown spectroscopy analysis of soil samples. Spectrochim Acta B At Spectrosc 62(12):1582–1589

Chernetsova ES, Morlock GE, Revelsky IA (2011) DART mass spectrometry and its applications in chemical analysis. Russ Chem Rev 80(3):235

Claessens M et al (2013) New techniques for the detection of microplastics in sediments and field collected organisms. Mar Pollut Bull 70(1–2):227–233

Cody RB, Laramée JA, Durst HD (2005) Versatile new ion source for the analysis of materials in open air under ambient conditions. Anal Chem 77(8):2297–2302

Contreras JA et al (2008) Hand-portable gas chromatograph-toroidal ion trap mass spectrometer (GC-TMS) for detection of hazardous compounds. J Am Soc Mass Spectrom 19(10):1425–1434

Cox R et al (2000) The forensic analysis of soil organic by FTIR. Forensic Sci Int 108(2):107–116

Davey NG et al (2014) Measurement of spatial and temporal variation in volatile hazardous air pollutants in Tacoma, Washington, using a mobile membrane introduction mass spectrometry (MIMS) system. J Environ Sci Health A 49(11):1199–1208

Durickovic I, Marchetti M (2014) Raman spectroscopy as polyvalent alternative for water pollution detection. IET Sci Meas Technol 8(3):122–128

Eckenrode BA (2001) Environmental and forensic applications of field-portable GC-MS: an overview. J Am Soc Mass Spectrom 12(6):683–693

Eide I, Zahlsen K (2005) A novel method for chemical fingerprinting of oil and petroleum products based on electrospray mass spectrometry and chemometrics. Energy Fuel 19(3):964–967

Fenech C et al (2012) The potential for a suite of isotope and chemical markers to differentiate sources of nitrate contamination: a review. Water Res 46(7):2023–2041

Fenech C et al (2013) An SPE LC–MS/MS method for the analysis of human and veterinary chemical markers within surface waters: an environmental forensics application. Environ Pollut 181:250–256

Frysinger GS, Gaines RB, Reddy CM (2002) GC× GC--a new analytical tool for environmental forensics. Environ Forensic 3(1):27–34

Gaines RB et al (1999) Oil spill source identification by comprehensive two-dimensional gas chromatography. Environ Sci Technol 33(12):2106–2112

Gallego J et al (2016) Insights into a 20-ha multi-contaminated brownfield megasite: an environmental forensics approach. Sci Total Environ 563:683–692

Gauch HG Jr (2012) Scientific method in brief. Cambridge University Press, New York

Giannoukos S, Brkic B, Taylor S (2016) Analysis of chlorinated hydrocarbons in gas phase using a portable membrane inlet mass spectrometer. Anal Methods 8(36):6607–6615
Godduhn A, Duffy LK (2003) Multi-generation health risks of persistent organic pollution in the far north: use of the precautionary approach in the Stockholm convention. Environ Sci Pol 6(4):341–353
Grabic R et al (2012) Multi-residue method for trace level determination of pharmaceuticals in environmental samples using liquid chromatography coupled to triple quadrupole mass spectrometry. Talanta 100:183–195
Hadley PW, Petrisor IG (2013) Incremental sampling: challenges and opportunities for environmental forensics. Environ Forensic 14(2):109–120
Harig R, Matz G, Rusch P (2002) Scanning infrared remote sensing system for identification, visualization, and quantification of airborne pollutants. In: Proc. SPIE 4574, Instrumentation for Air Pollution and Global Atmospheric Monitoring, 83 (February 12, 2002). doi:https://doi.org/10.1117/12.455146
Harvey SD et al (2002) Blind field test evaluation of Raman spectroscopy as a forensic tool. Forensic Sci Int 125(1):12–21
Hendricks PI et al (2014) Autonomous in situ analysis and real-time chemical detection using a backpack miniature mass spectrometer: concept, instrumentation development, and performance. Anal Chem 86(6):2900–2908
Hildenbrand ZL et al (2016) Point source attribution of ambient contamination events near unconventional oil and gas development. Sci Total Environ 573:382–388
Hoang CV et al (2013) Monitoring the presence of ionic mercury in environmental water by plasmon-enhanced infrared spectroscopy. Sci Rep 3:1175
Hollas JM (2004) Modern spectroscopy, 4th edn. Wiley, New York
Howard PH, Muir DC (2010) Identifying new persistent and bioaccumulative organics among chemicals in commerce. Environ Sci Technol 44:2277
Jjunju FPM et al (2015) Analysis of polycyclic aromatic hydrocarbons using desorption atmospheric pressure chemical ionization coupled to a portable mass spectrometer. J Am Soc Mass Spectrom 26(2):271–280
Khanmohammadi M, Garmarudi AB, de la Guardia M (2012) Characterization of petroleum-based products by infrared spectroscopy and chemometrics. TrAC Trends Anal Chem 35:135–149
Lay-Ekuakille A et al (2013) Experimental infrared measurements for hydrocarbon pollutant determination in subterranean waters. Rev Sci Instrum 84(1):015103
Lega M et al (2012) Using advanced aerial platforms and infrared thermography to track environmental contamination. Environ Forensic 13(4):332–338
Lega M et al (2014) Remote sensing in environmental police investigations: aerial platforms and an innovative application of thermography to detect several illegal activities. Environ Monit Assess 186(12):8291–8301
Lenz R et al (2015) A critical assessment of visual identification of marine microplastic using Raman spectroscopy for analysis improvement. Mar Pollut Bull 100(1):82–91
Li D-W et al (2014) Recent progress in surface enhanced Raman spectroscopy for the detection of environmental pollutants. Microchim Acta 181(1-2):23–43
Mach PM, Wright KC, Verbeck GF (2015) Development of multi-membrane near-infrared diode mass spectrometer for field analysis of aromatic hydrocarbons. J Am Soc Mass Spectrom 26(2):281–285
Mach PM et al (2015) Vehicle-mounted portable mass spectrometry system for the covert detection via spatial analysis of clandestine methamphetamine laboratories. Anal Chem 87(22):11501–11508
Mächler L, Brennwald MS, Kipfer R (2012) Membrane inlet mass spectrometer for the quasi-continuous on-site analysis of dissolved gases in groundwater. Environ Sci Technol 46(15):8288–8296
Meier-Augenstein W (1999) Applied gas chromatography coupled to isotope ratio mass spectrometry. J Chromatogr A 842(1-2):351–371

Michel K et al (2004) Monitoring of pollutant in waste water by infrared spectroscopy using chalcogenide glass optical fibers. Sensors Actuators B Chem 101(1):252–259
Miranda L et al (2013) Calibration of membrane inlet mass spectrometric measurements of dissolved gases: differences in the responses of polymer and nano-composite membranes to variations in ionic strength. Talanta 116:217–222
Mohamed E et al (2016) Near infrared spectroscopy techniques for soil contamination assessment in the Nile Delta. Eurasian Soil Sci 49(6):632–639
Na N et al (2007) Direct detection of explosives on solid surfaces by mass spectrometry with an ambient ion source based on dielectric barrier discharge. J Mass Spectrom 42(8):1079–1085
Nelson RK et al (2006) Tracking the weathering of an oil spill with comprehensive two-dimensional gas chromatography. Environ Forensic 7(1):33–44
O'Leary AE et al (2015) Combining a portable, tandem mass spectrometer with automated library searching - an important step towards streamlined, on-site identification of forensic evidence. Anal Methods 7(8):3331–3339
Pies C et al (2008) Characterization and source identification of polycyclic aromatic hydrocarbons (PAHs) in river bank soils. Chemosphere 72(10):1594–1601
Ram NM, Wiest MA, Davis CP (2005) Allocating cleanup costs at hazardous waste sites. Environ Sci Technol 39(6):128A–135A
Santos FJ, Galceran MT (2002) The application of gas chromatography to environmental analysis. TrAC Trends Anal Chem 21(9-10):672–685
Santos FJ, Galceran MT (2003) Modern developments in gas chromatography–mass spectrometry-based environmental analysis. J Chromatogr A 1000(1-2):125–151
Sharma SK, Misra AK, Sharma B (2005) Portable remote Raman system for monitoring hydrocarbon, gas hydrates and explosives in the environment. Spectrochim Acta A Mol Biomol Spectrosc 61(10):2404–2412
Sharpe SW et al (2004) Gas-phase databases for quantitative infrared spectroscopy. Appl Spectrosc 58(12):1452–1461
Slater G (2003) Stable isotope forensics--when isotopes work. Environ Forensic 4(1):13–23
Smith PA et al (2004) Detection of gas-phase chemical warfare agents using field-portable gas chromatography–mass spectrometry systems: instrument and sampling strategy considerations. TrAC Trends Anal Chem 23(4):296–306
Smith PA et al (2010) Field-portable gas chromatography with transmission quadrupole and cylindrical ion trap mass spectrometric detection: chromatographic retention index data and ion/molecule interactions for chemical warfare agent identification. Int J Mass Spectrom 295(3):113–118
Syage JA et al (2001) Field-portable, high-speed GC/TOFMS. J Am Soc Mass Spectrom 12(6):648–655
Takats Z, Wiseman JM, Cooks RG (2005) Ambient mass spectrometry using desorption electrospray ionization (DESI): instrumentation, mechanisms and applications in forensics, chemistry, and biology. J Mass Spectrom 40(10):1261–1275
Takats Z et al (2004) Mass spectrometry sampling under ambient conditions with desorption electrospray ionization. Science 306(5695):471–473
Tudorachi L (2014) Linking chemistry to law in environmental forensics. Environ Forensic 15(3):213–218
Van Cauwenberghe L et al (2013) Microplastic pollution in deep-sea sediments. Environ Pollut 182:495–499
Vandenabeele P, Edwards H, Jehlička J (2014) The role of mobile instrumentation in novel applications of Raman spectroscopy: archaeometry, geosciences, and forensics. Chem Soc Rev 43(8):2628–2649
Visotin A, Lennard C (2016) Preliminary evaluation of a next-generation portable gas chromatograph mass spectrometer (GC-MS) for the on-site analysis of ignitable liquid residues. Aust J Forensic Sci 48(2):203–221
Vítek P et al (2012) Evaluation of portable Raman spectrometer with 1064 nm excitation for geological and forensic applications. Spectrochim Acta A Mol Biomol Spectrosc 86:320–327

de Vos J et al (2011) Comprehensive two-dimensional gas chromatography time of flight mass spectrometry (GC × GC-TOFMS) for environmental forensic investigations in developing countries. Chemosphere 82(9):1230–1239
Wells JM, Roth MJ, Keil AD, Grossenbacher JW, Justes DR, Patterson GE, Barket DJ Jr (2008) Implementation of DART and DESI ionization on a fieldable mass spectrometer. J Am Soc Mass Spectrom 19(10):1419–1424
White R (2011) Environmental law enforcement: the importance of global networks and collaborative practices. Aust Policing 3(1):12
White R (2012) Environmental forensic studies and toxic towns. Curr Issues Crim Just 24:105
Wiley JS, Shelley JT, Cooks RG (2013) Handheld low-temperature plasma probe for portable "point-and-shoot" ambient ionization mass spectrometry. Anal Chem 85(14):6545–6552
Yim UH et al (2012) Oil spill environmental forensics: the Hebei Spirit oil spill case. Environ Sci Technol 46(12):6431–6437

Consortia and Collaboration in Environmental Sciences

12

Linda M. Hooper-Bùi, Brian Matherne, and George Washburn

Abstract

Consortia are a voluntary partnership that builds knowledge and community. The idea of consortia originated from Japan and British business models. Changes in the laws in the USA in the 1980s allowed for business consortia to form in the USA. Academic consortia often use shared resources to work on large projects or solve big problems. Best practices are (1) defining metrics of success and failure in the planning stages, (2) bringing the best scientists to the group and trusting them to participate and communicate, (3) sharing data and discussing ownership and authorship often, and (4) creating policies that reduce the impact of Rogue Researchers and Rogue Writers. Inclusivity should be intentional from the outset of the consortium and revisited periodically.

12.1 Introduction

A consortium is broadly defined as an association, a voluntary partnership, or a union which builds knowledge as well as community. Here we will explore how partnerships in science can share resources to achieve goals that would not be achievable individually. In a sense, a research consortium should be able to conduct more and/or higher quality of research than smaller groups or individual collaborations. Consortia in scientific research are becoming more common and are typically formed to solve problems that are so large they can't be tackled by individuals or smaller collaborations. Another main reason to form research consortia rather than working in small, siloed groups is that the activities of a healthy consortium can be

L. M. Hooper-Bùi (✉) · B. Matherne · G. Washburn
Department of Environmental Sciences, Louisiana State University, Baton Rouge, LA, USA
e-mail: Lindabui@lsu.edu

© Springer International Publishing AG, part of Springer Nature 2018

W. Burggren, B. Dubansky (eds.), *Development and Environment*,
https://doi.org/10.1007/978-3-319-75935-7_12

reciprocally strengthening for the members. The purpose of this chapter is to collect best practices of successful collaborations and consortia and also to deliver some warnings of how consortia can go wrong.

The senior author (hereafter LMHB) has participated in and led several consortia for the past 8 years. Consortia formation and function cannot be discussed here without discussing methods of collaboration. Several groups have identified development of best practices among business and education consortia, but best practices of scientific research consortia are needed. After a short history of consortia, the literature exploring the strengths and weaknesses of consortia will be presented and discussed. Part of collaboration and consortia is collaborative writing of reports and publications and producing projects as outcomes; therefore two case studies are presented to address this. Much of what is presented are education and business consortia challenges and practices adjusted to fit research consortia. Best practices will be identified.

12.2 History of Consortia

Japan pioneered the use of research consortia in business with technology research associations (TRA) that were allowed via the 1961 Act on Mining and Manufacturing Industry in Japan (Aldrich and Sasaki 1995). The Japanese TRA model imitates the British Research Associations which were started during World War I (Aldrich and Sasaki 1995). A case study from Japan (Bransetetter and Sakakibara 2002) examined government-sponsored research consortia resulting in technology patents, and the empirical data suggest that Japanese consortia focused on fundamental research. Japanese research consortia were developed to internalize externalities resulting from corporate research and development (Bransetetter and Sakakibara 2002; Katz 1986).

Prior to 1984, business consortia were rare in the USA. In 1984, the National Cooperative Research Act allowed US research consortia to form and gave antitrust protection for registered research and development business consortia (Aldrich and Sasaki 1995); business consortia in the USA were rare before because of the Sherman Act (1890) that made such entities illegal. In business both in Japan and the USA, consortia help two or more, often disparate, entities achieve a common goal with shared resources (McKeen 2016). This utilizes individual strengths and allows the individual corporations to achieve goals they wouldn't be able to achieve alone (McKeen 2016). This model can be translated to science.

The solidification of consortia in university research occurred in 1973. Aldrich and Sasaki (1995) report in that year the US National Science Foundation (NSF) began a program supporting university research connected with industry (Industry-University Cooperative Research Centers or IUCRCs), and this program is flourishing today. In fact, existing IUCRCs are eligible for support as collaborative research "clusters" for contributing to Grand Challenges (Wang and Kurose 2016 https://www.nsf.gov/pubs/2017/nsf17035/nsf17035.jsp).

Interestingly, Aldrich and Sasaki (1995) show that Japanese and US consortia conduct themselves quite differently with different values. Whereas, the goals of consortia tend to be similar, US consortia tend to have broader focus of research than Japanese consortia which also shun university partnerships. Another major difference is Japanese companies tend to utilize "member firms" and the US business consortia add usage of shared facilities to the Japanese model (Aldrich and Sasaki 1995). Aldrich and Sasaki (1995) note in business consortia, the Japanese cooperation is typically supported by government and in the US business consortia funding often is via membership dues.

In business, Aldrich and Sasaki (1995) found consortia activities are numerous. In both the USA and Japan, there is a focus on technical feasibility, whereas in the USA the most common activity is idea generation and in Japan applied product development, and prototype and pilot plants also dominate. Rarely do interim manufacturing and full commercialization fall within consortia activities (Aldrich and Sasaki 1995).

12.2.1 Modern Consortia

Academic consortia are connections of universities that share resources (Dotolo and Stràndness 1999) and work on large projects or problems. University consortia naturally formed in the 1960s and are often centered around teaching (Baus and Ramsbottom 1999); these were followed by consortia coming back into vogue in the 1990s with the rise of technology and also the cost of education (Baus and Ramsbottom 1999). Typically, research consortia are associated with massive projects where the consortium facilitates pool resources or coordination of efforts. Formation of consortia around a shared agenda can be larger than working in smaller collaborations. Sometimes the project is so big it requires a consortium to tackle the issue at hand. Bench to Bassinet (B2B) offers a great example of a consortium taking on a huge topic (https://benchtobassinet.com/About/ProgramOverview.aspx). B2B works on questions about how the heart forms and reasons why children are born with heart defects or disease. Additionally, research consortia can be assembled to create a community that has common interests and goals. Because of the fact that consortia are either focused on a large project or have numerous members or both, they can be very complex (Baus and Ramsbottom 1999). It is the organization, energetic cooperation, and excessive communication that can keep the complexity from being overwhelming. Another example of research consortia is those that are formed after the Deepwater Horizon oil drilling disaster (see Chap. 10). Those consortia received funding from the Gulf of Mexico Research Initiative. An example of an interdisciplinary consortium might be one which would form to study macroecology at continental scales (Heffernan et al. 2014).

Another type of consortia is the Sloan Consortium (Sloan-C), which is organized to help institutions and organizations produce quality online educational experiences and tools (Moore 2005). In 1993, the academic-focused Sloan coined the term Asynchronous Learning Network (ALN) which describes people as learning continuously in the place they are and at various times (Moore 2005). Moore (2005) emphasizes that ALNs are people oriented not technology oriented which allows for "anytime-anywhere learning." One of the keys to ALN is self-study with interconnectivity of the people involved. Research consortia are similar in focus as they form around people and the skills they bring to solve a large problem.

12.2.2 Where Are Successful Consortia Occurring in Science? Some Examples from a Variety of Disciplines

In science, there are mechanisms for this such as NSF's linked collaborative proposals. The US National Institutes of Health (NIH) allows for an unlimited number of principal investigators (PIs) on a project as long as they are qualified and fit the definition of PI. NIH does not recognize formal or informal leadership (see best practices) within the funded group of PIs and views them all as equally responsible (https://grants.nih.gov/grants/multi_pi/faq.htm#2963). The US National Aeronautics and Space Agency (NASA) also has consortium-led proposals (https://www.nasa.gov/pdf/418826main_Space%20Grant%202010%20Solicitation%20Rev%20B%5B1%5D.pdf). This is a standardized state-based practice that "through the establishment of state-based consortia, consisting of universities, university systems, associations, government agencies, industries, and informal education organizations involved in aerospace activities, lead institutions provide leadership and support for program objectives in their state and nationally by interacting with other universities, by broadening joint activities with NASA and aerospace-related industries, and by providing public service functions, such as support to elementary and secondary schools and to the public." NASA's consortia require a comprehensive strategic plan in addition to a description of the management plan, structure, and operational policies and procedures (https://www.nasa.gov/pdf/418826main_Space%20Grant%202010%20Solicitation%20Rev%20B%5B1%5D.pdf). Funded NASA consortia are suggested to have an advisory or executive committee or board that can be internal or external.

Business consortia often have a coordinator or project manager that is paid to organize and support the activities of the consortium (Baus and Ramsbottom 1999); scientific consortia also have project managers or coordinators. Baus and Ramsbottom (1999) recommend that decision-makers be not only at the organization and planning meetings but also at operational meetings. Whether internally or externally funded, a consortium must be able to sustain itself after initial funding. Even successful consortia will fall apart if their initial funding isn't sustained. One of the ways to achieve sustainability is to have a single tactical trajectory and to depoliticize the decision-making. Directing the decisions that focus on what is best for each member will allow conclusions to stand up to scrutiny. Every step requires open and extensive communication (Baus and Ramsbottom 1999).

McKeen (2016) sets out these hurdles for business consortia. (1) One pitfall is individuals have a particular job to do within their company, and they don't have time for consortium activities (Mckeen 2016). Hurdle 1 can be viewed as having competing priorities; however, participation in scientific consortia often enhances a scientific career. (2) Participation in a consortium requires demonstration of the consortium's value, i.e., in business, investment of resources by consortium members can be significant and the businesses need to see return on investment (ROI). Successful business consortia are typically very focused so that ROI is more easily trackable. In science, the demonstration of value comes at the funding stage which usually has to be repeated over the life of the consortium. (3) All business consortia require funding for running the organization and its efforts; scientific consortia do also. The funding mechanism for scientific consortia is more affable to the concept when the justification is properly crafted. (4) The consortium leadership and structure should be designed for most productive use of collaborations and resources in both business and scientific research consortia. (5) Within both types of consortia, consortium member renewal rates vary and succession planning should be part of the organization.

In business, consortia in the USA most commonly have a $1–2 million (1989 dollars) budget (Aldrich and Sasaki 1995). In Japan, when yen are converted to dollars, 22% of consortia are run for less than $0.5 million, but 18% have $10–19.9 million (Aldrich and Sasaki 1995). Research consortia need to dedicate dollars to administration of the consortia and also communication within the consortia (more below). This can be viewed as a challenge as the dollars spent on administration and on communication are not available for research.

12.2.3 Best Practices

Newly formed consortia form a leadership team or an executive committee which decides on its own organization and the organization of the consortium. This team sets the tone and the culture for the consortium. NIH funding requires a leadership plan for research consortia that specifically describes the roles and responsibility of each PI, plan for handling disputes, and specific manner for determination of direction of research (https://grants.nih.gov/grants/multi_pi/faq.htm#2963).

From the outset, the consortium members should decide upon metrics of successful consortia. Particularly, reliable metrics should be determined, and metrics should determine from the beginning what the end of the consortium should look like. Whereas most consortia are very productive and healthy, imagining a "hemorrhaging" consortium can be uncomfortable and can put a damper on initial excitement, but it's necessary and instructive. Defining metrics of consortium success and failure from the outset can aid in diagnosing consortium failure and allow the consortium to be repaired rather than disbanded. Although consortium failure can result in the activities greatly slowing down, other signs of a failing consortium can be spending a lot of money with very few results or it can be members spending consortium time on non-consortium activities or the cohesiveness of the group shuts down and communication stops. (Some examples are provided later in this chapter.) Failure to examine what "terminal illness" looks like for the consortium can be terminal in itself. For example, if infighting is too strong and the executive or leadership committee doesn't have broad support, if ghost writing is a common practice (ghost writing is discussed below), and if egos are ruling the consortium instead of a culture of collaboration, the consortium is weak and may be destined for failure. A good plan with strong leaders will bring an attitude of bringing the best scientists to the table, and then trusting them to do the very best science would make the consortium successful. The trust aspect takes out the need for micromanaging, but it doesn't negate strong leadership. The best fix for these problems is to prevent them with a strong team that trusts each other.

The number of consortium members (individuals, universities, or companies) is an important decision. After the Deepwater Horizon oil spill research, consortia were formed having principal investigators numbering from 10 to 50+ and were funded by the Gulf of Mexico Research Initiative (GoMRI). Typically, business consortia members vary from 2 to 50+ with most consortia having 10–30 members (Aldrich and Sasaki 1995). The research example given follows the business numbers. However one consortia leader LMHB spoke with confidence that one research consortium with 50 principal investigators was too large.

One of the oft-cited best practices in consortia is to decide what success looks like. Deciding what failure looks like is also valuable for knowing when to repair or disband a damaged consortium. For some consortia, disbanding is planned when the money runs out. That's a sign of weak leadership in the consortium. Reliable metrics that determine success can be a driving force in the progress of the consortium and can also help to correct issues that emerge. A source of consortium failure is lack of clear communication or lopsided communication. For example, successful research consortia leadership creates a collaborative discussion to develop a consortial policy on authorship or patents. One of the added values of research consortia can be the sharing of data or specimens collected by an individual or subset of the consortium. Authorship is the subject of numerous publications (Primack et al. 2014; Heffner 1981), and consortia leaders are encouraged to explore the peer-reviewed literature that most closely pertains to their discipline. From that literature, it is recommended that the leaders craft a coauthor agreement as in Primack et al. (2014). Strong consortia have established rules (culture) and communication that start from the beginning and carry throughout the functioning of the consortium.

Interconnectedness of the research of members can help, as does data sharing. Data sharing is a form of healthy communication within a consortium. Ensuring transparent data sharing processes via OneDrive, Google Docs, or Dropbox can eliminate communication gaps which are a cause of failure for consortia. One challenge with data sharing within an academic consortium is institutional requirements for use of Google Docs, Dropbox, and OneDrive. Some organizations do not allow the use of Dropbox because the "data ownership" issue in the user agreement. If these differ among consortium members, that is a hurdle that must be overcome.

Green et al. (2005) evaluation of the cancer research consortium indicated that "ability to have open dialogue" was the chief way to progress performance of consortia. It might be helpful to the consortium to designate a capable staff member as the director of communications (Baus and Ramsbottom 1999). Although this costs money that is not directly associated with results, an effective communication strategy will hold a consortium together, and lack of communication will kill the consortium. In Green et al.'s (2005) cancer consortium analysis, they cited high quality of communication as one of the pillars. A method to reduce the inclination toward secrecy is maintaining a healthy, continuous flow of communication. Research consortia should consider monthly conference calls where the members of the consortium contribute to the agenda that is ultimately fashioned by the executive committee. Ideally these calls would be via GoToMeeting or Adobe Connect, but audio-only calls are better than silence. Face-to-face meetings are also ideal but likely not practical for most consortia.

Sloan-C describes five pillars of quality for education and business consortia that the authors adapted to the formation and function of research consortia (Moore 2005). (1) Learning effectiveness is one metric that must be measured. In research, this is measured by new knowledge being gained and also positive feedback from members of the advisory council associated with the research consortium. (2) Moore (2005) also presents cost effectiveness and institutional commitment as a critical principle. Effectiveness of the research is often measured in outputs such as publications and presentations. The quality of these can be measured through citations and sometimes impact factors of journals. Additionally, public outreach and adoption of new ideas or technology are other metrics of effectiveness. Dissemination of the results or conclusions to mass media such as *The Atlantic*, *The Economist*, or *The New Yorker* and other similar publications that are read by decision-makers. Lastly, securing additional funding is another metric applicable to research consortia. (3) Access is a metric from education and business (Moore 2005) that is applicable to research consortia. Creation of international connectedness and new opportunities increases research capacity through consortia. (4 and 5) Satisfaction among employees and students are two quality principles cited among members' successful business and education that should be measured. A healthy competition among members of the research consortium and competition created outside the consortium by the formation of the consortium is a metric to be examined. Green et al. (2005) identified five "domains": (1) extent of collaboration and quality of communication, (2) performance of projects and infrastructure, (3) data quality, (4) scientific productivity, and (5) impact on member organizations. Not all these principles apply to every research consortium and some are difficult to measure, but the values should be examined upon formation of consortia and during the function of consortia.

12.2.4 Funding

Members of a consortium can compete with each other or choose to cooperate. There are several ways to cooperate in terms of getting funding where competition among members is minimized and the strengths of the members are maximized. The first is to individually apply for funding with linked proposals, and the second is to apply as a consortium. When each strategy is undertaken depends upon the granting agency and the urgency. Linked proposals are a good strategy when time is short and will remove the obstacle of subcontractual agreements between institutions that can take months to put into place. Some funding agencies have funds for consortia and other funds for individual investigators (that are rarely individual, just small groups). Others only have mechanisms for funding individuals or small groups, so consortia may have the option of linked proposals. Lastly, other grant-awarding agencies fund only consortia. When linking proposals, it is important to make sure each proposal can stand on its own if the other(s) are not funded, but each proposal will emphasize that if all are funded, the funder will receive "more bang for their buck" because of the interwoven efforts.

Funding agencies might be concerned about the track record or longevity of a consortium (Peterson 1999). Having a solid communication plan and succession plan in place will help with credibility of the consortium. Another concern some agencies express is consortia that appear to be assembled in response to a request for proposals (RFP) (Peterson 1999); however, the seven consortia that assembled after the BP drilling disaster were justified and initially funded by GoMRI. In the case of these consortia, longevity is still an issue. Some of the initially formed consortia didn't receive subsequent funding, and whether the remaining consortia will continue after the GoMRI funds (10 years) have been expended in 2019 will be telling.

Another aspect of successful consortia is they allow members to play to other member's strengths. "Everybody is a genius. But if you judge a *fish* by its ability to *climb* a *tree*, it will live its whole life believing that it is stupid." This is a quote attributed to Albert Einstein. The question that follows is, "What is your genius?" For example, one researcher might be very good at organizing large projects and communicating among members of the consortium, while other's strength might be more detail oriented. A researcher that holds onto publications and edits them endlessly can be balanced by a researcher who is a better judge of when the writing is done and submission is timely. The main point is to recognize people's strengths and enhance those strengths to minimize weakness of the research group.

Inset 1: A major activity in consortia is collaboration, especially collaborative writing. A role a consortium can play is allowing for exploration or collaboration with unique training opportunities for students. Integrating students into all aspects of consortial life is important. Typically, students are brought in after a proposal is funded and much of the initial work is started. Below is an example where a small consortium brought in a student in the idea generation and writing stages.

12.3 Case Study 1: Case Study in Collaboration—Collaborative Writing by a Student and His Mentor

From the beginning of a consortium or collaboration, there is a massive amount of collaborative writing to be done. Students often don't experience collaborative writing because the thesis or dissertation is largely a solo activity of their own writing and their own thoughts. It's important to provide students opportunities that will mimic the types of writing they will be doing as professionals. A better model might be to include the students, early in their tenure as students, in collaborative writing—as equals—as I did for this case study within this chapter. Brian W. Matherne (BWM) participated in writing a collaborative grant proposal with his graduate advisor, myself, and a professor from another department; in this case, the professors included him just as they would any colleague. Collaborative writing and development of research ideas can be a transformative activity for a graduate student. I asked BWM to write about this activity, and in doing so, I'm furthering the model of collaborative writing among professors and students. This inset serves as an example of collaborative writing with a student and professor for a project or grant proposal. The next inset is about collaborative writing for reports.

BWM As a graduate student, I am still developing my research and writing skills. Being aware of this can often lead to insecurities as I am trying to find my voice as a young scientist. This lack of self-confidence is a double-edged sword. At times, it can be crippling as I felt like my ideas will be ultimately ripped apart by my mentors. However, it is this same lack of confidence that drove me to put forth my best work. It forced me to become obsessive about writing as my only hope was to impress my colleagues. As I became more confident, this sense of vulnerability fades. You begin to trust that your work is valuable and that you in fact are an integral part of the larger effort.

LMHB I think the three professors kept possible insecurities on BWM's part in mind while we were collaboratively writing; I know I did. The group was intentional and careful to create a culture of "friendly challenges and edits," not just for BWM but for all of us. There was a lot of enthusiasm for the collaborative and friendly nature of the group. BWM had done previous research and an extensive literature search on the subject of the proposal, and that earned a great deal of respect among the group. Additionally, he had worked outside of the university for a while before his PhD. So, the three professors trusted his knowledge and skills probably more than he trusted himself.

Additionally, BWM was involved in all aspects of idea generation. His previous experience provided valuable insights, and he enhanced the project proposal from the beginning.

BWM Throughout the process of collaborative proposal writing, I discovered I no longer felt that I had to hang on to every word or phrase I wrote; it allowed me to explore the greater depth of my own research and our proposed research. As I begin to see my ideas molded into the project, I found new details to contribute. It was interesting to see the collective effort come together like pieces of a puzzle. A previously conceived notion I had was, that ideally, each collaborator writes a section in a manner that will allow disparate sections written by others to flow together. However, I learned that sometimes my ideas were furthered and elucidated by other members of the group. Writing separately then turned into a weaving together activity of each idea. The other participants also provided inspiration which not only allowed someone to expand upon my original thoughts but also fosters new discussion points from all participants.

LMHB Collaborative writing with students can help the professor and the students to cultivate and grow their ideas and also hone their writing skills. Each person brings strengths to the task. As BWM writes below, our small collaboration quickly understood the strengths and carefully cultivated them. The trust and collegiality among our group were evident, and people seemed to leave their egos at the door when we were writing together.

BWM To work with others, it is important for me and others to understand the strengths and weaknesses of our colleagues. This allowed our group to divide and conquer. Additionally, it enables you to focus on what you do well and to trust that other members of your team can help to fill gaps in your knowledge. Working as a team allows each person to shine as the group achieves their common goal.

LMHB I asked BWM to talk about the top three things he learned.

BWM

1. *No idea is too small*: I learned to value the importance of my ideas.
2. *If at first you don't succeed, you probably learned something anyway*: Learning from mistakes can actually be more productive than stressing over doing something perfectly the first time.
3. *Many hands make light work*: Embrace the diversity within your team. Allowing each person to work to their strengths makes the collaborative process more effective.
4. *Don't be so myopic*: It is also important to have a fresh set of eyes on the editing process; otherwise instead of proposing a study on blue crab, you might propose a study on blue crap.

LMHB I have to admit that I laughed when I read BWM's point #2 (and #4) above. Upon writing this chapter, we recently found out our proposal wasn't funded. BWM, as the student in the group, no doubt learned a lot. Other members of the group also learned that we work well together and more about the strengths of the individuals that could only come through in writing a proposal together. One thing you will note is the common word "trust."

Interestingly, throughout the whole proposal writing project, the group only met in person once. That was at the beginning. Everything was done online via email. In another proposal writing experience recently that LMHB was involved in with ~15 investigators, the group never met or Skyped. We simply wrote it through Google Docs. Previously, on GoMRI proposals, the three executive board members (LMHB included) wrote collaboratively and only met once to go over budget numbers in detail. The rest was done through Skype or via email.

12.4 Case Study 2: Collaborative Writing Among Colleagues

LMHB I asked George Washburn (GW) to write about a collaborative writing project where there was a deliverable. Part of collaboration and consortia is collaborative writing of reports and publications and producing projects as outcomes.

GW The group project that I was a part of consisted of five graduate students, guided by an experienced professor. The project assigned was to design and create an environmental awareness campaign. All of our work was being turned into the professor of the class for a grade, and then eventually the product was delivered to an agency to use as an educational campaign. The group had enthusiastic participants; we had strong, animated discussions about the material. We agreed on the amount of work to be done and by whom with each person embracing their particular roles.

LMHB I see a challenge in that they discussed roles but did not identify a leader. They were professional and treated each other as colleagues.

GW In this process, we were all equals, and did not have a traditional leader, which helped us all contribute to the discussion.

LMHB George makes good points about how their equality leads to contributions by everyone. However, you will see the downfall of not having a designated leader.

GW Partially through the project we realized that before turning the work into our professor, one person would do final edits and adjust the voice and tone of the whole work. This was an excellent idea for group writing. Essentially, that person would be the de facto leader. We ran into trouble when we met to decide on a mutually agreed-upon deadline for the next round of work and how the work was going to be divided up. We discovered that despite all of us having very similar levels of educations and backgrounds, we wrote very differently. When our parts combined, the manuscripts read like a word salad. It took one final editor to tie all the work together, so that it was not glaringly obvious when the portions of the work jumped between authors. Having only one final editor was also designed to eliminate the need for everyone to feel the need to critique and edit each other's work. Many writers felt that if something is written out of their style, it was wrong and needed to be changed. Our group struggled with this, until we added the final editor.

LMHB Designating a final editor for a consortium proposal, manuscript, or deliverable should occur at the start of the writing process. It doesn't have to be the lead author. If you have a skilled editor in the group, that person should be designated the final editor.

GW The largest hurdle for our group was not the agreement on a deadline but sticking to the set deadline. We agreed to send our collective work to our editor (who was a student member of the group), and then he would send the material off to the professor. This was agreed upon both verbally and in written communications, so there were no issues of misunderstandings. One of the members of the group continually ignored this deadline, and would continue to change and edit work past the deadlines, often at late at night.

LMHB I term this person the rogue writer. More on this below.

GW This led to the creation of multiple copies of the final draft and general confusion over how much work had been done and had been changed. As a group, this caused great strife because members would finish their work and send it to the editor, only the next day to wake up and find more problems. This led to trust issues among the group, which exacerbated the earlier mentioned problem of everyone wanting to edit each other's writing styles.

To remedy this, we sat down as a group and discussed what was happening. We discovered that it was a matter of a member believing they could fix many of the supposed problems in the manuscripts and that they could just send it in when they were done. Effectively they thought they could override the editor. After discussing this, we all had to agree on a permanent deadline for everything, and if we were not able to finish in time, we had to ask for more time. This initial hurdle led to the group having to trust each other in order to finish our work.

LMHB The group had to have a frank discussion with the rogue writer and implore them to understand their position in the group. Once that was clear, the group moved forward on the project.

12.5 Characteristics of Rogue Researchers and Rogue Writers (and What to Do about Them)

Research consortia are valuable, efficient mechanisms for conducting diverse research. Most of the time, participation in consortia is productive and positive for an individual. However, on rare occasions in research consortia, there are people who act as, at the least, productivity killers and, most extreme, unethical practitioners. They are the rogue researcher and the rogue writer. Although difficult to imagine and discuss, knowing the characteristics of problematic participants can help alleviate the problems they create while their problems are small. A rogue researcher (RR) takes group resources and uses them in a way that is outside the ethics or identified goals of the group. Sometimes it is difficult to identify the RR as they work in secret, and the impact is only noticed in the results. In science, it might be the group member that takes group data and publishes it without the group knowledge or someone who manipulates (or outright changes) the data they are entrusted to. It appears the key to RR behavior is secrecy and lack of communication. To thwart the RR, data transparency and frequent communication are necessary. Sharing of raw data as soon as it is generated is ideal. And, strong consortium leaders who set out a policy on authorship and data use from the beginning will help thwart RR. Sadly, sometimes, there is nothing that can be done with RR, and they either have to be tolerated or eased out of the group effort.

The rogue writer (RW) doesn't respect deadlines and submits group material as their own or without consulting the consortium (ghost authorship), someone who edits outside of the collaboration or after deadlines and agrees in person but electronically or on their own does something completely different. The ghost authorship alluded to above can happen in the wake of the actions of the rogue writer. This form of ghost authorship happens when researchers who design the original experiments, collect, analyze, and interpret data or otherwise contribute in important ways are then not named as coauthors (another definition is authors that write the paper but are not included as an author). This can happen in situations where the data collected are openly shared but authorship is not discussed.

Rogues are productivity killers. People with these characteristics need to be in mind when setting up the consortium in order to minimize the damage they cause. The RW writes and edits his/her own work past the deadline (or won't submit it to the group) or is someone who writes and edits after the group has agreed the work is finished. Another RW act is to take a paper they are not the lead on and do something different with it without the knowledge of the other authors. For example, the RW might present the contents of the paper at a conference without the coauthors' knowledge or nominate that paper for an award and inform the coauthors afterward. Creating a culture of constant communication can help, but sometimes RWs just act. If the RW edits after the manuscript is complete, those edits should not be allowed or incorporated. Again, creating a culture of communication and trying to eliminate secrecy is the best policy to reign in the RW. Creating policies about authorship and collaboration within the consortium from the beginning can help reign in the RW through enforcement of existing policies. The lack of policies creates an opening for RW to act.

12.6 Consortial Inclusivity

One thing that hasn't been discussed among business, educational, or research consortia literature is including underrepresented minorities (URM) in the consortium. Isolation of URM is a large factor in reduced retention (Daley et al. 2006). Inclusion of URM in a consortium has many benefits for the consortium and also for the URM faculty member. Just as diversity of disciplines and perspectives creates stronger consortia, diversity of people will follow. In forming consortia, a best practice would be to overtly seek qualified URM and determine if they fit into the consortium. This very act will reduce the isolation which is typical among URM (Daley et al. 2006).

12.7 Summary

The words trust, transparency, and communication are common words in this chapter. Others are organization and succession planning. The keys to successful, productive, and long-term consortia are (1) an organized group (consortium) with a defined leadership structure agreed upon by the members, (2) deciding from the outset what success is and also what the scenario would be for disbanding the consortium, (3) strong and frequent communication within the consortium by the leadership and among members, (4) shared resources and shared data in an open-access manner to reduce secrecy, (5) initial creation of policies regarding use of shared data and authorship, (6) trust that the members will do their best work and trust that the members will work for the betterment of the consortium as a whole, and (7) seeking diversity among consortia members for the best ideas and diversity of perspectives.

References

Aldrich HE, Sasaki T (1995) R&D consortia in the United State and Japan. Res Policy 24:301–316

Baus F, Ramsbottom CA (1999) Starting and sustaining a consortium. In: Dotolo LG, Strandness JT (eds) Best practices in higher education consortia: how institutions can work together, New directions for higher education, no. 106 vol XXVII. Jossey-Bass Inc, San Francisco, p 125

Bransetetter LG, Sakakibara M (2002) When do research consortia work well and why? Evidence from Japanese panel data. Am Econ Rev 92:143–159

Daley S, Wingard DL, Reznik V (2006) Improving the retention of underrepresented minority faculty in academic medicine. J Natl Med Assoc 98:1434–1440

Dotolo LG, Strándness JT (1999) Best practices in higher education consortia: how institutions can work together: new directions for higher education, number 106. Wiley, Hoboken, p 125

Green SM, Hart G, Wagner EH (2005) Measuring and improving performance in multicenter research consortia. JNCI Monogr 35:26–32

Heffernan JB, Soranno PA, Angilletta MJ Jr, Buckley LB, Gruner DS, Keitt TH, Kellner JR, Kominoski JS, Rocha AV, Xiao J, Harms TK, Goring SJ, Koenig LE, McDowell WH, Powell H, Richardson AD, Stow CA, Vargas R, Weathers KC (2014) Macrosystems ecology: understanding patterns and processes at continental scales. Front Ecol Environ 12:5–14. https://doi.org/10.1890/120017

Heffner A (1981) Funded research, multiple authorship, and subauthorship collaboration in four disciplines. Scientometrics 3:5–12. https://doi.org/10.1007/BF02021860

Katz ML (1986) An analysis of cooperative research and development. RAND J Econ 17:527–243

McKeen S (2016) Best practices for a successful consortium. https://www.linkedin.com/pulse/5-best-practices-successful-consortium-shannon-mckeen. Last accessed 18 Aug 2017

Moore JC (2005) The Sloan Consortium quality framework and the five pillars. http://www.mit.jyu.fi/OPE/kurssit/TIES462/Materiaalit/Sloan.pdf. Last accessed 23 Aug 2017

Peterson LM (1999) Consortial fundraising. In: Dotolo LG, Strandness JT (eds) Best practices in higher education consortia: how institutions can work together, New directions for higher education, no. 106 vol XXVII. Jossey-Bass Inc, San Francisco, p 125

Primack RB, Cigliano JA, Parsons C (2014) Three scientific authors share their experiences and solutions and ask you for yours (with a sample agreement for co-authors). Biological Conservation. https://www.elsevier.com/connect/co-authors-gone-bad-how-to-avoid-publishing-conflicts

Wang G, Kurose J (2016) IUCRC clusters for grand challenges, a collaborative opportunity for research to address grand challenges. NSF 17-035. https://www.nsf.gov/pubs/2017/nsf17035/nsf17035.jsp. Last accessed 28 Aug 2017

Part III

Environmental Effects and Experimental Outcomes

Challenges of Endocrine Disruption and Cardiac Development

13

Graciel Diamante and Daniel Schlenk

Abstract

The endocrine system plays a crucial role in the survival of animals. It regulates many signaling processes that control the cardiovascular, immune, reproductive, and central nervous systems. In addition to its role in maintaining the functions of major organ systems, signaling by the endocrine system is critical for normal development. For example, estrogen signaling has been shown to regulate cellular physiology by influencing its proliferation rate. With the diverse array of pathways that the endocrine system regulates, disruption by exogenous agents can impair vital functions, not just at the cellular and organismal levels but also at the population level. The integrated role of various biological systems and the interaction between organs can make it difficult to assess the effects of endocrine-disrupting compounds (EDCs). Adding to the challenges of studying the effects of EDCs is the complex series of signaling events that occur in a precise spatio-temporal manner during embryogenesis. However, using model organisms with a well-characterized developmental pattern such as zebrafish can facilitate the study of EDCs on physiological endpoints with apical relevance, such as cardiovascular development. Consequently, this review will focus on the developmental challenge of EDC exposure on cardiac development in aquatic organisms.

13.1 Introduction

The endocrine system is a network of cellular signaling tissues that secrete hormones into the circulatory system. When a hormone binds to its receptor on/in a target tissue, it typically initiates a signaling cascade. Activation of the hormone

G. Diamante (✉) · D. Schlenk
University of California, Riverside, CA, USA
e-mail: Gdiam001@ucr.edu; Daniel.Schlenk@ucr.edu

© Springer International Publishing AG, part of Springer Nature 2018
W. Burggren, B. Dubansky (eds.), *Development and Environment*,
https://doi.org/10.1007/978-3-319-75935-7_13

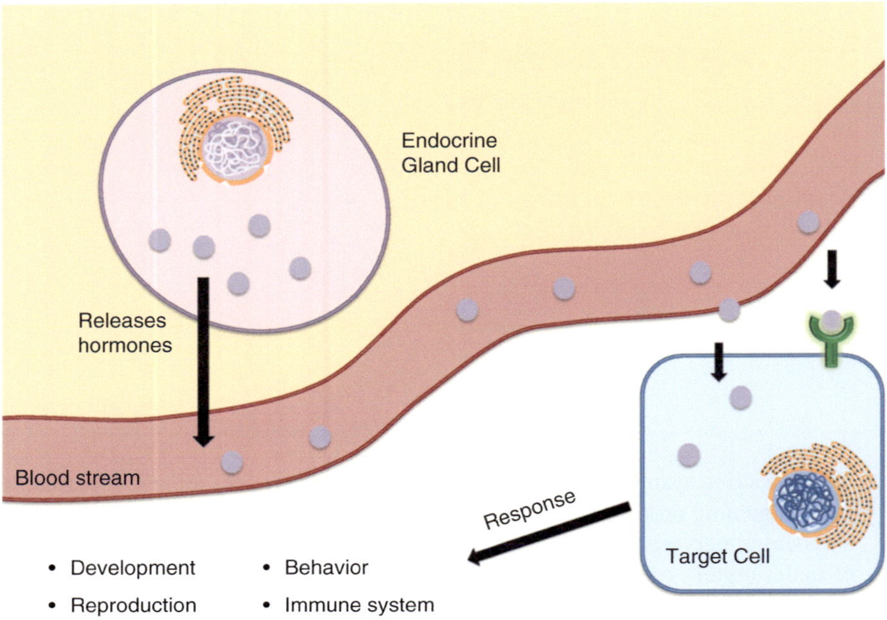

Fig. 13.1 The endocrine system is a network of glands that synthesize and secrete hormones as chemical messages. These hormones are released into the circulatory system to reach a target organ. Upon binding to a specific receptor, the hormones can induce signal transduction events that regulate many physiological functions

receptor can have multiple downstream physiological effects, including regulation of the immune system, reproduction, behavior, and development (Fig. 13.1).

The ability of the endocrine system to control cellular processes requires close coordination with the central nervous system (CNS). The primary link between the CNS and the endocrine system is the hypothalamus. It is organized into groups of neuronal bodies that can receive signals from the brain or afferent signals from the viscera, which then initiate the release of specific factors to the anterior or posterior portions of the pituitary gland. The magnocellular neurons release oxytocin (OT), vasopressin (AVP), and neurophysins (NP) to the posterior pituitary. The parvocellular neurons release corticotropin-releasing hormone (CRH), dopamine (DA), growth hormone-releasing hormone (GHRH), gonadotropin-releasing hormone (GnRH), and thyrotropin-releasing hormone (TRH) to the anterior pituitary (Fig. 13.2). The hypothalamus also releases inhibitory factors such as the gonadotropin-inhibitory hormone (GnIH) and somatostatin (SS) (Ubuka et al. 2013). The hormones produced by pituitary cells are typically released into the circulation, eventually reaching a diverse array of targets which elicit either the synthesis or release of a multitude of downstream factors.

Endocrine signaling can be negatively influenced by a host of factors including contaminants (DeRosa et al. 1998; Crisp et al. 1998; Sumpter 1998). Given the

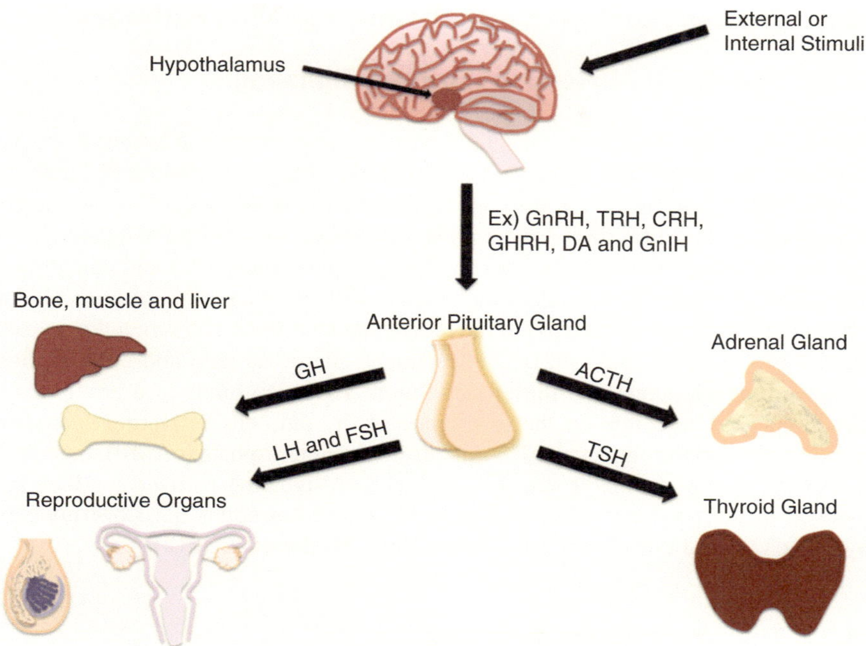

Fig. 13.2 Hormone signaling and synthesis begins at the hypothalamus. Specialized cells in the hypothalamus release hormones that either inhibit or promote release of other hormones from the anterior lobe of the pituitary. The hormones that target the anterior pituitary include corticotropin-releasing hormone (CRH), dopamine (DA), growth hormone-releasing hormone (GHRH), gonadotropin-releasing hormone (GnRH), thyroid-releasing hormone (TRH), and gonadotropin-inhibitory hormone (GnIH). These hormones then initiate or inhibit the synthesis of hormones from the pituitary gland such as luteinizing hormone (LH), follicle-stimulating hormone (FSH), thyroid-stimulating hormone (TSH), growth hormone (GH), and adrenocorticotropic hormone (ACTH) that have unique functions in specific target organs

importance of the endocrine system to reproduction, most studies have focused on this apical endpoint. However, endocrine-disrupting chemicals (EDCs) have also been shown to have adverse developmental, neurological, and more recently cardiovascular effects (Panzica et al. 2007; Waye and Trudeau 2011; Frye et al. 2012; Gao and Wang 2014). In an attempt to better regulate endocrine endpoints, the US Environmental Protection Agency established the Endocrine Disruptor Screening Program (EDSP) to identify chemicals that adversely affect the estrogen, androgen, thyroid, and steroid biosynthesis pathways in a biological system. Consequently, background on these four pathways will be initially the focus of this chapter with subsequent emphasis on steroid signaling. The goal of this review is to highlight the role of the different endocrine pathways in regulating the cardiovascular system and to present the challenges in evaluating EDCs during development.

13.2 Thyroid Hormone Synthesis and Signaling Pathways

13.2.1 Thyrotropin-Releasing Hormone Signaling

Thyroid signaling is initiated when TRH reaches the pituitary and targets thyrotrophic cells that synthesize and release the glycoprotein, thyroid-stimulating hormone (TSH). TRH binds to a subfamily of $G_{q/11}$ protein-coupled receptors, activating phospholipase C (PLC) that cleaves phosphatidylinositol 4,5-bisphosphate (PIP_2) into two secondary messengers: inositol 1,4,5-trisphosphate (IP3) and diacylglycerol (DAG). Each of the secondary messengers will cause an influx of intracellular Ca^{2+} initiating the release of TSH to the bloodstream where it will target the thyroid gland (Hsieh and Martin 1992). TSH regulates energy storage and consumption throughout development and during adulthood. TSH also binds to a group of G_s protein-coupled receptors on the membrane of thyroid cells activating adenylate cyclase (AC) and forming cyclic 3′5′-adenosine monophosphate (cAMP). Increases in cAMP can activate the protein kinase A (PKA) signaling pathways, which are involved in the production and secretion of thyroid hormones 3,3′3,5′-triiodo-L-thyronine (T_3) and L-thyroxine (T_4) (Brent 1994; Mullur et al. 2014).

13.2.2 Thyroid Hormone Synthesis

T_3 and T_4 are both synthesized in the follicles of the thyroid gland, derived from iodinated tyrosine. Blood-borne iodide is transported to the lumen of the thyroid follicle where it is oxidized to iodine by thyroperoxidase. Diet is another source of iodine, which is absorbed in the small intestine and transported to the thyroid. As a part of this process, thyroglobulin (Tg) is synthesized in the endoplasmic reticulum of thyroid gland cells and further processed in the Golgi apparatus. Tg is secreted into the colloid space, where the tyrosine residues undergo iodination (Kim and Arvan 1991, 1993). Iodinated Tg will reenter the cell by receptor-mediated endocytosis. After Tg is internalized, the endosome will fuse with lysosomes allowing the formation of T_3 and T_4 through the cleavage of Tg (Fig. 13.3). The resulting thyroid hormones will then diffuse into the bloodstream (Kostrouch et al. 1991; Seljelid et al. 1970; Brix and Herzog 1994; Dunn et al. 1991). Approximately 85% of the hormone released is T_4 and is predominantly converted to the more biologically active thyroid hormone T_3, by 5′-monodeiodinase in multiple organs such as the liver (Klein and Danzi 2007). In the bloodstream, T_4 and T_3 are transported by several thyroid hormone-binding plasma proteins to ensure the supply of thyroid hormones to the cells (Herzog 1983; Kostrouch et al. 1993). Thyroxine-binding globulin, transthyretin, and albumin are plasma proteins that transport T_4 and T_3 in the bloodstream (Ingenbleek and Young 1994). The degradation and synthesis of thyroid hormones are highly regulated by available iodine, which influence the synthesis of Tg. TSH can increase the transport of iodine into the cells leading to a rise of thyroid hormone levels, which control TSH release through negative feedback (Breen et al. 1997).

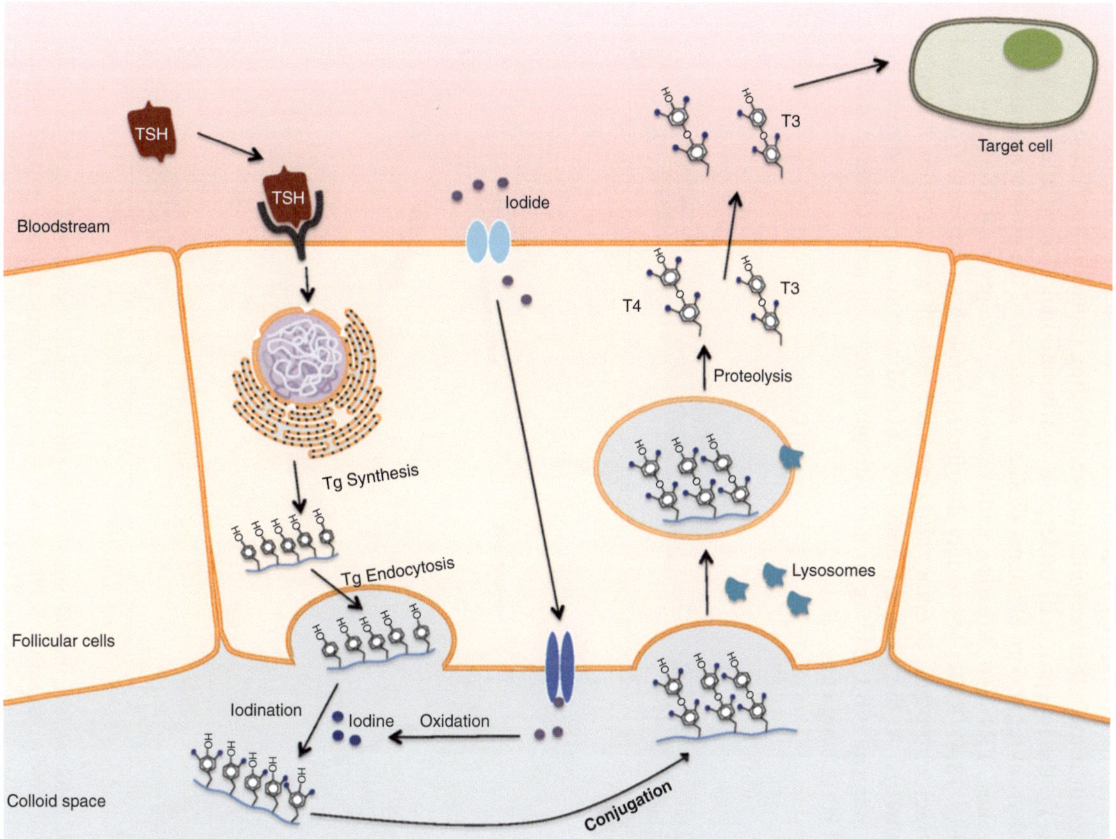

Fig. 13.3 Diagram of the major steps involved in thyroid hormone (T3 and T4) synthesis and secretion. Thyroid hormones are formed in follicles lined by a single layer of follicular cells. Thyroglobulin (Tg) is formed in the endoplasmic reticulum and then secreted into the colloid where it undergoes iodination. Iodinated residues of tyrosine are then conjugated. Processed thyroglobulin will reenter the cell by endocytosis and are broken down by proteases to yield T3 and T4

13.2.3 Thyroid Hormone Signaling

Thyroid hormones play an important role in growth, development, and metabolism. T_3 and T_4 bind to a superfamily of thyroid hormone receptors (TRs) that act as transcription factors. Unbound thyroid receptor is associated with a Co-repressor protein that prevents it from initiating transcription. Upon binding of T_3 or T_4, the receptor will undergo a conformational change allowing it to bind to thyroid regulatory elements (TREs), upstream of the promoter region of target genes. TRs also dimerize with retinoid X receptors (RXRs) and bind TREs (Fig. 13.4). RXRs are widely distributed throughout different tissues and can interact with other receptors allowing TR crosstalk with additional regulatory pathways. At the plasma membrane, thyroid hormones can bind to receptors on integrin avB3 leading to downstream effects on ion exchange such as Na^+ and H^+ (Cheng et al. 2010).

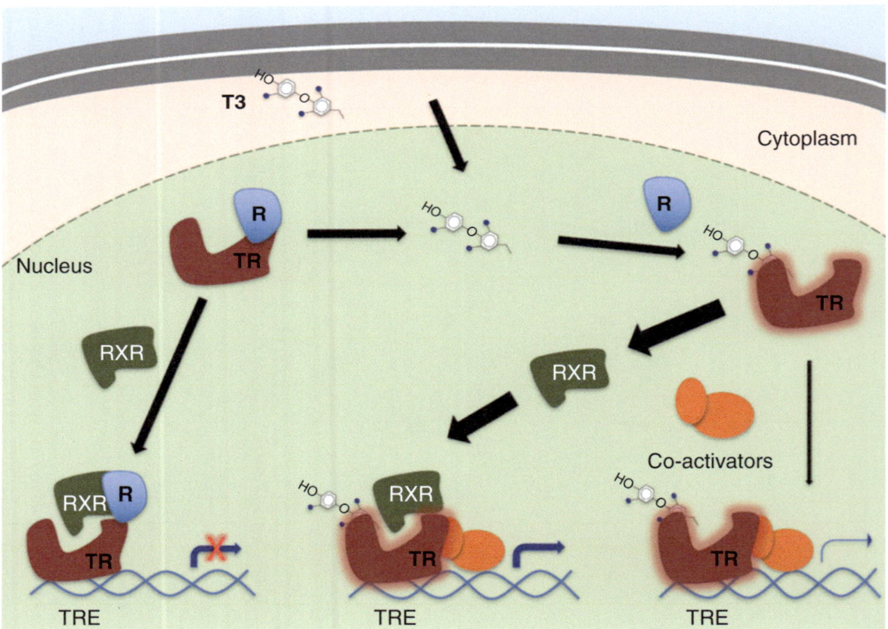

Fig. 13.4 In the nucleus, thyroid hormones can bind to thyroid receptors (TRs), which recruit coactivator protein. An activated TR can then activate the transcription of thyroid response element (TRE) containing genes. TRs form heterodimers with retinoid X receptors (RXRs) and bind to TREs located in the promoter regions of target genes. In the absence of the ligand, the RXR-TR heterodimer interacts with repressor proteins (R) to inhibit target gene transcription

13.3 Steroid Hormone Synthesis and Signaling Pathways

13.3.1 Gonadotropin-Releasing Hormone

Another factor released from the hypothalamus is GnRH, which regulates steroid synthesis and signaling. To mediate its role in reproduction, GnRH can bind to a $G_{q/11}$ protein-coupled receptor, found in gonadotrophic cells of the pituitary gland. Binding of GnRH to the membrane-bound receptor will cause the G-protein to undergo a conformational change resulting in the activation of phospholipase C leading to the mobilization of Ca^{2+}. Influx of Ca^{2+} ions initiates a cascade of signaling events including the production and secretion of two glycoproteins, luteinizing hormone (LH) and follicle-stimulating hormone (FSH). Each hormone plays a major role in reproduction by inducing steroidogenesis, spermatogenesis, folliculogenesis, and ovulation. The increased levels of sex steroid hormones in the blood can cause a negative feedback decreasing the release of GnRH from the hypothalamus, which will affect the secretion of LH and FSH from the anterior pituitary gland.

FSH and LH have gender-specific roles in regulating reproduction. In females, FSH stimulates the growth of granulosa cells, which regulates follicular maturation and the production of estradiol (E2) by inducing CYP19 (Padmanabhan et al. 1991). In addition to enhancing proliferation of Sertoli cells in the testis, FSH stimulates spermatogenesis by inducing androgen-binding protein (ABP) in Sertoli cells (Hansson et al. 1975). ABP aids in the intracellular transport of hormones such as testosterone, dihydrotestosterone, and E2 (Meachem et al. 1996). Under the control of circulating testosterone and E2, FSH is also regulated by several transforming growth factor β (TGF-β) proteins: activin and inhibin (Gregory and Kaiser 2004). Activin increases the secretion of FSH, and inhibin diminishes the release of FSH in females (Ling et al. 1986; De Jong 1988).

In females, an increase in circulating LH levels enhances ovulation and development of the corpus luteum, which secretes progesterone during the luteal phase of the ovarian cycle. In theca cells of the ovary, LH signaling initiates the conversion of cholesterol to androgens, which are precursors for estrogen synthesis via cytochrome P450 aromatase (CYP19). In males, LH is involved in the production of testosterone in the testis (Hu et al. 2010). Increased levels of testosterone will activate a negative feedback loop by inhibiting the release of GnRH and ultimately LH in the hypothalamus.

13.3.2 Steroid Hormone Synthesis

Production of steroid hormones can occur in several different tissues utilizing multiple forms of the cytochrome p450 monooxygenases or the hydroxysteroid dehydrogenase families of proteins (Payne and Hales 2004). Cytochrome p450 enzymes are a group of heme-containing enzymes that are found in all organisms. They are not only important for steroid synthesis but also catalyze the biotransformation of endogenous as well as exogenous chemicals (Klingenberg 1958; Omiecinski et al. 2011). Hydroxysteroid dehydrogenases belong to a family of alcohol dehydrogenases with tissue-specific expression throughout development (Voutilainen and Miller 1986; Kayes-Wandover and White 2000; Mellon and Griffin 2002; Payne and Hales 2004). Steroid hormones are derived from the parent compound cholesterol. Intermediary products include aldosterone, cortisol, estrogens, and androgens. The production of estrogens and androgens will be the main pathway described in this chapter.

The first step in steroidogenesis occurs in the mitochondrial membrane, where the side-chain moiety from cholesterol is cleaved, forming pregnenolone (Miller 2002, 2008). Pregnenolone can either be hydroxylated by cytochrome p450 17 (CYP17) to form 17-hydroxypregnenolone or is converted into progesterone by 3β-hydroxysteroid dehydrogenase (3β-HSD). 17-Hydroxypregnenolone and progesterone are both precursors for androstenedione but under different pathways. 17-Hydroxypregnenolone is converted via CYP17 to dehydroepiandrosterone (DHEA), which is converted into androstenedione by an oxidoreductase known as 3β-HSD. Alternatively, progesterone may undergo hydroxylation by CYP17 into 17-hydroxyprogesterone, which is subsequently converted into androstenedione then into estrone or testosterone by CYP19 or by 17β-HSD, respectively. Estrone and testosterone are both precursors for E2. Estrone can be hydrolyzed to E2 by 17β-HSD, and testosterone can be converted into E2 by CYP19 (Fig. 13.5). The

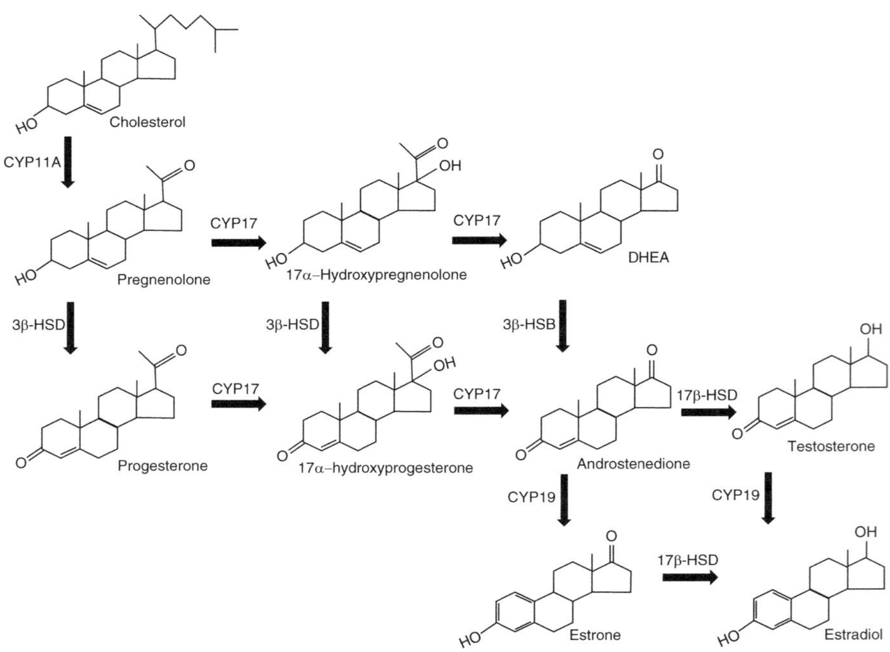

Fig. 13.5 Diagram of the steroid biosynthesis pathway along with the enzymes involved in each step. 17β-Hydroxysteroid dehydrogenase (17β-HSD), 3β-hydroxysteroid dehydrogenase (3β-HSD), cytochrome P450 family 11 subfamily A (cyp11A), cytochrome P450 family 17 (cyp17), and cytochrome P450 family 19 (cyp19)

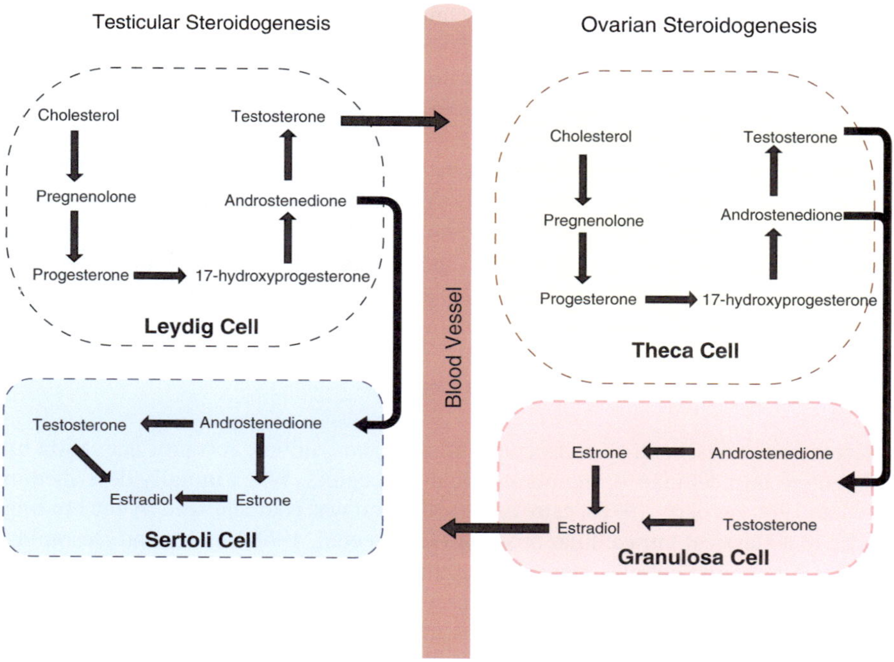

Fig. 13.6 Conversion of cholesterol to the primary sex steroid hormones occurs in the testis and the ovary. This process requires both the theca cells and granulosa cells of the ovaries and the Leydig cells and Sertoli cells of testis

conversion of cholesterol to androstenedione and testosterone primarily occurs in either the theca cells and granulosa cells of the ovaries or in the Leydig cells of the testis (Fig. 13.6). Conversion of androstenedione to estrone and testosterone to estradiol occurs in the granulosa cell of the ovaries (Fig. 13.6).

Estrogens and androgens are synthesized in both males and females. However, differences in concentration and hormone receptor expression contribute to the difference between the sexes. In males, androgens are the main steroid hormones that regulate reproduction and physiology. In Leydig cells, testosterone is rapidly created from androstenedione due to the high expression of 17β-HSD (Andersson et al. 1996; O'Shaughnessy et al. 2000). However, estradiol is also present in males and is produced in mammalian testis (Stumpf 1969). Estrogen has an essential role in reproduction as well as Leydig cell development and function (Khan et al. 1998; Abney 1999). In addition to its roles in normal male physiology, estrogens have a role in regulating steroidogenesis (Melner and Abney 1980a, b; Kalla et al. 1980). Studies have shown that estradiol can inhibit enzymes important for steroid production such as 17α-hydroxylase in males (Brinkmann et al. 1980; Abney 1999). The biosynthesis of androgens in females occurs in the ovary and in the adrenal gland (Lebbe and Woodruff 2013). The production of androgens in theca cells is primarily under LH regulation (Baird et al. 1981). Low levels of LH have been shown to result in the increased production of androgens (Campbell et al. 1998). Maintaining the appropriate amounts of androgens in female is important in sexual development and bone formation.

Since steroid hormones play a significant role in vertebrate physiology and development, there are essential mechanisms that have evolved to help maintain hormonal homeostasis. Steroidogenic acute regulatory (StAR) protein alters the production of steroid hormones through the manipulation of cholesterol transport in the cell (Miller and Strauss 1999, Miller 2007). Steroidogenesis can also be regulated transcriptionally through the expression of the above steroidogenic enzymes (Hiroi et al. 2004). Recent evidence has shown that these enzymes can also be epigenetically regulated (Martinez-Arguelles and Papadopoulos 2010).

13.3.3 Signaling of Steroid Hormones

Steroid hormones, such as androgens and estrogens, initiate receptor activation by binding to intracellular- or membrane-bound receptors. When initially described in the late 1960s to early 1970s, estrogenic activation was characterized by the binding of E2 to a classical intracellular receptor (Jensen et al. 1968; Jensen and Desombre 1973). These intracellular steroid receptors belong to a superfamily of nuclear receptors that can either stimulate or inhibit the transcription of specific gene targets through genomic interactions. In addition to nuclear receptor activation, steroid hormones can also induce non-genomic signals that initiate the production of secondary messengers. Increases in secondary messengers can also control the expression of genes through a variety of additional downstream signal transduction pathways. The primary difference between genomic and non-genomic pathways is that the latter does not require the entry of hormones into the cell but rather mediates their action through receptors at the cell membrane.

13.3.3.1 Signaling of Androgens

Androgens initiate activity by binding to cytosolic androgen receptors (ARs) (Chang et al. 1988a, b; Gobinet et al. 2002). One of the most well-studied androgens that bind to ARs is testosterone (Benten et al. 1999). After the binding of testosterone to an AR, the receptor undergoes a conformational change exposing the DNA-binding domain. The ligand-receptor complex then moves to the nucleus and activates the expression of genes by binding to androgen response elements (AnRE) (Jenster et al. 1991; Chang et al. 1995). Once bound to the DNA, the complex can recruit coactivators that initiate chromatin structure modifications, which allow RNA polymerases access to initiate transcription (Wang et al. 2005; McKenna and O'Malley 2002) (Fig. 13.7).

AR is expressed in several tissues of all vertebrates (De Waal et al. 2008; Gustafsson and Pousette 1975; Chang et al. 1988b). The expression of AR is dependent on age and reproductive stage of an organism. In rats, aside from the reproductive organs of both male and female, AR is expressed in the pancreas, liver, thyroid gland, and heart in varying degrees (Gustafsson and Pousette 1975; Takeda et al. 1990; Kimura et al. 1993). ARs regulate the expression of enzymes important for

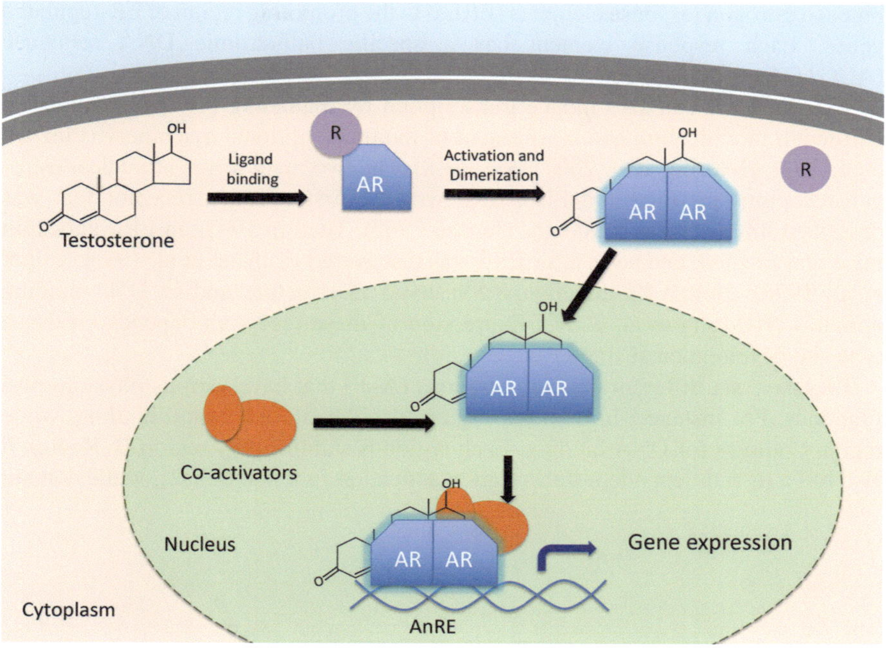

Fig. 13.7 In the absence of its ligand, androgen receptors (ARs) are bound to repressors (R) that inhibit its transport to the nucleus. Testosterone-bound ARs form dimers that translocate into the nucleus to initiate the transcription of AR-regulated genes. Activated ARs will recruit coactivators and bind to androgen response element (ARE) upstream the promoter region of target genes

metabolism, cell differentiation, growth, and detoxification (Sahlin et al. 1994; Hossain et al. 2008). Androgens can also initiate rapid non-genomic effects through secondary messengers (Foradori et al. 2008). Studies in mouse have shown that testosterone can induce an influx of extracellular Ca^{2+} following activation of the phospholipase C pathway and the downstream effectors, calmodulin and extracellular signal-regulated kinase (ERK) (Benten et al. 1999; Mellström and Naranjo 2001; Estrada et al. 2003).

13.3.3.2 Signaling of Estrogens

Estrogens can induce biological effects by binding to G-protein-coupled receptors or nuclear receptors (Hall et al. 2001; Prossnitz et al. 2008). Of the different estrogens, E2 is the strongest endogenous agonist for estrogen receptors (Kuiper et al. 1997; Heldring et al. 2007). The classical pathway of estrogen signaling regulates gene transcription by binding to the nuclear estrogen receptor (ER). In the absence of a ligand, heat shock proteins inhibit ERs. When E2 binds to the receptor, it will induce a conformational change causing the release of the inhibitory factors and initiating dimerization with another molecule of ligand-ER complex. The activation of ER also results in the exposure of the DNA-binding domain in the receptor that binds to estrogen response elements (EREs) in the promoting region of ER-regulated genes. Each response element has a specific palindromic DNA sequence (AGGTCAnnnTGACCT).

Activated ER can also induce transcription by regulating proteins involved in chromatin modification and recruitment of the transcriptional machinery (Pinzone et al. 2004; Heldring et al. 2007) (Fig. 13.8). The two main forms of nuclear estrogen receptors are ERα and ERβ. ERα was the first estrogen receptor that was described in the 1960s in humans (Jensen 1962). Then in 1993, an ERα knockout mice was created, and soon after ERβ was discovered (Lubahn et al. 1993; Kuiper et al. 1996). Both ERs are highly conserved in structure and in DNA-binding domains (Heldring et al. 2007). Expression of these receptors depends on tissue type and developmental stage.

ERs have six different structural regions (A–F) that have various physiological functions. For instance, in humans, region *E* is the site for ligand binding, while region *C* allows for DNA binding. Each region is stabilized by region *D*. Region *E* also has a ligand-dependent transcription activation function (AF-2), while domain

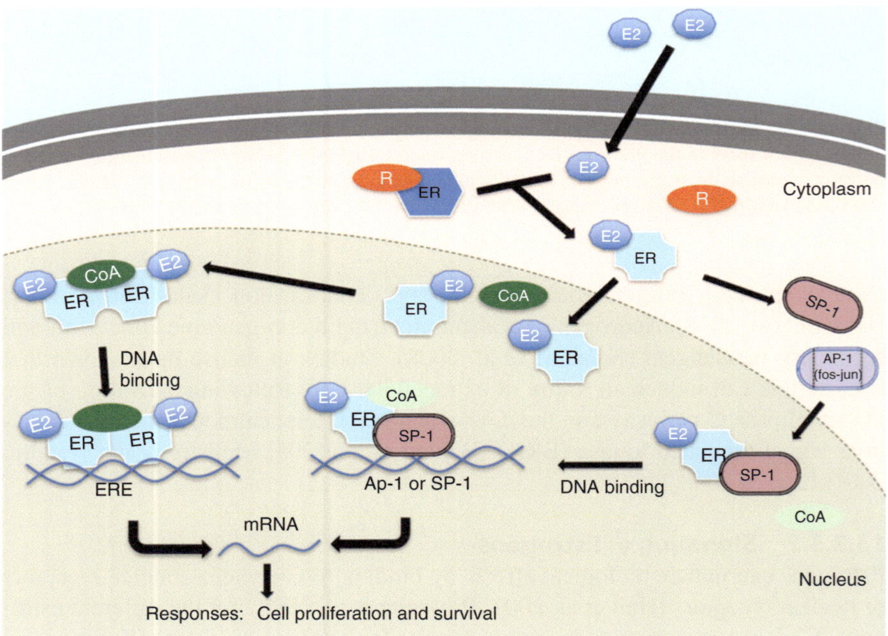

Fig. 13.8 The classical genomic activity of estrogens is mediated through the signaling of nuclear estrogen receptors (ERs). Unbound ER is bound to repressor (R) proteins. 17β-Estradiol (E2) binding induces formation of ER dimers, which relocate into the nucleus. After dimerization, ER recruits coactivators (CoA) and binds to estrogen response elements (EREs) on DNA. ERs also associate with other transcription factors such as specificity protein 1 (Sp1) and activating protein-1 (AP-1) to induce transcription of target genes

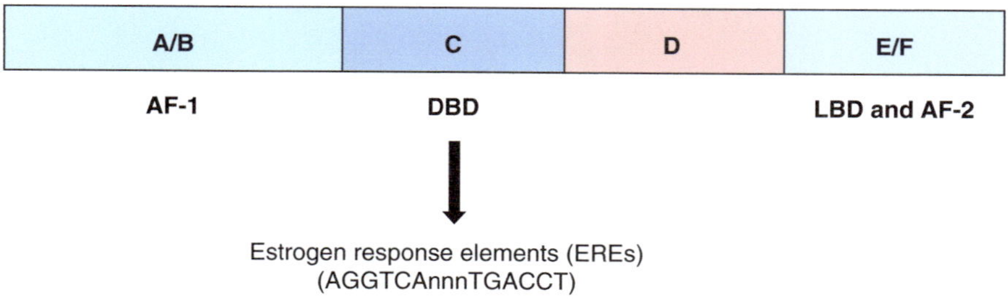

Fig. 13.9 Functional domains of the nuclear estrogen receptors. Both ERα and ERβ have four main functional domains. Domain C harbors a DNA-binding domain (DBD), while the ligand-binding domain (LBD) is located in E/F. The two transcriptional activation functions (AF-1 and AF-2) of the ERs are in A/B and E/F

A/B has a ligand-independent transcription activation function (AF-1) (Fig. 13.9). AF-2 binds a family of coactivator proteins called p160s, which includes GRIP1, TIF1, and SRC-1. These coactivators then recruit other transcriptional factors such as CREB-binding proteins and p300. Together these proteins can induce histone modification and allow the recruitment of the transcription machinery complex. Unlike AF-2, less is known about the activity for AF-1 (Webb et al. 1998).

Genes lacking ERE sequences can still be regulated by E2, further showing the diversity of estrogen signaling. ERs can also be transcriptional regulators by interacting with other proteins to induce expression. For example, ER can interact with activation protein 1 (AP-1) and affect the expression of genes that contain AP-1 binding sites (Fig. 13.8). AP-1 is a complex of proteins that consist of different combinations of FOS and JUN proteins. AP-1 binding sites are found in promoters of genes essential for many cellular processes like cell division and death. Interaction between ER and AP-1 sites has been shown to induce transcription of ovalbumin, cyclin D1, and IGF genes (Gaub et al. 1990; Umayahara et al. 1994; Liu et al. 2002). ER and AP-1 protein interactions are dependent on the receptor subtype and the AF domain (Webb et al. 1999).

There are also genes that only contain imperfect sequences or half of the ERE sequence in their promoting regions (Duan et al. 1998; Petz and Nardulli 2000). Genes with unique EREs include low-density lipoprotein receptor (*LDLR*) and *c-fos* and are regulated through an ER-Sp1 complex (Weisz and Rosales 1990). Studies in breast cancer cells have shown that ERα-Sp1 can induce genes involved in cell cycle progression and proliferation (Kim et al. 2005). Similar to the signaling pathways induced by ER/AP-1, the actions of the ER/Sp1 complex are also dependent on receptor type. E2 signaling is also involved in the regulatory functions of NF-KB and GATA1. Interaction of ER with NF-KB has been shown to have a role in bone metabolism by affecting the transcription of interleukin 6 (*il-6*) (Stein and Yang 1995).

Similar to other steroid hormones, the effects of estrogen can also occur through non-genomic interactions mediated through secondary messengers such as Ca^{2+} and cAMP (Fig. 13.10) (Szego and Davis 1967; Pietras and Szego 1975). E2 membrane

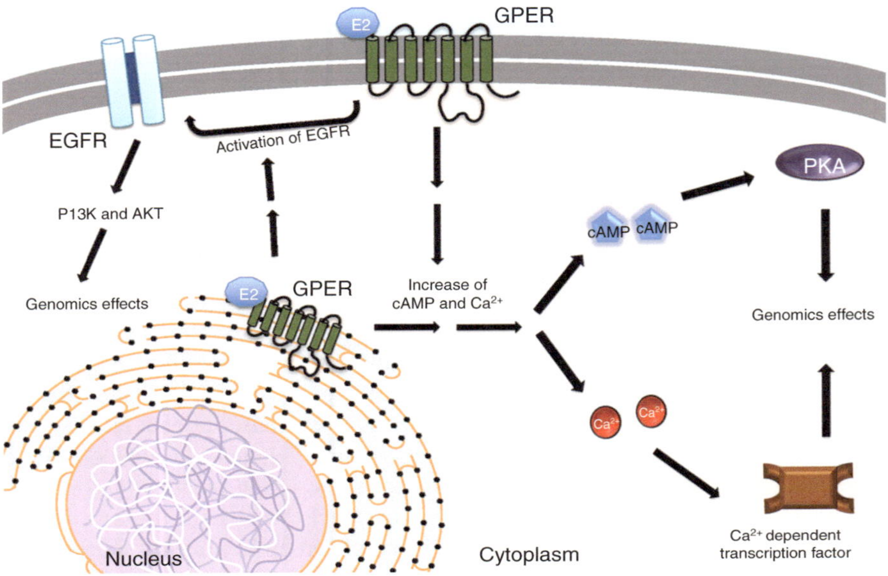

Fig. 13.10 Rapid non-genomic signaling of G-protein-coupled estrogen receptors (GPER) can also regulate gene expression. Binding of 17β-estradiol (E2) to GPER can activate the epidermal growth factor receptor (EGFR) leading to the initiation of other signal transduction pathways such as PI3K/AKT that can have downstream genomic effects. Activation of GPER can also lead to an increase in the conversion of adenosine triphosphate (ATP) to cyclic adenosine monophosphate (cAMP) as well as the release of Ca^{2+} in the cytosol. Accumulation of cAMP in the cytoplasm can activate protein kinase A (PKA), which can phosphorylate transcription factors to regulate transcription. Increased levels of Ca^{2+} can lead to the activation of different transcription factors

signaling is complex and quite controversial. These rapid changes in levels of Ca^{2+} and cAMP can be induced by E2 binding to membrane-bound receptors such as a transmembrane, G-protein-coupled estrogen receptor (GPER) (Revelli et al. 1998; Revankar et al. 2005; Thomas et al. 2005) (Fig. 13.10). GPER has been found on the plasma membrane as well as the outer surfaces of intracellular organelles such as the endoplasmic reticulum (Thomas et al. 2005; Revankar et al. 2005; Funakoshi et al. 2006). Although rapid production of secondary messengers is the predominant effect produced, GPER activation alters downstream gene expression (Vivacqua et al. 2012). Estrogen-induced increases of cAMP have been observed in uterine, mammary gland, and breast cancer cells (Szego and Davis 1967; Aronica et al. 1994). In breast cancer cells, E2 was shown to bind to GPER and activate adenylyl cyclase (AC) and increase cAMP levels (Thomas et al. 2005). A study in vascular smooth muscle cells has shown that E2 can promote apoptosis by inhibition of PKA via GPER (Ding et al. 2009). GPER signaling can also activate the phosphoinositide 3-kinase (PI3K) pathway (Ding et al. 2009). PI3K/AKT regulates cellular proliferation and suppression of apoptosis (Datta et al. 1997, 1999). GPER-induced apoptosis in vascular cells has been shown to be via ERK signaling (Ding et al. 2009) and involves EGFR transactivation (Daub et al. 1996; Filardo et al. 2000). The EGF pathway is important in regulating meiosis, mitosis, and cellular differentiation (Carpenter and Cohen 1979). Binding of GPER has been shown to induce the mobilization of Ca^{2+} ions into the cytosol by its interaction with EGFR (Fig. 13.10) (Revankar et al. 2005).

While differences are present between the genomic and non-genomic regulatory pathways of estrogens, overlap also occurs. For example, cyclin D1 (*ccnd1*) expression is regulated by AP-1, but studies have shown that E2 can induce *ccnd1* transcription via cAMP, which was linked to the activation of PKA (Castro-Rivera et al. 2001). The ability of estrogen to regulate the transcription of genes such as *ccnd1* by several mechanisms further shows the complexity of estrogen-signaling pathways. Additionally, the interaction of ERα and GPER is required for estrogen-induced pathways in ovarian cancer cells (Albanito et al. 2007).

13.4 Role of the Endocrine System on the Cardiovascular System

Hormones play a key role in a diverse number of signaling pathways in different organ systems like the cardiovascular system. Thyroid hormone receptors are expressed in many different tissues such as the brain, heart, kidney, and liver (Cheng et al. 2010). In the heart, changes in the concentration of both T_3 and T_4 can cause changes in cardiac output, myocyte contraction, as well as oxygen consumption (Kahaly and Dillmann 2005; Biondi et al. 2002; Danzi and Klein 2002). However, in cardiac myocytes, T_3 initiates most of the signaling because of limited deiodination in these cells (Everts et al. 1996). T_3 can initiate signaling in cardiac myocytes by binding to TR or RXR and regulate the expression of multiple genes involved in cardiac function (Brent 1994). Expression of cardiac genes such as myosin heavy chain, sarcoplasmic reticulum Ca^{2+} ATPase (SERCA), and phospholamban (PLB) is regulated by TR (Hartong et al. 1994; Kiss et al. 1994; Klein and Danzi 2007). Myosins are important parts of muscle fibers and aid in muscle contraction. During muscle relaxation, cytosolic Ca^{2+} is remobilized into the sarcoplasmic reticulum by SERCA under PLB regulation. Thyroid hormones can also mediate physiological effects by non-genomic pathways in cardiac cells. T_3 can alter membrane channels that control ion influxes such as Ca^{2+} essential for proper cardiac function (Davis and Davis 2002). In mice, treatment of vascular endothelial cells with T_3 led to the phosphorylation of protein kinase B (AKT) and endothelial nitric oxide synthase (eNOS), a protein involved in synthesis of nitric oxide (NO), which regulates contraction, heart rate, and myocyte contractility (Hiroi et al. 2006).

Steroid hormones have also been shown to have cardiac effects. In addition to reproductive tissues, testosterone has also been shown to impact macrophages, cardiac myocytes, and neuroblastoma cells (Ceballos et al. 1999; Estrada et al. 2006a, b). Androgen-induced Ca^{2+} influxes cause cell proliferation and cell death by apoptosis or necrosis (Estrada et al. 2006b). In cardiac myocytes, activation of a G-protein-coupled receptor by testosterone causes an increase in intracellular Ca^{2+} (Vicencio et al. 2006). In rabbits, the non-genomic effects of androgens have been shown to affect the relaxation of coronary arteries and the aorta in rabbit (Yue et al. 1995). Xiao et al. (2015) showed that testosterone could have a protective role in cardiac myocytes by decreasing cell death via AR. Although androgens have been linked to cardiac health, the role of androgen signaling on the heart has not been fully characterized (Liu et al. 2003; Kaushik et al. 2010).

As with androgens, estrogens can regulate nonreproductive organ systems, including the cardiovascular system (Murphy 2011), often having a protective role against cardiovascular diseases (Mendelsohn and Karas 1999; Deschamps and Murphy 2009). For example, estrogen therapy has been linked to decreasing cardiac arrhythmias in postmenopausal women (Cagnacci et al. 1992), and estrogens can affect blood flow by regulating the metabolism of cholesterol and reduce arterial plaque formation (Chow 1995). The role of the estrogen pathway on cardiovascular function is complex (Meyer et al. 2009, 2011). Activation of GPER also prevents the opening of mitochondria permeability transition pores, which may reduce intracellular Ca^{2+} levels during ischemia-reperfusion injury (Hausenloy and Yellon 2003; Bopassa et al. 2010). Contraction of cardiac myocytes is dependent upon rapid fluctuations in intracellular Ca^{2+}. Studies in mice have shown that the number of L-type Ca^{2+} channels is maintained by E2 (Johnson et al. 1997). L-Type Ca^{2+} channels are one of the most abundant calcium channels found in cardiac cells and transport Ca^{2+} into the cytosol (Catterall 1991). Signaling pathways via estrogen can also impact the cardiovascular system by regulating the expression of nitric oxide synthase (NOS), a protein that catalyzes the production of NO (Chambliss and Shaul 2002). NO targets vascular endothelial cells and modulates myocardial contractility.

13.5 Endocrine-Disrupting Chemicals

The ability of hormones to affect an important network like the cardiovascular system shows the complex and integral role of the endocrine system in normal organismal physiology. Biological processes that are required for normal development and function of organ systems can be influenced not only by endogenous molecules but as well as exogenous agents. Environmental contaminants can alter the endocrine system signaling (DeRosa et al. 1998; Crisp et al. 1998; Sumpter 1998). Endocrine-disrupting chemicals (EDCs) include a host of compounds such as pesticides, steroid hormones, and pharmaceutical agents, which can enter water bodies through municipal discharge of wastewater, as well as agricultural, industrial, and urban runoff (Ternes et al. 1999; Purdom et al. 1994; Larsson et al. 1999). The presence of EDCs in the environment has become a major concern due to the potential risks they pose to exposed populations.

EDCs can alter endocrine pathways by mimicking endogenous hormones and binding to specific receptors, either as an agonist or as an antagonist. The synthesis or clearance of endogenous hormones may also be altered by EDCs. Moreover, some EDCs can have multiple modes of action. For example, bisphenol-A (BPA), an industrial plasticizer, acts on several endocrine pathways by binding to estrogen, androgen, and thyroid receptors (Kuiper et al. 1997; Moriyama et al. 2002; Zoeller et al. 2005; Sohoni and Sumpter 1998; Lee et al. 2003). As described above each of these pathways have roles in the cardiovascular system. Epidemiological evidence has correlated exposure to BPA with increased cardiovascular risks (Melzer et al. 2010). Identification of the specific pathway affected by toxicants such as BPA can be especially challenging, due to the significant amount of crosstalk between endocrine pathways.

Hormones have also been suggested to play a role in cardiac development, and exposure to EDCs during embryogenesis can have significant effects that may impair the fitness of the organism. The heart serves a critical role in a developing organism. Consequently, impairment of endocrine pathways at sensitive life stages may significantly alter cardiac development and cause multiple lethal and sublethal effects that may not be observed until later life stages. Due to the complex nature of the evaluating EDCs, researchers have established and benefitted from the use of zebrafish as a model organism.

13.6 Zebrafish as a Model for Cardiac Development

Zebrafish (*Danio rerio*) have been a useful model in better understanding developmental toxicology. Traditionally, toxicants were screened for developmental effects using rodents, which bring certain limitations owing to their high cost of use and protracted developmental time, both of which limit the ability to evaluate the impacts of toxicants at sensitive stages. Zebrafish can be easily maintained and cultured in a laboratory setting. Due to transparent embryos, visual analysis of organogenesis can be measured throughout embryonic development. The zebrafish development toxicity assay (ZEDTA) has evaluated the teratogenicity of hundreds of compounds with high predictive ability (Ball et al. 2014). Large-scale mutagenesis screens have allowed the classification of important genes involved in the development of many organs such as the heart (Haffter et al. 1996; Driever et al. 1996; Chen et al. 1996). To further explore zebrafish development, small molecule screens have also been conducted to investigate both time and dose effects (Peterson et al. 2000; Mathew et al. 2007).

Many of the genes involved in development are highly conserved between zebrafish and humans allowing it to serve not only as a general vertebrate model but also one that can be utilized in human health biomedical research (Lam et al. 2006). There are several transgenic strains that have greatly contributed to developmental biology research (Veldman and Lin 2008; Ni et al. 2012). One of the first strains developed was a GFP labeled *gata1* gene that allows the study of erythrocyte development (Long et al. 1997). Transgenic zebrafish models have also been generated to study human diseases. A transgenic zebrafish line containing the mutated human cardiac sodium channel gene *scn5a* was developed as an arrhythmia model (Huttner et al. 2013).

The development of zebrafish is rapid; embryos hatch approximately 72 h after fertilization. The release of eggs by fecund females is driven by light, and the eggs are then externally fertilized immediately. Zebrafish development is divided into several stages: zygotic, cleavage, blastula, gastrula, segmentation, pharyngula, and the hatching period (Kimmel et al. 1995). Zebrafish are not dependent on a functional cardiovascular system for the first 5 days of its development, allowing phenotypic analysis after genetic alterations or chemical exposure. Although the cardiovascular system of zebrafish is different from mammals, early cardiac development has many similarities. The many complex processes that occur during cardiovascular development in zebrafish have been extensively reviewed (Stainier 2001; Staudt and Stainier 2012).

Myocardial and endocardial progenitor cell formation is one of the earliest events in cardiac development, and many of the genes and pathways involved are highly conserved among vertebrates. Cardiac progenitor cells that will later differentiate into myocardial and endocardial cells are derived from mesodermal cells (Fig. 13.11). At around 3 h postfertilization (hpf), cardiac progenitor cells are found on the marginal zone of the embryo. One of the earliest genes to be expressed in cardiac cells and is thought to be a key player in cardiac differentiation is *gata5*. It has been shown to initiate and can regulate expression of nkx2.5 during cardiac cell specification (Reiter et al. 1999). Nkx2.5 is crucial for myocardial differentiation in zebrafish (Chen and Fishman 1996). Several mutation studies have identified that *nodal* and *bmp* signaling also affects nkx2.5 expression. Mutations on *nodal* and *bmp* result in a decrease in nkx2.5 expression which could be restored by forcible expression of GATA5 (Reiter et al. 2001; Kishimoto et al. 1997). NKX2.5 homologs have been identified in pre-cardiac mesoderm cells of other organisms such as mice and frogs (Lints et al. 1993; Tonissen et al. 1994). Another important gene in cardiac differentiation is helix-loop-helix transcription factor gene (*hand2*) (Yelon et al. 2000). *Hand2* also contains the binding site for GATA transcription factors in its promoter region (McFadden et al. 2000). Hand2-deficient embryos result in a decrease in myocyte cell number (Yelon et al. 2000). The differentiation of endocardial cells is less understood compared to that of myocardial cells. One of the genes that are important for endocardial cell formation is *cloche* (*clo*). Mutations in this gene have been shown to block endocardial and blood cell differentiation (Stainier et al. 1995).

Starting from mid- to late gastrula and during the segmentation period, cardiac precursor cells move toward anterior lateral plate mesoderm (ALPM) (~16 hpf) and then fuse in the midline (~19 hpf) (Fig. 13.11). When cardiac cells reach the ALPM, they continue to further differentiate (De Pater et al. 2009). The myocardial cells will further differentiate into ventricular or atrial myocytes by expression of different myosin heavy chain genes (De Pater et al. 2009). *Grinch*, which encodes a G-protein-coupled receptor, is expressed in the marginal zone and the ALPM, while its ligand apelin is found in the midline (Scott et al. 2007; Zeng et al. 2007). *Grinch-Apelin* signaling has been shown to be important for cardiac cell migration during gastrulation. Fibronectin is an essential factor in cell migration and cardiac fusion. Mutation in the fibronectin gene (*natter*) has been demonstrated

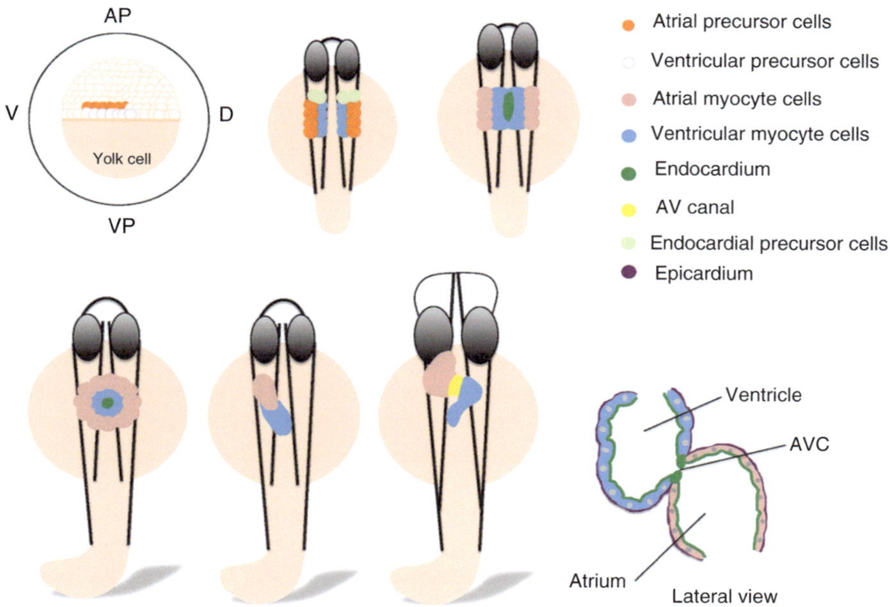

Fig. 13.11 Cardiac progenitor cells are located bilaterally in the lateral marginal zone. The cardiac progenitor cells move dorsally toward the midline to end up in the anterior lateral plate mesoderm (ALPM). The pre-endocardial and myocardial cells migrate toward the midline. When the bilateral heart fields fuse at the midline, they form a cardiac disc structure with the endocardial cells forming the inner layer and the ventricular and atrial myocytes at the periphery of the disc. Cardiac morphogenesis transforms the cardiac disc into a cardiac tube. When the linear heart tube forms, the venous pole is located at the anterior left, and the arterial pole is fixed at the midline. The heart tube will then loop with two distinct chambers. The heart will continue to develop and form an epicardium and valve leaflets at the atrioventricular canal (AVC). Ventral side (V), dorsal side (D)

to disrupt the movement of cells to the midline (Trinh and Stainier 2004). Levels of fibronectin are balanced and regulated by *hand2* (Garavito-Aguilar et al. 2010). A gene mutation known as Miles apart (*mil*), and its ligand regulator two of hearts (*toh*), is reported crucial for cell movement to the midline (Kupperman et al. 2000; Osborne et al. 2008). Endocardial cells will migrate toward the midline with the aid of the transcription factor *tal1* (Bussmann et al. 2007). VEGF signaling is also required for endocardial movement in order to coat the interior heart tube. Interaction between the endocardial and myocardial cells is also important. Endocardial cells may play a role in myocardial cell migration to form the outer layers of the cardiac cone (Staudt and Stainier 2012).

After the fusion of the two bilateral cardiac primordia (~19 hpf), the heart forms a disclike structure in the midline that will form the primitive heart tube (~26 hpf). The myocardial disc will rotate forming the venous pole at the left side of the midline, while the arterial pole remains at the midline (Fig. 13.11). The myocardial cells originating from the right field will involute toward the ventral side into the anterior region of the embryo. Mutations to *has* and *nok* have been shown to alter cell polarity and interfere with heart tube formation (Rohr et al. 2006, 2008). Both *nodal* and *bmp* signaling also help regulate this well-organized process (Bakkers 2011). Bmp4 is asymmetrically expressed with high amounts on the left field (Chen et al. 1997). Along with the left rotation of the disc, the heart subsequently undergoes extension.

After formation of the cardiac tube, at ~36 hpf, the heart tube loops to form a two-chambered organ with an inner and outer curvature (Fig. 13.11). The ventricle will bend to the right side of the midline, while the future atrium remains in place. This second rotation is also regulated by asymmetric gene expression. The venous and arterial poles will rotate at different degrees. Then the chambers will obtain their curvature by undergoing a process called cardiac ballooning. Atrial natriuretic peptide type A (ANPa) expression sets the outer curvature of the myocardium (Auman et al. 2007). Several molecular events are required for ventricular specification, which will form the major pumping chamber. Alterations in several genes such as *gata5* and *hand2* specifically produce altered ventricle formation (Reiter et al. 1999; Yelon et al. 2000).

Another important step in heart development is the atrioventricular canal (AVC) and valve formation. This starts at ~36 hpf and continues until about 4 days postfertilization when the valve leaflets are formed. The AVC is formed at the border between the two cardiac chambers. AVC specification is observed when restriction of *bmp4* and *versican* expression is limited to the AVC myocardial cells, and *notch1b* is limited to the endocardial cells (Westin and Lardelli 1997; Walsh and Stainier 2001). Cardiac valves are essential in maintaining the direction of blood flow from the ventricle to the atrium. Valve development starts with the formation of an endocardium cushion in the atrioventricular canal. Endothelial cells will take a cuboidal shape and start to express *dm-grasp* (Beis et al. 2005; Bakkers 2011). The molecular signaling of valve formation involves communication between the myocardial and endocardial layers via *nodal*, *nfat*, and *tgf-ß* pathways. After hatching, the endocardial cushion will enlarge and differentiate to form the atrioventricular valve leaflets. Blood flow also plays a role in valve development. Molecular changes that occur during valvulogenesis can be stimulated by physical changes of blood flow. The sheer stress produced from the oscillating flow induces the transcription of *klf2*, which regulates *notch* expression and enhances valve formation (Vermot et al. 2009; Timmerman et al. 2004).

As the heart is developing from a tube to a looped structure, it contains layers of two different cell types (the myocardial and endocardial cells) (Fig. 13.11). However, a third layer arises from the surrounding cells called the epicardium. This cell layer forms a surface around the entire heart. Bmp4 signaling from the myocardium layer induces epicardial specification (Liu and Stainier 2010). Expression of *wt1*, *tbx18*, and *tcf21* are the molecular markers for the epicardial cell population (Serluca 2008). The epithelial layer that surrounds the heart protects it from injury. After injury, the epicardial cells respond by increasing the expression of *tbx18* causing epithelial to mesenchymal transition, which will move into the injured myocardium layer forming a vascular network to promote repair (Lepilina et al. 2006; Kim et al. 2010).

Contractions are first observed at the venous pole of the developing heart (~24 hpf). Contractions initially occur at an irregular rhythm until the heart tube is fully formed. As the heart continues to develop, the contractions continue to increase and occur in a sequential pattern. The cells that maintain this pulse are located at the inner atrial node. *Islet-1* regulates the activity of pacemaker cells, which is seen at 48 hpf (Tessadori et al. 2012). Before the heart starts to contract in a peristaltic pattern, there is a decrease in pulse in the canal myocardium. This delay disappears in zebrafish that lack an endothelium suggesting that this event is initiated from signals from the endocardial cells (Milan et al. 2006). Transcription factor *tbx2b*, which is expressed in the myocardium of the AV canal, is also required for the conduction delay as well as marking fast- and slow-conducting cells. Bmp4 can regulate expression of *tbx2b*, which are both regulated by Wnt signaling in the AV canal myocardium (Verhoeven et al. 2011).

13.7 Cardiotoxicity in the Zebrafish

The growing use of zebrafish and detailed studies on its cardiovascular development have greatly contributed in characterizing the effects of cardiac toxicants. 2,3,7,8-Tetrachlorodibenzo-*p*-dioxin (TCDD), a polychlorinated dibenzo-*p*-dioxin congener, is known to target the cardiovascular system. TCDD mediates its toxicity by binding with high affinity to the aryl hydrocarbon receptor (AHR) in the cytoplasm. Activated AHR will move to the nucleus and form a dimer with the aryl hydrocarbon receptor nuclear translocator (ARNT). The AHR/ARNT dimer can then bind to xenobiotic response elements (XRE) and regulate transcription of specific genes (Denison et al. 1988a, b; Schmidt and Bradfield 1996; Rowlands and Gustafsson 1997). AHR regulates the genes involved in chemical metabolism such as CYP1A (Fujisawa-Sehara et al. 1987; Telakowski-Hopkins et al. 1988; Denison et al. 1988a) (Fig. 13.12).

AHR also has an important role in the development of several organs, including the heart (Gonzalez and Fernandez-Salguero 1998; Antkiewicz et al. 2006). Disruption of the AHR pathway results in cardiovascular defects. AHR-null mice resulted in an increase in heart weight that is associated with myocyte hypertrophy (Thackaberry et al. 2002; Lund et al. 2003). Furthermore, over-activation of AHR by TCDD causes heart size defects and dysregulation in myocyte proliferation and heart rate (Hornung et al. 1999; Walker and Catron 2000; Thackaberry et al. 2005). Other effects that include looping malformations, pericardial edema, and decreases in myocyte number have also been associated with TCDD toxicity (Henry et al. 1997; Antkiewicz et al. 2006; Carney et al. 2006). Further evidence that the developmental effects of TCDD were mediated through AHR comes from the observation that malformations in wildtype mice were reversed with a loss of function mutation of the *ahr* and *arnt* genes (Walisser et al. 2004). However, there is growing evidence that TCDD toxicity isn't dependent on activation of *CYP1A* (Carney et al. 2004).

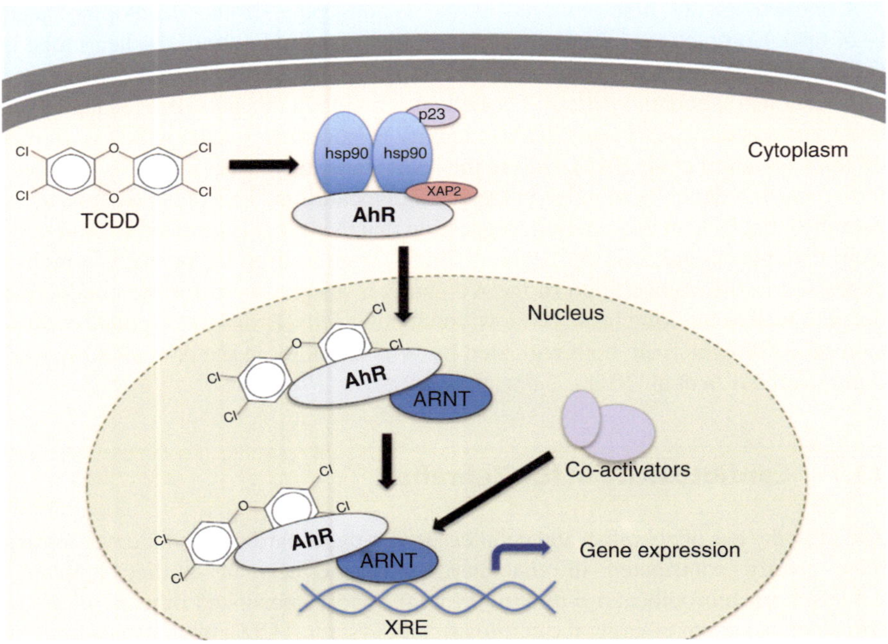

Fig. 13.12 In the absence of bound ligand, aryl hydrocarbon receptor (AHR) resides in a complex with heat shock protein 90 (HSP90), the co-chaperone protein X-associated protein 2 (XAP2), and p23. When bound to an agonist ligand such as 2,3,7,8-tetrachlorodibenzo-p-dioxin (TCDD), the AHR complex translocates to the nucleus, resulting in HSP90 displacement. AHR binds to the AHR nuclear translocator (ARNT), leading to AHR-ARNT heterodimer formation. This heterodimer is capable of binding to a xenobiotic response element (XRE). Both AHR and ARNT can recruit coactivators, leading to the transcription of a wide variety of genes such as cytochrome P450 1A1 (CYP1A1)

Since the genome is well annotated for zebrafish, many researchers have used this model to uncover other AHR targets during development. Handley-Goldstone et al. (2005) showed exposure to TCDD can alter the expression of genes that form the sarcomere, which is important for contraction. To further elucidate specific targets of AHR/ARNT, transcriptional changes were also temporally assessed in heart tissue of zebrafish after TCDD exposure (Carney et al. 2006). The genes that were dysregulated were involved in xenobiotic metabolism, cell division and proliferation, as well as ion transport. Along with the gene alterations, there was a significant reduction in cardiac function, heart size, and looping defects (Carney et al. 2006).

Another group of compounds that have been shown to activate the AHR and cause cardiac defects are polycyclic aromatic hydrocarbons (PAHs). PAHs are by-products formed from incomplete combustion of organic matter. Mixtures of PAHs can also be found in fossil fuels such as crude oil and coal. Many studies have shown that exposure to crude oil can result in several malformations and sublethal effects (Incardona et al. 2013, 2014, 2015). Some of the abnormal phenotypes that have been observed after exposure to PAH mixtures are pericardial edema, yolk sac edema, and curved body axis (Incardona et al. 2004; Heintz et al. 1999). Effects on heart rate (bradycardia) have also been described after PAH exposure (Incardona et al. 2004). To determine the specific compounds that cause toxicity, zebrafish embryos were exposed to a host of individual PAHs with subsequent evaluation of development (Incardona et al. 2004, 2006, 2011). These studies showed that individual PAHs can induce toxicity with differing severity. Three-ring PAHs such as phenanthrene and dibenzothiophene caused more severe effects compared to pyrene, as well as different deformities. Phenanthrene and dibenzothiophene showed effects on cardiac conduction, while pyrene resulted in anemia and loss of peripheral circulation showing that ring number is an important characteristic in PAH toxicity (Incardona et al. 2004).

Studies have also been done to evaluate the molecular mechanism of cardiac toxicity of individual PAHs. Research on PAHs has been primarily focused on AHR activation (Jayasundara et al. 2015). However, growing evidence suggests that certain PAHs can induce toxicity independent of AHR (Incardona et al. 2005; Goodale et al. 2013). AHR morpholinos were able to prevent the cardiac toxicity seen in zebrafish embryos treated with pyrene, but not phenanthrene or dibenzothiophene (Incardona et al. 2005). This further shows that other pathways may also be involved in PAH cardiac-induced toxicity and warrants further investigation especially those that are known to interact with the AHR pathway.

Compounds that alter the endocrine system can also cause cardiotoxic effects. Estrogenic compounds have also been shown to change Ca^{2+} homeostasis in cardiomyocyte cells. E2 and BPA can alter Ca^{2+} regulation, promoting cardiac arrhythmias in rodents (Yan et al. 2011). Levels of BPA measured in the urine from adults in the USA were associated with increased risk of myocardial infarction (Lang et al. 2008), coronary artery disease (Melzer et al. 2010), and hypertension (Shankar et al. 2012). Zebrafish embryos exposed to BPA (15 mg/L) caused pericardial edema, indicating an effect on the development of the cardiovascular system (Duan et al. 2008). Although research has established the role of the endocrine system on cardiovascular maintenance and function, its specific role in development remains unclear and warrants further investigation. Exposure to 17β-estradiol (3 μM) has also shown to result in pericardial edema as well as other defects (Chandrasekar et al. 2010). GPER has been suggested to have a role in heart development in zebrafish. Exposure to the GPER agonist G1 resulted in cardiac malformations (Jayasinghe and Volz 2012; Diamante et al. 2017). However, minimal studies have been done to further elucidate this as a potential pathway for cardiac toxicity. However, examining estrogen signaling can be multifaceted because of the great deal of crosstalk between several pathways. In addition, the AHR and ER signaling have been shown to interact, leading to another potential mechanism for cardiotoxicity (Matthews and Gustafsson 2006). Activated AHR has been shown to recruit ERs to promoter regions and to affect normal estrogen signaling (Matthews et al. 2005). The exact role of ER to AHR-regulated genes needs to be further examined.

Conclusion

The endocrine system plays a vital role in many cellular processes that are important for maintaining normal physiological functions. As described in this chapter, the thyroid, androgen, and estrogen pathways not only have the ability to signal to their specific receptors but can crosstalk with other pathways that add to the complexity of the signaling cascades that are induced. Therefore, studying the molecular effects of EDCs on an organism can become quite complicated, even more so when assessing their effects during development. With the numerous pathways that are endocrine regulated along with the additional complexity of cellular processes occurring during development, assessing toxicity at these intricate times brings about more difficulties. Therefore, despite numerous studies showing the effects of hormones on development, the specific molecular pathways responsible for these phenotypic effects remain unclear. Using the zebrafish as a model organism may help bridge the gaps.

The zebrafish has become a popular model for developmental and toxicological research. It has played a role in obtaining the current knowledge we have about the cardiotoxic effects of dioxins and PAHs. However, there are still many compounds that need to be further studied to better assess their potential to negatively affect embryo development. Estrogens have been well studied in terms of their effect on reproduction, but their role in embryo development and cardiovascular functions are now being examined. Using zebrafish as a tool can aid in the study of the role of estrogens on heart development. These findings will be essential in elucidating the mechanisms of toxicity of xenoestrogens. Future research needs to be able to match the molecular events of toxicity to a specific time point in the very complex timeline of development. A detailed mechanistic understanding of the normal role of endocrine system and the ability of EDCs to induce cardiac effects during development may help to address the toxicity of other environmental contaminants.

References

Abney TO (1999) The potential roles of estrogens in regulating Leydig cell development and function: a review. Steroids 64(9):610–617

Albanito L, Madeo A, Lappano R, Vivacqua A, Rago V, Carpino A, Oprea TI, Prossnitz ER, Musti AM, Andò S, Maggiolini M (2007) G protein-coupled receptor 30 (GPR30) mediates gene expression changes and growth response to 17beta-estradiol and selective GPR30 ligand G-1 in ovarian cancer cells. Cancer Res 67(4):1859–1866

Andersson S, Geissler WM, Wu L, Davis DL, Grumbach MM, New MI, Schwarz HP, Blethen SL, Mendonca BB, Bloise W, Witchel SF, Cutler GB Jr, Griffin JE, Wilson JD, Russel DW (1996) Molecular genetics and pathophysiology of 17 beta-hydroxysteroid dehydrogenase 3 deficiency. J Clin Endocrinol Metab 81(1):130–136

Antkiewicz DS, Peterson RE, Heideman W (2006) Blocking expression of AHR2 and ARNT1 in zebrafish larvae protects against cardiac toxicity of 2,3,7,8-tetrachlorodibenzo-p-dioxin. Toxicol Sci 94(1):175–182

Aronica SM, Kraus WL, Katzenellenbogen BS (1994) Estrogen action via the cAMP signaling pathway: stimulation of adenylate cyclase and cAMP-regulated gene transcription. Proc Natl Acad Sci U S A 91(18):8517–8521

Auman HJ, Coleman H, Riley HE, Olale F, Tsai HJ, Yelon D (2007) Functional modulation of cardiac form through regionally confined cell shape changes. PLoS Biol 5(3):e53
Baird DT, Swanston IA, McNeilly AS (1981) Secretion of androgens and estrogens by the preovulatory follicle in the ewe. Biol Reprod 24:1013–1025
Bakkers J (2011) Zebrafish as a model to study cardiac development and human cardiac disease. Cardiovasc Res 91(2):279–288
Ball JS, Stedman DB, Hillegass JM, Zhang CX, Panzica-Kelly J, Coburn A, Enright BP, Tornesi B, Amouzadeh HR, Hetheridge M, Gustafson AL, Augustine-Rauch KA (2014) Fishing for teratogens: a consortium effort for a harmonized zebrafish developmental toxicology assay. Toxicol Sci 139(1):210–219
Beis D, Bartman T, Jin SW, Scott IC, D'Amico LA, Ober EA, Verkade H, Frantsve J, Field HA, Wehman A, Baier H, Tallafuss A, Bally-Cuif L, Chen JN, Stainier DY, Jungblut B (2005) Genetic and cellular analyses of zebrafish atrioventricular cushion and valve development. Development 132(18):4193–4204
Benten WP, Lieberherr M, Stamm O, Wrehlke C, Guo Z, Wunderlich F (1999) Testosterone signaling through internalizable surface receptors in androgen receptor-free macrophages. Mol Biol Cell 10(10):3113–3123
Biondi B, Palmieri EA, Lombardi G, Fazio S (2002) Effects of thyroid hormone on cardiac function: the relative importance of heart rate, loading conditions, and myocardial contractility in the regulation of cardiac performance in human hyperthyroidism. J Clin Endocrinol Metab 87(3):968–974
Bopassa JC, Eghbali M, Toro L, Stefani E (2010) A novel estrogen receptor GPER inhibits mitochondria permeability transition pore opening and protects the heart against ischemia-reperfusion injury. Am J Physiol Heart Circ Physiol 298(1):16–23
Breen JJ, Hickok NJ, Guff JA (1997) The rat TSH gene contains distinct response elements for regulation by retinoids and thyroid hormone. Mol Cell Endocrinol 131:137–146
Brent GA (1994) The molecular basis of thyroid hormone action. N Engl J Med 331(13):847–853
Brinkmann AO, Leemborg FG, Roodnat EM, De Jong FH, Van der Molen HJ (1980) A specific action of estradiol on enzymes involved in testicular steroidogenesis. Biol Reprod 23(4):801–809
Brix K, Herzog V (1994) Extrathyroidal release of thyroid hormones from thyroglobulin by J774 mouse macrophages. J Clin Invest 93(4):1388–1396
Bussmann J, Bakkers J, Schulte-Merker S (2007) Early endocardial morphogenesis requires Scl/Tal1. PLoS Genet 3(8):e140
Cagnacci A, Soldani R, Puccini E, Fioretti P, Melis GB (1992) Lipid-independent therapeutic properties of transdermal 17 B-estradiol on cardiovascular diseases. Acta Obstet Gynecol Scand 71(8):639–641
Campbell BK, Baird DT, Webb R (1998) Effects of dose of LH on androgen production and luteinization of ovine theca cells cultured in a serum-free system. J Reprod Fertil 112(1):69–77
Carney SA, Chen J, Burns CG, Xiong KM, Peterson RE, Heideman W (2006) Aryl hydrocarbon receptor activation produces heart-specific transcriptional and toxic responses in developing zebrafish. Mol Pharmacol 70(2):549–561
Carney SA, Peterson RE, Heideman W (2004) 2,3,7,8-Tetrachlorodibenzo-p-dioxin activation of the aryl hydrocarbon receptor/aryl hydrocarbon receptor nuclear translocator pathway causes developmental toxicity through a CYP1A-independent mechanism in zebrafish. Mol Pharmacol 66(3):512–521
Carpenter G, Cohen S (1979) Epidermal growth factor. Annu Rev Biochem 48:193–216
Castro-Rivera E, Samudio I, Safe S (2001) Estrogen regulation of cyclin D1 gene expression in ZR-75 breast cancer cells involves multiple enhancer elements. J Biol Chem 276(33):30853–30861
Catterall WA (1991) Functional subunit structure of voltage-gated calcium channels. Science 253(5027):1499–1500
Ceballos G, Figueroa L, Rubio I, Gallo G, Garcia A, Martinez A, Yañez R, Perez J, Morato T, Chamorro G (1999) Acute and nongenomic effects of testosterone on isolated and perfused rat heart. J Cardiovasc Pharmacol 33(5):691–697

Chambliss KL, Shaul PW (2002) Estrogen modulation of endothelial nitric oxide synthase. Endocr Rev 23(5):665–686
Chandrasekar G, Archer A, Gustafsson JA, Andersson Lendahl M (2010) Levels of 17b-estradiol receptors expressed in embryonic and adult zebrafish following in vivo treatment of natural or synthetic ligands. PLoS One 5(3):e9678
Chang C, Saltzman A, Yeh S, Young W, Keller E, Lee HJ, Wang C, Mizokami A (1995) Androgen receptor: an overview. Crit Rev Eukaryot Gene Expr 5(2):97–125
Chang CS, Kokontis J, Liao ST (1988a) Molecular cloning of human and rat complementary DNA encoding androgen receptors. Science 240(4850):324–326
Chang CS, Kokontis J, Liao ST (1988b) Structural analysis of complementary DNA and amino acid sequences of human and rat androgen receptors. Proc Natl Acad Sci U S A 85(19):7211–7215
Chen JN, Fishman MC (1996) Zebrafish tinman homolog demarcates the heart field and initiates myocardial differentiation. Development 122(12):3809–3816
Chen JN, Haffter P, Odenthal J, Vogelsang E, Brand M, van Eeden FJ, Furutani-Seiki M, Granato M, Hammerschmidt M, Heisenberg CP, Jiang YJ, Kane DA, Kelsh RN, Mullins MC, Nüsslein-Volhard C (1996) Mutations affecting the cardiovascular system and other internal organs in zebrafish. Development 123:293–302
Chen JN, van Eeden FJ, Warren KS, Chin A, Nüsslein-Volhard C, Haffter P, Fishman MC (1997) Left-right pattern of cardiac BMP4 may drive asymmetry of the heart in zebrafish. Development 124(21):4373–4382
Cheng SY, Leonard JL, Davis PJ (2010) Molecular aspects of thyroid hormone actions. Endocr Rev 31(2):139–170
Chow MS (1995) Benefit/risk of estrogen therapy in cardiovascular disease: current knowledge and future challenges. J Clin Pharmacol 35(9 Suppl):11S–17S
Crisp TM, Clegg ED, Cooper RL, Wood WP, Anderson DG, Baetcke KP, Hoffmann JL, Morrow MS, Rodier DJ, Schaeffer JE, Touart LW, Zeeman MG, Patel YM (1998) Environmental endocrine disruption: an effects assessment and analysis. Environ Health Perspect 106(Suppl 1):11–56
Danzi S, Klein I (2002) Thyroid hormone-regulated cardiac gene expression and cardiovascular disease. Thyroid 12(6):467–472
Datta SR, Brunet A, Greenberg ME (1999) Cellular survival: a play in three Akts. Genes Dev 13(22):2905–2927
Datta SR, Dudek H, Tao X, Masters S, Fu H, Gotoh Y, Greenberg ME (1997) Akt phosphorylation of BAD couples survival signals to the cell-intrinsic death machinery. Cell 91(2):231–241
Daub H, Weiss FU, Wallasch C, Ullrich A (1996) Role of transactivation of the EGF receptor in signalling by G-protein-coupled receptors. Nature 379(6565):557–560
Davis PJ, Davis FB (2002) Nongenomic actions of thyroid hormone on the heart. Thyroid 12(6):459–466
De Jong FH (1988) Inhibin. Physiol Rev 68(2):555–607
De Pater E, Clijsters L, Marques SR, Lin YF, Garavito-Aguilar ZV, Yelon D, Bakkers J (2009) Distinct phases of cardiomyocyte differentiation regulate growth of the zebrafish heart. Development 136(10):1633–1641
De Waal PP, Wang DS, Nijenhuis WA, Schulz RW, Bogerd J (2008) Functional characterization and expression analysis of the androgen receptor in zebrafish (Danio rerio) testis. Reproduction 136(2):225–234
Denison MS, Fisher JM, Whitlock JP Jr (1988a) Inducible, receptor-dependent protein-DNA interactions at a dioxin-responsive transcriptional enhancer. Proc Natl Acad Sci U S A 85(8):2528–2532
Denison MS, Fisher JM, Whitlock JP Jr (1988b) The DNA recognition site for the dioxin-Ah receptor complex. nucleotide sequence and functional analysis. J Biol Chem 263(33):17221–17224
DeRosa C, Richter P, Pohl H, Jones DE (1998) Environmental exposures that affect the endocrine system: public health implications. J Toxicol Environ Health B Crit Rev 1(1):3–26
Deschamps AM, Murphy E (2009) Activation of a novel estrogen receptor, GPER, is cardioprotective in male and female rats. Am J Physiol Heart Circ Physiol 297(5):1806–1813

Diamante G, Menjivar-Cervantes N, Leung MS, Volz DC, Schlenk D (2017) Contribution of G protein-coupled estrogen receptor 1 (GPER) to 17β-estradiol-induced developmental toxicity in zebrafish. Aquat Toxicol 186:180–187

Ding Q, Gros R, Limbird LE, Chorazyczewski J, Feldman RD (2009) Estradiol-mediated ERK phosphorylation and apoptosis in vascular smooth muscle cells requires GPR 30. Am J Physiol Cell Physiol 297(5):C1178–C1187

Driever W, Solnica-Krezel L, Schier AF, Neuhauss SC, Malicki J, Stemple DL, Stainier DY, Zwartkruis F, Abdelilah S, Rangini Z, Belak J, Boggs C (1996) A genetic screen for mutations affecting embryogenesis in zebrafish. Development 123:37–46

Duan R, Porter W, Safe S (1998) Estrogen-induced c-fos protooncogene expression in MCF-7 human breast cancer cells: role of estrogen receptor Sp1 complex formation. Endocrinology 139(4):1981–1990

Duan Z, Zhu L, Zhu L, Kun Y, Zhu X (2008) Individual and joint toxic effects of pentachlorophenol and bisphenol A on the development of zebrafish (Danio rerio) embryo. Ecotoxicol Environ Saf 71(3):774–780

Dunn AD, Crutchfield HE, Dunn JT (1991) Proteolytic processing of thyroglobulin by extracts of thyroid lysosomes. Endocrinology 128(6):3073–3080

Estrada M, Espinosa A, Müller M, Jaimovich E (2003) Testosterone stimulates intracellular calcium release and mitogen-activated protein kinases via a G protein-coupled receptor in skeletal muscle cells. Endocrinology 144(8):3586–3597

Estrada M, Uhlen P, Ehrlich BE (2006a) Ca2+ oscillations induced by testosterone enhance neurite outgrowth. J Cell Sci 119(Pt 4):733–743

Estrada M, Varshney A, Ehrlich BE (2006b) Elevated testosterone induces apoptosis in neuronal cells. J Biol Chem 281(35):25492–25501

Everts ME, Verhoeven FA, Bezstarosti K, Moerings EP, Hennemann G, Visser TJ, Lamers JM (1996) Uptake of thyroid hormones in neonatal rat cardiac myocytes. Endocrinology 137(10):4235–4242

Filardo EJ, Quinn JA, Bland KI, Frackelton AR Jr (2000) Estrogen-induced activation of Erk-1 and Erk-2 requires the G protein-coupled receptor homolog, GPR30, and occurs via transactivation of the epidermal growth factor receptor through release of HB-EGF. Mol Endocrinol 14(10):1649–1660

Foradori CD, Weiser MJ, Handa RJ (2008) Non-genomic actions of androgens. Front Neuroendocrinol 29(2):169–181

Frye CA, Bo E, Calamandrei G, Calzà L, Dessì-Fulgheri F, Fernández M, Fusani L, Kah O, Kajta M, Le Page Y, Patisaul HB, Venerosi A, Wojtowicz AK, Panzica GC (2012) Endocrine disrupters: a review of some sources, effects, and mechanisms of actions on behaviour and neuroendocrine systems. J Neuroendocrinol 24(1):144–159

Fujisawa-Sehara A, Sogawa K, Yamane M, Fujii-Kuriyama Y (1987) Characterization of xenobiotic responsive elements upstream from the drug-metabolizing cytochrome P-450c gene: a similarity to glucocorticoid regulatory elements. Nucleic Acids Res 15(10):4179–4191

Funakoshi T, Yanai A, Shinoda K, Kawano MM, Mizukami Y (2006) G protein-coupled receptor 30 is an estrogen receptor in the plasma membrane. Biochem Biophys Res Commun 346(3):904–910

Gao X, Wang HS (2014) Impact of bisphenol a on the cardiovascular system – epidemiological and experimental evidence and molecular mechanisms. Int J Environ Res Public Health 11(8):8399–8413

Garavito-Aguilar ZV, Riley HE, Yelon D (2010) Hand2 ensures an appropriate environment for cardiac fusion by limiting Fibronectin function. Development 137(19):3215–3220

Gaub MP, Bellard M, Scheuer I, Chambon P, Sassone-Corsi P (1990) Activation of the ovalbumin gene by the estrogen receptor involves the fos-jun complex. Cell 63(6):1267–1276

Gobinet J, Poujol N, Sultan C (2002) Molecular action of androgens. Mol Cell Endocrinol 198:15–24

Gonzalez FJ, Fernandez-Salguero P (1998) The aryl hydrocarbon receptor: studies using the AHR-null mice. Drug Metab Dispos 26(12):1194–1198

Goodale BC, Tilton SC, Corvi MM, Wilson GR, Janszen DB, Anderson KA, Waters KM, Tanguay RL (2013) Structurally distinct polycyclic aromatic hydrocarbons induce differential transcriptional responses in developing zebrafish. Toxicol Appl Pharmacol 272(3):656–670

Gregory SJ, Kaiser UB (2004) Regulation of gonadotropins by inhibin and activin. Semin Reprod Med 22(3):253–267

Gustafsson J, Pousette K (1975) Demonstration and partial characterization of cytosol receptors for testosterone. Biochemistry 14(14):3094–3101

Haffter P, Granato M, Brand M, Mullins MC, Hammerschmidt M, Kane DA, Odenthal J, van Eeden FJ, Jiang YJ, Heisenberg CP, Kelsh RN, Furutani-Seiki M, Vogelsang E, Beuchle D, Schach U, Fabian C, Nüsslein-Volhard C (1996) The identification of genes with unique and essential functions in the development of the zebrafish, Danio rerio. Development 123:1–36

Hall JM, Couse JF, Korach KS (2001) The multifaceted mechanisms of estradiol and estrogen receptor signaling. J Biol Chem 276(40):36869–36872

Handley-Goldstone HM, Grow MW, Stegeman JJ (2005) Cardiovascular gene expression profiles of dioxin exposure in zebrafish embryos. Toxicol Sci 85(1):683–693

Hansson V, Weddington SC, Petrusz P, Ritzen EM, Nayfeh SN (1975) French FS (1975) FSH stimulation of testicular androgen binding protein (ABP): comparison of ABP response and ovarian augmentation. Endocrinology 97(2):469–473

Hartong R, Wang N, Kurokawa R, Lazar MA, Glass CK, Apriletti JW, Dillmann WH (1994) Delineation of three different thyroid hormone-response elements in promoter of rat sarcoplasmic reticulum Ca2+ATPase gene. Demonstration that retinoid X receptor binds 5′ to thyroid hormone receptor in response element 1. J Biol Chem 269(17):13021–13029

Hausenloy DJ, Yellon DM (2003) The mitochondrial permeability transition pore: its fundamental role in mediating cell death during ischaemia and reperfusion. J Mol Cell Cardiol 35(4): 339–341

Heintz RA, Short JW, Rice SD (1999) Sensitivity of fish embryos to weathered crude oil: II. Increased mortality of pink salmon (Oncorhynchus gorbuscha) embryos incubating downstream from weathered Exxon Valdez crude oil. Environ Toxicol Chem 18(3):494–503

Heldring N, Pike A, Andersson S, Matthews J, Cheng G, Hartman J, Tujague M, Ström A, Treuter E, Warner M, Gustafsson JA (2007) Estrogen receptors: how do they signal and what are their targets. Physiol Rev 87(3):905–931

Henry TR, Spitsbergen JM, Hornung MW, Abnet CC, Peterson RE (1997) Early life stage toxicity of 2,3,7,8-tetrachlorodibenzo-p-dioxin in zebrafish (Danio rerio). Toxicol Appl Pharmacol 142(1):56–68

Herzog V (1983) Transcytosis in thyroid follicle cells. J Cell Biol 97(3):607–617

Hiroi H, Christenson LK, Strauss JF (2004) Regulation of transcription of the steroidogenic acute regulatory protein (StAR) gene: temporal and spatial changes in transcription factor binding and histone modification. Mol Cell Endocrinol 215(1–2):119–126

Hiroi Y, Kim HH, Ying H, Furuya F, Huang Z, Simoncini T, Noma K, Ueki K, Nguyen NH, Scanlan TS, Moskowitz MA, Cheng SY, Liao JK (2006) Rapid nongenomic actions of thyroid hormone. Proc Natl Acad Sci U S A 103(38):14104–14109

Hornung MW, Spitsbergen JM, Peterson RE (1999) 2,3,7,8-Tetrachlorodibenzo-p-dioxin alters cardiovascular and craniofacial development and function in sac fry of rainbow trout (Oncorhynchus mykiss). Toxicol Sci 47(1):40–51

Hossain MS, Larsson A, Scherbak N, Olsson PE, Orban L (2008) Zebrafish androgen receptor: isolation, molecular, and biochemical characterization. Biol Reprod 78(2):361–369

Hsieh KP, Martin TF (1992) Thyrotropin-releasing hormone and gonadotropin-releasing hormone receptors activate phospholipase C by coupling to the guanosine triphosphate-binding proteins Gq and G11. Mol Endocrinol 6(10):1673–1681

Hu J, Zhang Z, Shen WJ, Azhar S (2010) Cellular cholesterol delivery, intracellular processing and utilization for biosynthesis of steroid hormones. Nutr Metab (Lond) 7:47

Huttner IG, Trivedi G, Jacoby A, Mann SA, Vandenberg JI, Fatkin D (2013) A transgenic zebrafish model of a human cardiac sodium channel mutation exhibits bradycardia, conduction-system abnormalities and early death. J Mol Cell Cardiol 61:123–132

Incardona JP, Carls MG, Holland L, Linbo TL, Baldwin DH, Myers MS, Peck KA, Tagal M, Rice SD, Scholz NL (2015) Very low embryonic crude oil exposures cause lasting cardiac defects in salmon and herring. Sci Rep 5:13499

Incardona JP, Carls MG, Teraoka H, Sloan CA, Collier TK, Scholz NL (2005) Aryl hydrocarbon receptor-independent toxicity of weathered crude oil during fish development. Environ Health Perspect 113(12):1755–1762

Incardona JP, Collier TK, Scholz NL (2004) Defects in cardiac function precede morphological abnormalities in fish embryos exposed to polycyclic aromatic hydrocarbons. Toxicol Appl Pharmacol 196(2):191–205

Incardona JP, Day HL, Collier TK, Scholz NL (2006) Developmental toxicity of 4-ring polycyclic aromatic hydrocarbons in zebrafish is differentially dependent on AH receptor isoforms and hepatic cytochrome P4501A metabolism. Toxicol Appl Pharmacol 217(3):308–321

Incardona JP, Gardner LD, Linbo TL, Brown TL, Esbaugh AJ, Mager EM, Stieglitz JD, French BL, Labenia JS, Laetz CA, Tagal M, Sloan CA, Elizur A, Benetti DD, Grosell M, Block BA, Scholz NL (2014) Deepwater horizon crude oil impacts the developing hearts of large predatory pelagic fish. Proc Natl Acad Sci U S A 111(15):E1510–E1518

Incardona JP, Linbo TL, Scholz NL (2011) Cardiac toxicity of 5-ring polycyclic aromatic hydrocarbons is differentially dependent on the aryl hydrocarbon receptor 2 isoform during zebrafish development. Toxicol Appl Pharmacol 257(2):242–249

Incardona JP, Swarts TL, Edmunds RC, Linbo TL, Aquilina-Beck A, Sloan CA, Gardner LD, Block BA, Scholz NL (2013) Exxon Valdez to Deepwater horizon: comparable toxicity of both crude oils to fish early life stages. Aquat Toxicol 142-143:303–316

Ingenbleek Y, Young V (1994) Transthyretin (prealbumin) in health and disease: nutritional implications. Annu Rev Nutr 14:495–533

Jayasinghe BS, Volz DC (2012) Aberrant ligand-induced activation of G protein-coupled estrogen receptor 1 (GPER) results in developmental malformations during vertebrate embryogenesis. Toxicol Sci 125(1):262–273

Jayasundara N, Van Tiem Garner L, Meyer JN, Erwin KN, Di Giulio RT (2015) AHR2-mediated transcriptomic responses underlying the synergistic cardiac developmental toxicity of PAHs. Toxicol Sci 143(2):469–481

Jensen EV (1962) On the mechanism of estrogen action. Perspect Biol Med 6:47–54

Jensen EV, Desombre ER (1973) Estrogen-receptor interaction: estrogenic hormones effect transformation of specific receptor proteins to a biologically active form. Science 182:126–134

Jensen EV, Suzuki T, Kawashima T, Stumpf WE, Jungblut PW, DeSombre ER (1968) A two-step mechanism for the interaction of estradiol with rat uterus. Proc Natl Acad Sci U S A 59(2):632–638

Jenster G, van der Korput HA, van Vroonhoven C, van der Kwast TH, Trapman J, Brinkmann AO (1991) Domains of the human androgen receptor involved in steroid binding, transcriptional activation, and subcellular localization. Mol Endocrinol 5(10):1396–1404

Johnson BD, Zheng W, Korach KS, Scheuer T, Catterall WA, Rubanyi GM (1997) Increased expression of the cardiac L-type calcium channel in estrogen receptor-deficient mice. J Gen Physiol 110(2):135–140

Kahaly GJ, Dillmann WH (2005) Thyroid hormone action in the heart. Endocr Rev 26(5):704–728

Kalla NR, Nisula BC, Menard R, Loriaux DL (1980) The effect of estradiol on Leydig cell testosterone biosynthesis. Endocrinology 106:35–39

Kaushik M, Sontineni SP, Hunter C (2010) Cardiovascular disease and androgens: a review. Int J Cardiol 142(1):8–14

Kayes-Wandover KM, White PC (2000) Steroidogenic enzyme gene expression in the human heart. J Clin Endocrinol Metab 85(7):2519–2525

Khan SA, Ball RB, Hendry WJ (1998) Effects of neonatal administration of diethylstilbestrol in male hamsters: disruption of reproductive function in adults after apparently normal pubertal development. Biol Reprod 58(1):137–142

Kim J, Wu Q, Zhang Y, Wiens KM, Huang Y, Rubin N, Shimada H, Handin RI, Chao MY, Tuan TL, Starnes VA, Lien CL (2010) PDGF signaling is required for epicardial function and blood vessel formation in regenerating zebrafish hearts. Proc Natl Acad Sci U S A 107(40):17206–17210

Kim K, Barhoumi R, Burghardt R, Safe S (2005) Analysis of estrogen receptor {α}-Sp1 interactions in breast cancer cells by fluorescence resonance energy transfer. Mol Endocrinol 19(4):843–854
Kim PS, Arvan P (1991) Folding and assembly of newly synthesized thyroglobulin occurs in a pre-Golgi compartment. J Biol Chem 266(19):12412–12418
Kim PS, Arvan P (1993) Hormonal regulation of thyroglobulin export from the endoplasmic reticulum of cultured thyrocytes. J Biol Chem 268(7):4873–4879
Kimmel CB, Ballard WW, Kimmel SR, Ullmann B, Schilling TF (1995) Stages of embryonic development of the zebrafish. Dev Dyn 203(3):253–310
Kimura N, Mizokami A, Oonuma T, Sasano H, Nagura H (1993) Immunocytochemical localization of androgen receptor with polyclonal antibody in paraffin-embedded human tissues. J Histochem Cytochem 41(5):671–678
Kishimoto Y, Lee KH, Zon L, Hammerschmidt M, Schulte-Merker S (1997) The molecular nature of zebrafish swirl: BMP2 function is essential during early dorsoventral patterning. Development 124(22):4457–4466
Kiss E, Jakab G, Kranias EG, Edes I (1994) Thyroid hormone-induced alterations in phospholamban protein expression: regulatory effects on sarcoplasmic reticulum Ca2+ transport and myocardial relaxation. Circ Res 75(2):245–251
Klein I, Danzi S (2007) Thyroid disease and the heart. Circulation 116(15):1725–1735
Klingenberg M (1958) Pigments of rat liver microsomes. Arch Biochem Biophys 75(2):376–386
Kostrouch Z, Bernier-Valentin F, Munari-Silem Y, Rajas F, Rabilloud R, Rousset B (1993) Thyroglobulin molecules internalized by thyrocytes are sorted in early endosomes and partially recycled back to the follicular lumen. Endocrinology 132(6):2645–2653
Kostrouch Z, Munari-Silem Y, Rajas F, Bernier-Valentin F, Rousset B (1991) Thyroglobulin internalized by thyrocytes passes through early and late endosomes. Endocrinology 129(4): 2202–2211
Kuiper GG, Carlsson B, Grandien K, Enmark E, Häggblad J, Nilsson S, Gustafsson JA (1997) Comparison of the ligand binding specificity and transcript tissue distribution of estrogen receptors alpha and beta. Endocrinology 138(3):863–870
Kuiper GG, Enmark E, Pelto-Huikko M, Nilsson S, Gustafsson JA (1996) Cloning of a novel receptor expressed in rat prostate and ovary. Proc Natl Acad Sci U S A 93(12):5925–5930
Kupperman E, An S, Osborne N, Waldron S, Stainier DY (2000) A sphingosine-1-phosphate receptor regulates cell migration during vertebrate heart development. Nature 406:192–195
Lam SH, Winata CL, Tong Y, Korzh S, Lim WS, Korzh V, Spitsbergen J, Mathavan S, Miller LD, Liu ET, Gong Z (2006) Transcriptome kinetics of arsenic-induced adaptive response in zebrafish liver. Physiol Genomics 27(3):351–361
Lang IA, Galloway TS, Scarlett A, Henley WE, Depledge M, Wallace RB, Melzer D (2008) Association of urinary bisphenol A concentration with medical disorders and laboratory abnormalities in adults. JAMA 300(11):1303–1310
Larsson DGJ, Adolfsson-Erici M, Parkkonen J, Pettersson M, Berg AH, Olsson PE, Förlin L (1999) Ethinyloestradiol – an undesired fish contraceptive? Aquat Toxicol 45(2–3):91–97
Lebbe M, Woodruff TK (2013) Involvement of androgens in ovarian health and disease. Mol Hum Reprod 19(12):828–837
Lee HJ, Chattopadhyay S, Gong EY, Ahn RS, Lee K (2003) Antiandrogenic effects of bisphenol A and nonylphenol on the function of androgen receptor. Toxicol Sci 75(1):40–46
Lepilina A, Coon AN, Kikuchi K, Holdway JE, Roberts RW, Burns CG, Poss KD (2006) A dynamic epicardial injury response supports progenitor cell activity during zebrafish heart regeneration. Cell 127(3):607–619
Ling N, Ying SY, Ueno N, Shimasaki S, Esch F, Hotta M, Guillemin R (1986) Pituitary FSH is released by a heterodimer of the beta-subunits from the two forms of inhibin. Nature 321(6072):779–782
Lints TJ, Parsons LM, Hartley L, Lyons I, Harvey RP (1993) Nkx-2. 5: a novel murine homeobox gene expressed in early heart progenitor cells and their myogenic descendants. Development 119(3):419–431

Liu J, Stainier DY (2010) Tbx5 and Bmp signaling are essential for proepicardium specification in zebrafish. Circ Res 106(12):1818–1828
Liu MM, Albanese C, Anderson CM, Hilty K, Webb P, Uht RM, Price RH Jr, Pestell RG, Kushner PJ (2002) Opposing action of estrogen receptors alpha and beta on cyclin D1 gene expression. J Biol Chem 277(27):24353–24360
Liu PY, Death AK, Handelsman DJ (2003) Androgens and cardiovascular disease. Endocr Rev 24(3):313–340
Long Q, Meng A, Wang H, Jessen JR, Farrell MJ, Lin S (1997) GATA-1 expression pattern can be recapitulated in living transgenic zebrafish using GFP reporter gene. Development 124(20):4105–4111
Lubahn DB, Moyer JS, Golding TS, Couse JF, Korach KS, Smithies O (1993) Alteration of reproductive function but not prenatal sexual development after insertional disruption of the mouse estrogen receptor gene. Proc Natl Acad Sci U S A 90(23):11162–11166
Lund AK, Goens MB, Kanagy NL, Walker MK (2003) Cardiac hypertrophy in aryl hydrocarbon receptor null mice is correlated with elevated angiotensin II, endothelin-1, and mean arterial blood pressure. Toxicol Appl Pharmacol 193(2):177–187
Martinez-Arguelles DB, Papadopoulos V (2010) Epigenetic regulation of the expression of genes involved in steroid hormone biosynthesis and action. Steroids 75(7):467–476
Mathew LK, Sengupta S, Kawakami A, Andreasen EA, Löhr CV, Loynes CA, Renshaw SA, Peterson RT, Tanguay RL (2007) Unraveling tissue regeneration pathways using chemical genetics. J Biol Chem 282(48):35202–35210
Matthews J, Gustafsson JA (2006) Estrogen receptor and aryl hydrocarbon receptor signaling pathways. Nucl Recept Signal 4:e016
Matthews J, Wihlén B, Thomsen J, Gustafsson JA (2005) Aryl hydrocarbon receptor-mediated transcription: ligand-dependent recruitment of estrogen receptor alpha to 2,3,7,8-tetrachlorodi benzo-p-dioxin-responsive promoters. Mol Cell Biol 25(13):5317–5328
McFadden DG, Charité J, Richardson JA, Srivastava D, Firulli AB, Olson EN (2000) A GATA-dependent right ventricular enhancer controls dHAND transcription in the developing heart. Development 127(24):5331–5341
McKenna NJ, O'Malley BW (2002) Combinatorial control of gene expression by nuclear receptors and coregulators. Cell 108(4):465–474
Meachem SJ, McLachlan RI, de Kretser DM, Robertson DM, Wreford NG (1996) Neonatal exposure of rats to recombinant follicle stimulating hormone increases adult Sertoli and spermatogenic cell numbers. Biol Reprod 54(1):36–44
Mellon SH, Griffin LD (2002) Neurosteroids: biochemistry and clinical significance. Trends Endocrinol Metab 13(1):35–43
Mellström B, Naranjo JR (2001) Mechanisms of Ca(2+)-dependent transcription. Curr Opin Neurobiol 11(3):312–319
Melner MH, Abney TO (1980a) The direct effect of 17β-estradiol on LH-stimulated testosterone production in hypophysectomized rats. J Steroid Biochem 13(2):203–210
Melner MH, Abney TO (1980b) Depletion of the cytoplasmic estrogen receptor in gonadotropin-desensitized testes. Endocrinology 107(5):1620–1626
Melzer D, Rice NE, Lewis C, Henley WE, Galloway TS (2010) Association of urinary bisphenol a concentration with heart disease: evidence from NHANES 2003/06. PLoS One 5(1):e8673
Mendelsohn ME, Karas RH (1999) The protective effects of estrogen on the cardiovascular system. N Engl J Med 340(23):1801–1811
Meyer MR, Haas E, Prossnitz ER, Barton M (2009) Non-genomic regulation of vascular cell function and growth by estrogen. Mol Cell Endocrinol 308(1–2):9–16
Meyer MR, Prossnitz ER, Barton M (2011) The G protein-coupled estrogen receptor GPER/GPR30 as a regulator of cardiovascular function. Vasc Pharmacol 55(1–3):17–25
Milan DJ, Giokas AC, Serluca FC, Peterson RT, MacRae CA (2006) Notch1b and neuregulin are required for specification of central cardiac conduction tissue. Development 133(6):1125–1132
Miller WL (2002) Androgen biosynthesis from cholesterol to DHEA. Mol Cell Endocrinol 198(1–2):7–14

Miller WL (2007) StAR search—what we know about how the steroidogenic acute regulatory protein mediates mitochondrial cholesterol import. Mol Endocrinol 21(3):589–601
Miller WL (2008) Steroidogenic enzymes. Endocr Dev 13:1–18
Miller WL, Strauss JF (1999) Molecular pathology and mechanism of action of the steroidogenic acute regulatory protein, StAR. J Steroid Biochem Mol Biol 69(1–6):131–141
Moriyama K, Tagami T, Akamizu T, Usui T, Saijo M, Kanamoto N, Hataya Y, Shimatsu A, Kuzuya H, Nakao K (2002) Thyroid hormone action is disrupted by bisphenol A as an antagonist. J Clin Endocrinol Metab 87(11):5185–5190
Mullur R, Liu YY, Brent GA (2014) Thyroid hormone regulation of metabolism. Physiol Rev 94(2):355–382
Murphy E (2011) Estrogen signaling and cardiovascular disease. Circ Res 109(6):687–696
Ni TT, Lu J, Zhu M, Maddison LA, Boyd KL, Huskey L, Ju B, Hesselson D, Zhong TP, Page-McCaw PS, Stainier DY, Chen W (2012) Conditional control of gene function by an invertible gene trap in zebrafish. Proc Natl Acad Sci U S A 109(38):15389–15394
O'Shaughnessy PJ, Baker PJ, Heikkilä M, Vainio S, McMahon AP (2000) Localization of 17beta-hydroxysteroid dehydrogenase/17-ketosteroid reductase isoform expression in the developing mouse testis--androstenedione is the major androgen secreted by fetal/neonatal leydig cells. Endocrinology 141(7):2631–2637
Omiecinski CJ, Vanden Heuvel JP, Perdew GH, Peters JM (2011) Xenobiotic metabolism, disposition, and regulation by receptors: from biochemical phenomenon to predictors of major toxicities. Toxicol Sci 120(Suppl 1):S49–S75
Osborne N, Brand-Arzamendi K, Ober EA, Jin SW, Verkade H, Holtzman NG, Yelon D, Stainier DY (2008) The spinster homolog, two of hearts, is required for sphingosine 1-phosphate signaling in zebrafish. Curr Biol 18(23):1882–1888
Padmanabhan V, Sairam MR, Hassing JM, Brown MB, Ridings JW, Beitins IZ (1991) Follicle-stimulating hormone signal transduction: role of carbohydrate in aromatase induction in immature rat sertoli cells. Mol Cell Endocrinol 79(1–3):119–128
Panzica GC, Viglietti-Panzica C, Mura E, Quinn MJ Jr, Lavoie E, Palanza P, Ottinger MA (2007) Effects of xenoestrogens on the differentiation of behaviorally-relevant neural circuits. Front Neuroendocrinol 28(4):179–200
Payne AH, Hales DB (2004) Overview of steroidogenic enzymes in the pathway from cholesterol to active steroid hormones. Endocr Rev 25(6):947–970
Peterson RT, Link BA, Dowling JE, Schreiber SL (2000) Small molecule developmental screens reveal the logic and timing of vertebrate development. Proc Natl Acad Sci U S A 97(24):12965–12969
Petz LN, Nardulli AM (2000) Sp1 binding sites and an estrogen response element half-site are involved in regulation of the human progesterone receptor A promoter. Mol Endocrinol 14(7):972–985
Pietras RJ, Szego CM (1975) Endometrial cell calcium and oestrogen action. Nature 253(5490):357–359
Pinzone JJ, Stevenson H, Strobl JS, Berg PE (2004) Molecular and cellular determinants of estrogen receptor alpha expression. Mol Cell Biol 24(11):4605–4612
Prossnitz ER, Arterburn JB, Smith HO, Oprea TI, Sklar LA, Hathaway HJ (2008) Estrogen signaling through the transmembrane G protein-coupled receptor GPR30. Annu Rev Physiol 70:165–190
Purdom CE, Hardiman PA, Bye VVJ, Eno NC, Tyler CR, Sumpter JP (1994) Estrogenic effects of effluents from sewage treatment works. Chem Ecol 8:275–285
Reiter JF, Alexander J, Rodaway A, Yelon D, Patient R, Holder N, Stainier DY (1999) Gata5 is required for the development of the heart and endoderm in zebrafish. Genes Dev 13(22):2983–2995
Reiter JF, Verkade H, Stainier DY (2001) Bmp2b and Oep promote early myocardial differentiation through their regulation of gata5. Dev Biol 234(2):330–338
Revankar CM, Cimino DF, Sklar LA, Arterburn JB, Prossnitz ER (2005) A transmembrane intracellular estrogen receptor mediates rapid cell signaling. Science 307(5715):1625–1630

Revelli A, Massobrio M, Tesarik J (1998) Nongenomic actions of steroid hormones in reproductive tissues. Endocr Rev 19(1):3–17
Rohr S, Otten C, Abdelilah-Seyfried S (2008) Asymmetric involution of the myocardial field drives heart tube formation in zebrafish. Circ Res 102(2):e12–e19
Rohr S, Bit-Avragim N, Abdelilah-Seyfried S (2006) Heart and soul/prkci and nagie oko/mpp5 regulate myocardial coherence and remodeling during cardiac morphogenesis. Development 133:107–115
Rowlands JC, Gustafsson JA (1997) Aryl hydrocarbon receptor-mediated signal transduction. Crit Rev Toxicol 27(2):109–134
Sahlin L, Norstedt G, Eriksson H (1994) Androgen regulation of the insulin-like growth factor-I and the estrogen receptor in rat uterus and liver. J Steroid Biochem Mol Biol 51(1–2):57–66
Schmidt JV, Bradfield CA (1996) Ah receptor signaling pathways. Annu Rev Cell Dev Biol 12:55–89
Scott IC, Masri B, D'Amico LA, Jin SW, Jungblut B, Wehman AM, Baier H, Audigier Y, Stainier DY (2007) The g protein-coupled receptor agtrl1b regulates early development of myocardial progenitors. Dev Cell 12(3):403–413
Seljelid R, Reith A, Nakken KF (1970) The early phase of endocytosis in rat thyroid follicle cells. Lab Investig 23:595–605
Serluca FC (2008) Development of the proepicardial organ in the zebrafish. Dev Biol 315(1):18–27
Shankar A, Teppala S, Sabanayagam C (2012) Urinary bisphenol a levels and measures of obesity: results from the national health and nutrition examination survey 2003–2008. ISRN Endocrinol 2012:965243
Sohoni P, Sumpter JP (1998) Several environmental oestrogens are also anti-androgens. J Endocrinol 158(3):327–339
Stainier DY (2001) Zebrafish genetics and vertebrate heart formation. Nat Rev Genet 2(1):39–48
Stainier DY, Weinstein BM, Detrich HW, Zon LI, Fishman MC (1995) Cloche, an early acting zebrafish gene, is required by both the endothelial and hematopoietic lineages. Development 121(10):3141–3150
Staudt D, Stainier D (2012) Uncovering the molecular and cellular mechanisms of heart development using the zebrafish. Annu Rev Genet 46:397–418
Stein B, Yang MX (1995) Repression of the interleukin-6 promoter by estrogen receptor is mediated by NF-kappa B and C/EBP beta. Mol Cell Biol 15(9):4971–4979
Stumpf WE (1969) Nuclear concentration of 3H-estradiol in target tissues. Dry-mount autoradiography of vagina, oviduct, ovary, testis, mammary tumor, liver and adrenal. Endocrinology 85(1):31–37
Sumpter JP (1998) Xenoendorine disrupters--environmental impacts. Toxicol Lett 102-103:337–342
Szego CM, Davis JS (1967) Adenosine 3′,5′-monophosphate in rat uterus: acute elevation by estrogen. Proc Natl Acad Sci U S A 58(4):1711–1718
Takeda H, Chodak G, Mutchnik S, Nakamoto T, Chang C (1990) Immunohistochemical localization of androgen receptors with mono- and polyclonal antibodies to androgen receptor. J Endocrinol 126(1):17–25
Telakowski-Hopkins CA, King RG, Pickett CB (1988) Glutathione S-transferase Ya subunit gene: identification of regulatory elements required for basal level and inducible expression. Proc Natl Acad Sci U S A 85(4):1000–1004
Ternes TA, Stumpf M, Mueller J, Haberer K, Wilken RD, Servos M (1999) Behavior and occurrence of estrogens in municipal sewage treatment plants--I. Investigations in Germany, Canada and Brazil. Sci Total Environ 225(1–2):81–90
Tessadori F, van Weerd JH, Burkhard SB, Verkerk AO, de Pater E, Boukens BJ, Vink A, Christoffels VM, Bakkers J (2012) Identification and functional characterization of cardiac pacemaker cells in zebrafish. PLoS One 7(10):e47644
Thackaberry EA, Gabaldon DM, Walker MK, Smith SM (2002) Aryl hydrocarbon receptor null mice develop cardiac hypertrophy and increased hypoxia-inducible factor-1alpha in the absence of cardiac hypoxia. Cardiovasc Toxicol 2(4):263–274

Thackaberry EA, Jiang Z, Johnson CD, Ramos KS, Walker MK (2005) Toxicogenomic profile of 2,3,7,8-tetrachlorodibenzo-p-dioxin in the murine fetal heart: modulation of cell cycle and extracellular matrix genes. Toxicol Sci 88(1):231–241

Thomas P, Pang Y, Filardo EJ, Dong J (2005) Identity of an estrogen membrane receptor coupled to a G protein in human breast cancer cells. Endocrinology 146(2):624–632

Timmerman LA, Grego-Bessa J, Raya A, Bertrán E, Pérez-Pomares JM, Díez J, Aranda S, Palomo S, McCormick F, Izpisúa-Belmonte JC, de la Pompa JL (2004) Notch promotes epithelial-mesenchymal transition during cardiac development and oncogenic transformation. Genes Dev 18(1):99–115

Tonissen KF, Drysdale TA, Lints TJ, Harvey RP, Krieg PA (1994) XNkx-2. 5, a Xenopus gene related to Nkx-2. 5 and tinman: evidence for a conserved role in cardiac development. Dev Biol 162(1):325–328

Trinh LA, Stainier DY (2004) Fibronectin regulates epithelial organization during myocardial migration in zebrafish. Dev Cell 6(3):371–382

Ubuka T, Son YL, Bentley GE, Millar RP, Tsutsui K (2013) Gonadotropin-inhibitory hormone (GnIH), GnIH receptor and cell signaling. Gen Comp Endocrinol 190:10–17

Umayahara Y, Kawamori R, Watada H, Imano E, Iwama N, Morishima T, Yamasaki Y, Kajimoto Y, Kamada T (1994) Estrogen regulation of the insulin-like growth factor I gene transcription involves an AP-1 enhancer. J Biol Chem 269(23):16433–16442

Veldman MB, Lin S (2008) Zebrafish as a developmental model organism for pediatric research. Pediatr Res 64(5):470–476

Verhoeven MC, Haase C, Christoffels VM, Weidinger G, Bakkers J (2011) Wnt signaling regulates atrioventricular canal formation upstream of BMP and Tbx2. Birth Defects Res A Clin Mol Teratol 91(6):435–440

Vermot J, Forouhar AS, Liebling M, Wu D, Plummer D, Gharib M, Fraser SE (2009) Reversing blood flows act through klf2a to ensure normal valvulogenesis in the developing heart. PLoS Biol 7(11):e1000246

Vicencio JM, Ibarra C, Estrada M, Chiong M, Soto D, Parra V, Diaz-Araya G, Jaimovich E, Lavandero S (2006) Testosterone induces an intracellular calcium increase by a nongenomic mechanism in cultured rat cardiac myocytes. Endocrinology 147(3):1386–1395

Vivacqua A, Romeo E, De Marco P, De Francesco EM, Abonante S, Maggiolini M (2012) GPER mediates the Egr-1 expression induced by 17β-estradiol and 4-hydroxitamoxifen in breast and endometrial cancer cells. Breast Cancer Res Treat 133(3):1025–1035

Voutilainen R, Miller WL (1986) Developmental expression of genes for the steroidogenic enzymes P450scc (20,22-desmolase), P450c17 (17 alpha-hydroxylase/17,20-lyase), and P450c21 (21-hydroxylase) in the human fetus. J Clin Endocrinol Metab 63(5):1145–1150

Walisser JA, Bunger MK, Glover E, Bradfield CA (2004) Gestational exposure of Ahr and Arnt hypomorphs to dioxin rescues vascular development. Proc Natl Acad Sci U S A 101(47):16677–16682

Walker MK, Catron TF (2000) Characterization of cardiotoxicity induced by 2,3,7, 8-tetrachlorodibenzo-p-dioxin and related chemicals during early chick embryo development. Toxicol Appl Pharmacol 167(3):210–221

Walsh EC, Stainier DY (2001) UDP-glucose dehydrogenase required for cardiac valve formation in zebrafish. Science 293(5535):1670–1673

Wang Q, Carroll JS, Brown M (2005) Spatial and temporal recruitment of androgen receptor and its coactivators involves chromosomal looping and polymerase tracking. Mol Cell 19(5):631–642

Waye A, Trudeau VL (2011) Neuroendocrine disruption: more than hormones are upset. J Toxicol Environ Health B Crit Rev 14(5–7):270–291

Webb P, Nguyen P, Shinsako J, Anderson C, Feng W, Nguyen MP, Chen D, Huang SM, Subramanian S, McKinerney E, Katzenellenbogen BS, Stallcup MR, Kushner PJ (1998) Estrogen receptor activation function 1 works by binding p160 coactivator proteins. Mol Endocrinol 12(10):1605–1618

Webb P, Nguyen P, Valentine C, Lopez GN, Kwok GR, McInerney E, Katzenellenbogen BS, Enmark E, Gustafsson JA, Nilsson S, Kushner PJ (1999) The estrogen receptor enhances AP-1

activity by two distinct mechanisms with different requirements for receptor transactivation functions. Mol Endocrinol 13(10):1672–1685
Weisz A, Rosales R (1990) Identification of an estrogen response element upstream of the human c-fos gene that binds the estrogen receptor and the AP-1 transcription factor. Nucleic Acids Res 18(17):5097–5106
Westin J, Lardelli M (1997) Three novel notch genes in zebrafish: implications for vertebrate notch gene evolution and function. Dev Genes Evol 207(1):51–63
Xiao FY, Nheu L, Komesaroff P, Ling S (2015) Testosterone protects cardiac myocytes from superoxide injury via NF-κB signalling pathways. Life Sci 133:45–52
Yan S, Chen Y, Dong M, Song W, Belcher SM, Wang HS (2011) Bisphenol A and 17β-estradiol promote arrhythmia in the female heart via alteration of calcium handling. PLoS One 6(9):e25455
Yelon D, Ticho B, Halpern ME, Ruvinsky I, Ho RK, Silver LM, Stainier DY (2000) The bHLH transcription factor hand2 plays parallel roles in zebrafish heart and pectoral fin development. Development 127(12):2573–2582
Yue P, Chatterjee K, Beale C, Poole-Wilson PA, Collins P (1995) Testosterone relaxes rabbit coronary arteries and aorta. Circulation 91(4):1154–1160
Zeng XX, Wilm TP, Sepich DS, Solnica-Krezel L (2007) Apelin and its receptor control heart field formation during zebrafish gastrulation. Dev Cell 12(3):391–402
Zoeller RT, Bansal R, Parris C (2005) Bisphenol-A, an environmental contaminant that acts as a thyroid hormone receptor antagonist in vitro, increases serum thyroxine, and alters RC3/neurogranin expression in the developing rat brain. Endocrinology 146(2):607–612

Adaptation in Polluted Waters: Lessons from Killifish

14

Elias M. Oziolor and Cole W. Matson

Abstract

The world is changing at a pace that has forced us to reevaluate the capabilities and rate of the evolutionary process. As industrial development continues to increase and as we improve analytical environmental chemistry methods, we have observed a growing number of contaminated environments. This increasing pressure from chronically released and persistent contaminants has amplified the importance of evolutionary toxicology. In this chapter, we aim to convey the principles of this relatively young field through a review of evolutionary adaptation in Atlantic and Gulf killifish (*Fundulus heteroclitus* and *F. grandis*). Killifish have been widely studied with regard to evolutionary toxicology and anthropogenic adaptation. In the past 20 years, researchers have identified multiple populations of both *F. heteroclitus* and *F. grandis*, which have adapted to resist high levels of various classes of industrial contaminants, including polychlorinated biphenyls (PCBs), polycyclic aromatic hydrocarbons (PAHs), and polychlorinated dibenzo-*p*-dioxins and furans (PCDD/Fs). Heritability of resistance has been confirmed, with offspring showing a similar capacity to resist the cardiovascular teratogenesis associated with early life stage exposures to dioxin-like compounds (DLCs). As more populations have been discovered and studied, different aspects of the acquired resistance have been revealed, but also unique attributes of the independently adapted populations have been identified. Populations adapted to PAHs exhibit high resistance to many classes of contaminants, while populations adapted to PCB contamination are only partially resistant to PAHs. PAH-adapted populations of *F. heteroclitus* and PCB-adapted populations of *F. grandis* are more tolerant to oxidative stress, while PCB-adapted populations of *F. heteroclitus* are more sensitive. Even geographically close adapted popula-

E. M. Oziolor · C. W. Matson (✉)
Department of Environmental Science, Center for Reservoir and Aquatic Systems Research (CRASR), Institute for Biomedical Studies, Baylor University, Waco, TX, USA
e-mail: cole_matson@baylor.edu

© Springer International Publishing AG, part of Springer Nature 2018

W. Burggren, B. Dubansky (eds.), *Development and Environment*,
https://doi.org/10.1007/978-3-319-75935-7_14

tions in *F. grandis* differ in their levels of cross-resistance to pesticides. These are only some of the ways in which populations of *Fundulus* with phenotypically similar genetic adaptations create a much more varied and intricate story to be explored in evolutionary toxicology. Here we summarize considerations in the field of evolutionary toxicology through a case study of anthropogenic adaptation in *Fundulus*, including its environmental and evolutionary relevance.

14.1 Ecology-Development Interactions

Organismal development is a complicated and time-sensitive process. Developmental processes have been carefully studied to establish their evolution and mechanisms (Amsterdam et al. 2004; Keller et al. 2003). To compound the complexity of the developmental process, many steps during early development are strongly influenced by gene-environment interactions (Gilbert 2001). For example, natural and anthropogenic stressors have both been known to cause divergent cardiac (Billiard et al. 2006), limb (Stocum 2000), and jaw (Burkhart et al. 1998) development in early life exposures (Gilbert 2001). The interaction between ecology and development has become more apparent as natural populations of the same species can vary strongly in their developmental phenotype based on their ecology (Sultan 2007). The strong linkage between ecological parameters and development has led to the synthesis of a relatively new and rapidly expanding field of eco-devo (Sultan 2007). This has become particularly important as the realities of a changing world have forced us to consider the persistence and survival of populations under rapidly changing ecological conditions (Chown and Gaston 2008; Sultan 2007).

Physiological and genetic variability in developmental responses to ecological conditions can dictate the ability of a species to persist in a given environment (Chown and Gaston 2008; Sultan 2007). Climate change is causing environments to experience more frequent extremes in the occurrence of natural stressors (temperature, precipitation, drought, pH) (Parmesan and Yohe 2003; Poloczanska et al. 2013), in addition to the continual increase in anthropogenic contamination (Pal et al. 2014). As many of these stressors are known to impact the developmental process, such changes are predicted to cause populations to either adapt (Visser 2008) or struggle and perish (Berg et al. 2010) in the novel environment. Some of the most crucial predictions of the persistence of natural populations in a changing world will come from understanding the genetic, physiological (Auld et al. 2010; Berg et al. 2010; Sultan 2007), and nongenetic (Meaney 2010) ability of individuals and populations to deal with novel extremes. To improve our understanding of how populations are affected by changing environmental factors, we need to better understand population genetic responses to long-term ecological change in the environment.

14.2 Gene-Environment Interactions

When studying adaptation in natural populations, one must consider the primary prerequisites for localized evolution in response to stressors. For differentiation among populations, there must be an ecological parameter that drives selection and a barrier to gene flow that keeps populations separate (Rundle and Nosil 2005). A common example is the evolution of sexual selection in the three-spined stickleback (*Gasterosteus aculeatus*), where males and females optimize mating signal emission and acceptance to foster genetic barriers between populations in different environments (Boughman 2001). On the other hand, anthropogenic contaminants have also been known to drive localized adaptation by allowing only adaptive genotypes to successfully complete development in contaminated aquatic environments (Di Giulio and Clark 2015; Oziolor and Matson 2015; Wirgin et al. 2011; Wirgin and Waldman 2004). These processes involve complex habitat-population interactions which can be influenced by a milieu of factors.

Three of the main considerations for local adaptation, which we will return to with specific examples, are selective pressure, population size, and migration rates (Kawecki and Ebert 2004). Strong selective pressures can result in the over-representation of adaptive genotypes that allow for proper development in natural populations (Coleman and Chisholm 2010). On the other hand, larger population sizes can maintain higher levels of genetic variability, providing a greater likelihood that adaptive genotypes are found within the population (Lynch and Conery 2003). And lastly, barriers to gene flow and lack of migration between populations exposed to strong selective pressures can result in selective sweeps for adaptive genotypes (Kawecki and Ebert 2004; Lenormand 2002). Population adaptation to strong selective pressures can alter the gene pool of populations, by reducing genetic diversity, and possibly leave it compromised to future ecological challenges (Bickham 2011; Bickham et al. 2000). Thus, localized adaptive events in response to selective pressures allow populations to maintain a presence in an adverse environment (i.e., evolutionary rescue). However, it is important to remember that this adaptive response is often accompanied by significant and long-lasting reductions in population size. On the other hand, adaptive genes can be maintained in isolated populations in the absence of selective pressure (Stearns 1993) and thus may compromise the overall ecological health of the ecosystems affected if there are any fitness costs associated with them (Bickham 2011; Meyer and Di Giulio 2003).

14.3 Evolutionary Toxicology

The field of evolutionary toxicology came about as a consequence of the increasing concentrations and continual increase in releases of anthropogenic contaminants in the environment (Bickham and Smolen 1994). With increasingly contaminated environments, wild populations of several species have been found to have adapted to resist the negative effects of the contaminants they were exposed to (De Coninck et al. 2014; Oziolor and Matson 2015; Theodorakis and Shugart 1997). Given these trends, it became necessary to establish the new field of evolutionary toxicology (Bickham and Smolen 1994) and state principles and best practices by which to abide when looking for evolutionary responses to anthropogenic compounds (Bickham 2011). Bickham (2011) summarized these "cornerstones" as (1) changes in genetic diversity, (2) changes in genotypic frequencies caused by contaminant-induced selection, (3) changes in dispersal patterns or gene flow between populations, and (4) changes in genotypic frequencies caused by increased mutation. With climate change in mind, the increase in anthropogenic contamination has been predicted to interact with changing environmental variables and force more populations to deal with increasing toxicity and thus higher selective pressures coming from contamination (Noury et al. 2006). Some of the first studies of population-wide change were focused on behavior or development in populations that differ in contamination history (Van Veld and Westbrook 1995; Van Veld et al. 1992; Vogelbein et al. 1990; Weis et al. 1981a, b; Zhou et al. 1996). With significant improvements in genetic and genomic technologies, the study of the toxicant-population relationships and the genotypes involved in adaptive selection has become one of the main ways to study evolutionary responses (Hawkins et al. 2010; Kimberly and Salice 2015; Klerks et al. 2011). Evolutionary toxicology exhibits a novel way in which environmental variables can interact with development and force population-wide changes in a changing world (Gilbert 2001).

14.4 The Origin of *Fundulus* Adaptation to Pollution Research

The genus *Fundulus* has a broad geographic range in North America, particularly in the eastern United States and all around the Gulf of Mexico coastline (introduced in Europe and elsewhere) (Bernardi et al. 1995; Gonzalez et al. 2009), as well as being found in both freshwater and saltwater environments (Whitehead 2010). In this review, we will focus on two of the species in the genus *Fundulus*—*F. heteroclitus* and *F. grandis*. This focus stems from the large number of papers that have described populations of these species that have adapted to locally polluted aquatic environments. Note that when referring to *Fundulus* throughout the remainder of chapter, we are specifically referring to the *F. heteroclitus*-*F. grandis* group. A thorough review of the potential of *Fundulus* as a teleost model organism is given by Burnett et al. (2007), but there are several key features of *Fundulus* that make them particularly useful for evolutionary toxicology. Several *Fundulus* species have been shown to have high site fidelity (Alldredge et al. 2011; Nelson et al. 2014; Skinner et al. 2005, 2012). High site fidelity increases the likelihood that the current population represents the product of any multigenerational effects resulting from local contamination. Populations of *Fundulus* generally maintain large sizes (Adams et al. 2006), allowing for higher genetic variability in populations. Some *Fundulus* species have also been known to have short generation turnover times, by reaching sexual maturity 1 year after hatching and surviving less than 4 years in the wild (Meredith and Lotrich 1979; Valiela et al. 1977). A short generation turnover allows for the more recent introductions of anthropogenic contamination to affect many generations of a population in a short amount of time, thus facilitating population adaptation (Bickham 2011). In addition to their life history advantages, *Fundulus* also have a well-documented embryonic development, which allows for teratogenesis to be easily observed (Armstrong and Child 1965). Thus, as a very suitable and environmentally relevant model species, *Fundulus* have become an exemplar in evolutionary toxicology.

There have been multiple aquatic species found to adapt to the chronic toxicity of anthropogenic compounds (Oziolor and Matson 2015; Oziolor et al. 2016a; Yuan et al. 2006), but *Fundulus* is the most commonly studied genus. We will focus on the progression of evolutionary toxicology research on *F. heteroclitus* and *F. grandis*, how studies have improved with the advent of technology, and how research on these species can continue to improve our understanding of the long-term impacts of anthropogenic contamination on natural populations.

The first instances of population divergence in *F. heteroclitus* were found to be in response to methylmercury (meHg) contamination (Weis et al. 1981b). As meHg was found to be responsible for several teratogenic malformations during killifish development (Weis and Weis 1977), Weis et al. (1981a) used a reference population from an unpolluted site and a population from a meHg-polluted location to determine if multigenerational exposure to meHg had altered their ability to deal with that stressor. After exposing both populations early in development to meHg, they observed a lower frequency of craniofacial, cardiovascular, and skeletal deformities in embryos from the meHg-polluted site (Weis et al. 1981a). These initial findings suggested that chronic contamination in one population had caused it to adapt to resist the teratogenic effects of meHg in polluted waters. The authors also found that differences between the tolerance of adapted and reference populations were very pronounced during embryonic stages of development, but were not significant during the larval stage (Khan and Weis 1987). The same line of research led to the discovery of decreased feeding rates and possible higher susceptibility to predation in larvae of adapted populations of *Fundulus heteroclitus* (Smith and Weis 1997; Weis and Khan 1990; Weis and Weis 1995; Zhou et al. 1996). This first set of studies on *Fundulus* adaptation set the stage for future evolutionary toxicology investigations.

When adaptation to anthropogenic contamination occurs, it is often due to a selective sweep, causing an abundance of adaptive genotypes in the population exposed to contamination (Bickham 2011). The primary concern after such an event is to understand the selective pressure driving the selective sweep. The investigation of meHg tolerance in *Fundulus* shows that tolerant larvae are more tolerant to meHg than Hg itself (Weis and Khan 1990). This is opposite to the toxicity of those compounds to reference populations, thus suggesting that the selective pressure that caused adaptation was likely meHg (Weis and Khan 1990). Further, such a selective process can cause genetic bottlenecks and can possibly reduce the fitness of adapted populations in the absence of the selective stressor (Bickham 2011). In this case, *Fundulus* larvae were observed to be more susceptible to predation and less proficient at catching prey, suggesting that the adaptive process may have resulted in other fitness costs (Smith and Weis 1997; Weis and Khan 1990; Zhou et al. 1996). To further understand the full impacts of this adaptation on multiple systems, the mechanism of adaptation needed to be investigated. In this case, the authors attempted to link tolerance to metallothionein, a protein responsible for sequestration of metals in the cell, but there were no differences between the levels of this enzyme in reference and adapted populations (Weis 1984). We use this initial and more simple adaptation to contamination example in *Fundulus* to show you the origins of adaptation studies in these species. These studies have served as a building block to expand and improve on evolutionary toxicology investigations in this group of fish. From this point on, we will present evolutionary studies in *Fundulus* populations that have been subjected to complex contaminant mixtures, affecting multiple molecular pathways. These investigations are some of the most diverse and thorough studies of adaptation to anthropogenic contamination and serve as a foundation in the field of evolutionary toxicology.

14.5 The Aryl Hydrocarbon Receptor (AHR) Pathway and Development

The next set of studies of *Fundulus* adaptation focused on the resistance of embryos to the cardiac teratogenesis associated with early life exposure to aryl hydrocarbon receptor (AHR) agonists. In fish, the AHR has undergone duplication events, and in some fish it is split into four isoforms—AHR1a/b and AHR2a/b (Aluru et al. 2015; Karchner et al. 2005; Karchner et al. 1999). AHR2 is a cytosolic receptor that readily binds to DLCs such as PCDD/Fs, PCBs, and PAHs (Billiard et al. 2006; Fukunaga et al. 1995; Hankinson 1995; Karchner et al. 1999, 2002; Schmidt and Bradfield 1996; Van den Berg et al. 1998). Before AHR2 complexes with a DLC, it is also bound by heat shock protein (HSP) 90, p23 and aryl hydrocarbon receptor-interacting protein (AIP), and following DLC binding it dissociates from these factors and moves across the nuclear membrane (Hahn 2001, 2002; Karchner et al. 2002). Once inside the nucleus, AHR2 binds with aryl hydrocarbon nuclear translocator (ARNT) and induces the expression of multiple genes to initiate biotransformation and ultimately excretion of the contaminant from the cell (Hahn 2001, 2002; Karchner et al. 2002). A number of these inducible genes are cytochrome p450s (*cyps*) and in particular *CYP1A*, which codes for a monooxygenase responsible for initiating biotransformation of aryl hydrocarbons (Stegeman 1993). When constitutive expression of this pathway is present in developing fish embryos, generally resulting from DLC exposure, it can lead to cardiac teratogenesis, which alters proper development and prevents the embryos from successfully hatching (Incardona et al. 2004; Matson et al. 2008a). It is still unclear whether constitutive AHR expression in the absence of a xenobiotic can induce cardiac teratogenesis. However, multiple populations of both *Fundulus grandis* and *Fundulus heteroclitus* from highly DLC-polluted environments have been shown to be resistant to the cardiac teratogenesis which these compounds induce in reference *Fundulus* populations (Elskus et al. 1999; Nacci et al. 1999; Ownby et al. 2002; Oziolor et al. 2014). We will focus on these studies to illustrate the types of approaches used to study evolutionary toxicology.

14.6 Finding and Characterizing Resistance

As the AHR pathway is responsible for the biotransformation and elimination of many anthropogenic contaminants (e.g., DLCs), it is thought to be protective of the negative effects of these compounds. Contrary to that notion, knockdown of this pathway actually protects from the teratogenic effects of DLCs in fish (Carney et al. 2004; Clark et al. 2010). Interestingly, a recalcitrant AHR pathway has been documented in *Fundulus* populations collected from contaminated waters in five separate locations and two different species (Elskus et al. 1999; Nacci et al. 1999; Oziolor et al. 2014; Van Veld and Westbrook 1995; Whitehead et al. 2012). These include some of the first studies showing that wild populations of *Fundulus* had adapted on a molecular level to resist the negative effects of DLCs. Many studies began by documenting depressed CYP1A activity (downstream-regulated enzyme of AHR pathway) in F_1 embryos from wild-caught fish (Meyer and Di Giulio 2002; Nacci et al. 1999; Oziolor et al. 2014) and correlating this depressed signal with higher resistance to cardiac teratogenesis when embryos were exposed to DLCs. In combination with the understanding that AHR suppression leads to protection, this suggests population-wide adaptation in response to chronic contamination in these populations. However, it does not conclusively confirm evolution.

To more definitively point toward a genetic and heritable explanation of the AHR recalcitrance and resistance to DLCs in *Fundulus*, researchers utilized several complementary approaches. In order to rule out maternal depuration of mRNA or DLC offloading into the embryo, which could alter their response to further DLC challenge, researchers performed reciprocal crosses (Meyer and Di Giulio 2003; Meyer et al. 2002; Nacci et al. 1999; Oziolor et al. 2014). They crossed resistant females with reference males, and vice versa, and tested their AHR responsiveness and resistance to DLCs (Meyer and Di Giulio 2003; Meyer et al. 2002; Nacci et al. 1999; Oziolor et al. 2014). By showing that both parents contribute equally to the resistance and recalcitrance of AHR, the authors confirmed that these traits were biparentally inherited and likely genetic. In order to further reduce the possibility of this being an epigenetic trait, authors raised F_1 embryos from resistant populations in clean conditions and used F_2 generation offspring for the same DLC challenges (Clark et al. 2013a; Meyer and Di Giulio 2002; Nacci et al. 1999). This is the first aspect in which adapted populations from separate environments differ. In northern adapted populations from New Bedford Harbor (NBH) (exposed to PCBs), the resistance to cardiac teratogenesis and recalcitrance of the AHR pathway are strongly inherited in F_2 embryos (Nacci et al. 1999), while southern adapted populations from the Atlantic Woods (AW) (exposed to PAHs) are more complicated. Ownby et al. (2002) observed no loss of resistance in F_2 embryos, whereas other studies suggest that F_2 and F_3 generations of AW fish lose their resistance and the recalcitrance of the AHR pathway (Meyer and Di Giulio 2002, 2003; Meyer et al. 2002). Other studies have gone on to suggest a disconnect in which resistance to a DLC challenge is sustained in F_2 generations, but the recalcitrance of the AHR pathway is reduced (Clark et al. 2013a). The consensus from these studies is that because of the biparental inheritance and likely multigenerational heritability, the resistant phenotype in *Fundulus* populations in contaminated environments is likely to be genetic and the pathways involved in it merit further exploration.

14.7 Altered Pathways and Systems

The first step toward understanding the impacts of adaptation on populations is by exploring the altered pathways. The NBH-adapted killifish population did not show differences in their basal expression of AHR pathway components (AHR2 and ARNT). On the other hand, this population showed a difference in AHR1, which was more broadly expressed in wild-caught resistant populations, but not in their offspring (Powell et al. 2000). Further analyses of the sensitivity of the AHR pathway in NBH fish revealed a recalcitrance to the induction of CYP1A activity and protein content in multiple tissues in NBH embryos and adults challenged with a DLC (Bello et al. 2001). While CYP1A activity and protein levels were suppressed, they were still inducible in NBH, while AW fish showed almost no inducible expression of CYP1A when challenged with a DLC (Meyer et al. 2003b). Similarly, CYP1A activity in *Fundulus grandis* populations from the Houston Ship Channel (HSC) (exposed to PCBs, PCDD/Fs, and PAHs) was highly suppressed when presented with PCB challenge but less suppressed when the challenge was a PAH mixture (Oziolor et al. 2014). In addition to *CYP1A* mRNA, AW fish also presented with repressed expression of aryl hydrocarbon receptor repressor (AHRR), CYP1B1, and CYP1C1 when presented with DLC challenges (Meyer et al. 2003b; Wills et al. 2010b). However, the expression of AHR2 itself was not altered in this adapted phenotype (Wills et al. 2010b). These results show that in both NBH and AW populations, the responses of multiple downstream-regulated members of the AHR pathway are recalcitrant. This finding is supported by the fact that AHR2 knockdown provided reference population embryos significant protection from the cardiovascular teratogenic impacts of DLC challenge, while CYP1A knockdown increased their sensitivity (Clark et al. 2010; Matson et al. 2008a). These studies support the likely role of suppression of the entire AHR pathway in the observed protection from the developmental cardiovascular deformities associated with DLC exposures in wild populations of *Fundulus* but still leave the mechanism of this suppression unknown.

14.8 Selection Driving Forces

An important aspect to searching for the molecular signatures of adaptation to contaminants is an understanding of the selective pressures that toxicants in the environment must present in order to drive adaptation. Despite the fact that many DLCs have a common mode of toxicity (the induction of the AHR pathway), their presence in complex mixtures in the environment complicates the question of which contaminants drive selective pressure. Studies of several *Fundulus* populations have concluded that there is a relationship between level of contamination in the environment and the populations' level of resistance to the negative impacts of that contamination (Nacci et al. 2002a, 2010; Oziolor and Matson 2015). Such correlations are difficult to draw because of the complexity of contamination present at most locations. With complex PCB mixtures present in the HSC (Howell et al. 2011; Lakshmanan et al. 2010; Oziolor et al. 2014) and NBH (Nacci et al. 2010), and a complex PAH mixture at AW (Clark et al. 2013b), the disparity between likely selective pressures becomes obvious. The diversity of contaminant concentrations among adapted sites can also be explanatory of some of the differences in the way molecular pathways have been affected in those populations.

When there is a disparity in chemical complexity, it is important to understand the effects of the predominant classes of contaminants on the adapted population to further elucidate possible mechanisms through which they could drive selection. When considering PCBs as a selective pressure, their DLC properties are one of their dominant features. Their DLC features explain why correlations between "total PCB" levels at contaminated sites and sensitivity to model DLC compounds of populations at those sites are very often drawn to explain selective potential (Nacci et al. 2010). On the other hand, some PCBs have a strong estrogenic potential, an area of research which has been explored recently as a focal area of investigation into adaptive responses (Greytak et al. 2010). It was determined that NBH-adapted populations have a lower aromatase (AroB or CYP19A) and estrogen receptor α (ERα) response to estrogen potentiation (Greytak et al. 2010). This difference led the researchers to investigate the architecture of the *esr 1* gene, which codes for ERα, and found two deletions specific to the NBH-adapted populations (Cotter et al. 2015). These deletions had a significant impact on two major splice variants of ERα, which explained lower responsiveness of this receptor to estrogen (Cotter et al. 2016). Even though the sensitivity of adapted populations recovered after several generations, this is likely adaptation, which allowed populations to maintain proper estrogen signaling in the presence of heavy estrogenic contamination (Cotter et al. 2016). Lastly, some PCB congeners are non-coplanar, which does not give them DLC properties. However, these non-coplanar PCBs can induce other pathways in fish. For example, researchers identified increased responsiveness, but lower basal levels, of pregnane X receptor (PXR), CYP3A, and P-glycoprotein in NBH populations of *F. heteroclitus* (Grans et al. 2015). These were hypothesized to cross-react with the altered AHR pathway in adapted fish.

The class of contaminants more predominant at AW are PAHs, and they have a highly divergent set of chemical properties. Like PCBs, they can cause cardiac teratogenesis through the activation of the AHR pathway (Incardona et al. 2004), but they can also cause chromosomal damage (through DNA adducts or oxidative stress) (Ashurst et al. 1983; Harbeitner et al. 2013; Jung et al. 2011; Nacci et al. 2002b). The following discussion will begin by exploring the differences in sensitivity of adapted populations of *Fundulus* in AHR activation in response to PAHs. The other effects of these complex mixtures will be discussed further below.

Primary investigations revealed that PAH and PCB dual exposures do not produce an additive effect in either causing cardiac teratogenesis or inducing the AHR pathway (Wassenberg et al. 2002). On the other hand, when a PAH acting as a CYP1A inducer was combined with a PAH that is a posttranslational CYP1A inhibitor, it caused a synergistic increase in cardiac teratogenesis in *F. heteroclitus* (Wassenberg and Di Giulio 2004a). In addition, a CYP1A inhibiting PAH added to even a complex mixture of PAH-contaminated sediment-derived pore-water caused a similar synergistic effect (Wassenberg and Di Giulio 2004b). These studies suggest that the composition and ratios of PAHs in contaminated environments can have a strong and nonadditive influence on the level of teratogenesis (Billiard et al. 2008; Wassenberg et al. 2005). Thus, when dealing with complex PAH mixtures, we need to understand their composition and ratios in the environment. The ratios and level of contamination in AW site showed large levels of both CYP1A inducing and inhibitory PAHs, which suggests that the teratogenic potential at that site is very strong (Jung et al. 2011). Studies of AW F_1 embryos also show that they are resistant to the synergistic toxicity observed in reference killifish (Wills et al. 2009).

Understanding selective pressures and the complex contamination profiles of aquatic environments is not an easy task. In addition, given the geographic breadth of *Fundulus* adaptation, it becomes even more important to be able to compare the adaptive responses and selective pressures among adapted populations. Many studies have relied on information derived for the polluted sites by environmental chemists, who are interested in sources and complexity of contamination (Howell et al. 2008, 2011; Lakshmanan et al. 2010), as well as contractors hired to monitor a known contaminated site (Anchor 2010). A recent study showed that there is a disparity between the congeners studied for each location and thus suggested that for better comparability, a standard group of congeners should be included (Oziolor et al. 2016a). By understanding the chemical composition and the effects of the classes of contaminants present at contaminated sites, we can better predict the potential for adaptation in wild populations (Nacci et al. 2002a). By comparing multiple environments and populations, we can increase the accuracy of such prediction (Oziolor et al. 2016a).

14.9 Fitness Costs and Cross-Resistances

The occurrence of the evolution of resistance to anthropogenic contamination reveals that these environments have imposed strong selective pressure on resident *Fundulus* populations. There are several ways that such selection can impose threats to the future health of populations that have undergone evolutionary rescue:

(1) Genetic bottlenecks reduce genetic diversity, leaving a population with a lower capacity to deal with future stressors.
(2) Genetic hitchhiking can cause genes that are physically linked to adaptive genotypes, but not adaptive themselves, to be overrepresented in populations.
(3) Changes in molecular pathway function can impair interacting pathways.
(4) Altered functionality of the AHR pathway itself may be a cost in the absence of the selective pressure that caused the fixation of genes coding for a nonresponsive pathway.

A significant challenge in studying these fitness costs stems from the fact that the polluted habitats that we are studying are never ecologically identical to reference sites. Often, polluted sites are closer to urban environments, which are more frequently impacted by eutrophication, hypoxia, and lower salinity in addition to anthropogenic contamination (Johnson and Munshi-South 2017). With these differences in mind, we will explore some of the fitness costs and cross-resistances studied in adapted *Fundulus* populations.

DLCs have the potential of causing oxidative stress in fish populations. Cells respond to some congeners by uncoupling CYP1A metabolism (inhibiting CYP1A activity, while still activating the AHR pathway) of those compounds and activating a response to oxidative stress—*Nrf2* (Schlezinger et al. 2006). Given this potential, studies of AW-adapted populations aimed to understand whether the chronic exposure of killifish to PAH contamination has driven a selective process resulting in them being able to resist oxidative damage. Studies identified a higher resistance to oxidative stress, as well as more responsive phase II biotransformation enzymes, also involved in mediating oxidative stress (Bacanskas et al. 2004; Meyer et al. 2003a). Similar investigations in HSC adapted Gulf killifish populations also showed an increased resistance to oxidative stress compared to reference populations (Oziolor et al. 2016b). On the other hand, early life exposures in NBH killifish show that adapted populations were more susceptible to oxidative stress during early developmental windows (Harbeitner et al. 2013). These studies may suggest that oxidative stress pathways are either affected differentially in adapted *Fundulus* populations or that the molecular interactions between the AHR and Nrf pathways are affecting their responsiveness to oxidative stress in these populations variably during different developmental stages.

When metabolized, PAHs can elicit additional toxicity through the formation of DNA adducts or induce formation of reactive oxygen species (ROS), resulting in DNA damage in cells (Ashurst et al. 1983). Studies comparing wild-caught AW adapted and reference populations revealed that basal DNA damage in adapted fish was significantly higher (Jung et al. 2009; Jung et al. 2011). On the other hand, wild-caught adults challenged with benzo[a]pyrene (BaP), whose metabolite (BaP diolepoxide (BPDE)) is known to form DNA adducts, revealed that adapted AW population fish were less susceptible to further DNA damage induction (Jung et al. 2009). The same resistance to DNA damage induction was found in F_1 generation larvae from adapted populations (Jung et al. 2009). Interestingly, the basal level of BPDE adducts in the AW populations was elevated (Jung et al. 2011). This higher basal DNA damage, higher level of BPDE adducts, but protection from further damage could be suggestive that in order to deal with an increased level of DNA damaging PAHs in the field, AW populations have increased DNA repair mechanisms. This was also suggested by the finding that the AW-adapted population was less likely to develop hepatic lesions following early life exposure to BaP, a known carcinogen (Wills et al. 2010a). On the other hand, PAHs are still metabolized in resistant populations of *F. heteroclitus* (Wills et al. 2009). Embryos from resistant populations show a lower level of transformation of BaP but also reveal a higher production of the 9,10-diol BaP metabolite (Wills et al. 2009). Thus, the protection from DNA damage observed in adapted killifish may be a result of both lower rate of biotransformation and also differential rate of production of specific harmful PAH metabolites.

Observed AHR recalcitrance would be predicted to increase the susceptibility of adapted populations to other anthropogenic contaminants that depend on the AHR pathway for detoxification (Clark and Di Giulio 2012; Oziolor et al. 2016b; Tang et al. 2002). When AW-adapted population embryos and larvae were challenged with pesticides that are either bioactivated or detoxified by CYP1A (as part of the AHR pathway), they showed higher resistance to both (Clark and Di Giulio 2012). It is logical that a contaminant bioactivated by CYP1A might have lower toxicity when CYP1A is downregulated, but it is mechanistically unclear why compounds that are detoxified by CYP1A are less toxic to resistant fish (Clark and Di Giulio 2012; Oziolor et al. 2016b). The likely explanation for this resistance is that there are multiple pathways capable of detoxifying these pesticides, and while CYP1A is the primary enzyme, other enzymes could be upregulated in adapted populations as a compensatory mechanism for the downregulation of the AHR and CYP1A (Clark and Di Giulio 2012; Oziolor et al. 2016b; Tang et al. 2002). These examples illustrate that there can be multiple pathways altered by the adaptation to anthropogenic contamination in *Fundulus* populations, which allow them to thrive and deal with multiple classes of stressors, sometimes against mechanistic hypotheses predicting fitness costs.

Another possible fitness cost exploration stems from the interaction between the AHR pathway and the hypoxia inducible factor (HIF). HIF is a protein regulated by O_2 tension and is designed to increase the ability of an organism to tolerate hypoxic conditions by increasing angiogenesis, anaerobic respiration, vasodilation, and erythropoiesis (Wang et al. 1995). As part of its normal response to hypoxia, HIF binds with ARNT to further bind DNA and induce the expression of a set of genes (Chan et al. 1999). Thus, AHR and HIF share ARNT as a binding partner, which has been seen to decrease the inducibility of the AHR pathway (Chan et al. 1999). It has been shown in zebrafish that some PAHs can interact with hypoxia resulting in severe teratogenesis (Matson et al. 2008b). In addition, hypoxia has been shown to reduce the potential of the AHR to respond to contaminant exposures (Fleming et al. 2008). If the reduction in AHR responsiveness in adapted populations is mediated by ARNT modification, this could lead to a lower ability of these populations to deal with hypoxic stress. Initial studies suggested that multiple generations of adapted AW populations were more susceptible to hypoxic challenge compared to a reference population (Meyer and Di Giulio 2003). On the other hand, studies of HSC Gulf killifish populations showed an increased resting metabolic rate and lower ability to deal with hypoxic stress of adapted populations compared to one reference, but not to a second reference population (Oziolor et al. 2016b). In these tests, one reference population had a much longer time to loss of equilibrium (LOE) due to hypoxic exposure, while the other reference population reached LOE at the same time as xenobiotic-resistant populations (Oziolor et al. 2016b). This suggests that in Gulf killifish, hypoxic responses between populations were likely affected by variability in an alternate environmental factor, rather than by the adaptation to anthropogenic pollution (Oziolor et al. 2016b). Thus, it can be quite difficult to accurately identify a fitness cost or cross-resistance to a stressor in adapted populations without either molecular evidence or specific understanding of the environmental variables that could be acting on the trait in different populations.

14.10 Current and Future Directions: Omics

The complicated mechanistic basis of adaptation in *Fundulus* has led to investigations utilizing transcriptomic and genomic approaches. Some of the initial investigations attempted to observe changes in gene expression in wild-caught populations of adapted *Fundulus*. While there were some significant differences in transcription between an adapted population and its proximate reference population (Meyer et al. 2005), there were few genes that were consistently differentially expressed in geographically separated adapted populations (Oleksiak 2008), and their function in conferred protection was not obvious. While transcriptomic studies were not able to establish any large differences in normal development in adapted populations (Bozinovic and Oleksiak 2010, 2011), when embryos were challenged with AHR agonists, there were a small but significant percentage of genes that were differentially expressed during development (Bozinovic et al. 2013; Oleksiak et al. 2011). On the other hand, analysis of killifish from three separate adapted populations challenged with a DLC revealed a common mechanism of adaptation in all three populations (Whitehead et al. 2012). This study shows that despite some transcriptional differences, a common suppression of the expression of AHR-related genes is observed in all populations and in multiple generations (Whitehead et al. 2012). Thus, transcriptomic studies suggest that the responsiveness of a few genes during a developmental challenge and in adults could be explanatory of the resistance of embryos to cardiac teratogenesis. However, they may also suggest that normal development is not affected by this adaptation in chronically polluted populations.

To further understand the aberrant transcriptomic response, studies attempted to find genes under selection in adapted populations of *F. heteroclitus*. One study used amplified fragment length polymorphisms (AFLP) to identify signatures of selection in adapted populations, but they did not find similar AFLPs responsible for those signatures between the three examined adapted populations (Williams and Oleksiak 2008). Further investigation was able to identify stronger signatures of selection using single nucleotide polymorphisms (SNPs) from both coding and noncoding regions (Williams and Oleksiak 2011a). Some of the SNPs associated with the AHR pathway (in particular *CYP1A*) were commonly shared among all three adapted populations (Williams and Oleksiak 2011a). More targeted approaches to follow up on the genetic signatures of selection found in these experiments led researchers to explore AHR-related genes (*ahr1*, *ahr2*, *ahrr*, and *CYP1A*) to search for variability that explained the refractory AHR phenotype. There was no strong evidence of reduced variability in the *CYP1A* promoter region (Williams and Oleksiak 2011b), but on the other hand, authors demonstrated strong genetic structure in all three other loci. In addition, there were SNPs in AHR2 that were suggested to be under selection and strongly associated with a resistant phenotype (Reitzel et al. 2014).

With great strides and lower costs in omics, current investigations have provided a reference genome for *F. heteroclitus* and are underway in *F. grandis* (Reid et al. 2017). While studies of genomic variation have shown no bottlenecks in resistant populations and have identified regions of selection, the genetic variation underlying resistance in *F. heteroclitus* was only recently uncovered through low-coverage population resequencing. A study by Reid et al. (2016) recently explored the genomic variation in four pairs of *F. heteroclitus* populations with differing contaminant exposure histories. The study identified regions of high divergence between populations with acquired resistance to contaminant effects and revealed low nucleotide diversity around those loci, consistent with rapid selective sweeps. In addition to revealing putative targets of selection that are related to estrogen regulation, ryanodine receptor function, and immune system modulation, Reid et al. (2016) identified a set of deletions in AHR1 and AHR2 that have swept almost to fixation in one resistant population, and found at low frequencies in the other resistant populations. While this deletion likely prevents the AHR pathway from functioning properly, multiple tolerant populations also had increased copy number of the downstream-regulated gene for CYP1A. It is possible that an increase of copy number of this gene is a possibly compensatory mechanism to allow basal expression of *CYP1A*, which may be necessary for some endogenous processes. One of the strongest advantages of the approach taken by Reid et al. (2016) is also the top-to-bottom approach that identifies regions of selection, independent of the main pathway affected. In this manner, this study has identified targets that can inform future investigations into possible fitness costs and broader impacts of anthropogenic selection in *Fundulus*. This study was the first to identify the specific mechanism of protection in resistant *Fundulus* populations and provides an idea of the landscape of selection that gave rise to this adaptation. Once resistant genotypes are known, the alterations to genes can be experimentally replicated using the already established CRISPR-Cas assay in *F. heteroclitus* (Aluru et al. 2015). Verifying experimentally that the altered genes confer protection in reference fish will allow us to examine potential fitness costs and cross-resistances, independent of environmental influence.

Conclusion

The *Fundulus* adaptation story is complicated but is also one of the most well-studied instances of evolution in response to anthropogenic contamination. The studies started by documenting resistance to several classes of anthropogenic contaminants in multiple populations of two *Fundulus* species. They linked this adaptation to a specific molecular pathway and investigated the likely effects and co-affected pathways. Investigations also focused on the properties of chemical contaminants found in those environments and how they could explain variable responses between resistant and reference populations. Possible fitness costs and cross-resistances were identified in resistant populations, but some were linked to environmental variables rather than the adapted phenotype. Lastly, omics approaches have been used to confirm that a common mechanism of adaptation was responsible for conferring resistance in adapted populations: recalcitrance of the AHR pathway caused by mutations in target AHR pathway related genes, as well as a deletion in AHR1/AHR2 in adapted populations. The investigations presented in this case study can serve as an example of a logical and hypothesis-driven succession of experiments that get progressively closer to answering questions about the effects of anthropogenic compounds on natural populations and our ability to predict their occurrence and consequences.

References

Adams SM, Lindmeier JB, Duvernell DD (2006) Microsatellite analysis of the phylogeography, Pleistocene history and secondary contact hypotheses for the killifish, Fundulus heteroclitus. Mol Ecol 15:1109–1123

Alldredge P, Gutierrez M, Duvernell D, Schaefer J, Brunkow P, Matamoros W (2011) Variability in movement dynamics of topminnow (Fundulus notatus and F. olivaceus) populations. Ecol Freshw Fish 20:513–521

Aluru N, Karchner SI, Franks DG, Nacci D, Champlin D, Hahn ME (2015) Targeted mutagenesis of aryl hydrocarbon receptor 2a and 2b genes in Atlantic killifish (Fundulus heteroclitus). Aquat Toxicol 158:192–201

Amsterdam A, Nissen RM, Sun ZX, Swindell EC, Farrington S, Hopkins N (2004) Identification of 315 genes essential for early zebrafish development. Proc Natl Acad Sci U S A 101:12792–12797

Anchor QEA (2010) Sediment and surface water contamination of potential concern delineation data report Patrick bayou superfund site, Deer Park

Armstrong PB, Child JS (1965) Stages in normal development of *Fundulus heteroclitus*. Biol Bull 128:143

Ashurst SW, Cohen GM, Nesnow S, Digiovanni J, Slaga TJ (1983) Formation of benzo[a]pyrene DNA adducts and their relationship to tumor initiation in mouse epidermis. Cancer Res 43:1024–1029

Auld JR, Agrawal AA, Relyea RA (2010) Re-evaluating the costs and limits of adaptive phenotypic plasticity. Proc R Soc B Biol Sci 277:503–511

Bacanskas LR, Whitaker J, Di Giulio RT (2004) Oxidative stress in two populations of killifish (*Fundulus heteroclitus*) with differing contaminant exposure histories. Mar Environ Res 58:597–601

Bello SM, Franks DG, Stegeman JJ, Hahn ME (2001) Acquired resistance to Ah receptor agonists in a population of Atlantic killifish (*Fundulus heteroclitus*) inhabiting a marine superfund site: in vivo and in vitro studies on the inducibility of xenobiotic metabolizing enzymes. Toxicol Sci 60:77–91

Berg MP, Kiers ET, Driessen G, van der Heijden M, Kooi BW, Kuenen F, Liefting M, Verhoef HA, Ellers J (2010) Adapt or disperse: understanding species persistence in a changing world. Glob Change Biol 16:587–598

Bernardi G, Fernandez-Delgado C, Gomez-Chiarri M, Powers DA (1995) Origin of a Spanish population of Fundulus heteroclitus inferred by cytochrome b sequence analysis. J Fish Biol 47:737–740

Bickham JW (2011) The four cornerstones of evolutionary toxicology. Ecotoxicology 20:497–502

Bickham JW, Sandhu S, Hebert PDN, Chikhi L, Athwal R (2000) Effects of chemical contaminants an genetic diversity in natural populations: implications for biomonitoring and ecotoxicology. Mutat Res 463:33–51

Bickham JW, Smolen MJ (1994) Somatic and heritable effects of environmental genotoxins and the emergence of evolutionary toxicology. Environ Health Perspect 102:25–28

Billiard SM, Meyer JN, Wassenberg DM, Hodson PV, Di Giulio RT (2008) Nonadditive effects of PAHs on early vertebrate development: mechanisms and implications for risk assessment. Toxicol Sci 105:5–23

Billiard SM, Timme-Laragy AR, Wassenberg DM, Cockman C, Di Giulio RT (2006) The role of the aryl hydrocarbon receptor pathway in mediating synergistic developmental toxicity of polycyclic aromatic hydrocarbons to zebrafish. Toxicol Sci 92:526–536

Boughman JW (2001) Divergent sexual selection enhances reproductive isolation in sticklebacks. Nature 411:944–948

Bozinovic G, Oleksiak MF (2010) Embryonic gene expression among pollutant resistant and sensitive Fundulus heteroclitus populations. Aquat Toxicol 98:221–229

Bozinovic G, Oleksiak MF (2011) Genomic approaches with natural fish populations from polluted environments. Environ Toxicol Chem 30:283–289

Bozinovic G, Sit TL, Di Giulio R, Wills LF, Oleksiak MF (2013) Genomic and physiological responses to strong selective pressure during late organogenesis: few gene expression changes found despite striking morphological differences. BMC Genomics 14:15

Burkhart JG, Helgen JC, Fort DJ, Gallagher K, Bowers D, Propst TL, Gernes M, Magner J, Shelby MD, Lucier G (1998) Induction of mortality and malformation in Xenopus laevis embryos by water sources associated with field frog deformities. Environ Health Perspect 106:841–848

Burnett KG, Bain LJ, Baldwin WS, Callard GV, Cohen S, Di Giulio RT, Evans DH, Gomez-Chiarri M, Hahn ME, Hoover CA, Karchner SI, Katoh F, MacLatchy DL, Marshall WS, Meyer JN, Nacci DE, Oleksiak MF, Rees BB, Singer TD, Stegeman JJ, Towle DW, Van Veld PA, Vogelbein WK, Whitehead A, Winn RN, Crawford DL (2007) *Fundulus* as the premier teleost model in environmental biology: opportunities for new insights using genomics. Comp Biochem Physiol D Genomics Proteomics 2:257–286

Carney SA, Peterson RE, Heideman W (2004) 2,3,7,8-tetrachlorodibenzo-p-dioxin activation of the aryl hydrocarbon receptor/aryl hydrocarbon receptor nuclear translocator pathway causes developmental toxicity through a CYP1A-independent mechanism in zebrafish. Mol Pharmacol 66:512–521

Chan WK, Yao G, Gu YZ, Bradfield CA (1999) Cross-talk between the aryl hydrocarbon receptor and hypoxia inducible factor signaling pathways – demonstration of competition and compensation. J Biol Chem 274:12115–12123

Chown SL, Gaston KJ (2008) Macrophysiology for a changing world. Proc R Soc B Biol Sci 275:1469–1478

Clark BW, Bone AJ, Di Giulio RT (2013a) Resistance to teratogenesis by F1 and F2 embryos of PAH-adapted Fundulus heteroclitus is strongly inherited despite reduced recalcitrance of the AHR pathway. Environ Sci Pollut Res Int 21(24):13898–13908

Clark BW, Cooper EM, Stapleton HM, Di Giulio RT (2013b) Compound- and mixture-specific differences in resistance to polycyclic aromatic hydrocarbons and PCB-126 among *Fundulus heteroclitus* subpopulations throughout the Elizabeth River Estuary (Virginia, USA). Environ Sci Technol 47:10556–10566

Clark BW, Di Giulio RT (2012) *Fundulus heteroclitus* adapted to PAHs are cross-resistant to multiple insecticides. Ecotoxicology 21:465–474

Clark BW, Matson CW, Jung D, Di Giulio RT (2010) AHR2 mediates cardiac teratogenesis of polycyclic aromatic hydrocarbons and PCB-126 in Atlantic killifish (*Fundulus heteroclitus*). Aquat Toxicol 99:232–240

Coleman ML, Chisholm SW (2010) Ecosystem-specific selection pressures revealed through comparative population genomics. Proc Natl Acad Sci 107:18634–18639

Cotter KA, Nacci D, Champlin D, Chuprin J, Callard GV (2015) Cloning of multiple ERalpha mRNA variants in killifish (Fundulus heteroclitus), and differential expression by tissue type, stage of reproduction, and estrogen exposure in fish from polluted and unpolluted environments. Aquat Toxicol 159:184–197

Cotter KA, Nacci D, Champlin D, Yeo AT, Gilmore TD, Callard GV (2016) Adaptive significance of ER alpha splice variants in killifish (Fundulus heteroclitus) resident in an estrogenic environment. Endocrinology 157:2294–2308

De Coninck DIM, Janssen CR, De Schamphelaere KAC (2014) An approach to assess the regulatory relevance of microevolutionary effects in ecological risk assessment of chemicals: a case study with cadmium. Environ Toxicol Chem 33:453–457

Di Giulio RT, Clark BW (2015) The Elizabeth River story: a case study in evolutionary toxicology. J Toxicol Env Health B Crit Rev 18:259–298

Elskus AA, Monosson E, McElroy AE, Stegeman JJ, Woltering DS (1999) Altered CYP1A expression in Fundulus heteroclitus adults and larvae: a sign of pollutant resistance? Aquat Toxicol 45:99–113

Fleming CR, Matson CW, Hinton DE, Di Giulio RT (2008) Developmental and molecular interactions between hypoxia and aryl hydrocarbon receptor (AHR) pathways in Zebrafish (Danio rerio) and Atlantic killifish (Fundulus heteroclitus). Mar Environ Res 66:72–72

Fukunaga BN, Probst MR, Reisz-Porszasz S, Hankinson O (1995) Identification of functional domains of the aryl hydrocarbon receptor. J Biol Chem 270:29270–29278

Gilbert SF (2001) Ecological developmental biology: developmental biology meets the real world. Dev Biol 233:1–12

Gonzalez I, Levin M, Jermanus S, Watson B, Gilg MR (2009) Genetic assessment of species ranges in *Fundulus heteroclitus* and *F. grandis* in northeastern Florida salt marshes. Southeast Nat 8:227–244

Grans J, Wassmur B, Fernandez-Santoscoy M, Zanette J, Woodin BR, Karchner SI, Nacci DE, Champlin D, Jayaraman S, Hahn ME, Stegeman JJ, Celander MC (2015) Regulation of pregnane-X-receptor, CYP3A and P-glycoprotein genes in the PCB-resistant killifish (Fundulus heteroclitus) population from New Bedford Harbor. Aquat Toxicol 159:198–207

Greytak SR, Tarrant AM, Nacci D, Hahn ME, Callard GV (2010) Estrogen responses in killifish (*Fundulus heteroclitus*) from polluted and unpolluted environments are site- and gene-specific. Aquat Toxicol 99:291–299

Hahn ME (2001) Dioxin toxicology and the aryl hydrocarbon receptor: insights from fish and other non-traditional models. Mar Biotechnol 3:S224–S238

Hahn ME (2002) Aryl hydrocarbon receptors: diversity and evolution. Chem Biol Interact 141:131–160

Hankinson O (1995) The aryl-hydrocarbon receptor complex. Annu Rev Pharmacol Toxicol 35:307–340

Harbeitner RC, Hahn ME, Timme-Laragy AR (2013) Differential sensitivity to pro-oxidant exposure in two populations of killifish (*Fundulus heteroclitus*). Ecotoxicology 22:387–401

Hawkins RD, Hon GC, Ren B (2010) Next-generation genomics: an integrative approach. Nat Rev Genet 11:476–486

Howell NL, Rifai HS, Koenig L (2011) Comparative distribution, sourcing, and chemical behavior of PCDD/Fs and PCBs in an estuary environment. Chemosphere 83:873–881

Howell NL, Suarez MP, Rifai HS, Koenig L (2008) Concentrations of polychlorinated biphenyls (PCBs) in water, sediment, and aquatic biota in the Houston Ship Channel, Texas. Chemosphere 70:593–606

Incardona JP, Collier TK, Scholz NL (2004) Defects in cardiac function precede morphological abnormalities in fish embryos exposed to polycyclic aromatic hydrocarbons. Toxicol Appl Pharmacol 196:191–205

Johnson MTJ, Munshi-South J (2017) Evolution of life in urban environments. Science 358(6363):eaam8327

Jung D, Cho Y, Collins LB, Swenberg JA, Di Giulio RT (2009) Effects of benzo a pyrene on mitochondrial and nuclear DNA damage in Atlantic killifish (*Fundulus heteroclitus*) from a creosote-contaminated and reference site. Aquat Toxicol 95:44–51

Jung DW, Matson CW, Collins LB, Laban G, Stapleton HM, Bickham JW, Swenberg JA, Di Giulio RT (2011) Genotoxicity in Atlantic killifish (*Fundulus heteroclitus*) from a PAH-contaminated superfund site on the Elizabeth River, Virginia. Ecotoxicology 20:1890–1899

Karchner SI, Franks DG, Hahn ME (2005) AHR1B, a new functional aryl hydrocarbon receptor in zebrafish: tandem arrangement of *ahr1b* and *ahr2* genes. Biochem J 392:153–161

Karchner SI, Franks DG, Powell WH, Hahn ME (2002) Regulatory interactions among three members of the vertebrate aryl hydrocarbon receptor family: AHR repressor, AHR1, and AHR2. J Biol Chem 277:6949–6959

Karchner SI, Powell WH, Hahn ME (1999) Identification and functional characterization of two highly divergent aryl hydrocarbon receptors (AHR1 and AHR2) in the teleost *Fundulus heteroclitus* – evidence for a novel subfamily of ligand-binding basic helix loop helix-Per-ARNT-Sim (bHLH-PAS) factors. J Biol Chem 274:33814–33824

Kawecki TJ, Ebert D (2004) Conceptual issues in local adaptation. Ecol Lett 7:1225–1241

Keller R, Davidson LA, Shook DR (2003) How we are shaped: the biomechanics of gastrulation. Differentiation 71:171–205

Khan AT, Weis JS (1987) Effect of methylmercury on egg and juvenile viability in two populations of killifish Fundulus heteroclitus. Environ Res 44:272–278

Kimberly DA, Salice CJ (2015) Evolutionary responses to climate change and contaminants: evidence and experimental approaches. Curr Zool 61:690–701

Klerks PL, Xie L, Levinton JS (2011) Quantitative genetics approaches to study evolutionary processes in ecotoxicology; a perspective from research on the evolution of resistance. Ecotoxicology 20:513–523

Lakshmanan D, Howell NL, Rifai HS, Koenig L (2010) Spatial and temporal variation of polychlorinated biphenyls in the Houston Ship Channel. Chemosphere 80:100–112
Lenormand T (2002) Gene flow and the limits to natural selection. Trends Ecol Evol 17:183–189
Lynch M, Conery JS (2003) The origins of genome complexity. Science 302:1401–1404
Matson CW, Clark BW, Jenny MJ, Fleming CR, Hahn ME, Di Giulio RT (2008a) Development of the morpholino gene knockdown technique in *Fundulus heteroclitus*: a tool for studying molecular mechanisms in an established environmental model. Aquat Toxicol 87:289–295
Matson CW, Timme-Laragy AR, Di Giulio RT (2008b) Fluoranthene, but not benzo a pyrene, interacts with hypoxia resulting in pericardial effusion and lordosis in developing zebrafish. Chemosphere 74:149–154
Meaney MJ (2010) Epigenetics and the biological definition of gene x environment interactions. Child Dev 81:41–79
Meredith WH, Lotrich VA (1979) Production dynamics of a tidal creek population of fundulus-heteroclitus (Linnaeus). Estuar Coast Mar Sci 8:99–118
Meyer J, Di Giulio R (2002) Patterns of heritability of decreased EROD activity and resistance to PCB 126-induced teratogenesis in laboratory-reared offspring of killifish (Fundulus heteroclitus) from a creosote-contaminated site in the Elizabeth River, VA, USA. Mar Environ Res 54:621–626
Meyer JN, Di Giulio RT (2003) Heritable adaptation and fitness costs in killifish (*Fundulus beteroclitus*) inhabiting a polluted estuary. Ecol Appl 13:490–503
Meyer JN, Nacci DE, Di Giulio RT (2002) Cytochrome P4501A (CYP1A) in killifish (*Fundulus heteroclitus*): heritability of altered expression and relationship to survival in contaminated sediments. Toxicol Sci 68:69–81
Meyer JN, Smith JD, Winston GW, Di Giulio RT (2003a) Antioxidant defenses in killifish (Fundulus heteroclitus) exposed to contaminated sediments and model prooxidants: short-term and heritable responses. Aquat Toxicol 65:377–395
Meyer JN, Volz DC, Freedman JH, Di Giulio RT (2005) Differential display of hepatic mRNA from killifish (Fundulus heteroclitus) inhabiting a superfund estuary. Aquat Toxicol 73:327–341
Meyer JN, Wassenberg DM, Karchner SI, Hahn ME, Di Giulio RT (2003b) Expression and inducibility of aryl hydrocarbon receptor pathway genes in wild-caught killifish (Fundulus heteroclitus) with different contaminant-exposure histories. Environ Toxicol Chem 22:2337–2343
Nacci D, Coiro L, Champlin D, Jayaraman S, McKinney R, Gleason TR, Munns WR, Specker JL, Cooper KR (1999) Adaptations of wild populations of the estuarine fish *Fundulus heteroclitus* to persistent environmental contaminants. Mar Biol 134:9–17
Nacci DE, Champlin D, Coiro L, McKinney R, Jayaraman S (2002a) Predicting the occurrence of genetic adaptation to dioxinlike compounds in populations of the estuarine fish *Fundulus heteroclitus*. Environ Toxicol Chem 21:1525–1532
Nacci DE, Champlin D, Jayaraman S (2010) Adaptation of the estuarine fish *Fundulus heteroclitus* (Atlantic killifish) to polychlorinated biphenyls (PCBs). Estuar Coasts 33:853–864
Nacci DE, Kohan M, Pelletier M, George E (2002b) Effects of benzo a pyrene exposure on a fish population resistant to the toxic effects of dioxin-like compounds. Aquat Toxicol 57:203–215
Nelson TR, Sutton D, DeVries DR (2014) Summer movements of the Gulf killifish (Fundulus grandis) in a northern Gulf of Mexico salt marsh. Estuar Coasts 37:1295–1300
Noury P, Geffard O, Tutundjian R, Garric J (2006) Non destructive in vivo measurement of ethoxyresorufin biotransformation by zebrafish prolarva: development and application. Environ Toxicol 21:324–331
Oleksiak MF (2008) Changes in gene expression due to chronic exposure to environmental pollutants. Aquat Toxicol 90:161–171
Oleksiak MF, Karchner SI, Jenny MJ, Franks DG, Welch DBM, Hahn ME (2011) Transcriptomic assessment of resistance to effects of an aryl hydrocarbon receptor (AHR) agonist in embryos of Atlantic killifish (Fundulus heteroclitus) from a marine superfund site. BMC Genomics 12:263
Ownby DR, Newman MC, Mulvey M, Vogelbein WK, Unger MA, Arzayus LF (2002) Fish (*Fundulus heteroclitus*) populations with different exposure histories differ in tolerance of creosote-contaminated sediments. Environ Toxicol Chem 21:1897–1902

Oziolor E, Matson C (2015) Evolutionary toxicology: population adaptation in response to anthropogenic pollution. In: Riesch R, Tobler M, Plath M (eds) Extremophile fishes. Springer, Berlin, pp 247–277
Oziolor EM, Bigorgne E, Aguilar L, Usenko S, Matson CW (2014) Evolved resistance to PCB- and PAH-induced cardiac teratogenesis, and reduced CYP1A activity in gulf killifish (*Fundulus grandis*) populations from the Houston Ship Channel, Texas. Aquat Toxicol 150:210–219
Oziolor EM, De Schamphelaere K, Matson CW (2016a) Evolutionary toxicology: meta-analysis of evolutionary events in response to chemical stressors. Ecotoxicology 25:1–9
Oziolor EM, Dubansky B, Burggren WW, Matson CW (2016b) Cross-resistance in gulf killifish (*Fundulus grandis*) populations resistant to dioxin-like compounds. Aquat Toxicol 175:222–231
Pal A, He YL, Jekel M, Reinhard M, Gin KYH (2014) Emerging contaminants of public health significance as water quality indicator compounds in the urban water cycle. Environ Int 71:46–62
Parmesan C, Yohe G (2003) A globally coherent fingerprint of climate change impacts across natural systems. Nature 421:37–42
Poloczanska ES, Brown CJ, Sydeman WJ, Kiessling W, Schoeman DS, Moore PJ, Brander K, Bruno JF, Buckley LB, Burrows MT, Duarte CM, Halpern BS, Holding J, Kappel CV, O'Connor MI, Pandolfi JM, Parmesan C, Schwing F, Thompson SA, Richardson AJ (2013) Global imprint of climate change on marine life. Nat Clim Chang 3:919–925
Powell WH, Bright R, Bello SM, Hahn ME (2000) Developmental and tissue-specific expression of AHR1, AHR2, and ARNT2 in dioxin-sensitive and -resistant populations of the marine fish *Fundulus heteroclitus*. Toxicol Sci 57:229–239
Reid NM, Jackson CE, Gilbert D, Minx P, Montague MJ, Hampton TH, Helfrich LW, King BL, Nacci DE, Aluru N, Karchner SI, Colbourne JK, Hahn ME, Shaw JR, Oleksiak MF, Crawford DL, Warren WC, Whitehead A (2017) The landscape of extreme genomic variation in the highly adaptable Atlantic killifish. Genome Biol Evol 9:659–676
Reid NM, Proestou DA, Clark BW, Warren WC, Colbourne JK, Shaw JR, Karchner SI, Hahn ME, Nacci D, Oleksiak MF, Crawford DL, Whitehead A (2016) The genomic landscape of rapid repeated adaptation to human-altered environments. Science 354:1305–1308
Reitzel AM, Karchner SI, Franks DG, Evans BR, Nacci D, Champlin D, Vieira VM, Hahn ME (2014) Genetic variation at aryl hydrocarbon receptor (AHR) loci in populations of Atlantic killifish (*Fundulus heteroclitus*) inhabiting polluted and reference habitats. BMC Evol Biol 14:6
Rundle HD, Nosil P (2005) Ecological speciation. Ecol Lett 8:336–352
Schlezinger JJ, Struntz WDJ, Goldstone JV, Stegeman JJ (2006) Uncoupling of cytochrome P450 1A and stimulation of reactive oxygen species production by co-planar polychlorinated biphenyl congeners. Aquat Toxicol 77:422–432
Schmidt JV, Bradfield CA (1996) Ah receptor signaling pathways. Annu Rev Cell Dev Biol 12:55–89
Skinner MA, Courtenay SC, Parker WR, Curry RA (2005) Site fidelity of mummichogs (Fundulus heteroclitus) in an Atlantic Canadian estuary. Water Qual Res J Canada 40:288–298
Skinner MA, Courtenay SC, Parker WR, Curry RA (2012) Stable isotopic assessment of site fidelity of mummichogs, Fundulus heteroclitus, exposed to multiple anthropogenic inputs. Environ Biol Fish 94:695–706
Smith GM, Weis JS (1997) Predator-prey relationships in mummichogs (Fundulus heteroclitus (L)): effects of living in a polluted environment. J Exp Mar Biol Ecol 209:75–87
Stearns S (1993) The evolutionary links between fixed and variable traits. Acta Palaeontol Pol 38:215–232
Stegeman JJ (1993) The cytochromes P450 in fish. Elsevier, Amsterdam and New York
Stocum DL (2000) Frog limb deformities: an "eco-devo" riddle wrapped in multiple hypotheses surrounded by insufficient data. Teratology 62:147–150
Sultan SE (2007) Development in context: the timely emergence of eco-devo. Trends Ecol Evol 22:575–582
Tang J, Cao Y, Rose RL, Hodgson E (2002) In vitro metabolism of carbaryl by human cytochrome P450 and its inhibition by chlorpyrifos. Chem Biol Interact 141:229–241

Theodorakis CW, Shugart LR (1997) Genetic ecotoxicology II: population genetic structure in mosquitofish exposed in situ to radionuclides. Ecotoxicology 6:335–354

Valiela I, Wright JE, Teal JM, Volkmann SB (1977) Growth, production and energy transformations in the salt-marsh killifish Fundulus heteroclitus. Mar Biol 40:135–144

Van den Berg M, Birnbaum L, Bosveld ATC, Brunstrom B, Cook P, Feeley M, Giesy JP, Hanberg A, Hasegawa R, Kennedy SW, Kubiak T, Larsen JC, van Leeuwen FXR, Liem AKD, Nolt C, Peterson RE, Poellinger L, Safe S, Schrenk D, Tillitt D, Tysklind M, Younes M, Waern F, Zacharewski T (1998) Toxic equivalency factors (TEFs) for PCBs, PCDDs, PCDFs for humans and wildlife. Environ Health Perspect 106:775–792

Van Veld PA, Vogelbein WK, Smolowitz R, Woodin BR, Stegeman JJ (1992) Cytochrome P450IA1 in hepatic lesions of a teleost fish (Fundulus heteroclitus) collected from a polycyclic aromatic hydrocarbon-contaminated site. Carcinogenesis 13:505–507

Van Veld PA, Westbrook DJ (1995) Evidence for depression of cytochrome P4501A in population of chemically resistant mummichog (*Fundulus heteroclitus*). Environ Sci 3:221–234

Visser ME (2008) Keeping up with a warming world; assessing the rate of adaptation to climate change. Proc R Soc B Biol Sci 275:649–659

Vogelbein WK, Fournie JW, Van Veld PA, Huggett RJ (1990) Hepatic neoplasms in the mummichog *Fundulus heteroclitus* from a creosote-contaminated site. Cancer Res 50:5978–5986

Wang GL, Jiang B-H, Rue EA, Semenza GL (1995) Hypoxia-inducible factor 1 is a basic-helix-loop-helix-PAS heterodimer regulated by cellular O-2 tension. Proc Natl Acad Sci U S A 92:5510–5514

Wassenberg DM, Di Giulio RT (2004a) Synergistic embryotoxicity of polycyclic aromatic hydrocarbon aryl hydrocarbon receptor agonists with cytochrome P4501A inhibitors in *Fundulus heteroclitus*. Environ Health Perspect 112:1658–1664

Wassenberg DM, Di Giulio RT (2004b) Teratogenesis in *Fundulus heteroclitus* embryos exposed to a creosote-contaminated sediment extract and CYP1A inhibitors. Mar Environ Res 58:163–168

Wassenberg DM, Nerlinger AL, Battle LP, Di Giulio RT (2005) Effects of the polycyclic aromatic hydrocarbon heterocycles, carbazole and dibenzothiophene, on in vivo and in vitro CYP1A activity and polycyclic aromatic hydrocarbon-derived embryonic deformities. Environ Toxicol Chem 24:2526–2532

Wassenberg DM, Swails EE, Di Giulio RT (2002) Effects of single and combined exposures to benzo(a)pyrene and 3,3′ 4,4′ 5-pentachlorobiphenyl on EROD activity and development in *Fundulus heteroclitus*. Mar Environ Res 54:279–283

Weis JS, Khan AA (1990) Effects of mercury on feeding behavior of the mummichog, *Fundulus heteroclitus* from a polluted habitat. Mar Environ Res 30:243–249

Weis JS, Weis P (1995) Swimming performance and predator avoidance by mummichog (Fundulus heteroclitus) larvae after embryonic or larval exposure to methylmercury. Can J Fish Aquat Sci 52:2168–2173

Weis JS, Weis P, Heber M, Vaidya S (1981a) Methylmercury tolerance of killifish (*Fundulus heteroclitus*) embryos from a polluted vs non-polluted environment. Mar Biol 65:283–287

Weis P (1984) Metallothionein and mercury tolerance in the killifish, fundulus-heteroclitus. Mar Environ Res 14:153–166

Weis P, Weis JS (1977) Methylmercury teratogenesis in the killifish, Fundulus heteroclitus. Teratology 16:317–325

Weis P, Weis JS, Heber MA (1981b) Variation in teratogenic response to methylmercury by killifish (*Fundulus heteroclitus*) embryos – differences within and between populations. Anat Rec 199:A270–A271

Whitehead A (2010) The evolutionary radiation of diverse osmotolerant physiologies in killifish (*Fundulus sp.)* Evolution 64:2070–2085

Whitehead A, Pilcher W, Champlin D, Nacci D (2012) Common mechanism underlies repeated evolution of extreme pollution tolerance. Proc R Soc B Biol Sci 279:427–433

Williams LM, Oleksiak MF (2008) Signatures of selection in natural populations adapted to chronic pollution. BMC Evol Biol 8:282

Williams LM, Oleksiak MF (2011a) Ecologically and evolutionarily important SNPs identified in natural populations. Mol Biol Evol 28:1817–1826

Williams LM, Oleksiak MF (2011b) Evolutionary and functional analyses of cytochrome P4501A promoter polymorphisms in natural populations. Mol Ecol 20:5236–5247

Wills LP, Jung D, Koehrn K, Zhu SQ, Willett KL, Hinton DE, Di Giulio RT (2010a) Comparative chronic liver toxicity of Benzo a pyrene in two populations of the Atlantic killifish (*Fundulus heteroclitus*) with different exposure histories. Environ Health Perspect 118:1376–1381

Wills LP, Matson CW, Landon CD, Di Giulio RT (2010b) Characterization of the recalcitrant CYP1 phenotype found in Atlantic killifish (*Fundulus heteroclitus*) inhabiting a superfund site on the Elizabeth River, VA. Aquat Toxicol 99:33–41

Wills LP, Zhu SQ, Willett KL, Di Giulio RT (2009) Effect of CYP1A inhibition on the biotransformation of benzo[a]pyrene in two populations of *Fundulus heteroclitus* with different exposure histories. Aquat Toxicol 92:195–201

Wirgin I, Roy NK, Loftus M, Chambers RC, Franks DG, Hahn ME (2011) Mechanistic basis of resistance to PCBs in Atlantic tomcod from the Hudson River. Science 331:1322–1325

Wirgin I, Waldman JR (2004) Resistance to contaminants in North American fish populations. Mutat Res 552:73–100

Yuan ZP, Courtenay S, Chambers RC, Wirgin I (2006) Evidence of spatially extensive resistance to PCBs in an anadromous fish of the Hudson River. Environ Health Perspect 114:77–84

Zhou T, Scali R, Weis P, Weis JS (1996) Behavioral effects in mummichog larvae (*Fundulus heteroclitus*) following embryonic exposure to methylmercury. Mar Environ Res 42:45–49

Case Study: Gill Plasticity in Larval Fishes

15

Michael A. Sackville and Colin J. Brauner

Abstract

Adult teleosts possess an impressive degree of plasticity with respect to gas exchange and ion regulation. The gill is central to this plasticity, serving as the dominant site for both of these processes. Indeed, reversible changes to gill morphology can modify the oxygen and ion transport cascades such that flux is maintained despite changes to supply and demand. Gill plasticity thus provides clear selective benefits, allowing teleosts to thrive in habitats with fluctuating levels of oxygen and ions, and likely contributing to their broad distribution. However, this differs markedly in early life stages. Here, we use rainbow trout as a case study to review gill plasticity during larval development with respect to gas exchange and ion regulation. Surprisingly, the larval gill exhibits little plasticity relative to the adult form. Gill morphology is mostly fixed early on, and plasticity appears limited to changes in convection and protein function (hemoglobin isoform, ion affinity, Na^+, K^+ ATPase activity, tight junction permeability). Larvae thus have a reduced capacity to maintain trans-epithelial gas and ion flux in hypoxic and ion-poor water, respectively. Interestingly, larval development is slowed by hypoxia, but not by reduced ion uptake in ion-poor water. We propose that this limited gill plasticity is associated with selection for rapid development and minimal endogenous reserves when larvae rear in a stable environment on a fixed energy budget. As a consequence, this limited gill plasticity may contribute to the survival bottleneck often associated with larval development. However, these ideas are untested and based on few studies. We therefore conclude this review with several research directions aimed to provide a more comprehensive understanding of gill plasticity in larval rainbow trout specifically, and teleosts generally.

M. A. Sackville · C. J. Brauner (✉)
Department of Zoology, University of British Columbia, Vancouver, BC, Canada
e-mail: Mikesack@zoology.ubc.ca; Brauner@zoology.ubc.ca

© Springer International Publishing AG, part of Springer Nature 2018

W. Burggren, B. Dubansky (eds.), *Development and Environment*,
https://doi.org/10.1007/978-3-319-75935-7_15

15.1 Introduction

Adult teleost fishes possess an impressive and well-documented degree of plasticity with respect to gas exchange and ion regulation. The gill is central to this plasticity, serving as the dominant site for both of these processes. Indeed, reversible changes to the gill can modify the oxygen and ion transport cascades such that flux demands are met despite changes in environmental availability. Gill plasticity thus provides clear selective benefits, allowing teleosts to thrive in habitats with fluctuating levels of oxygen and ions, and likely contributing to their widespread distribution.

Here, we use rainbow trout as a case study to review the plasticity of the gill during larval development with respect to gas exchange and ion regulation. Perhaps surprisingly, the larval gill exhibits very little plasticity relative to the adult form. Gill morphology is mostly fixed early on, and plasticity appears limited to changes in convection and protein function. Although it remains untested, this relative absence of plasticity may result from selection for rapid development and minimal endogenous reserves when larvae rear in a stable environment on a fixed energy budget. As a consequence, this limited plasticity may contribute to the survival bottleneck often associated with larval development.

15.2 Gills of Adult Fishes Are Plastic

As stated above, plasticity of the adult fish gill plays a central role in the plasticity of gas exchange and ion regulation. Reversible changes to the gill can modify the oxygen and ion transport cascades to uphold the necessary levels of trans-epithelial flux when challenged by the environment. As the presence and absence of these changes in larval fishes are the focus of this review, establishing a simple framework to clearly discuss and describe exactly how they sustain trans-epithelial flux is essential. This is done below using a modified Fick equation based on Fick's law of diffusion.

Our modified Fick equation is first organized to describe trans-epithelial oxygen flux, where "F" is the rate of flux, "A" is epithelial area, "t" is thickness, "K" is permeability, and "ΔP" is the partial pressure gradient across the epithelium ($P_{\text{external}} - P_{\text{internal}}$). This iteration of the equation is commonly used to describe trans-epithelial gas flux in biological systems (Cameron 1989). All examples are taken from adult fish, and ion flux will be addressed subsequently.

$$F = A \cdot K \cdot \Delta P \cdot t^{-1} \tag{15.1}$$

When the partial pressure of environmental oxygen (P_{external}) decreases, flux decreases because of the associated reduction in "ΔP." Maintaining oxygen flux (F) in this instance therefore requires compensatory changes to other components in the equation. For example, structural remodeling of the gill can increase "A" by expanding lamellar surface area and decrease "t" by thinning of the lamellar epithelium. Indeed, scaleless carp exposed to hypoxia undergo dramatic changes to gill morphology in as little as 24 h, nearly doubling lamellar surface area and halving whole-blood diffusion distance (Figs. 15.1 and 15.2; Matey et al. 2008). Flux can also be maintained by modifying ventilation and perfusion to increase bulk flow of respiratory media through the gills, effectively increasing "ΔP." For example, rainbow trout exposed to ~25–35% of normal atmospheric oxygen increase ventilation volume anywhere from two- to sixfold (the hypoxic ventilatory response; Davis and Cameron 1971; Holeton and Randall 1967; Perry et al. 2009). Although not directly associated with the gill, "ΔP" can also be manipulated by increasing hemoglobin oxygen affinity, effectively reducing P_{internal} on the blood side of the gill epithelium (Soivio et al. 1980).

Trans-epithelial ion flux is affected by similar principles as oxygen, but important differences exist. Foremost is that ions often move across a single cell layer through transporters and channels powered by active transport, whereas oxygen mostly diffuses freely across epithelia down its partial pressure gradient. The ion flux machinery is associated with specialized gill cells called "ionocytes" (also commonly referred to as "chloride cells" and "mitochondrion-rich cells"; Hwang

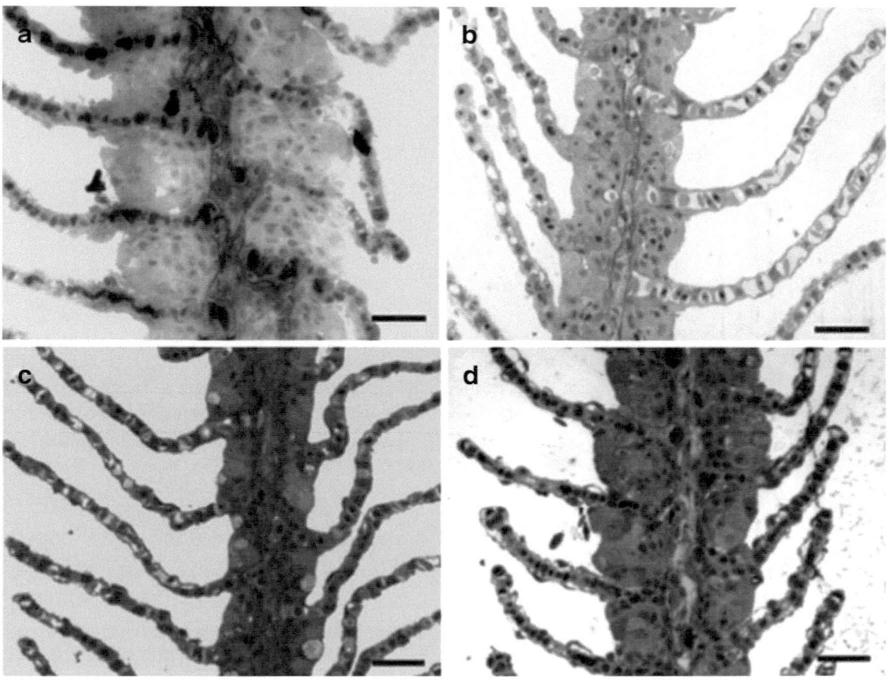

Fig. 15.1 General view of the scaleless carp gill filaments imaged with light microscopy in fish exposed to normoxia (control; **a**), following exposure to 0.3 mg O_2 l^{-1} hypoxia for 12 h (**b**), 24 h (**c**), and 12 h recovery in normoxia (**d**). Scale bars, 20 μm. Reproduced from Matey et al. (2008) with permission from the Journal of Experimental Biology

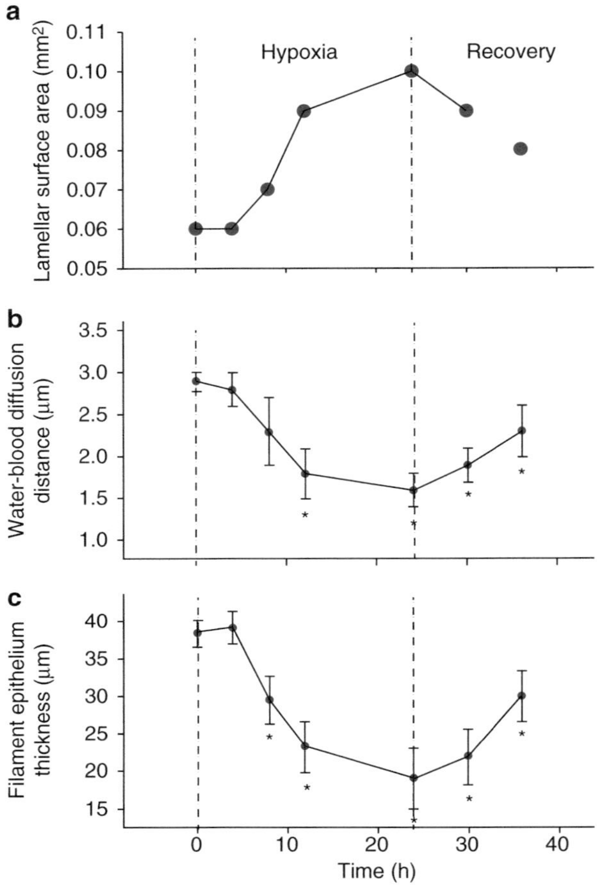

Fig. 15.2 Changes in (**a**) protruding lamellar surface area, (**b**) water-blood diffusion distance, and (**c**) filament epithelial thickness in the gills of the scaleless carp during exposure to hypoxia (0.3 mg O_2 l^{-1}) for 24 h followed by recovery in normoxia for 12 h. An asterisk represents statistically significant difference from the normoxic control values. Data are means ± s.e.m. (N = 7). Reproduced from Matey et al. (2008) with permission from the Journal of Experimental Biology

et al. 2011). Thus, changes to ion flux kinetics are largely achieved through ionocyte modification. When conceptually reorganizing our Fick equation for ion flux (Eq. (15.2)), these differences must be taken into account. Accordingly, "K/t" from Eq. (15.1) are replaced with a coefficient of permeability for the epithelium (p), and "ΔP" with transporter affinity (Km) and environmental ion concentration [i]. Please note that Eq. (15.2) demonstrates conceptual relationships among factors that affect flux and should not be used for quantitative flux calculations.

$$F = A \cdot p \cdot [i] \cdot \mathrm{Km}^{-1} \tag{15.2}$$

When environmental ion concentration [i] decreases enough to limit flux, F can be restored by increasing transporter affinity (decreasing Km), increasing ion permeability (p), and/or increasing area for ion flux (A). Increasing ion permeability in this case is effectively increasing maximum flux capacity (also known as J_{max}), which is achieved by increasing the total activity of transporter machinery. Transporter machinery activity can be increased in many ways, some of which include increasing ionocyte number (Greco et al. 1995, 1996), transporter number per ionocyte (Tipsmark et al. 2002), or switching to higher activity transporters/machinery (posttranslational modification, isoform switching, etc; Flemmer et al. 2010; Richards et al. 2003; Tipsmark et al. 2010). As an example, rainbow trout acclimated to softwater double the epithelial surface area covered by ionocytes through extensive cell proliferation and expansion (Fig. 15.3; Greco et al. 1995, 1996). Theoretically, ion flux can also be modified by changes to ventilation and perfusion to maintain suitable levels of counter-ion availability. To our knowledge this has yet to be shown, likely because the respiratory demands placed on ventilation and perfusion may exceed the ionic demands in most situations examined. A comprehensive review of the many ways ion flux kinetics can be modified is beyond the scope of this chapter, but excellent sources exist (Hwang et al. 2011).

The modifications listed above highlight a central role for the adult teleost gill in plasticity of gas exchange and ion regulation and, ultimately, the ability for teleost fishes to thrive in variable habitats. This central role should come as no surprise because the adult gill is the dominant site for both of these processes. As we shall see in the following sections, larval trout initially possess very underdeveloped gills and, perhaps unsurprisingly, exhibit relatively little plasticity in gill development with respect to gas exchange and ion regulation.

15.3 Larval Development and the Gill

In this review, we define larval development as the period between hatch and the completion of yolk sac absorption. Although this period lasts little more than 30 days for trout reared at 10 °C in normoxic conditions, dramatic shifts in life history occur. Larval trout begin this period as sedentary organisms, feeding endogenously as they develop in the protection of their gravel redd. However, upon completion of this period, trout emerge as free-swimming juvenile fish ready to feed exogenously.

This dramatic shift in life history is accompanied and facilitated by equally dramatic shifts in morphology and physiology (reviewed by Finn and Kapoor 2008). Here, we focus specifically on the development of the gill with respect to gas exchange and ion regulation during this period. Gill development under normal levels of oxygen and ions will be described first, followed by a review of what few studies examine gill development under reduced levels of oxygen and variable ion concentrations. As we shall see, larval gill development is relatively fixed when

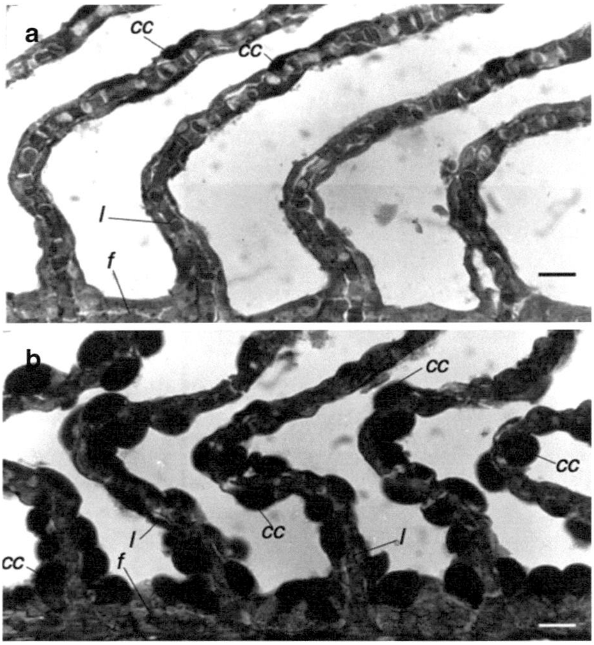

Fig. 15.3 Representative light micrographs of trout gills showing filament and lamellae from (**a**) control fish and (**b**) fish exposed to softwater conditions for 2 weeks. Ionocytes (also known as chloride cells *CC*), which are stained black, have proliferated over the entire surface of the gill epithelium of the softwater fish, *l* lamellae, *f* filament. The CCs of the control fish are smaller and far less numerous. Scale bars, 15 μm. Reproduced from Greco et al. (1995) with permission from the Journal of Experimental Biology

subjected to changes in these environmental conditions. This is a case study of rainbow trout, but where data are limited, examples are drawn from related fishes.

15.4 Larval Gill Development in Normoxic, Ion-Sufficient Water at 10 °C

The larval trout gill is severely underdeveloped at hatch. Gill arches are present but have only small filamental buds that often lack lamellae (González et al. 1996). Consequently, the vast majority of gas exchange and ion regulation occurs across the skin. This is made possible by the small body size of larval trout, which initially provides a sufficiently high surface area to volume ratio (SA:V) to meet flux demands cutaneously. As fish develop, however, this cutaneous capacity progressively diminishes due to the morphological constraints associated with growth. During growth, body mass generally scales in proportion to length cubed, yet surface area only scales in proportion to length squared. This mismatch in scaling results in a reduction of mass-specific body surface area as trout get larger. As illustrated by our Fick equation (Eq. (15.1)), this reduction in mass-specific body surface area (*A*), coupled with a thickening of the dermis (*t*), leads to the eventual insufficiency of the body surface to maintain flux demands (*F*). As we shall see, gill development during this time supplants the skin as the dominant site of gas exchange and ion regulation, ultimately facilitating the morphological changes to body shape and dermis associated with growth. We first discuss the morphological changes that underlie gill development with respect to gas exchange and ion regulation, and then follow with direct measures of flux. As we shall see, the gill supplants the skin as the dominant site of ion flux before gas flux. Specific drivers for this sequence of gill recruitment remain unknown, but possibilities are discussed.

15.4.1 Lamellae as a Morphological Index of Gas Flux Capacity

Estimates of morphological diffusive capacity in larval trout can provide valuable insight as to how gas exchange develops during this phase and the role that gill development plays in this process. Measurements of mass-specific surface area suggest that the gill plays only a small role at hatch relative to the body surface, representing only ~5% of total gill and body area (Rombough 1999). This value increases to ~35% by ysa but still suggests the skin retains a primary role in gas exchange at this stage. According to our Fick equation (Eq. (15.1)), thickness (*t*) is also an important factor driving morphological diffusive capacity. Thus, a more accurate estimate would be the anatomical diffusion factor (ADF; Perry 1990), which is defined as the mass-specific area per harmonic mean diffusion distance ($cm^2\ g^{-1}\ \mu m^{-1}$). Measurements of ADF are lacking for larval trout, but excellent data do exist for Atlantic salmon. Wells and Pinder (1996) report body and gill ADFs for developing larval salmon that suggest a slightly different story for diffusive capacity than the surface area data from trout. Total ADF in larval salmon is maintained at a constant ~3 $cm^2\ g^{-1}\ \mu m^{-1}$ from hatch to ysa. As with mass-specific surface area in trout, the relative contribution of the skin in larval salmon at hatch to this total ADF is high at ~92%, the majority of which occurs at the thin, well-vascularized epithelium of the yolk sac. However, the shift during development is more substantial, dropping to ~32% by ysa. These results suggest that the gills play a much greater role in gas exchange at ysa than would be predicted by changes in surface area alone (~68% vs ~35%) and that an increased role of the gill helps to maintain a high and stable total diffusive capacity throughout development as cutaneous capacity decreases.

During this shift from skin to gill, the most dramatic changes to ADF also occur at the gill. Although dermal thickening and the mismatched scaling of body mass with surface area reduce cutaneous ADF to ~1/3 its initial value by ysa, total ADF is maintained by an almost ninefold increase in branchial ADF (Wells and Pinder 1996). This increase in branchial ADF is caused by extensive proliferation and expansion of filaments and their associated lamellae. In larval Atlantic salmon, total filament numbers rise from ~250 to 650 and lamellae from near 0 to ~10,000. This results in an approximate 20-fold increase in total gill area, with filaments and lamellae making surprisingly similar contributions (~44% vs 56% total branchial area, respectively). Despite similar contributions to total gill area by both of these structures, lamellae prove to be the primary driver behind increased branchial ADF. Indeed, lamellar harmonic mean diffusion distance is approximately fivefold thinner than that of filaments at ysa and ~40-fold thinner than that of the outer body dermis. By the completion of ysa, lamellae account for over 90% of branchial ADF and more than 60% of total ADF. It should be noted that filaments are still an important structure for gas exchange, as filament size and number determine available space for lamellar development. However, lamellar proliferation and expansion appear to be the main drivers for increased branchial ADF, which allows cutaneous ADF to drop during development without compromising total ADF. This relationship is clearly illustrated in Fig. 15.4.

15.4.2 Ionocyte Distribution as a Morphological Index of Ion Flux Capacity

In much the same way that ADF can be used to estimate the relative partitioning of gas exchange capacity between skin and gills, ionocyte distribution can be used to estimate the relative partitioning of ion flux capacity. As previously stated, ionocytes are specialized epithelial cells associated with the molecular machinery that largely drives trans-epithelial ion flux. Thus, ionocyte distribution and density are often used as proxies to determine where ion flux occurs. Interestingly, ionocyte distribution in larval trout follows a pattern very similar to ADF during development

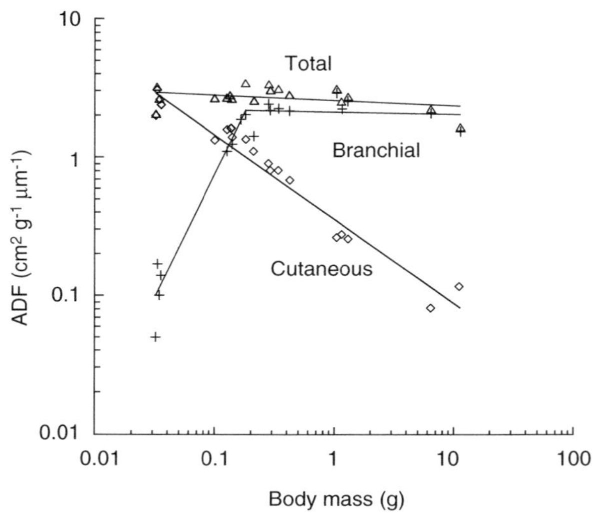

Fig. 15.4 Allometry of the anatomical diffusion factor (ADF) in developing Atlantic salmon. Total (open triangle), cutaneous (open diamond), and branchial (plus). Reproduced from Wells and Pinder (1996) with permission from the Journal of Experimental Biology

in Atlantic salmon, where increasing numbers of branchial ionocytes help maintain a constant total mass-specific ionocyte density despite diminishing contributions from cutaneous surfaces (Rombough 1999). At hatch, only 20% of the total 70,000 ionocytes are found on the gill, with the remaining 80% found cutaneously. As with ADF, most cutaneous ionocytes initially occur on the thin, well-vascularized epithelium of the yolk sac, likely because it favors trans-epithelial ion flux by allowing single cells to contact the external environment and circulatory system simultaneously. Indeed, ~60% of all ionocytes at hatch are found on the yolk sac, and they occur at a much higher density than other cutaneous surfaces (~800 cells mm^{-2} on yolk sac vs 100 cells mm^{-2} on head and trunk). As the yolk sac is absorbed during development, the number of cutaneous ionocytes decreases from ~55,000 to 40,000, but total ionocyte number actually increases from ~70,000 to 170,000. This rise in total ionocyte number is driven exclusively by increasing numbers of branchial ionocytes, which allow larval trout to maintain a stable mass-specific density of 1500 cells mg^{-1} during growth despite diminishing cutaneous contributions. By ysa, 75% of all ionocytes are found on the gill and occur almost exclusively on the filaments. Furthermore, ionocyte density on filaments remains fairly stable throughout development at approximately 2000 cells mm^{-2}. Thus, the rising number of branchial ionocytes is driven primarily by proliferation and expansion of filaments in much the same way that branchial ADF is driven primarily by proliferation and expansion of lamellae.

Our modified Fick equation describing trans-epithelial ion flux (Eq. (15.2)) suggests that flux capacity can also be affected by changes in activity of transport machinery. Thus, measurements of activity for key enzymes implicated in ionocyte machinery can also provide estimates of ion flux capacity. Activity of Na^+/K^+-ATPase (NKA) is often used in this way and has been measured in the skin and gills of larval trout during development. NKA activity of gill tissue increases at least threefold with development, while activity of cutaneous surfaces remains low and constant (Fu et al. 2010; Zimmer et al. 2014). This result may seem unsurprising given the pattern of ionocyte proliferation, but enzyme activity is expressed as a function of protein mass. Thus, if ionocyte density on the gills is stable throughout development as previously shown (Rombough 1999), increasing levels of NKA activity suggest that ion flux capacity per individual cell might also be increasing.

15.4.3 Morphological Indices of Gas and Ion Flux Capacities Accurately Predict Direct Measurements of Trans-epithelial Flux

The morphological indices identified above can be useful predictors for how flux partitions between gills and the skin during development. However, they provide only best estimates of actual trans-epithelial flux in vivo. Among other factors, no control for convection and small errors in measurements of gill dimensions can result in very poor estimations of how flux actually partitions in living fish. Although often more difficult to acquire, direct measurements of flux are thus necessary to validate the accuracy of these morphological proxies and to determine the extent to which they influence flux in vivo.

Direct measurements of oxygen and sodium flux across the gills and skin of larval trout do provide strong support for the predictive power of our morphological estimates of capacity (Fu et al. 2010; Zimmer et al. 2014). In resting, anesthetized animals, percent oxygen uptake at the gills increases from ~20% at hatch to ~55% at ysa. These values are slightly higher than would be predicted by ADF at hatch (10%) and slightly lower than would be predicted at ysa (70%). However, recalling our Fick equation for gas exchange (Eq. (15.1)), flux is affected by more than just "*A*" and "*t*" as predicted by ADF. Variation in partial pressure gradients (ΔP) could be driving these modest discrepancies. Indeed, differences in local oxygen demand of different tissues could result in ΔPs that create a pattern of oxygen flux different from that predicted by ADF alone, especially during development when the importance of the circulatory system to oxygen delivery is ever changing. Similarly, differences in ventilation between the skin and gills could also be affecting ΔP and the observed mismatch between ADF and flux. It should also be noted that estimates of ADF were made in Atlantic salmon and direct measures of flux were made in rainbow trout. Although relative skin to gill surface area partitioning is similar between studies (Rombough 1999; Wells and Pinder 1996), possible species differences in epithelial thickness could result in different ADFs.

For sodium uptake, percent contribution of the gills increases from ~10% at hatch to ~100% by ysa. These values are lower than would be predicted by ionocyte distribution at hatch (20%) and higher than would be predicted at ysa (75%). Much like the modest mismatch observed between ADF and oxygen uptake, our Fick equation can help identify candidates for these observed differences. Variation in the activity of flux machinery at individual cells ("*p*"; Eq. (15.2)) as well as convection to enhance counter-ion availability could be responsible ([*i*]; Eq. (15.2)).

15.4.4 Ion Flux Shifts from Skin to Gill Before Gas Flux

Direct measurements of flux also reveal an interesting pattern for the relative timing of transition from skin to gills for oxygen and sodium. The gill has long been viewed as an organ primarily for gas exchange and secondarily for ion regulation. During development, however, the gill takes on a primary role for sodium uptake well before oxygen. In resting, anesthetized animals, the gill assumes 50% of sodium uptake at only ~15 days post-hatch (dph) but 50% of oxygen uptake at ~28 dph (Fig. 15.5; Brauner and Rombough 2012). Even at 30 dph, which coincides with ysa, the gill is only responsible for ~55% of oxygen uptake, yet ~100% of sodium uptake occurs branchially. This pattern is consistent across studies of trout and holds true for measures of fluxes, ADF, and ionocyte distribution in all species of larval teleosts examined to date, both in freshwater and seawater (Japanese flounder, Hiroi et al. 1998 and Hwang 1990; killifish, Katoh et al. 2000; tilapia, Li et al. 1995; brown trout, Pisam et al. 2000; zebra fish, Rombough 2002; European sea bass, Varsamos et al. 2002).

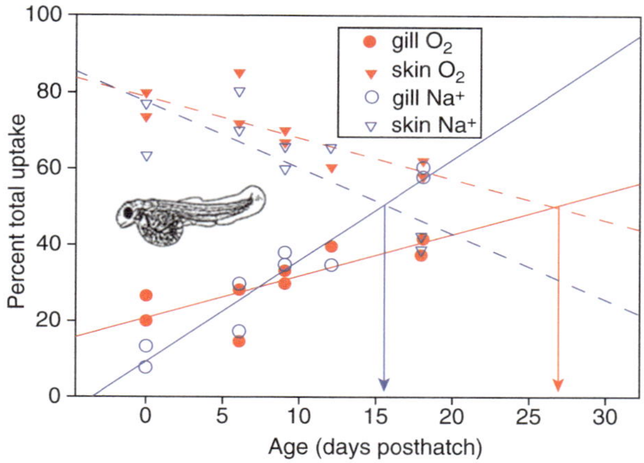

Fig. 15.5 Relative contributions of the gill and skin to total O_2 and Na^+ uptake in rainbow trout larvae. Vertical arrows indicate the age at which 50% of total Na^+ (15.5 dph) or O_2 (27 dph) uptake has shifted to the gills (after Fu et al. 2010; data for soft and hard water combined). Reproduced from Brauner and Rombough (2012) with permission from Respiratory Physiology & Neurobiology

The reasons for this pattern of skin to gill transition remain unknown. Trans-epithelial ion flux often requires an epithelial thickness of one cell layer; thus, thickening of the outer dermis might be more limiting to sodium uptake than oxygen. Additionally, branchial capacity for ion flux might simply increase before that of gases because filaments must exist for lamellae to proliferate. Alternatively, this pattern could change when fluxes are measured under periods of variable demand and/or supply. These scenarios and others will be further explored in subsequent sections.

15.4.5 Summary of Larval Gill Development in Normoxic, Ion-Sufficient Water at 10 °C

Morphological indices provide impressively accurate estimates of how gas and ion flux partitions between the skin and gills during larval development. Differences between estimates and direct measurements of flux are modest and, although untested, can be easily reconciled within the principles of our Fick equations. This suggests that morphology and ionocyte distribution are good indicators of the relative contributions of the gill to gas and ion flux during development. Furthermore, the accuracy of these indices emphasizes a central role for morphological change in dictating branchial capacity of both gas and ion flux. Indeed, dramatic compensatory proliferation of branchial structures maintains total mass-specific capacity during development despite diminishing cutaneous contributions imposed by the morphological constraints associated with growth. Diffusive capacity for gas exchange appears closely linked to lamellar proliferation, while ionocyte number appears more closely linked to filamental proliferation. Additionally, increases to branchial ion flux capacity may also occur independently of changes to gill morphology, as increasing levels of branchial NKA activity suggest increases in flux capacity of individual ionocytes. Under resting, control conditions, all of these changes appear to result in an earlier and more complete shift from the skin to gill for trans-epithelial flux of ions than gases.

15.5 Larval Gill Development in Hypoxia

Low environmental oxygen, or hypoxia, can occur naturally during larval trout development. Variation in water flow, redd depth, and clutch size has been implicated in reducing oxygen in the gravel redd to as low as 20% of normal atmospheric levels (Coble 1961; Peterson and Quinn 1996; Youngson et al. 2004). This ecological relevance of hypoxia has resulted in many studies investigating its effects on larval development (Ciuhandu et al. 2007; Holeton 1971; Rombough 1988b). However, surprisingly few have specifically examined gill development with respect to gas exchange under these conditions.

When exposed to moderate hypoxia, larval trout generally appear unable to mount a plastic response sufficient to maintain normoxic levels of oxygen consumption. The resulting reduction in aerobic metabolism slows larval development overall. As we shall see, this pattern of slowed development also appears to extend to the gill early on, but there is some evidence for increasing plasticity to enhance oxygen uptake as gill structures form.

15.5.1 Morphological Plasticity at the Gill

To our knowledge, only one study has investigated the effects of chronic hypoxia on larval gill development in a salmonid. McDonald and McMahon (1977) reared arctic char following hatch in normoxia (121 mmHg, 75% saturation) and hypoxia (33 mmHg, 20% saturation) at 6.5 °C for 47 days. The number and area of filaments and lamellae were measured at 2, 15, 38, and 47 dph in both treatments. As in other fishes, hypoxia appears to slow development overall. Larvae were smaller than those reared in normoxia at each time point (68 vs 74 mg at 15 dph, 76 vs 90 mg at 38 dph, 79 vs 99 mg at 47 dph), and while yolk sac absorption was over 90% complete in normoxia by 47 dph, it was only 50% complete in hypoxia.

Up until 38 dph, gill development in hypoxia also appears to be merely slowed, showing little evidence for plasticity. Hypoxic larvae had ~20% fewer filaments by 38 dph, which resulted in significantly less total filament area relative to normoxic larvae (14.7 vs 22.1 mm^2). Less filament area in turn provided less substrate for proliferation of lamellae, which led to 50% fewer lamellae (1221 vs 2439). Fewer lamellae in turn resulted in hypoxic larvae having half the total lamellar area of normoxic larvae at 38 dph (3.0 vs 6.2 mm^2) and about 60% the mass-specific lamellar area (~40 vs 70 $mm^2\ g^{-1}$).

As previously discussed, lamellae are the primary gas exchange structures of the gill; thus, these findings suggest a lower branchial diffusive capacity in hypoxia at 38 dph. The observed differences in gill morphology between treatments up to this point are likely due to a reduction in overall developmental rate rather than plasticity associated specifically with gill morphology. However, comparisons between stage-matched individuals and measurements of lamellar thickness would be required to definitively rule out hypoxia-induced plasticity at 38 dph or earlier.

Despite the absence of these additional metrics, hypoxia-induced plasticity of the gill is clearly evident by 47 dph. Filament number (433 vs 554), area (23.6 vs 29.0 mm^2), and lamellar number (2296 vs 3686) are all significantly lower in hypoxic larvae at 47 dph, which is consistent with the pattern of slowed development at 38 dph. Yet, total lamellar area is similar between treatments at 47 dph (~8 vs 9 mm^2, hypoxia vs normoxia; no significant difference), and mass-specific surface area is actually slightly higher in hypoxia (~9.8 vs 9.1 cm^2 g^{-1}; n.s.). Hypoxic larvae appear to make up the difference in total lamellar area by increasing the size of existing individual lamellae, rather than by proliferation of new additional lamellae. Indeed, mean single lamellar surface area of hypoxic larvae at 47 dph was 30% larger than that of normoxic larvae (0.0033 vs 0.0024 mm^2). Additionally, this difference does not appear to be linked to a faster rate of lamellar expansion. Mean single lamellar surface area in hypoxia is similar to that in normoxia at 15 dph (0.0013 vs 0.0016 mm^2, nonsignificant) and 38 dph (0.0023 vs 0.0024 mm^2), but while lamellae continue to expand until at least 47 dph in hypoxia, expansion in normoxia stops at 38 dph. Thus, it appears that larval gills may develop at relatively fixed rates regardless of treatment and that plasticity may manifest as the continuation of normal lamellar expansion beyond normoxic requirements. Again, stage-matched comparisons and measurements of lamellar thickness would be welcome here, ideally providing a more accurate assessment of the true nature and extent of this hypoxia-induced plasticity. Regardless, larvae in both treatments possess similar lamellar surface areas at 47 dph despite hypoxic larvae being less developed (50 vs 90% ysa) and smaller (79 vs 99 mg) than those reared in normoxia.

15.5.2 Other Plasticity at the Gill

Larval trout reared in hypoxia also exhibit other forms of plasticity aside from morphological change to enhance oxygen uptake at the gill. Behaviorally, larvae increase ventilation and heart rate immediately following hatch (Holeton 1971). As in adults, this modified convection helps maintain a higher ΔP across the gill and increases bulk flow of oxygen to tissues. Larval trout also retain an embryonic isoform of hemoglobin later into development when reared in hypoxia (Bianchini and Wright 2013). The exact function of retaining this embryonic isoform in hypoxia remains unknown, as in vitro measurements of oxygen affinity appear similar to juvenile hemoglobin. However, allosteric modulation may result in higher oxygen affinity in vivo, and the reduced rate of erythrocyte turnover may provide valuable energetic savings at a time when oxygen availability is low (Bianchini and Wright 2013).

15.5.3 Impacts of Plasticity on Performance and Development in Hypoxia

These different levels of plasticity affect larvae at different points in development. Changes in convection and possible hemoglobin function have effects immediately following hatch, while changes in gill morphology appear to occur later as lamellae fully form. In larval steelhead reared in normoxia at 9 °C, P_{crit} does progressively decline from ~70% atmospheric oxygen at hatch to 33% at ysa, suggesting that oxygen uptake during hypoxia may improve with development (Rombough 1988a). However, the relative contributions of these different responses toward the enhancement of oxygen uptake in hypoxia remain unknown, as does their effect on the timing of transition for oxygen uptake from the skin to gills. Furthermore, the slower rate of development by larvae reared in 20% atmospheric oxygen suggests that these changes are still unable to maintain normoxic levels of oxygen uptake in chronic, moderate hypoxia – especially early on. Possible constraints and/or drivers underlying this insufficiency of plasticity will be discussed in Sect. 15.7.

15.6 Larval Gill Development in Variable Ion Compositions

The ionic composition of freshwater can be both complex and variable. As an example, a comparison between the ionic profiles of Vancouver and Hamilton city water reveals differences of more than tenfold in concentration for at least five different ions (Hamilton vs Vancouver in mM; [Na^+] 0.6 vs 0.06, [Cl^-] 0.8 vs 0.05, [Ca^{2+}] 1.8 vs 0.03, [Mg^{2+}] 0.3 vs 0.007, [K^+] 0.05 vs 0.004; from Fu et al. 2010; Zimmer et al. 2014). Factors such as local geology, rainfall, and tidal fluctuations can all lead to similar levels of natural variation across temporal and spatial scales in the gravel redd. An appreciation for this complexity and variability is important because multiple ions play very different and important biological roles in the development of larval trout. Furthermore, subtle changes in the concentrations of these ions can drastically affect rates of trans-epithelial ion flux, epithelial integrity, and even the buffering of important chemical reactions in the water that can affect trans-epithelial flux of ions *and* gases (for reviews, please see Hwang et al. 2011; Hwang and Lin 2013).

An attempt to summarize the effects of variation in all ions and their possible combinations is well beyond the scope of this case study. Additionally, the majority of existing work has focused primarily on variation in environmental sodium and its trans-epithelial flux. This popularity of sodium likely stems from its relative ease of measurement (concentration and flux), role as the dominant extracellular cation, and involvement of its trans-epithelial flux in maintaining homeostasis for more than just sodium balance (i.e., ammonia excretion, acid-base balance). Thus, here we also limit our review primarily to variation in environmental sodium and its trans-epithelial flux.

15.6.1 Morphological Plasticity at the Gill

To our knowledge, only one study has investigated the effects of environmental ion composition on larval trout gill morphology with respect to ion regulation. Shen and Leatherland (1978) reared larval trout from hatch to ysa at 10 °C in 13 ppt seawater, 11 ppt seawater, and distilled water. They found no difference in the number and structural appearance of ionocytes between treatments at any point in development.

15.6.2 Other Plasticity at the Gill

Despite this absence of morphological plasticity, there are indications of plasticity at other levels. Gallagher (2014) reared larval trout from hatch to ysa at 10 °C in three concentrations of sodium (0.06, 0.43, and 1.7 mM [Na^+]). As early as 11 dph, measurements of sodium uptake kinetics revealed a general pattern of increasing sodium affinity with decreasing sodium concentration. Similarly, Fu et al. (2010) found that larvae reared in 0.06 mM [Na^+] at 10 °C had 25% higher levels of gill NKA activity than those reared in 2.3 mM [Na^+] by 18 dph (~2.5 vs 2.0 μmol ADP mg^{-1} h^{-1}). Both studies suggest that the ion flux machinery at individual ionocytes can be modified to enhance sodium uptake in ion-poor conditions.

Larval trout also appear to possess a degree of behavioral plasticity for enhancing sodium uptake. We (Sackville M.A. and Brauner C.J., unpublished observations) reared trout in 2.3 mM [Na^+] from hatch to ysa at 10 °C. Subsets of larvae were transferred to 0.06 mM [Na^+] every 3 dph to examine the effects of decreasing environmental sodium on ventilation frequency, sodium uptake, and oxygen consumption. Prior to 24 dph, larvae increased ventilation frequency by ~20% following exposure to low sodium for 24 h. This increased ventilation was accompanied by a twofold increase in the rate of sodium uptake between 1 and 24 h postexposure, but no change in oxygen consumption. In adult trout, the proliferation of ionocytes on lamellae following a 2-week acclimation to ion-poor water impairs gas exchange by increasing the lamellar diffusion distance (Greco et al. 1995, 1996). Ventilation in adults then increases by ~35% to maintain routine levels of gas exchange, which yields similar results to Lee et al. (2016, unpublished). Although not measured by Lee et al. (2016, unpublished), the shorter time course of the exposure/response relative to Greco et al. (1995) and the absence of plasticity in ionocyte proliferation observed during development (Shen and Leatherland 1978) suggest that lamellar thickening is likely not occurring here. Thus, it appears that larval trout may modify convection to enhance trans-epithelial ion flux in much the same way that adult trout do for gas exchange.

15.6.3 Impacts of Plasticity on Performance and Development in Variable Ion Compositions

As with the hypoxia-induced plasticity discussed in Sect. 15.4, the relative contributions that these different forms of ionoregulatory plasticity make toward enhancing trans-epithelial ion flux remain unknown. Similarly, ionoregulatory plasticity also appears unable to maintain rates of trans-epithelial ion flux as environmental ion concentration decreases. Indeed, all studies to date show sustained reductions in sodium uptake when larval trout are reared in more dilute water. Fu et al. (2010) show a twofold difference in sodium uptake between 0.06 and 2.3 mM Na, which was subsequently corroborated by Zimmer et al. (2014) and Lee et al. (2016, unpublished).

We also saw in Sect. 15.4 that the insufficiency of hypoxia-induced plasticity to maintain normoxic levels of oxygen uptake results in a slowing of larval development overall. Interestingly, the insufficiency of ionoregulatory plasticity to maintain levels of sodium uptake in more dilute water appears to have no such effect. Despite large differences in sodium uptake, no differences in overall developmental rate, whole-body sodium content, or timing of skin-to-gill transition for sodium flux were observed between any of the environmental ion concentrations tested in the studies listed above. This suggests that rates of sodium loss are also reduced to maintain a similar rate of net sodium uptake (Fu et al. 2010). Ion loss in adult trout occurs largely across the gill epithelium, and epithelial "leakiness" can be modified by changes in tight junction complexes to mitigate this loss (for review see Chasiotis et al. 2012). Tight junction complexes in the gills of larval trout are thus another likely source of ionoregulatory plasticity worthy of future exploration.

15.6.4 Summary of Larval Gill Development in Variable Ion Compositions

Plasticity of gill development with respect to ion regulation follows a pattern similar to that of gas exchange. Morphological plasticity is largely absent, and enhancement of trans-epithelial ion flux appears limited to changes in convection and the kinetic properties of existing ionocytes. The capacity of ionoregulatory plasticity also appears unable to maintain levels of trans-epithelial ion flux as environmental availability decreases. However, the reduced levels of ion uptake do not slow larval development in the same way as reduced levels of oxygen uptake in hypoxia. Possible constraints and/or drivers underlying this insufficiency of plasticity and the differences between ion- and hypoxia-induced effects are discussed in the following section (Sect. 15.7).

15.7 Limited Morphological Plasticity and the Differential Response to Hypoxia and Ion-Poor Water

In Sect. 15.3, we saw that gill morphology serves as a good indicator for flux capacity of both gases and ions. Lamellae correlate well with the capacity for branchial gas flux and filamental distribution of ionocytes with the capacity for branchial ion flux. Perhaps unsurprisingly, changes to gill morphology are a primary source of plasticity for gas and ion flux in adult trout, and many examples are provided in Sect. 15.2.

As seen in Sects. 15.5 and 15.6, gill morphology of larval trout exhibits very little plasticity, especially early in development. Plasticity of gas and ion flux at the gill is instead limited primarily to changes in convection and protein function (hemoglobin isoform, ion affinity, NKA activity, tight junction permeability). Furthermore, these remaining sources of plasticity appear unable to maintain levels of trans-epithelial gas and ion flux as environmental availability decreases. Interestingly, larval development is slowed overall by the resulting reduction of oxygen uptake in hypoxia, but not by the reduction of sodium uptake in dilute water.

Here, we attempt to explain (1) the drivers underlying this apparent lack of plasticity in gill morphology during larval development and (2) why reduced levels of oxygen uptake slow development when ion uptake does not.

15.7.1 Why Is Morphological Plasticity of the Gill Limited During Larval Development?

The absence of morphological plasticity at the gill during larval development is likely the product of three related factors: (a) the gill develops at a near-maximum rate, (b) endogenous feeding limits the energetic scope for plasticity, and (c) the osmorespiratory compromise places reciprocal costs on gas and ion flux when either is enhanced.

15.7.1.1 The Gill Develops at a Near-Maximum Rate

Studies examining the energetics of larval development suggest that growth and metabolism are operating at close to maximum rates (Rombough 2011). Indeed, specific growth rates of many larval teleosts are reported as 10–30% day^{-1} (Rombough 2011), which is an order of magnitude greater than values reported for many adults. Furthermore, routine metabolic rates of larval trout are ~80% of maximum at hatch (Ninness et al. 2006), which is also much higher than the ~30% reported for adults (Chen et al. 2015).

This investment in rapid development is believed to provide selective benefits by moving fish through the vulnerable larval phase as quickly as possible. However, if the larval gill is also developing at a near-maximum rate, hypoxia-induced proliferation and expansion of gill structures would only become apparent as normoxic growth reaches completion later in development. This is actually what we see in Sect. 15.5, where hypoxia-induced lamellar expansion in arctic char appears to simply continue at a normal rate after normoxic expansion has ceased. This same principle can also be applied to ionocyte proliferation and maturation, helping to explain why NKA activity between larvae reared in high and low [Na^+] only differs after 18 dph (Fu et al. 2010).

15.7.1.2 Endogenous Feeding Limits the Energetic Scope for Plasticity

The yolk is a finite energy supply that fuels the larval phase of trout development. Use of this endogenous reserve has been shown to be exceptionally efficient. An estimated 80% is devoted to growth and development, with conversion efficiencies upward of 75% commonly observed in many larval teleosts. Another 5% of the reserve is lost to waste production, and the remaining 15% serves to fuel activity and all other processes (Rombough 2011).

Efficient use of the yolk reserve is believed to be driven partly by selection for increased fecundity in adult fish. If less endogenous energy is required to successfully fuel larval development, adults can increase reproductive output by producing a greater number of smaller eggs, rather than a smaller number of larger eggs. If generations of larvae are consistently reared in a stable, homogenous environment, selection for narrower margins of the energy budget fueling this phase of development is predicted. However, imparting less energy per egg limits the energetic scope that larvae have to deal with additional and unexpected costs that may fall outside the normal developmental trajectory.

This limited energetic scope might also help explain why larval trout exhibit limited plasticity in response to environmental variability. Mounting a plastic response requires energy. If the amount of energy required for plasticity exceeds the amount allocated in the endogenous budget, energy is taken from other processes. For example, larval trout reared in hatcheries routinely exhibit lower yolk-to-tissue conversion efficiencies because of the additional energy spent on righting behavior in the artificial substrate (Mead and Woodall 1968). Similar reductions in yolk-to-tissue conversion efficiency are also observed in larval pink salmon reared under climate change relevant levels of CO_2 (Ou et al. 2015), presumably due to increased costs associated with maintaining acid-base balance. In severe cases, this reallocation of energy away from other critical processes might prematurely deplete endogenous reserves, thus preventing the completion of larval development. This may be why larval plasticity is limited to "low-cost" modifications in convection and protein function early on, while "high-cost" modifications to gill morphology only appear closer to the completion of development and ysa.

15.7.1.3 The Osmorespiratory Compromise

The osmorespiratory compromise refers to the paradoxical relationship between gas exchange and ion/osmoregulation. In this relationship, enhancing gas exchange can negatively impact ion/osmoregulation, and enhancing ion/osmoregulation can negatively impact gas exchange. For example, lamellar expansion enhances oxygen uptake in adult trout by increasing the area for diffusion ("a" in Eq. (15.1); Sect. 15.2). However, this same increase in area also enhances the passive diffusion or "leak" of ions and water across the gill epithelium. This increased leak results in a greater hydromineral disturbance, which then requires an increased level of active ion/osmoregulation to maintain ion and water balance. Thus, increasing lamellar area to enhance oxygen uptake also incurs an ion/osmoregulatory cost. Similarly, adult trout acclimated to ion-poor water undergo ionocyte proliferation on lamellae to enhance active ion uptake (Fig. 15.3, Sect. 15.2; Greco et al. 1995). However, ionocytes are generally thicker than the pavement cells that typically make up the lamellar epithelium. As illustrated in Fig. 15.6, lamellar diffusion distance increases with ionocyte coverage (Greco et al. 1996). This increased diffusion distance decreases the diffusive capacity of the gills, which requires adult trout to increase ventilation to maintain routine levels of gas exchange (Greco et al. 1995, 1996). Thus, ionocyte proliferation to enhance ion uptake also incurs respiratory costs.

Morphological changes at the gill to enhance gas exchange or ion regulation thus incur additional reciprocal costs. If the energetic scope for plasticity is limited during larval development as suggested, these reciprocal costs may provide further rationale for the absence of morphological plasticity at the gill early on.

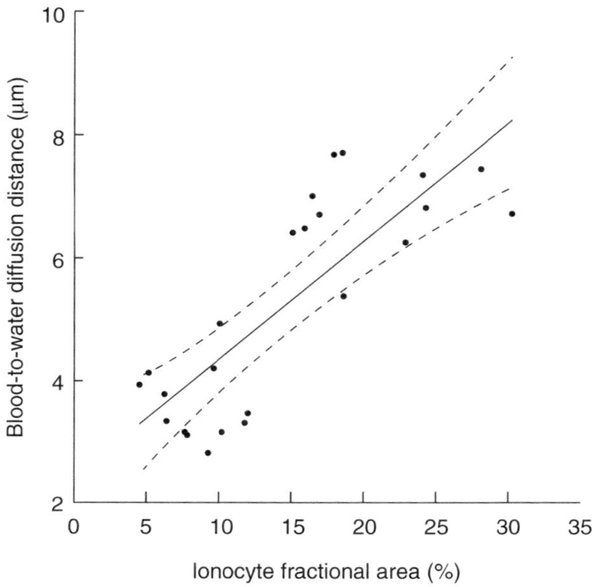

Fig. 15.6 Correlation between ionocyte fractional area and the blood-water diffusion distance ($n = 24$). Values are individual data points obtained from control and softwater fish at all sampling times [$r^2 = 0.81$; $y = 0.19x + 2.42$]. Broken lines indicate 95% confidence intervals. Reproduced from Greco et al. (1996) with permission from Cell and Tissue Research

15.7.2 Differential Effects/Responses to Hypoxia and Ion-Poor Water

As discussed in Sects. 15.5 and 15.6, larval trout reared in hypoxia experience reductions in oxygen uptake and develop more slowly. Conversely, larval trout reared in ion-poor water experience reductions in ion uptake yet develop at a normal rate. As proposed below, slowing metabolism and development may serve as an adaptive response to decreasing environmental availability of oxygen or ions. Thus, the absence of slowed development with decreasing ion concentration and uptake is somewhat surprising. The reasons for this different effect on developmental rate remain unknown, but as discussed below, fundamental differences between hypoxic and ionic disturbances may play a role.

In hypoxia, lower environmental oxygen reduces the rate of oxygen uptake by larval trout in the absence of compensation (reduced ΔP; Eq. (15.1)). Provided the oxygen demands of aerobic metabolism and the rates of other metabolic processes can be reduced without complication, life and development simply slow down. Developing more slowly may carry few negative consequences if emergence timing from the redd is not under strong selection and predation pressure is low. Furthermore, if larval development is under selection for smaller endogenous reserves and less energetic scope (Sect. 15.7.1.2), slowed development in hypoxia might actually be favored over using energetically costly forms of plasticity to maintain normoxic rates of oxygen uptake and development. Thus, in the appropriate context, developing more slowly might be an adaptive response to decreasing environmental oxygen.

Similar principles can be applied to the scenario of ion-poor water. Lower ion concentration can lower the rate of ion uptake in the absence of compensation (reduced [i]; Eq. (15.2)). Slowing metabolic rate and development can also lower the ion uptake demand associated with metabolism; thus, similar arguments for an adaptive response can be made. However, while oxygen demand is set exclusively by aerobic metabolism, ion demand is set by metabolism *and* passive ion leak across the gill epithelium. Furthermore, the gradient driving passive ion leak increases as environmental ion concentration decreases. Thus, depending on the degree of the ionic challenge, reducing metabolic rate and development might not be a viable option for reducing ion demand as it is for oxygen.

Although gross ion uptake is reduced in the studies reviewed here (Sect. 15.6), measurements of larval ion content suggest that net ion uptake is likely maintained, probably by reducing ion loss. Most surprising is that the reduction in ion loss to maintain net ion uptake appears to come without significant cost. Comparisons across the range of water ion concentrations reviewed here did not reveal any significant differences in metabolic rate, larval ion content, or skin-to-gill transition timing of gas/ion flux. However, measurements of yolk absorption rate and conversion efficiency were not made, both of which could reveal energetic costs. Furthermore, nonsignificant trends present in some of the reviewed studies suggest that small costs might be incurred over this range of ion concentrations. Indeed, Fu et al. (2010) and Zimmer and Wood (2015) found that ion and gas flux transitioned to the gill ~1–2 days earlier in the ion-poor water. Further, although Gallagher (2014) found no differences in sodium content between larvae reared in 0.06 and 1.7 mM [Na^+], Zimmer and Wood (2015) report ~20% lower sodium content between larvae reared in 0.6 and 60 mM [Na^+] at 18 dph. These differences range from nonsignificant to modest but may become more pronounced if measured across a broader range of ion concentrations.

15.8 Summary and Future Directions

Relative to the adult form, the larval trout gill exhibits very little plasticity during development with respect to gas exchange and ion regulation. Gill morphology is mostly fixed early on, and plasticity appears limited to changes in convection and protein function. This limited plasticity is likely produced by selection for rapid development and minimal endogenous reserves and may play a major role in the survival bottleneck associated with larval development. However, the conclusions presented here are based on very few studies, and the drivers proposed for the observed patterns are largely untested. More research is clearly required for a comprehensive understanding of developmental plasticity in rainbow trout specifically and teleosts generally. We thus conclude with suggestions to more completely address some of the knowledge gaps and patterns of interest identified by this review.

15.8.1 What Is the Full Extent of Larval Gill Plasticity?

Larval gill plasticity is clearly limited relative to the adult form, but it likely still plays an important role in determining how larvae cope with environmental variability. Surprisingly, the full extent of this plasticity is poorly characterized. For example, chronic hypoxic studies have generally tested only one oxygen tension

(33 mmHg), and measurements of hypoxia-induced changes to gill morphology have not been compared with stage-matched controls.

Comparing stage-matched larvae reared under multiple oxygen tensions until ysa would help characterize the full extent of larval gill plasticity. This would allow us to understand how different oxygen tensions impact larval development and how that might change with plasticity as development progresses. Repeating this experiment under multiple ion concentrations would be equally informative. This work would yield important insight into how different environmental conditions might impact the larval survival bottleneck at different points in development.

15.8.2 Are There Costs Associated with Reduced Rates of Gross Sodium Uptake?

As discussed in Sect. 15.6, larval trout reared in low-sodium water decrease the rate of gross sodium uptake by more than half without any apparent cost to sodium balance or oxygen uptake. It was also discussed how this reduction in gross sodium uptake is likely balanced by a reduction in sodium loss to maintain net sodium uptake. However, it remains unknown why higher rates of gross sodium uptake and loss are maintained in higher environmental concentrations if lower rates of gross uptake and loss bear no additional costs. One possible explanation is that a lower sodium uptake rate affects biological processes not measured or challenged in the experiments presented in this review. Indeed, only 10% of sodium taken up from the water at 2.3 mM [Na^+] is estimated as biologically incorporated (Fu et al. 2010), while the rest is believed to facilitate trans-epithelial flux of ammonia/ammonium (Zimmer et al. 2014) and acid-base relevant ions (Brauner and Rombough 2012). Zimmer and Wood (2015) found no difference in ammonia excretion between larvae reared in 0.6 and 60 mM [Na^+], but effects might be revealed over a broader range of concentrations. To our knowledge, the capacity for acid-base regulation in larval trout has not been tested as a function of environmental sodium and should thus be challenged across a range of concentrations as well.

15.8.3 Do Endogenous Reserves Determine the Scope for Plasticity?

Section 15.7.1.2 proposes selection for smaller endogenous reserves as a driver for limited larval gill plasticity. Examining the relationship between natural or experimental variation in relative yolk mass and metrics of plasticity like P_{crit} and ionic imbalance could help test this hypothesis. Furthermore, if a relationship between plasticity and endogenous reserves is found, natural populations of salmonids that exhibit high spawning site fidelity could provide insight into the selective forces at play. Variation in relative yolk mass across populations could be expressed as a function of historical environmental variability and other selective pressures hypothesized to affect the level of endogenous reserves and/or the degree of plasticity expressed during larval development.

15.8.4 Does Larval Gill Development Recapitulate Chordate Gill Evolution?

One of the more notable aspects of larval gill development is how the gill assumes a primary role in the flux of sodium long before oxygen (15 vs 28 dph, Sect. 15.4.4; Fu et al. 2010). This pattern challenges traditional dogma that defines the fish gill as an organ primarily for gas exchange and secondarily for ion regulation. As discussed in Sect. 15.4, the shift from the skin to gills by gas and ion flux during growth facilitates the reduction in cutaneous flux capacity that occurs as the dermis thickens and body SA:V declines. The earlier shift by sodium to the gill suggests that these morphological changes associated with growth might be more limiting to cutaneous ion flux than gas.

Similarly, early chordate evolution is characterized by changes to dermal thickness and body SA:V that accompany dramatic changes to organism size and shape. As in larval teleosts, these morphological changes are believed to be facilitated by a shift from cutaneous to branchial flux. However, oxygen uptake is widely accepted as the primary role of the gill in this scenario. Testing gill function in extant species representative of early chordates might reveal a pattern similar to larval teleosts where ion flux shifts to the gill before gases. Such a finding would also challenge traditional dogma, suggesting that early chordate evolution was shaped primarily by constraints associated with ion regulation rather than oxygen uptake, clearly an exciting area for further research.

References

Bianchini K, Wright PA (2013) Hypoxia delays hematopoiesis: retention of embryonic hemoglobin and erythrocytes in larval rainbow trout, *Oncorhynchus mykiss*, during chronic hypoxia exposure. J Exp Biol 216:4415–4425

Brauner CJ, Rombough PJ (2012) Ontogeny and palaeophysiology of the gill: new insights from larval and air-breathing fish. Respir Physiol Neurobiol 184:293–300

Cameron JN (1989) The respiratory physiology of animals. Oxford University Press, Oxford

Chasiotis H, Kolosov D, Bui P, Kelly SP (2012) Tight junctions, tight junction proteins and paracellular permeability across the gill epithelium of fishes: a review. Respir Physiol Neurobiol 184:269–281

Chen Z, Snow M, Lawrence CS, Church AR, Narum SR, Devlin RH, Farrell AP (2015) Selection for upper thermal tolerance in rainbow trout (*Oncorhynchus mykiss* Walbaum). J Exp Biol 218:803–812

Ciuhandu CS, Wright PA, Goldberg JI, Stevens ED (2007) Parameters influencing the dissolved oxygen in the boundary layer of rainbow trout (*Oncorhynchus mykiss*) embryos and larvae. J Exp Biol 210:1435–1445

Coble DW (1961) Influence of water exchange and dissolved oxygen in redds on survival of steelhead trout embryos. Trans Am Fish Soc 90:469–474

Davis JC, Cameron JN (1971) Water flow and gas exchange at the gills of rainbow trout, *Salmo gairdneri*. J Exp Biol 54:1–18

Finn RN, Kapoor BG (2008) Fish larval physiology. CRC Press, Boca Raton

Flemmer AW, Monette MY, Djurisic M, Dowd B, Darman R, Gimenez I, Forbush B (2010) Phosphorylation state of the Na+-K+-Cl- cotransporter (NKCC1) in the gills of Atlantic killifish (*Fundulus heteroclitus*) during acclimation to water of varying salinity. J Exp Biol 213:1558–1566

Fu C, Wilson JM, Rombough PJ, Brauner CJ (2010) Ions first: Na^+ uptake shifts from the skin to the gills before O_2 uptake in developing rainbow trout, *Oncorhynchus mykiss*. Proc Roy Soc B 277:1553–1560

Gallagher EJ (2014) The ontogeny of sodium balance of rainbow trout (*Oncorhynchus mykiss*) and pink salmon (*Oncorhynchus gorbuscha*) during post-embryonic development in freshwater. cIRcle: UBC's digital repository: electronic theses and dissertations. http://hdl.handle.net/2429/50906

González ME, Blánquez MJ, Rojo C (1996) Early gill development in the rainbow trout, *Oncorhynchus mykiss*. J Morphol 229:201–217

Greco AM, Gilmour KM, Fenwick JC, Perry SF (1995) The effects of softwater acclimation on respiratory gas transfer in the rainbow trout *Oncorhynchus mykiss*. J Exp Biol 198:2557–2567

Greco AM, Fenwick JC, Perry SF (1996) The effects of softwater acclimation on gill structure in the rainbow trout *Oncorhynchus mykiss*. Cell Tissue Res 285:75–82

Hiroi J, Kaneko T, Seikai T, Tanaka M (1998) Developmental sequence of chloride cells in the body skin and gills of Japanese flounder (*Paralichthys olivaceus*) Larvae. Zool Sci 15:455–460

Holeton GF (1971) Respiratory and circulatory responses of rainbow trout larvae to carbon monoxide and to hypoxia. J Exp Biol 55:683–694

Holeton GF, Randall DJ (1967) Changes in blood pressure in the rainbow trout during hypoxia. J Exp Biol 46:297–305

Hwang PP (1990) Salinity effects on development of chloride cells in the larvae of ayu (*Plecoglossus altivelis*). Mar Biol 107:1–7

Hwang PP, Lin LY (2013) Gill ionic transport, acid-base regulation, and nitrogen excretion. In: Evans DH, Claiborne JB, Currie S (eds) The physiology of fishes, 4th edn. CRC Press, Boca Raton

Hwang PP, Lee TH, Lin LY (2011) Ion regulation in fish gills: recent progress in the cellular and molecular mechanisms. Am J Physiol Regul Integr Comp Physiol 301:R28–R47

Katoh F, Shimizu A, Uchida K, Kaneko T (2000) Shift of chloride cell distribution during early life stages in seawater-adapted killifish, *Fundulus heteroclitus*. Zool Sci 17:11–18

Li J, Eygensteyn J, Lock R, Verbost P, Heijden A, Bonga S, Flik G (1995) Branchial chloride cells in larvae and juveniles of freshwater tilapia *Oreochromis mossambicus*. J Exp Biol 198:2177–2184

Matey V, Richards JG, Wang Y, Wood CM, Rogers J, Davies R, Murray BW, Chen X-Q, Du J, Brauner CJ (2008) The effect of hypoxia on gill morphology and ionoregulatory status in the Lake Qinghai scaleless carp, *Gymnocypris przewalskii*. J Exp Biol 211:1063–1074

McDonald DG, McMahon BR (1977) Respiratory development in Arctic char *Salvelinus alpinus* under conditions of normoxia and chronic hypoxia. Can J Zool 55:1461–1467

Mead RW, Woodall WL (1968) Comparison of sockeye salmon fry produced by hatcheries, artificial channels and natural spawning areas. Int Pac Salmon Fish Comm 20:1–41

Ninness MM, Stevens ED, Wright PA (2006) Energy expenditure during hatching in rainbow trout (*Oncorhynchus mykiss*) embryos. Can J Fish Aquat Sci 63:1405–1413

Ou M, Hamilton TJ, Eom J, Lyall EM, Gallup J, Jian A, Lee J, Close DA, Yun S-S, Brauner CJ (2015) Responses of pink salmon to CO_2-induced aquatic acidification. Nat Clim Chang 5:950–955

Perry SF (1990) Recent advances and trends in the comparative morphometry of vertebrate gas exchange organs. In: Boutilier RG (ed) Vertebrate gas exchange. Advances in comparative and environmental physiology, vol 6. Springer, Berlin

Perry SF, Jonz MG, Gilmour KM (2009) Oxygen sensing and the hypoxic ventilatory response. In: Richards JG, Farrell AP, Brauner CJ (eds) Hypoxia. Fish physiology, vol 27. Elsevier, Amsterdam

Peterson NP, Quinn TP (1996) Spatial and temporal variation in dissolved oxygen in natural egg pockets of chum salmon in Kennedy Creek, Washington. J Fish Biol 48:131–143

Pisam M, Massa F, Jammet C, Prunet P (2000) Chronology of the appearance of beta, A, and alpha mitochondria-rich cells in the gill epithelium during ontogenesis of the brown trout (*Salmo trutta*). Anat Rec 259:301–311

Richards JG, Semple JW, Bystriansky JS, Schulte PM (2003) Na+/K+-ATPase α-isoform switching in gills of rainbow trout (*Oncorhynchus mykiss*) during salinity transfer. J Exp Biol 206:4475–4486
Rombough PJ (1988a) Growth, aerobic metabolism, and dissolved oxygen requirements of embryos and alevins of steelhead, *Salmo gairdneri*. Can J Zool 66:651–660
Rombough PJ (1988b) Respiratory gas exchange, aerobic metabolism, and effects of hypoxia during early life. In: Fish physiology, vol 11. Elsevier, Amsterdam, pp 59–161
Rombough PJ (1999) The gill of fish larvae. Is it primarily a respiratory or an ionoregulatory structure? J Fish Biol 55:186–204
Rombough PJ (2002) Gills are needed for ionoregulation before they are needed for O_2 uptake in developing zebrafish, *Danio rerio*. J Exp Biol 205:1787–1794
Rombough PJ (2011) The energetics of embryonic growth. Respir Physiol Neurobiol 178:22–29
Shen AC, Leatherland JF (1978) Effect of ambient salinity on ionic and osmotic regulation of eggs, larvae, and alevins of rainbow trout (*Salmo gairdneri*). Can J Zool 56:571–577
Soivio A, Nikinmaa M, Westman K (1980) The blood oxygen binding properties of hypoxic *Salmo gairdneri*. J Comp Physiol B 136:83–87
Tipsmark CK, Madsen SS, Seidelin M, Christensen AS, Cutler CP, Cramb G (2002) Dynamics of Na+, K+, 2Cl− cotransporter and Na+, K+-ATPase expression in the branchial epithelium of brown trout (*Salmo trutta*) and Atlantic salmon (*Salmo salar*). J Exp Zool 293:106–118
Tipsmark CK, Mahmmoud YA, Borski RJ, Madsen SS (2010) FXYD-11 associates with Na+-K+-ATPase in the gill of Atlantic salmon: regulation and localization in relation to changed ion-regulatory status. Am J Physiol Regul Integr Comp Physiol 299:R1212–R1223
Varsamos S, Diaz JP, Charmantier G, Blasco C, Connes R, Flik G (2002) Location and morphology of chloride cells during the post-embryonic development of the European sea bass, *Dicentrarchus labrax*. Anat Embryol 205:203–213
Wells P, Pinder A (1996) The respiratory development of Atlantic salmon. I. Morphometry of gills, yolk sac and body surface. J Exp Biol 199:2725–2736
Youngson AF, Malcolm IA, Thorley JL, Bacon PJ, Soulsby C (2004) Long-residence groundwater effects on incubating salmonid eggs: low hyporheic oxygen impairs embryo development. Can J Fish Aquat Sci 61:2278–2287
Zimmer AM, Wood CM (2015) Ammonia first? The transition from cutaneous to branchial ammonia excretion in developing rainbow trout is not altered by exposure to chronically high NaCl. J Exp Biol 218:1467–1470
Zimmer AM, Wright PA, Wood CM (2014) What is the primary function of the early teleost gill? Evidence for Na^+/NH_4^+ exchange in developing rainbow trout (*Oncorhynchus mykiss*). Proc Biol Sci 281:20141422

Part IV

The Human Component

The Human Embryo and Its Environment: In Utero Challenges and Physiological Outcomes

16

Damian Hutter

Abstract

The human embryo is totally dependent on the mother. Anything that affects the mother might have a direct impact on the growing embryo. While intoxication due to maternal drug abuse in industrialized countries is rare, chronic or intermittent hypoxia due to various causes remains one of the highest risks during pregnancy for the unborn child. This chapter will focus on the different mechanisms for intrauterine hypoxia and their impact on the fetal cardiovascular system.

16.1 Introduction

During the intrauterine life of the mammalian fetus, gas exchange and metabolism are solely dependent on the umbilical blood flow provided by the host. Therefore exposure to hypoxic environments, critical injuries of vital organs, impaired placental function, or altered oxygen carrier transport systems in the host directly impact intrauterine growth and organ function. This may affect embryogenesis and organogenesis and subsequently lead to perinatal fetal demise, stillbirth, intrauterine growth restriction (IUGR), birth asphyxia, or prematurity.

In the United States, IUGR and prematurity (<37 weeks) complicate 12.5% of all deliveries (Ananth and Vintzileos 2006). It remains the leading cause of perinatal mortality and morbidity, accounting for as many as 75% of perinatal deaths. Long-term disabilities including cerebral palsy, hearing loss, retinopathies, and mental and cardiopulmonary disorders lead to emotional stress for the families and to tremendous costs for the public healthcare system (Ananth and Vintzileos 2006).

D. Hutter
Division of Pediatric Cardiology, University Hospital of Bern, Bern, Switzerland
e-mail: Damian.Hutter@insel.ch

© Springer International Publishing AG, part of Springer Nature 2018

W. Burggren, B. Dubansky (eds.), *Development and Environment*,
https://doi.org/10.1007/978-3-319-75935-7_16

This chapter will explain different mechanisms for intrauterine hypoxia and their impact on the fetal cardiovascular system.

16.2 What Happens During Normal Pregnancy of a Human Being

To understand when and how hypoxia may expose the growing fetus to lethal risk, we have to understand what is happening during a normal pregnancy.

16.2.1 Invasion of the Endometrium or "Hypoxic" Organogenesis

As soon as the blastocyst makes contact with the epithelium of the uterus, it initiates the process of placentation. It is followed by cytotrophoblast, syncytiotrophoblast, and extravillous trophoblast formation (Irving et al. 1995). During the uterine invasion, the extravillous trophoblast forms plugs to prevent maternal blood flow from entering the intervillous space creating a hypoxic environment (Jauniaux et al. 2003; Jauniaux et al. 1992; Burton et al. 1999). Therefore during organogenesis, oxygen tension in the placenta remains lower than in the endometrium. Oxygen tension will only reach similar levels approximately in the period from the 12th to the 13th week of gestation when maternal blood finally enters the intervillous spaces (Rodesch et al. 1992; Jauniaux et al. 2001). In addition during the first trimester of pregnancy, the growing embryo and the placenta are unprotected against oxidative damage. It is not until the 8th to 9th week of gestation that antioxidant enzymes such as mitochondrial superoxide dismutase will be expressed by the syncytiotrophoblast (Watson et al. 1998; Jauniaux et al. 2000). When maternal blood flow is finally entering the extravillous trophoblast spaces, the sudden fall in the vascular resistance of the uteroplacental circulation increases the uterine artery blood flow from 10 to 350 ml/min near term (Palmer et al. 1992). At term, the uteroplacental blood flow is about 20% of the total cardiac output of the mother (Palmer et al. 1992). Increased endothelial NO production and other vasodilators are thought to be responsible for maintaining uterine artery flow, while adrenergic vasoconstrictor response is reduced (Mateev et al. 2003; Xiao et al. 2002; Hu et al. 1999). By mid-gestation the human uterine artery has doubled its diameter and the increased flow is accommodated by hyperplasia of all cell layers (Cipolla and Osol 1994). This change in diameter is likely related to the increasing uteroplacental circulation secondary to hormonal stimuli and uterine expression of vascular endothelial growth factor (VEGF) (Mulvany et al. 1996; Keyes et al. 1996; Ni et al. 1997).

16.2.2 Normal Physiologic Alteration During Pregnancy

Normal pregnancy is accompanied by complex physiologic changes of the cardiopulmonary system. Thus, changes may cause significant hemodynamic burden to preexisting underlying cardiovascular disease, or it could be the origin for newly acquired cardiovascular disease. In both circumstances, the life of the pregnant woman and the fetus are at high risk (Siu et al. 2001; Abbas et al. 2005).

16.2.2.1 Maternal Cardiovascular System

In the first trimester, cardiac output increases by 20–25% and reaches its maximum by the late second and beginning of the third trimester (30–40% above the pre-pregnant condition). This is achieved by a rise in stroke volume (30%) and to a lesser extent by an accelerated heart rate (10–20 beats/min). The increased cardiac output is a response to the lower systemic blood pressure at the end of the first trimester (Miller 2000; Siu and Colman 2001). It is suggested that the gestational hormones, the circulating prostaglandins, the excessive release of human placental growth factors, and the low-resistance uteroplacental unit are responsible for this hemodynamic phenomena (Chapman et al. 1998; Magness 1998). A crucial role is played by the stimulated endothelium-derived NO synthase and the fall of endothelin-1 levels (Furuhashi et al. 1995; Cadnapaphornchai et al. 2001). The increase of total blood volume is related to plasma expansion (30–40 ml/kg body weight) rather than an increase in total red blood cells, which is accounting for the relative anemia during pregnancy. Together with the improved cardiac output and the low blood viscosity, a rightward shift on the hemoglobin-oxygen dissociation curve occurs (Abbas et al. 2005; Miller 2000; Siu and Colman 2001; Magness 1998).

16.2.2.2 Maternal Pulmonary System

With the maternal hemodynamic changes, the gas exchange has to adapt accordingly. The chest diameter increases as the diaphragm rises due to the larger uterus muscle. In contrast to the cardiovascular system, the ventilatory adaption is established by anatomical changes to maintain the natural lung capacity and to accommodate the required oxygen demands. Hormonal stimulation (progesterone) of the central nervous system initiates mild hyperventilation to reduce the alveolar CO_2 resulting in a compensated respiratory alkalosis of pregnancy (American College of Obstetricians and Gynecologists 1996).

The heart develops early post conception in a rapid sequence from its origins in the heart field to a complete, looped organ by 8 weeks of gestation. During most of the cardiac development, the oxygen saturation never exceeds 20% exposing the fetus at an increased risk for oxidative damage. The fetal lungs play no role in the oxygenation. Approximately 60–70% of the total cardiac output of the fetus bypasses the lungs related to the unique fetal circulation in which the right and left ventricles work in parallel and are connected with several physiological shunts. By the time the extravillous spaces of the trophoblast are starting to be filled with maternal blood, the fetal heart is ready to deliver the increasing oxygen and nutritional demands of the growing fetus. The oxygen saturation in the fetus increases to about 60%. To maintain an adequate circulation, the fetal heart adjusts continuously to the rise in circulatory blood volume and the pressure load of the fetus. Indeed as soon as the heart is formed, the cardiovascular system grows along with the fetus and adapts to the physiological circulatory changes before and after birth.

16.3 In Utero Hypoxia

Intrauterine hypoxia may be caused by a variety of conditions. It is important to distinguish between host-, placenta-, and fetus-related causes as each mechanism manifests differently and has different outcomes. Kingdom (Kingdom and Kaufmann 1997) and colleagues suggested to divide the hypoxic changes in the feto-placental unit into three subtypes: (1) preplacental hypoxia where the mother and the fetus experience hypoxia (i.e., high altitude, cardiopulmonary disease in the host, etc.); (2) uteroplacental hypoxia, where maternal oxygenation is normal but uteroplacental circulation is impaired (i.e., preeclampsia, placenta insufficiency, etc.); and (3) post-placental hypoxia where only the fetus is hypoxic which is reflected by reduced, absent, or reversed end-diastolic flow in the umbilical arteries. Another way to classify intrauterine hypoxia is by dividing early- and late-onset intrauterine growth restriction (IUGR). The late-onset IUGR group involves the preplacental and uteroplacental hypoxic group (Kingdom et al. 2000) and is characterized by asymmetric patterns of growth (Rosenberg 2008).

16.3.1 Preplacental Hypoxia

16.3.1.1 Host-Related Intrauterine Hypoxia: Preplacental Hypoxia

There are three major causes for host-related intrauterine hypoxia:

- *A hypoxic environment, which demands major cardiopulmonary adaption, related to high altitude*
- *Poor hemodynamic predisposition with limited ability to respond to the physiologic cardiovascular changes*
- *Poor ventilatory capability to respond sufficiently to the increased oxygen demands during pregnancy*

Additional factors like anemia, infections, or chronic inflammatory diseases in combination with one or several of the abovementioned causes further enhance fetal hypoxia and increase the risk for adverse intrauterine events.

16.3.1.2 Hypoxic Environment: High Altitude

Chronic hypoxia is believed to play a key role in the etiology of IUGR and onset of preeclampsia. High-altitude environments provide a perfect setting to mimic such conditions.

It is well known that by every gain of 1000 m altitude, the birth weight declines on average by 100 g (Jensen and Moore 1997; Giussani et al. 2001; Mortola et al. 2000). Interestingly, chronic hypoxia at high altitudes even has a greater effect on birth weight than maternal risk factors like low weight gain, smoking, primiparity, or preeclampsia (Jensen and Moore 1997). The fact that two thirds of fetal growth occurs in the third trimester of pregnancy indicates that a decline in birth weight must be related to changes in this late period of fetal life (Rosenberg 2008).

An epidemiologic investigation in birth weight showed that at 2744 m above sea level, the average birth weight is 240 g less than at 915 m above sea level which was difficult to explain by prematurity as gestational age was on average only 0.5 weeks shorter (Jensen and Moore 1997). Today we know that the growth of premature babies begins to slow down between the 25th and 31st week of gestation and is more affected in the abdominal than the head circumference (Krampl et al. 2000). A possible relationship between chronic hypoxia and preeclampsia or preeclamptic-like clinical conditions was suspected to contribute to the intrauterine growth restriction (Moore et al. 1982). Indeed, preeclamptic-like pregnancies are more frequent in high-altitude environments. As these pregnancies already carry an increased risk for IUGR and prematurity under normal conditions, they were suspected to be major contributors for the birth weight decline in high altitudes (Keyes et al. 2003). Why there is a higher incidence of preeclamptic-like pregnancies at high altitude remains unclear. The early suggested concepts with hypoxia leading to impaired trophoblast invasion into the uteroplacental arteries during the first trimester of pregnancy are no longer found to be valid explanations (Sibai et al. 2005; Caniggia et al. 2000a, b). The observation of systemic arterial hypertension and proteinuria in women with high-altitude residency may relate to maladaption to chronic hypoxia rather than classic preeclampsia (Keyes et al. 2003; Sibai et al. 2005; Palmer et al. 1999). While adapting to the hypoxic environment, the pregnant host has to deal with the oxygen and nutritional demands of its fetus. This is accommodated by increased alveolar ventilation. The ventilatory change improves the arterial oxygen saturation to levels similar to conditions at sea level (McAuliffe et al. 2001). The higher the sensitivity to hypoxia, the better the birth weight of the newborn (Moore 2001).

While oxygen uptake is well maintained by the ventilatory system, the effect of chronic hypoxia on the systemic and uteroplacental vascular system remains somewhat more demanding; cardiac output in high altitude is lower due to lower heart rate and smaller stroke volume (Kitanaka et al. 1989; Kametas et al. 2004). In addition, the total blood volume in pregnant women living permanently at high altitude is lower and it is already lower in pre-pregnant conditions. Near term, the arterial oxygen content in pregnant women is lower than in nonpregnant women, which may be the result of the physiologic anemia during pregnancy with greater plasma volume expansion in favor of red blood cell mass. These findings though are only observed in women living in high altitudes for less than three generations (McAuliffe et al. 2001). While ventilation seems not to interfere with cardiac output, systemic vascular resistance (SVR) does (Moore et al. 1986). Chronic hypoxic guinea pigs near term have lower cardiac output and increased SVR than her normoxemic controls (Harrison and Moore 1990). In fact pregnant women of high-altitude residency not only lack the physiological blood pressure fall at the beginning of the second trimester, but also the "normotensive"-considered individuals have higher blood pressures (Moore et al. 1982; Palmer et al. 1999). It is suspected that chronic hypoxia leads to a diminished response of endothelial nitric oxide synthase (eNOS), while the sympathetic nervous system (α_1-/α_2-adrenergic receptor) shows increased activity (Mateev et al. 2003, 2006; Hu et al. 1999; Cadnapaphornchai et al. 2001; Hu et al. 1996; Zhang et al. 2008). In addition, potent vasoconstrictors like endothelin-1 and the hypoxia-inducible factor (HIF) may be stimulated early in pregnancy by excessive generation of reactive oxygen species (ROS) pathways due to oxidative stress (James et al. 2006). It is believed that these early changes have in part an important impact on the severity in clinical manifestation of preeclampsia near term in affected high-altitude residents.

In hypoxic environments, the uterine artery has an inadequate growth response with reported smaller uterine artery diameter in pregnant high-altitude residents. These individuals have reported lower uterine artery blood flow (Zamudio et al. 1995). A direct association between uterine arterial flow and birth weight is supported by different studies conducted in women with different origins (Moore et al. 2001; Julian et al. 2009).

Hypoxic conditions as encountered in high-altitude environments impose a significant burden on the genetically unadapted host. The complex cardiopulmonary response triggered by the early placentation shows subtle physiologic changes to accommodate the most vital demands of the growing embryo/fetus and the pregnant host.

16.3.1.3 Poor Hemodynamic Predisposition with Limited Ability to Respond to the Physiologic Cardiovascular Changes: Underlying (Congenital) Cardiac Disease

Women with underlying (congenital) cardiac disease are at an increased risk to develop pregnancy complications (Siu et al. 2002). Patients at risk are classified as low, intermediate, or high with a 5%, 25%, and 75% chance of developing cardiac complications during pregnancy, respectively (Siu et al. 2001). The highest risk is observed in mothers with severe left-sided obstructive lesions (i.e., aortic stenosis, coarctation), pulmonary hypertension, and Marfan syndrome with aortic root dilatation, as well as individuals with symptoms of moderate to severe heart failure (NYHA functional class III and IV). Increasing maternal hypotension is the most important symptom associated with intrauterine growth restriction (20–25%) and prematurity (20–25%) (Hameed et al. 2001). Interestingly, mothers with a congenital heart disease do not belong to the high-risk group for an adverse maternal outcome. The live-birth rate is reportedly only 40–45% if the mother has cyanotic heart disease. Rates decrease another 10–15% when the maternal oxygen saturation drops below 85%. In addition, extreme prematurity affects 35–40% of these pregnancies (Presbitero et al. 1994). Fetal or neonatal death, brain hemorrhage secondary to maternal anticoagulation or to extreme prematurity, and IUGR are common findings in offspring of pregnant women with congenital or acquired heart disease (Siu et al. 2001, 2002; Sellman and Holman 2000).

16.3.1.4 Poor Ventilatory Capability to Respond Sufficiently to the Increased Oxygen Demands During Pregnancy: Underlying Pulmonary Disease

Chronic pulmonary disease may have similar maternal-fetal consequences as chronic exposure to hypoxia (American College of Obstetricians and Gynecologists 1996) (Sect. 16.3.1.2). Poorly controlled asthma is associated with preeclampsia, uterine hemorrhage, preterm delivery, and low birth weight (Kumar 2008; Guy et al. 2004). Among chronic lung diseases, cystic fibrosis (CF) and tuberculosis are the most common conditions: 5% of the world population is carrying the CF gene, and 30% of humans have been infected with *Mycobacterium tuberculosis* (Inge and Wilson 2008; Rowe et al. 2005). Pregnancy in cystic fibrosis patients seems to have a positive effect on maternal long-term survival (Goss et al. 2003), despite the increased maternal risk for infections and insulin resistance (Cheng et al. 2006; Hardin et al. 2005) and the increased fetal risk of prematurity and/or IUGR (Rosenberg 2008; Cheng et al. 2006; Odegaard et al. 2002). All pregnant mothers carry a higher risk for acute respiratory infections independent of their underlying health condition. One percent of the women experience symptoms of bronchitis or pneumonia during the course of pregnancy. Although thanks to current antibiotic treatment, maternal mortality has decreased dramatically, but it remains a relevant cause of maternal death (American College of Obstetricians and Gynecologists 1996). The major risk for the fetus lies in maternal respiratory failure and inappropriate antibiotic and/or antiviral treatment with the potential of adverse fetal effects (American College of Obstetricians and Gynecologists 1996; McColgin et al. 1992; Shapiro 2006).

16.3.1.5 Decreased Oxygen Delivery Capacity

Maternal hematological disorders may directly affect oxygen transfer. Iron deficiency anemia (IDA) is common in pregnancy (Allen 1997) and often related to malnutrition or micronutrient diets (Mahajan et al. 2008; Singla et al. 1997; Baker et al. 2009). IDA is associated with increased risk for IUGR and prematurity (Singla et al. 1997; Barker et al. 1990; Lao and Wong 1997; Williams et al. 1997).

Contrarily to IDA, the oxygen carrier capacity is inherently altered in hemoglobinopathies. Sickle cell disease, the most common hereditary blood disorder, is particularly common in Africans and Afro-Americans (Davies et al. 2000; Motulsky 1973). It may be present in combination with hemoglobin C or β-thalassemia (Hb S/C or Hb S/β). The most severe form Hb SS (homozygous HbS) is called sickle cell anemia, but any Hb S combination (Hb S/C or Hb S/β) can potentially cause vaso-occlusive crisis and hemolysis (ACOG Committee on Obstetrics 2007). The red blood cells assume an abnormal, rigid sickle shape with decreasing oxygen tension. The increase in blood viscosity leads to regional hypoperfusion, acidosis, and hemolysis (vaso-occlusive crisis). Patients with sickle cell disease and combination of it are at higher risk for maternal (i.e., preterm labor, preterm rupture of membranes, postpartum infections) and fetal complications (i.e., abortion, prematurity, IUGR, low birth weight, stillbirth) (Sun et al. 2001). Close fetal monitoring during pregnancy and prophylactic exchange transfusion seem to be often effective in abolishing life-threatening intrauterine hypoxic events (Koshy et al. 1988). Thalassemia is an inherited autosomal recessive blood disease which is particularly prevalent in Asians (α-form) and among Mediterranean people (β-form). The genetic defect results in reduced rate of synthesis of α- or β-globin chains which make up hemoglobin (ACOG Committee on Obstetrics 2007; Kazazian Jr. 1990). Homozygous individuals present with severe anemia (Cooley's anemia) and extramedullary erythropoiesis. Alpha-thalassemia major (Hb Bart's) is associated with hydrops fetalis, intrauterine death, and preeclampsia (Davies et al. 2000). β-Thalassemia is a result of a mutation in the β-globin gene causing deficient or absent β chain production with absence of hemoglobin. Death usually occurs in infancy unless periodic blood transfusions are initiated at a young age. The clinical picture of β-thalassemia varies in severity as it is dependent on the variable expression of Hb A. Pregnancy in thalassemia carriers is usually uncomplicated. Successful pregnancies in women with α- and β-thalassemia major have been reported (Ong et al. 1977) but were associated with a higher incidence of IUGR, low birth weight, and prematurity (Aessopos et al. 1999; Jensen et al. 1995).

16.3.2 Uteroplacental Hypoxia

Uteroplacental hypoxia has been related to abnormal placentation early in gestation and to placental vascular disease later in pregnancy. Abnormal placental implantation is a common finding in pregnancies complicated by IUGR, by gestational hypertension, and by preeclampsia.

Multiple studies were able to demonstrate an increased risk for both mother and fetus to develop ischemic heart disease later in life (Barker et al. 1989, 1990, 2000; Ramsay et al. 2003; Wilson et al. 2003; Haukkamaa et al. 2004; Barker 1996; Herrera et al. 2007; Martyn et al. 1995; Sartori et al. 1999; Ness and Sibai 2006). We will focus on the group with chronic uteroplacental hypoxia as it impacts the whole duration of pregnancy.

Preeclampsia is a complex multisystem disorder observed in human pregnancy. Maternal clinical manifestations range from mild arterial hypertension and proteinuria to fully established HELLP syndrome (*h*emolysis, *e*levated *l*iver enzymes, *l*ow *p*latelet count) or eclampsia (Sibai 2003, 2004; Douglas and Redman 1994) with severe arterial hypertension, proteinuria, and multi-organ involvement (pulmonary edema, CNS symptoms, oliguria, thrombocytopenia, liver failure) (Sibai 2003; National High Blood Pressure Education Program Working Group 2000).

Causes for its origin are largely unknown (Sibai et al. 2005) but may be the result of a systemic inflammatory response perhaps related to an immature maternal immune response. As preeclampsia occurs more commonly in primigravid women of young age (incidence of 2–7%), it is suggested that this might be related to limited sperm exposure (Dekker and Sibai 2001; Einarsson et al. 2003). Semen in the female genital tract may induce an inflammatory-like response by the seminal vesicle-derived transforming growth factor β1 (TGF β1) inhibiting a normal immune response for semi-allogenic conceptus (Robertson et al. 2002). Sperm contact usually causes mucosal alloimmunization which seems to be reduced after limited sperm exposure or after intracytoplasmic sperm injection (Wang et al. 2002; Peters et al. 2004). Superficial placentation is a common finding in preeclamptic pregnancies (Sibai et al. 2005). This may be enhanced by the so-called hypoxia-related syncytiotrophoblast shedding (release of microvillous membrane particles and fetal DNA into the maternal circulation) in the first trimester of pregnancy and may result in systemic inflammatory-like response in the mothers (Sibai et al. 2005; Neale et al. 2003; Saito and Sakai 2003). Key abnormalities of preeclampsia include a rise in systemic vascular resistance, endothelial dysfunction, and activation of the coagulation system with enhanced platelet aggregation (National High Blood Pressure Education Program Working Group 2000). Endothelial dysfunction is responsible for the impaired generation and activity of vasodilators such as prostacyclin and eNO and could explain surface-mediated platelet activation and fibrin formation in the uteroplacental circulation (Redman 1990).

Depending on the severity of the preeclampsia, the condition may lead to intrauterine hypoxia and/or oxidative stress in the fetus. Preeclampsia is associated with IUGR and prematurity (Ness and Sibai 2006). Fetal morbidity and mortality increases significantly when preeclampsia develops prior to 33 gestational weeks (Hauth et al. 2000; Hnat et al. 2002; Buchbinder et al. 2002; Zhang et al. 2003). Preeclamptic mothers and their offspring are considered at an increased risk for early-onset cardiovascular disease later in life.

16.3.3 Post-placental Hypoxia

A main consequence of chronic hypoxia is the failure of the fetus to achieve its genetically determined growth potential. About 10% of all babies grow poorly in utero and are born small for gestational age (Pollack and Divon 1992). IUGR is associated with distress and asphyxia and a six- to tenfold increased perinatal mortality (Pollack and Divon 1992). Frequent hypoxia-mediated complications include meconium aspiration, metabolic and hematologic disturbances, cognitive dysfunction, and cerebral palsy. Acute and chronic hypoxia is also associated with various obvious morphological and functional fetal cardiac changes (Meyer and Lubo 2007; Webster and Abela 2007; Zhang 2005) that either aim to compensate for the reduced oxygenation of vital organs or are the result of hypoxia-mediated fetal tissue damage.

Hemodynamic consequences: Irrespective of the underlying hypoxia mechanism, at an initial stage, the human fetus may be able to adapt with an increase in blood supply from the venous duct to the brain, myocardium, and upper body and a reduction in perfusion of the kidneys, gastrointestinal tract, and lower extremities. This redistribution of blood allows preferential delivery of nutrients and oxygen to the most vital organs. Cerebral vasodilatation to spare the brain hypoxic damage leads to a decrease in left ventricular afterload, while systemic arterial vasoconstriction of the lower body vessels increases right ventricular afterload. This mechanism may be verified by echocardiography demonstrating a low middle cerebral artery pulsatility index and a shift of cardiac output in favor of the left ventricle (Fouron et al. 1999). With further deterioration of the fetal oxygenation, this protective mechanism is overwhelmed by the decline in cardiac output and the emergence of fetal distress. The final stage is characterized by a decline in systolic and diastolic fetal cardiac function, secondary to myocardial ischemia (Fouron et al. 1991). Raised atrial contraction results in transmission of atrial pressure waves into the venous duct and umbilical vein, causing end-diastolic umbilical vein pulsation. At this stage, reduced or reversed end-diastolic flow velocity may also be found in pulmonary veins, and coronary blood flow may become visible with increased systolic-diastolic flow velocities ("heart sparing"). If not delivered, intrauterine death occurs usually within a few days (Gudmundsson et al. 1996).

Similar findings to humans have been reproduced in animals. In high altitude, e.g., the cardiac output of hypoxic fetal sheep heart is reduced (Browne et al. 1997a, b), but the hemoglobin level increased to maintain a near-normal oxygen delivery to the fetal myocardium. In addition, coronary blood flow is increased (Garcia et al. 2000; Kamitomo et al. 1993) although there is no change in capillary/muscle fiber ratio, capillary volume density, or capillary diameter (Thompson 2003). In addition, myocardial contractility of hypoxic lambs is reduced (Browne et al. 1997a, b).

While chronic hypoxia leads to disadvantageous fetal cardiac consequences, chronic anemia seems to have rather beneficial effects. The different responses of the fetal heart may be explained by the oxygen dissociation curve. It is far easier for the fetus to compensate for anemic conditions due to the higher oxygen affinity of fetal hemoglobin than it is to compensate for a hypoxic environment. In maternal anemia-mediated hypoxia, the fetus is able to increase the cardiac output and to increase the transplacental oxygen transfer by actively interfering with the iron metabolism of the mother.

Surviving babies seem to be particularly susceptible to the development of arterial hypertension and cardiovascular disease later in life. A direct association between low birth weight and early onset of essential arterial hypertension has first been articulated in the "fetal origins of adult disease hypothesis" (Barker 1992a, b, c; Barker and Martyn 1992). Barker's theory states that physiologic adaptations that enable a fetus to survive a period of intrauterine deprivation result in permanent reprogramming of the development of key organs and pathological consequences in postnatal life. In older children and adults, a low birth weight has been linked with increased arterial stiffness, systolic blood pressure, premature coronary heart disease, stroke, and diabetes (Barker et al. 1989, 1990; Bonamy et al. 2005; Johansson et al. 2005; Martin et al. 2000). Despite the strong epidemiologic evidence that support the "fetal origin" concept, we still do not know what mechanisms and pathways are involved in "fetal programming." As a matter of fact, chronic hypoxia increases oxidative stress, irrespective of the mechanism. This initiates a complex signaling cascade with upregulation of genes dealing with oxidative stress, while many genes with effects on cell communication, survival, and signaling are downregulated (Zhang 2005; Bae et al. 2003). This downregulation is believed to play a role in the long-term cardiovascular programming. This hypoxia-mediated cardiovascular imprint appears also to leave the adult heart more vulnerable against ischemia/reperfusion injury (Beere and Green 2001; Hutter et al. 1996; Jayakumar et al. 2001; Li et al. 2003; Snoeckx et al. 2001; Xi et al. 2001; Yellon et al. 1992).

16.3.3.1 Teratogenicity

Recently it has also been suggested that hypoxia early in gestation may be teratogenic to the human embryo. As such, maternal asthma exacerbation during the first trimester of pregnancy reportedly increased the risk for congenital malformations including the risk of cardiovascular malformations (Blais and Beauchesne 2004). As described above, maternal blood enters the intervillous space of the human placenta only after 10–12 gestational weeks (Burton et al. 1999; Jauniaux et al. 2000), and until this moment the placental metabolism is anaerobic. Yet, the human heart forms early in the period of anaerobic metabolism between day 15 and day 60 post-conception. Interestingly, if animal embryos are exposed to chronic hypoxia, cardiac malformations seem not to occur more common.

16.3.3.2 Cellular Effects of Hypoxia

That early fetal hypoxia triggers cardiac remodeling has been shown in rats that were randomized either into a hypoxic, nutrient-restricted, or control study cohort. Hypoxia was associated with enhanced apoptosis together with significant increase in binucleated myocytes (Bae et al. 2003). At the age of 4 months, hypoxia exposure was associated with increased heart weight/body weight ratio presumably due to hypertrophy of myocardium in the presence of slowed fetal growth, increased β-/α-myosin heavy chain ratio, and increased collagen I and III expression, while matrix metalloproteinase-2 activity was lower when compared with the other groups. The hemodynamic consequences of these changes are higher end-diastolic pressure related to less compliant left ventricle and a reduced capability to recover from ischemia. Interestingly, nutrient-restricted rats had similar findings at the age of 7 months (Xu et al. 2006).

Apoptosis is an active physiologic process that removes unwanted or defective cells by intrinsic programmed cell suicide (Zhang 2005). Even under normal conditions, billions of cells die each day due to natural apoptosis. The extrinsic pathway of cell apoptosis is initiated through binding of death-inducing ligands from the tumor necrosis factor (TNF) gene superfamily such as Fas, TNF-related apoptosis ligand (TRAIL), and TNF-α to the surface of the target cell. This activates initiator caspase-8, which in turn activates effector caspase-3 which leads to the controlled cell death by apoptosis where normal DNA molecules are cut into small fragments in a genetically programmed manner. The intrinsic pathway is activated via cytochrome c release from the mitochondria and is regulated by Bcl-2 family proteins with pro-apoptotic (e.g., Bax, Bak) or anti-apoptotic (e.g., Bcl-2, Bcl-XL) members. Overexpression of pro-apoptotic proteins leads to cytochrome c release from the mitochondria, which activates initiator caspase-9. Further signaling initiates effector caspase-3 to execute controlled cell death by apoptosis. Cross talk between the extrinsic and intrinsic pathway amplifies cytochrome c release from the mitochondria and accelerates apoptosis (Green and Reed 1998). Gene profiling of fetal rat hearts which were exposed to maternal hypoxic conditions shows an upregulation of genes related to oxidative stress and metabolism, while proliferation, communication, defense, and survival are downregulated (Zhang 2005; Bae et al. 2003). This is reflected in increased levels of caspase-3, caspase-8, and Fas (Bae et al. 2003; Tanaka et al. 1994; Yaniv et al. 2002). In contrast, anti-apoptotic protein levels are decreased (Bae et al. 2003; Kajstura et al. 1995). As pro-apoptotic proteins (e.g., Bax) likely need the extrinsic pathway for translocation from the cytosol to the mitochondria, their levels remain unaffected by chronic hypoxia (Bae et al. 2003; Murphy et al. 2000; Saikumar et al. 1998).

Another important role in the signaling process of apoptosis is played by nitric oxide (NO) which is primarily known as a vasodilator. NO is also able to inhibit or to induce controlled cell death. In rat hearts exposed to oxidative stress, many genes that affect cell communication, survival, and signaling were downregulated (Zhang 2005; Bae et al. 2003). This downregulation is believed to be partly responsible for the long-term consequences of intrauterine hypoxia and leaves a persistent cardiovascular "imprint" that leads to cardiovascular disease in later life. The transcription of the heat shock protein 70 (Hsp70) might be an example of this observed cardiac programming phenomenon. Hsp70 is a protein that protects against myocardial ischemia and stress (hyperthermia) (Snoeckx et al. 2001; Xi et al. 2001; Yellon et al. 1992) and inhibits apoptosis by preventing the formation of caspase-9. In chronic intrauterine hypoxia conditions, the expression of Hsp70 is downregulated (Beere and Green 2001). This effect persists into adulthood and may explain why some adult hearts are more vulnerable against ischemia/reperfusion injury (Beere and Green 2001; Hutter et al. 1996; Jayakumar et al. 2001; Li et al. 2003; Snoeckx et al. 2001; Xi et al. 2001; Yellon et al. 1992). The expression of endothelial nitric oxide is also important for the long-term cardioprotection of the cardiomyocytes. eNOS levels are also decreased in rat hearts that were exposed to intrauterine hypoxia (Thompson and Dong 2005). Similar changes were observed in the regulation of the β-adrenoreceptors (βAR) and the coupling G proteins. β_2AR and $G_s\alpha$ are upregulated in adult rat hearts that were exposed in utero to chronic hypoxia. This upregulation preserves cardiac contractility during states of hypoxia (Xiao 2000; Chesley et al. 2000), but the regulatory mechanism appears to be lost in adulthood presumably due to altered prenatal programming.

Conclusion

Hypoxia does not play a major role in the early development of structural cardiac malformations probably because early embryogenesis already takes place in anaerobic conditions. During the second and third trimester, oxygen becomes more important for the normal fetal organogenesis and growth. If hypoxia occurs at that stage, the fetus has a number of protective options. Immediate protection against oxidative stress is established by upregulation of the corresponding genes. Nitric oxide generation sets in for enhanced cell signaling for defense, platelet inhibition, and regulation of apoptosis. β_2AR and $G_s\alpha$ will be upregulated to maintain a sufficient cardiac output. With persistent hypoxia premature exit of cell cycle is initiated together with enhanced apoptosis, which results in fewer, but hypertrophied cardiomyocytes. This process probably aims for better energy efficiency during hypoxic conditions, but the ventricles become less compliant. Altered regulatory gene expression in response to in utero hypoxia appears to extend into adulthood and may mimic the changes that are found in adults with chronic heart failure. Hypoxia slows fetal growth. Growth restriction is currently considered an important risk factor of premature arterial hypertension and cardiovascular disease, probably secondary to endothelial dysfunction. Further investigations are needed to explore preventative strategies such as the early use of antioxidants and selective vasodilators to limit the effects of intrauterine hypoxia.

References

Abbas AE, Lester SJ, Connolly H (2005) Pregnancy and the cardiovascular system. Int J Cardiol 98(2):179–189

ACOG Committee on Obstetrics (2007) ACOG Practice Bulletin No. 78: hemoglobinopathies in pregnancy. Obstet Gynecol 109(1):229–237

Aessopos A et al (1999) Pregnancy in patients with well-treated beta-thalassemia: outcome for mothers and newborn infants. Am J Obstet Gynecol 180(2 Pt 1):360–365

Allen LH (1997) Pregnancy and iron deficiency: unresolved issues. Nutr Rev 55(4):91–101

American College of Obstetricians and Gynecologists (1996) ACOG technical bulletin. Pulmonary disease in pregnancy. Number 224--June 1996. American College of Obstetricians and Gynecologists. Int J Gynaecol Obstet 54(2):187–196

Ananth CV, Vintzileos AM (2006) Epidemiology of preterm birth and its clinical subtypes. J Matern Fetal Neonatal Med 19(12):773–782

Bae S et al (2003) Effect of maternal chronic hypoxic exposure during gestation on apoptosis in fetal rat heart. Am J Physiol Heart Circ Physiol 285(3):H983–H990

Baker PN et al (2009) A prospective study of micronutrient status in adolescent pregnancy. Am J Clin Nutr 89(4):1114–1124

Barker DJ (1992a) The fetal origins of adult hypertension. J Hypertens Suppl 10(7):S39-44

Barker DJ (1992b) The fetal origins of diseases of old age. Eur J Clin Nutr 46(Suppl 3):S3–S9

Barker DJ (1992c) Fetal growth and adult disease. Br J Obstet Gynaecol 99(4):275–276

Barker DJ (1996) Growth in utero and coronary heart disease. Nutr Rev 54(2 Pt 2):S1–S7

Barker DJ, Martyn CN (1992) The maternal and fetal origins of cardiovascular disease. J Epidemiol Community Health 46(1):8–11

Barker DJ et al (1989) Growth in utero, blood pressure in childhood and adult life, and mortality from cardiovascular disease. BMJ 298(6673):564–567

Barker DJ et al (1990) Fetal and placental size and risk of hypertension in adult life. BMJ 301(6746):259–262

Barker DJ et al (2000) Growth in utero and blood pressure levels in the next generation. J Hypertens 18(7):843–846

Beere HM, Green DR (2001) Stress management - heat shock protein-70 and the regulation of apoptosis. Trends Cell Biol 11(1):6–10

Blais L, Beauchesne MF (2004) Use of inhaled corticosteroids following discharge from an emergency department for an acute exacerbation of asthma. Thorax 59(11):943–947

Bonamy AK et al (2005) Preterm birth contributes to increased vascular resistance and higher blood pressure in adolescent girls. Pediatr Res 58(5):845–849

Browne VA et al (1997a) Cardiac beta-adrenergic receptor function in fetal sheep exposed to long-term high-altitude hypoxemia. Am J Phys 273(6 Pt 2):R2022–R2031

Browne VA et al (1997b) Activator calcium and myocardial contractility in fetal sheep exposed to long-term high-altitude hypoxia. Am J Phys 272(3 Pt 2):H1196–H1204

Buchbinder A et al (2002) Adverse perinatal outcomes are significantly higher in severe gestational hypertension than in mild preeclampsia. Am J Obstet Gynecol 186(1):66–71

Burton GJ, Jauniaux E, Watson AL (1999) Maternal arterial connections to the placental intervillous space during the first trimester of human pregnancy: the Boyd collection revisited. Am J Obstet Gynecol 181(3):718–724

Cadnapaphornchai MA et al (2001) Chronic NOS inhibition reverses systemic vasodilation and glomerular hyperfiltration in pregnancy. Am J Physiol Renal Physiol 280(4):F592–F598

Caniggia I et al (2000a) Hypoxia-inducible factor-1 mediates the biological effects of oxygen on human trophoblast differentiation through TGFbeta(3). J Clin Invest 105(5):577–587
Caniggia I et al (2000b) Oxygen and placental development during the first trimester: implications for the pathophysiology of pre-eclampsia. Placenta 21(Suppl A):S25–S30
Chapman AB et al (1998) Temporal relationships between hormonal and hemodynamic changes in early human pregnancy. Kidney Int 54(6):2056–2063
Cheng EY et al (2006) Aggressive prenatal care results in successful fetal outcomes in CF women. J Cyst Fibros 5(2):85–91
Chesley A et al (2000) The beta(2)-adrenergic receptor delivers an antiapoptotic signal to cardiac myocytes through G(i)-dependent coupling to phosphatidylinositol 3'-kinase. Circ Res 87(12):1172–1179
Cipolla M, Osol G (1994) Hypertrophic and hyperplastic effects of pregnancy on the rat uterine arterial wall. Am J Obstet Gynecol 171(3):805–811
Davies SC et al (2000) Screening for sickle cell disease and thalassaemia: a systematic review with supplementary research. Health Technol Assess 4(3):i–v. 1–99
Dekker G, Sibai B (2001) Primary, secondary, and tertiary prevention of pre-eclampsia. Lancet 357(9251):209–215
Douglas KA, Redman CW (1994) Eclampsia in the United Kingdom. BMJ 309(6966):1395–1400
Einarsson JI, Sangi-Haghpeykar H, Gardner MO (2003) Sperm exposure and development of pre-eclampsia. Am J Obstet Gynecol 188(5):1241–1243
Fouron JC et al (1991) Diastolic circulatory dynamics in the presence of elevated placental resistance and retrograde diastolic flow in the umbilical artery: a Doppler echographic study in lambs. Am J Obstet Gynecol 164(1 Pt 1):195–203
Fouron JC et al (1999) Relationship between flow through the fetal aortic isthmus and cerebral oxygenation during acute placental circulatory insufficiency in ovine fetuses. Am J Obstet Gynecol 181(5 Pt 1):1102–1107
Furuhashi N et al (1995) Maternal plasma endothelin levels and fetal status in normal and pre-eclamptic pregnancies. Gynecol Obstet Investig 39(2):88–92
Garcia FC et al (2000) Ca(2+) sensitivity of fetal coronary arteries exposed to long-term, high-altitude hypoxia. J Soc Gynecol Investig 7(3):161–166
Giussani DA et al (2001) Effects of altitude versus economic status on birth weight and body shape at birth. Pediatr Res 49(4):490–494
Goss CH et al (2003) The effect of pregnancy on survival in women with cystic fibrosis. Chest 124(4):1460–1468
Green DR, Reed JC (1998) Mitochondria and apoptosis. Science 281(5381):1309–1312
Gudmundsson S et al (1996) Venous Doppler in the fetus with absent end-diastolic flow in the umbilical artery. Ultrasound Obstet Gynecol 7(4):262–267
Guy ES, Kirumaki A, Hanania NA (2004) Acute asthma in pregnancy. Crit Care Clin 20(4):731–745. x
Hameed A et al (2001) The effect of valvular heart disease on maternal and fetal outcome of pregnancy. J Am Coll Cardiol 37(3):893–899
Hardin DS et al (2005) The metabolic effects of pregnancy in cystic fibrosis. Obstet Gynecol 106(2):367–375
Harrison GL, Moore LG (1990) Systemic vascular reactivity during high-altitude pregnancy. J Appl Physiol 69(1):201–206
Haukkamaa L et al (2004) Risk for subsequent coronary artery disease after preeclampsia. Am J Cardiol 93(6):805–808
Hauth JC et al (2000) Pregnancy outcomes in healthy nulliparas who developed hypertension. Calcium for Preeclampsia Prevention Study Group. Obstet Gynecol 95(1):24–28
Herrera EA et al (2007) High-altitude chronic hypoxia during gestation and after birth modifies cardiovascular responses in newborn sheep. Am J Physiol Regul Integr Comp Physiol 292(6):R2234–R2240
Hnat MD et al (2002) Perinatal outcome in women with recurrent preeclampsia compared with women who develop preeclampsia as nulliparas. Am J Obstet Gynecol 186(3):422–426

Hu XQ et al (1996) Effects of long-term high-altitude hypoxemia on alpha 1-adrenergic receptors in the ovine uterine artery. Am J Phys 270(3 Pt 2):H1001–H1007
Hu XQ et al (1999) Effect of chronic hypoxia on alpha-1 adrenoceptor-mediated inositol 1,4,5-trisphosphate signaling in ovine uterine artery. J Pharmacol Exp Ther 288(3):977–983
Hutter JJ et al (1996) Overexpression of heat shock protein 72 in transgenic mice decreases infarct size in vivo. Circulation 94(6):1408–1411
Inge LD, Wilson JW (2008) Update on the treatment of tuberculosis. Am Fam Physician 78(4):457–465
Irving JA et al (1995) Characteristics of trophoblast cells migrating from first trimester chorionic villus explants and propagated in culture. Placenta 16(5):413–433
James JL, Stone PR, Chamley LW (2006) The regulation of trophoblast differentiation by oxygen in the first trimester of pregnancy. Hum Reprod Update 12(2):137–144
Jauniaux E et al (1992) Doppler ultrasonographic features of the developing placental circulation: correlation with anatomic findings. Am J Obstet Gynecol 166(2):585–587
Jauniaux E et al (2000) Onset of maternal arterial blood flow and placental oxidative stress. A possible factor in human early pregnancy failure. Am J Pathol 157(6):2111–2122
Jauniaux E, Watson A, Burton G (2001) Evaluation of respiratory gases and acid-base gradients in human fetal fluids and uteroplacental tissue between 7 and 16 weeks' gestation. Am J Obstet Gynecol 184(5):998–1003
Jauniaux E et al (2003) Comparison of ultrasonographic and Doppler mapping of the intervillous circulation in normal and abnormal early pregnancies. Fertil Steril 79(1):100–106
Jayakumar J et al (2001) Heat shock protein 70 gene transfection protects mitochondrial and ventricular function against ischemia-reperfusion injury. Circulation 104(12 Suppl 1):I303–I307
Jensen GM, Moore LG (1997) The effect of high altitude and other risk factors on birthweight: independent or interactive effects? Am J Public Health 87(6):1003–1007
Jensen CE, Tuck SM, Wonke B (1995) Fertility in beta thalassaemia major: a report of 16 pregnancies, preconceptual evaluation and a review of the literature. Br J Obstet Gynaecol 102(8):625–629
Johansson S et al (2005) Risk of high blood pressure among young men increases with the degree of immaturity at birth. Circulation 112(22):3430–3436
Julian CG et al (2009) Augmented uterine artery blood flow and oxygen delivery protect Andeans from altitude-associated reductions in fetal growth. Am J Physiol Regul Integr Comp Physiol 296(5):R1564–R1575
Kajstura J et al (1995) Programmed cell death and expression of the protooncogene bcl-2 in myocytes during postnatal maturation of the heart. Exp Cell Res 219(1):110–121
Kametas NA et al (2004) Maternal cardiac function during pregnancy at high altitude. BJOG 111(10):1051–1058
Kamitomo M et al (1993) Effects of long-term, high-altitude hypoxemia on ovine fetal cardiac output and blood flow distribution. Am J Obstet Gynecol 169(3):701–707
Kazazian HH Jr (1990) The thalassemia syndromes: molecular basis and prenatal diagnosis in 1990. Semin Hematol 27(3):209–228
Keyes LE et al (1996) Pregnancy-stimulated growth of vascular smooth muscle cells: importance of protein kinase C-dependent synergy between estrogen and platelet-derived growth factor. J Cell Physiol 166(1):22–32
Keyes LE et al (2003) Intrauterine growth restriction, preeclampsia, and intrauterine mortality at high altitude in Bolivia. Pediatr Res 54(1):20–25
Kingdom JC, Kaufmann P (1997) Oxygen and placental villous development: origins of fetal hypoxia. Placenta 18(8):613–621. discussion 623–6
Kingdom J et al (2000) Development of the placental villous tree and its consequences for fetal growth. Eur J Obstet Gynecol Reprod Biol 92(1):35–43
Kitanaka T, Gilbert RD, Longo LD (1989) Maternal responses to long-term hypoxemia in sheep. Am J Phys 256(6 Pt 2):R1340–R1347
Koshy M et al (1988) Prophylactic red-cell transfusions in pregnant patients with sickle cell disease. A randomized cooperative study. N Engl J Med 319(22):1447–1452

Krampl E et al (2000) Fetal biometry at 4300 m compared to sea level in Peru. Ultrasound Obstet Gynecol 16(1):9–18
Kumar R (2008) Prenatal factors and the development of asthma. Curr Opin Pediatr 20(6):682–687
Lao TT, Wong WM (1997) Placental ratio--its relationship with mild maternal anaemia. Placenta 18(7):593–596
Li G et al (2003) Effect of fetal hypoxia on heart susceptibility to ischemia and reperfusion injury in the adult rat. J Soc Gynecol Investig 10(5):265–274
Magness RR (1998) Maternal cardiovascular and other physiologic responses to the endocrinology of pregnancy. In: Bazer FW (ed) The endocrinology of pregnancy. Humana Press, Totowas, pp 507–539
Mahajan S et al (2008) Nutritional anaemia dysregulates endocrine control of fetal growth. Br J Nutr 100(2):408–417
Martin H et al (2000) Impaired endothelial function and increased carotid stiffness in 9-year-old children with low birthweight. Circulation 102(22):2739–2744
Martyn CN et al (1995) Growth in utero, adult blood pressure, and arterial compliance. Br Heart J 73(2):116–121
Mateev S et al (2003) Chronic hypoxia opposes pregnancy-induced increase in uterine artery vasodilator response to flow. Am J Physiol Heart Circ Physiol 284(3):H820–H829
Mateev SN et al (2006) Chronic hypoxia augments uterine artery distensibility and alters the circumferential wall stress-strain relationship during pregnancy. J Appl Physiol 100(6):1842–1850
McAuliffe F et al (2001) Blood gases in pregnancy at sea level and at high altitude. BJOG 108(9):980–985
McColgin SW, Glee L, Brian BA (1992) Pulmonary disorders complicating pregnancy. Obstet Gynecol Clin N Am 19(4):697–717
Meyer K, Lubo Z (2007) Fetal programming of cardiac function and disease. Reprod Sci 14(3):209–216
Miller MD (2000) Anesthesia, 5th edn. Churchill Livingston, Elsevier
Moore LG (2001) Small babies and big mountains: John Lichty solves a Colorado mystery in Leadville. In: Reeves JT, Grover FT (eds) Attitudes on altitude. University of Colorado Press, Boulder
Moore LG et al (1982) The incidence of pregnancy-induced hypertension is increased among Colorado residents at high altitude. Am J Obstet Gynecol 144(4):423–429
Moore LG et al (1986) Maternal hypoxic ventilatory response, ventilation, and infant birth weight at 4,300 m. J Appl Physiol 60(4):1401–1406
Moore LG et al (2001) Tibetan protection from intrauterine growth restriction (IUGR) and reproductive loss at high altitude. Am J Hum Biol 13(5):635–644
Mortola JP et al (2000) Birth weight and altitude: a study in Peruvian communities. J Pediatr 136(3):324–329
Motulsky AG (1973) Frequency of sickling disorders in U.S. blacks. N Engl J Med 288(1):31–33
Mulvany MJ et al (1996) Vascular remodeling. Hypertension 28(3):505–506
Murphy KM, Streips UN, Lock RB (2000) Bcl-2 inhibits a Fas-induced conformational change in the Bax N terminus and Bax mitochondrial translocation. J Biol Chem 275(23):17225–17228
National High Blood Pressure Education Program Working Group (2000) Report of the National High Blood Pressure Education Program Working Group on high blood pressure in pregnancy. Am J Obstet Gynecol 183(1):S1–S22
Neale D et al (2003) Maternal serum of women with pre-eclampsia reduces trophoblast cell viability: evidence for an increased sensitivity to Fas-mediated apoptosis. J Matern Fetal Neonatal Med 13(1):39–44
Ness RB, Sibai BM (2006) Shared and disparate components of the pathophysiologies of fetal growth restriction and preeclampsia. Am J Obstet Gynecol 195(1):40–49
Ni Y et al (1997) Pregnancy augments uteroplacental vascular endothelial growth factor gene expression and vasodilator effects. Am J Phys 273(2 Pt 2):H938–H944

Odegaard I et al (2002) Maternal and fetal morbidity in pregnancies of Norwegian and Swedish women with cystic fibrosis. Acta Obstet Gynecol Scand 81(8):698–705
Ong HC, White JC, Sinnathuray TA (1977) Haemoglobin H disease and pregnancy in a Malaysian woman. Acta Haematol 58(4):229–233
Palmer SK et al (1992) Quantitative estimation of human uterine artery blood flow and pelvic blood flow redistribution in pregnancy. Obstet Gynecol 80(6):1000–1006
Palmer SK et al (1999) Altered blood pressure course during normal pregnancy and increased preeclampsia at high altitude (3100 meters) in Colorado. Am J Obstet Gynecol 180(5):1161–1168
Peters B et al (2004) Effect of heterosexual intercourse on mucosal alloimmunisation and resistance to HIV-1 infection. Lancet 363(9408):518–524
Pollack RN, Divon MY (1992) Intrauterine growth retardation: definition, classification, and etiology. Clin Obstet Gynecol 35(1):99–107
Presbitero P et al (1994) Pregnancy in cyanotic congenital heart disease. Outcome of mother and fetus. Circulation 89(6):2673–2676
Ramsay JE et al (2003) Microvascular dysfunction: a link between pre-eclampsia and maternal coronary heart disease. BJOG 110(11):1029–1031
Redman CW (1990) Platelets and the beginnings of preeclampsia. N Engl J Med 323(7):478–480
Robertson SA et al (2002) Transforming growth factor beta--a mediator of immune deviation in seminal plasma. J Reprod Immunol 57(1-2):109–128
Rodesch F et al (1992) Oxygen measurements in endometrial and trophoblastic tissues during early pregnancy. Obstet Gynecol 80(2):283–285
Rosenberg A (2008) The IUGR newborn. Semin Perinatol 32(3):219–224
Rowe SM, Miller S, Sorscher EJ (2005) Cystic fibrosis. N Engl J Med 352(19):1992–2001
Saikumar P et al (1998) Role of hypoxia-induced Bax translocation and cytochrome c release in reoxygenation injury. Oncogene 17(26):3401–3415
Saito S, Sakai M (2003) Th1/Th2 balance in preeclampsia. J Reprod Immunol 59(2):161–173
Sartori C et al (1999) Augmented vasoreactivity in adult life associated with perinatal vascular insult. Lancet 353(9171):2205–2207
Sellman JS, Holman RL (2000) Thromboembolism during pregnancy. Risks, challenges, and recommendations. Postgrad Med 108(4):71–72. 77–8, 81–4
Shapiro JM (2006) Critical care of the obstetric patient. J Intensive Care Med 21(5):278–286
Sibai BM (2003) Diagnosis and management of gestational hypertension and preeclampsia. Obstet Gynecol 102(1):181–192
Sibai BM (2004) Diagnosis, controversies, and management of the syndrome of hemolysis, elevated liver enzymes, and low platelet count. Obstet Gynecol 103(5 Pt 1):981–991
Sibai B, Dekker G, Kupferminc M (2005) Pre-eclampsia. Lancet 365(9461):785–799
Singla PN et al (1997) Fetal growth in maternal anaemia. J Trop Pediatr 43(2):89–92
Siu SC, Colman JM (2001) Heart disease and pregnancy. Heart 85(6):710–715
Siu SC et al (2001) Prospective multicenter study of pregnancy outcomes in women with heart disease. Circulation 104(5):515–521
Siu SC et al (2002) Adverse neonatal and cardiac outcomes are more common in pregnant women with cardiac disease. Circulation 105(18):2179–2184
Snoeckx LH et al (2001) Heat shock proteins and cardiovascular pathophysiology. Physiol Rev 81(4):1461–1497
Sun PM et al (2001) Sickle cell disease in pregnancy: twenty years of experience at Grady Memorial Hospital, Atlanta, Georgia. Am J Obstet Gynecol 184(6):1127–1130
Tanaka M et al (1994) Hypoxia induces apoptosis with enhanced expression of Fas antigen messenger RNA in cultured neonatal rat cardiomyocytes. Circ Res 75(3):426–433
Thompson LP (2003) Effects of chronic hypoxia on fetal coronary responses. High Alt Med Biol 4(2):215–224
Thompson LP, Dong Y (2005) Chronic hypoxia decreases endothelial nitric oxide synthase protein expression in fetal guinea pig hearts. J Soc Gynecol Investig 12(6):388–395
Wang JX et al (2002) Surgically obtained sperm, and risk of gestational hypertension and preeclampsia. Lancet 359(9307):673–674

Watson AL et al (1998) Susceptibility of human placental syncytiotrophoblastic mitochondria to oxygen-mediated damage in relation to gestational age. J Clin Endocrinol Metab 83(5):1697–1705
Webster WS, Abela D (2007) The effect of hypoxia in development. Birth Defects Res C Embryo Today 81(3):215–228
Williams LA, Evans SF, Newnham J (1997) Prospective cohort study of factors influencing the relative weights of the placenta and the newborn infant. BMJ 314(7098):1864–1868
Wilson BJ et al (2003) Hypertensive diseases of pregnancy and risk of hypertension and stroke in later life: results from cohort study. BMJ 326(7394):845
Xi L et al (2001) Whole body hyperthermia and preconditioning of the heart: basic concepts, complexity, and potential mechanisms. Int J Hyperth 17(5):439–455
Xiao RP (2000) Cell logic for dual coupling of a single class of receptors to G(s) and G(i) proteins. Circ Res 87(8):635–637
Xiao D et al (2002) Cortisol-mediated potentiation of uterine artery contractility: effect of pregnancy. Am J Physiol Heart Circ Physiol 283(1):H238–H246
Xu Y et al (2006) Hypoxia or nutrient restriction during pregnancy in rats leads to progressive cardiac remodeling and impairs postischemic recovery in adult male offspring. FASEB J 20(8):1251–1253
Yaniv G et al (2002) Hypoxia predisposes neonatal rat ventricular myocytes to apoptosis induced by activation of the Fas (CD95/Apo-1) receptor: Fas activation and apoptosis in hypoxic myocytes. Cardiovasc Res 54(3):611–623
Yellon DM et al (1992) The protective role of heat stress in the ischaemic and reperfused rabbit myocardium. J Mol Cell Cardiol 24(8):895–907
Zamudio S et al (1995) Effect of altitude on uterine artery blood flow during normal pregnancy. J Appl Physiol 79(1):7–14
Zhang L (2005) Prenatal hypoxia and cardiac programming. J Soc Gynecol Investig 12(1):2–13
Zhang J, Meikle S, Trumble A (2003) Severe maternal morbidity associated with hypertensive disorders in pregnancy in the United States. Hypertens Pregnancy 22(2):203–212
Zhang L, Xiao D, Hu X (2008) Effect of cGMP on pharmacomechanical coupling in the uterine artery of near-term pregnant sheep. J Pharmacol Exp Ther 327(2):425–431

Intergenerational Transfer of Biological Responses to Trauma: Impact of Psychosocial Stress in Fathers on Offspring

17

Mallory E. Bowers and Rachel Yehuda

Abstract

Maternal experiences and psychological disposition prior to and during pregnancy are associated with specific biological and behavioral phenotypes in their children. Less clear, however, is the contribution of paternal experiences to offspring disposition. Increasingly, animal models indicate that paternal experiences are as impactful to offspring development as are maternal experiences. Importantly, these experiences in fathers can precipitate biological changes that are concomitant with biological effects observed in offspring. Here, we examine the body of evidence across species that suggests paternal environmental exposures are associated with specific phenotypes in offspring and note potential mechanisms for the "transmission" of intergenerational experience. Finally, we acknowledge the caveats endemic to this line of research and suggest points of consideration for the field.

17.1 Introduction

Many psychiatric disorders, particularly mood disorders, result from an interaction between genes and environmental exposures. Genome-wide gene association, monozygotic twin, and candidate gene studies demonstrate the contribution of heritability and even point to specific alleles that could confer risk for mental illness

M. E. Bowers
Mental Health Care Center, James J. Peters Veterans Affairs Medical Center, Bronx, NY, USA

R. Yehuda (✉)
Department of Psychiatry, Icahn School of Medicine at Mount Sinai, New York, NY, USA

Department of Neuroscience, Icahn School of Medicine at Mount Sinai, New York, NY, USA
e-mail: Rachel.Yehuda@va.gov

© Springer International Publishing AG, part of Springer Nature 2018

W. Burggren, B. Dubansky (eds.), *Development and Environment*,
https://doi.org/10.1007/978-3-319-75935-7_17

(Almli et al. 2014; Goldberg et al. 1990; Smoller 2016). Nevertheless, data from these studies indicate that genetics may only confer a partial contribution to susceptibility. Therefore, it is critical to understand the role of environmental experience in precipitating the development of psychopathology. As part of a broader understanding of the impact of environment in the diathesis for mental disorders, investigators have focused on the role of parental experiences and their effects on offspring. It has been particularly informative to consider the contribution of parental events occurring prior to conception. It is not clear how large of a role paternal contributions have relative to other environmental factors.

It has recently been established that first- and second-generation offspring of persons exposed to trauma prior to conception exhibit phenotypic differences compared to peers sired by parents who have not been exposed to comparable trauma (Flory et al. 2011; Schulz 2010; Solomon et al. 1988; Yehuda et al. 2016). The phenomena by which parental experiences impact parental phenotypes, including psychopathology, which are then reflected by changes in offspring phenotypes is sometimes referred to as "intergenerational transmission" and can have multiple causes (Bowers and Yehuda 2016; Dias and Ressler 2014a). For instance, there can be long-term consequences from parent to offspring social learning and/or variation in parental rearing styles (Bandura 1962, 1978). Additionally, distinct offspring phenotypes could theoretically result from changes in offspring biology that were triggered by some experience-dependent alteration in parental biology. The putative mechanism for transmitting biological changes from parent to offspring is the gamete, i.e., spermatozoa in men and oocytes in women (Daxinger and Whitelaw 2012). More specifically, a mechanism of transmission could involve epigenetic marks in gametes. Epigenetic modifications in somatic tissues influence gene expression, and thus phenotypes, and are responsive to environmental factors, including parental stress. Although transcription is repressed in germ cells, gametic epigenetic marks can impact embryonic development and, potentially, long-term phenotypes.

Evidence for transmission in humans is supported by studies that observe significant correlations between parent and offspring biology. For instance, mothers who have higher circulating levels of the stress hormone, cortisol, during pregnancy give birth to children with comparably higher levels of cortisol (Gutteling et al. 2004, 2005). These findings may in part reflect genetics. Studies investigating an association between parental and offspring biology have largely focused on maternal effects (Bowers and Yehuda 2016). However, offspring phenotypes can also be influenced by paternal effects (Lehrner et al. 2014). Paternal studies are particularly intriguing because they offer an opportunity to noninvasively study germ cells.

Rodent studies have begun the necessary work of clarifying whether paternal experience, via sperm, can impact offspring outcomes. At least some rodent studies provide solid evidence that paternal stress induces epigenetic changes in sperm that trigger specific behavioral phenotypes in offspring (Bohacek and Mansuy 2015). Comparable biological data in humans is extremely preliminary and has largely examined intergenerational effects of substances such as pollutants and endocrine disruptors, rather than trauma and resultant mental illness (Skinner 2016; Vecoli et al. 2016). The matter of examining transmitted biological effects induced by extreme stress is complicated, however, by limited understanding of the primary effects of stress. For instance, long-term changes to the hypothalamic-pituitary-adrenal (HPA) axis resulting from trauma exposure and subsequent development of posttraumatic stress disorder (PTSD) are still being worked out (Yehuda 2002; Yehuda and Bierer 2008). This is the fundamental difference between studying transmitted biological effects of psychosocial stress exposure versus examining the effects of alcohol consumption or famine exposure, for instance.

Here, we examine the evidence for transmission of paternal psychosocial stress in rodents, focusing on sperm epigenetics as a mechanism for transmission. First we examine potential epigenetic markers in sperm that could be used to convey parental experience, including DNA methylation, small noncoding RNAs, and histone post-translational modifications. Some of these mechanisms are, to some extent, theoretical because whether and how these marks are maintained through embryogenesis is unknown. Additionally, we present evidence from rodent studies, which suggest that parental stress is associated with biological and behavioral phenotypes in offspring. Human studies are largely preliminary, and thus only offer correlative associations between epigenetic marks in sperm and environmental exposures as evidence that experience is transmitted. Because the field is nascent, we suggest several future directed lines of inquiry that address gaps in our current understanding. Although the question of whether parental experience shapes offspring outcomes via biological mechanisms is now gaining widespread acceptance within psychiatry, particularly among those who use rodent models, questions still remain as to the precise epigenetic marks (and the signaling cascades up- and downstream of these marks) that carry this signal.

17.2 Epigenetic Mechanisms of Transmission in Sperm

In considering a biological mechanism that is sensitive to environmental conditions and can exert long-lasting, but impermanent, phenotypic changes across generations, epigenetics are an appealing candidate (Feil and Fraga 2012; Zhang and Meaney 2010). Epigenetics encompass mechanisms that alter gene expression, including DNA methylation, histone modifications, and small noncoding RNAs. Here, we present a brief review of proposed epigenetic mechanisms and evidence for their role in transfer of ancestral experience (for more exhaustive reviews, see (Daxinger and Whitelaw 2012; Feng and Chen 2015; Yan 2014)).

17.2.1 DNA Methylation

DNA methylation and demethylation involve the transfer or removal of methyl groups to cytosines in cytosine-phosphate-guanine (CpG) sites. Approximately 70% of annotated gene promoters are associated with a CpG island, a 1000 base pair region with an elevated CpG composition (Klengel et al. 2014). Heightened promoter methylation is typically associated with downregulation of gene expression and has been shown to interfere with transcription factor binding (Klose and Bird 2006; Weaver et al. 2004). Although analysis of DNA methylation in somatic tissues has typically focused on gene promoter regions, intergenerational research has pivoted to analysis of imprinted genes in gametes because they are thought to escape reprogramming during embryonic development (Lane et al. 2003). In this way, changes to the methylation status of imprinted genes could be maintained in the newly formed embryo, potentially impacting embryonic development and other long-term phenotypic effects downstream of development. Imprinting center regions (ICR) control genetic imprinting via robust methylation according to allelic parent-of-origin. The process of establishing imprinting marks occurs in the parental developing germ line and is maintained in the offspring embryo until a second wave of DNA methylation reprogramming occurs during offspring's respective primordial germ cell development. IGF2, DLK1, and H19 are examples of imprinted genes that have been implicated in intergenerational studies (Ouko et al. 2009; Soubry et al. 2016). For instance, parental exposure to adverse nutritional conditions during the Dutch Hunger Winter has been associated with offspring aberrant methylation at the IGF2 imprint regulatory region almost half a century later (Heijmans et al. 2008; Schulz 2010).

17.2.2 RNA

There are four types of noncoding RNAs (ncRNA) found in sperm: repetitive elements, transcription start sites (TSS)/promoter-associated RNA, piRNAs, and miRNAs (Krawetz et al. 2011). miRNAs and piRNAs are involved in posttranscriptional regulation of gene expression and are among the most studied of the various types of sperm ncRNAs (Daxinger and Whitelaw 2012). Numerous miRNAs and piRNAs are uniquely present in sperm and are of unknown function (Pantano et al. 2015). Although the specific roles of these ncRNAs have yet to be determined, many of these miRNAs and piRNAs are likely critical to embryonic development, as studies show that they are transmitted to the oocyte upon fertilization (Kawano et al. 2012; Ostermeier et al. 2004; Smythies et al. 2014). For instance, the most abundant sperm miRNA, miR-34c, is essential for the first cleavage division in mouse zygotes (Liu et al. 2012).

17.2.3 Histone and Protamine Posttranslational Modifications

Modifications to histone proteins are an additional avenue for epigenetic regulation of phenotypes (Sweatt 2013). Histone modifications alter the degree of DNA compaction around octameric histone clusters (Campos and Reinberg 2009). During spermatogenesis in humans, approximately 85% of histones are replaced with highly basic protamines, which facilitate increased compaction of DNA and transcriptional silencing of genes (Goldberg et al. 1977). Consequently, less research has focused on posttranslational modifications (PTMs) of histones and protamines as mechanisms of paternal transmission. New data, however, are upending the assumption that histone and protamines play a circumscribed role in transmission. Protamines appear to be modified by several types of PTMs (previously, very little was known regarding PTMs of protamines), suggesting an alternative function besides compaction of DNA (Brunner et al. 2014; Chirat et al. 1993). Additionally, the approximately 15% histones that are not replaced by protamines are thought to keep their posttranslational modifications, with eventual transfer to the fertilized oocyte (Hammoud et al. 2009). Interestingly, remaining histones, and their accompanying histone tail modifications, appear to cluster predominantly around loci that contain genes that are important for development and, in fact, have been shown to play a role in embryogenesis via regulation of gene expression (Hammoud et al. 2009; Teperek et al. 2016).

Altogether, the epigenetic mechanisms in sperm that could play a role in intergenerational transmission overlap substantially with typical epigenetic modifications present in somatic tissues, with protamine posttranslation modification being a notable exception. As the field learns more about how epigenetic marks are transferred from gametes to embryo, it will become clearer which modifications are critical for transmission and in what ways these signals are shaped to affect expression of offspring phenotypes.

17.3 Can Paternal Experience Be Transmitted via Sperm?

17.3.1 Rodents

Rodent studies have begun to address the question of whether paternal experience is transmitted to offspring (Bohacek and Mansuy 2015). Mouse and rat studies are useful primarily in that fathers do not play a role in pup rearing, and isogenic strains eliminate the confound of genetic variation. However, it has become clear that, despite lack of paternal investment in offspring care, paternal experience can be indirectly transmitted behaviorally via interactions with mothers during mating (Mashoodh et al. 2012). More recent studies have begun to control for so-called maternal provisioning by using in vitro fertilization (IVF) or by manipulating the biology of the early embryo in order to recapitulate offspring behavior thought to be caused by paternal experience.

Rodent studies addressing the question of biological transmission have used parental stressor exposure, including maternal separation and stress (MSUS), social defeat, and chronic variable stress, as environmental insults. In the unpredictable maternal separation/maternal stress model (MSUS) used by Isabelle Mansuy, pups (F_1) are separated from mothers for 3 h, once per day, from postnatal day 1–14 (Bohacek et al. 2015). During maternal separation, dams are exposed to restraint stress or the forced swim test. As adults, F_1 and F_2 control and experimental males are bred to naïve wild-type females to generate F_2 and F_3 offspring. MSUS affects F_1, F_2, and F_3 behaviors on a number of tasks that can be clustered according to their assessment of behavioral flexibility/impulsivity, cognition, sociability, and anxiety-like and depressive-like behaviors (Bohacek et al. 2015; Franklin et al. 2010, 2011;

Gapp et al. 2014a, b). The extensive behavioral phenotyping of the Mansuy group demonstrates that MSUS affects behavior across generations, but that these behaviors are not necessarily expressed equally in males and females, or in every generation (Franklin et al. 2010; Weiss et al. 2011). For instance, F_1 MSUS males spend more time floating on the forced swim test compared to control males; however, this depressive-like behavior disappears in F_2 MSUS males and reappears in F_3 MSUS males (Franklin et al. 2010). Alternatively, F_1 MSUS females spend less time floating, F_2 MSUS females spend more time floating, and F_3 MSUS females spend an equal amount of time floating compared to controls (Franklin et al. 2010).

At least some parentally mediated offspring behaviors can be reversed by paternal environmental enrichment (EE) after weaning (and upon termination of MSUS) (Gapp et al. 2016). When F_1 control and MSUS males are placed in social groups in an enriched cage from weaning until adulthood, F_2 MSUS spend comparable time in the bright compartment of the light-dark box compared to F_2 controls (Gapp et al. 2016). Behavior on an active avoidance task is also reversed by environmental enrichment (Gapp et al. 2016).

Behavioral changes in offspring as a consequence of MSUS are accompanied by epigenetic alterations in parental sperm, including variation in DNA methylation and changes in small RNA content, and these epigenetic changes can be surprisingly consistent across tissues and generations. For instance, decreased methylation in the promoter region of the glucocorticoid receptor (GR or Nrc31) in F_1 MSUS sperm is also mirrored in the hippocampus F_2 MSUS offspring (Gapp et al. 2016). Both F_1 and F_2 MSUS subjects exhibit increased GR expression in the hippocampus, additionally (Gapp et al. 2016). Parallel intergenerational methylation changes are also observed in the promoter region of protein kinase C ɣ (Prkcc), an important regulator of synaptic plasticity, in F_1 MSUS sperm and F_2 MSUS hippocampus (Bohacek et al. 2015).

Studies also note divergent epigenetic changes across tissues and generations; however, these changes may still promote or induce similar functional outputs in that expression of specific genes are uniformly up- or downregulated. Although there are no differences in hippocampal promoter methylation of the mineralocorticoid receptor (MR) gene between F_2 MSUS and control subjects, acetylation of H4K5 (H4K5ac) and H3K14 (H3K14ac), dimethylation of H3K4 (H3K4me2), and trimethylation of H3K36 (H3K36me3) are significantly decreased at an intragenic region of the MR gene in F_2 MSUS hippocampus (Gapp et al. 2014b). Additionally, increased MR promoter methylation across a number of CpG sites in F_1 MSUS sperm is observed (Gapp et al. 2014b). Increased promoter methylation suggests that the MR gene could be primed for transcriptional repression, although, it should be noted here that transcription is already repressed in spermatozoa (Gapp et al. 2014b). Increased promoter methylation serves a functionally similar role as the observed histone posttranslation modifications adjacent to MR, in that both reduce MR expression.

Perhaps the strongest evidence suggesting that MSUS affects offspring behavior via the paternal germ line demonstrates that subjects derived from wild-type oocytes microinjected with RNAs from F_1 MSUS sperm recapitulate anxiety-like and depressive-like behaviors observed in F_2 MSUS offspring resulting from traditional breeding pairs (Gapp et al. 2014a). These data contradict the possibility that effects are conveyed via behaviorally naïve mothers and, moreover, highlight the importance of sperm RNAs in mediating transfer of ancestral experience.

As in MSUS offspring, male and female mice from social defeat stressed fathers exhibit increased depressive-like behavior on the forced swim test (Dietz et al. 2011). These offspring also exhibit greater anxiety-like behavior and more social avoidance in a social interaction test (Dietz et al. 2011). However, most of these effects are not recapitulated in IVF experiments, suggesting that an interaction between defeated fathers and naïve mothers mediates transfer of paternal experience to offspring (Dietz et al. 2011). IVF offspring of defeated fathers exhibit greater depressive-like behavior on the forced swim test as in offspring from natural breeding experiments, offering some evidence of direct paternal transmission of social defeat stress (Dietz et al. 2011).

In the chronic variable stress model, fathers are randomly exposed during puberty or adulthood to one of seven different stressors per day for 6 weeks prior to breeding. Offspring of fathers exposed to stress, either during adolescence or adulthood, display reduced HPA-axis stress responsivity in that they produce less corticosterone following a 15 min restraint; however, no behavioral changes are observed when testing mice on the Barnes Maze, paired pulse inhibition (PPI), tail suspension, and the light-dark box test. Male, but not female, of fathers exposed to social defeat stress exhibit higher baseline corticosterone, but do not exhibit differences after acute restraint stress (Dietz et al. 2011). Nine specific miRNAs are significantly increased in sperm from both paternal stress groups (Rodgers et al. 2013), and when these miRNAs are microinjected into wild-type zygotes, offspring exhibit the reduced HPA axis stress responsivity phenotype (Rodgers et al. 2015).

Aside from ancestral stress studies, there are few alternative behavioral models in rodents that are used to better understand a biological mechanism of paternal behavior transmission. These studies are limited to examination of cued fear conditioning, self-administration of cocaine, and ethanol exposure. Analysis of biological factors that transmit substance abuse behavior, however, is confounded by the epimutagenic properties of drugs and alcohol.

When F_0 male mice are fear conditioned using acetophenone, their F_1 and F_2 male offspring have an increased behavioral sensitivity to acetophenone, but not to other odors (Dias and Ressler 2014b). These behavioral effects are maintained with IVF and in cross-fostered F_1 males whose mothers were fear conditioned using acetophenone, suggesting that transmission is not specific to male germ line epigenetic changes (Dias and Ressler 2014b). Notably, the Olfr151 gene, which encodes an odorant receptor activated by acetophenone, is hypomethylated in sperm of fear-conditioned F_1 males and their behaviorally naïve F_1 offspring (Dias and Ressler 2014b).

Male offspring from fathers placed in ethanol ("E-sired") vapor chambers for 5 weeks prior to conception have significantly decreased preference and consumption of ethanol compared to control male offspring (Finegersh and Homanics 2014). This effect is not observed in female offspring (Finegersh and Homanics 2014). Ethanol exposure changes DNA methylation in the *bdnf* promoter of sperm and in the brain of naïve offspring (Finegersh and Homanics 2014). Male, but not female, offspring of sires that self-administered cocaine exhibit delayed acquisition and reduced maintenance of cocaine self-administration (Vassoler et al. 2013). The authors observe similar molecular changes in F_0 sperm and F_1 brain, specifically, an increase in the association between acetylated histone 3 and the BDNF promoter, which is thought to underlie an increase in BDNF mRNA and protein in F_1 frontal cortex (Vassoler et al. 2013). Uniquely, the authors also note molecular changes in testes of F_0 that self-administered cocaine (Vassoler et al. 2013).

17.3.2 Humans

There are almost no studies that address the question of whether paternal experience is transmitted to offspring via epigenetic changes to sperm. This stands in stark contrast, particularly given the ease in collecting semen specimens, to the number of studies that have examined the contribution of maternal experience to biological and behavioral characteristics of the offspring (Bowers and Yehuda 2016). Nevertheless, preliminary data support the rodent literature in the contention that paternal experience is transmitted via the germ line. A recent study finds some evidence for specific differential DNA modifications associated with major depressive disorder (MDD) that is common across tissues (Oh et al. 2015). DNA modification analysis was performed in white blood cells from monozygotic twins discordant for MDD, in the brain prefrontal cortex, and in sperm samples from bipolar (BP) affected individuals and control subjects (Oh et al. 2015). It is important to note, however, that tissue samples were from non-overlapping cohorts (Oh et al. 2015). Changes in DNA methylation are also observed in sperm from men addicted to alcohol and in sperm from men addicted to opioids (Chorbov et al. 2011; Ouko et al. 2009). Specifically, differentially methylated regions (DMR) such as H19 DMR and IG DMR methylation in sperm are correlated with alcohol consumption in men (Ouko et al. 2009). These DMRs are thought to fall within the recognition sequences for DNA methyltransferases (DNMTs), which could alter methylation patterns across neighboring genetic regions. In sperm-derived DNA from opioid addicts, methylation was significantly increased at CpG 2 of OPRM1 promoter (Chorbov et al. 2011), which controls expression of opioid receptors. Changes in expression of OPRM1 could alter sensitivity and tolerance to opioids, potentially altering opioid intake.

17.4 Conclusions and Future Directions

There are several issues the field must reconcile in order to advance the question of whether (and how) preconception paternal experience can be transmitted to offspring. This review has chosen to highlight several big-picture questions that will shape the direction of the field going forward. Some of these hurdles are addressable given the currently available methods, while others are technically out of reach and unanswerable at present.

In the short-term, greater widespread consideration should be given to the use of IVF and artificial insemination experiments in intergenerational rodent studies. Given the influence of maternal provisioning, cross-fostering studies are not completely sufficient in shoring up evidence for a strictly biological transmission of paternal experience. However, artificial reproductive technologies, such as IVF, are not without their own problems, particularly issues related to cost. Additionally, hormone induction of ovulation may influence biological markers involved in intergenerational transmission, confounding signals of interest. Thorough studies should include cross-fostering and either IVF or artificial insemination to evaluate biological and behavioral phenotypes across several different methods.

As noted by others, parental stress models are also vulnerable to underpowering, which could potentially inflate statistical significance. Again, researchers will have to balance cost concerns with the minimum number of animals needed to appropriately evaluate phenotypic differences. This issue of cost is further complicated by an ethical consideration in the use of research animals, as the true number of subjects required could violate the third "R" concept—reduction—according to IACUC.

Although there is better understanding of how sperm epigenetics can transmit to fertilized oocytes, how, exactly, somatic responses to stress are conveyed to sperm is still speculative. Unlike epimutagens, such as alcohol, chemotherapy, and endocrine disruptors, stress effects on sperm are indirect in that they must initiate some biological change in the body that is then transferred to gametes. Accordingly, we propose clarifying if (and how) steroid hormones produced by the stress-responsive (HPA) axis signal gametes as a research imperative for the intergenerational stress field. Steroid hormones, in particular cortisol, are an appealing target, as they are transported via the bloodstream in response to stress to coordinate physiological and behavioral responses for survival. Alternatively, exosomes could act as shuttles for epigenetic messages sent to sperm from tissues modified by extreme stress. These messages could be sent from nearby tissues, such as Sertoli cells, or secreted from as far away as the central nervous system (Daxinger and Whitelaw 2012; Vassoler et al. 2013).

Additionally, it is unclear whether transmission is a mechanism to enhance seemingly "adaptive" capabilities in offspring or whether changes to sperm biology, and thus offspring phenotype, are a circumstantial byproduct of stress-mediated biological changes intended to mobilize tissues critical for survival responses. At least some data suggest that transmission is advantageous. Offspring of mothers who were malnourished early in pregnancy have higher obesity rates, reflecting potential changes in offspring metabolism and/or feeding behavior that would be expedient during a time of relative food scarcity (Roseboom et al. 2006). A similar phenomenon is observed in offspring sired from fathers forced to consume ethanol, where affected offspring demonstrate differences in consumption, preference, and sensitivity compared to controls (Finegersh and Homanics 2014). Moreover, Dias and Ressler (2014b) demonstrate enhanced sensitivity to an odorant associated with an aversive stimulus, indicating that there may be some specificity to transmission. The question of whether ancestral transmission of experience is advantageous is likely better suited for the field of evolutionary biology to test, rather than for neuroscience or psychiatry.

As mentioned, researchers are pursuing studies designed to test the hypothesis that DNA methylation of gametic imprinting regions is a mechanism by which parental effects are transmitted to offspring. These studies have been developed based on data suggesting that gametic imprinting regions escape epigenetic reprogramming during embryogenesis (Lane et al. 2003). Intriguingly, imprinted genes appear to be disproportionally involved in development, and there is some evidence to suggest that paternal imprinting favors the production of larger offspring, and maternal imprinting favors smaller offspring (Plasschaert and Bartolomei 2014). Developmental-related genes are also now appreciated to contribute to learning-related plasticity in adulthood, e.g., *bdnf*, which is involved in neuron growth and differentiation, has also been shown to play a role in amygdala-dependent learning (Dias et al. 2014; Ehrlich and Josselyn 2016; Heldt et al. 2014; Maguschak and Ressler 2011, 2012). Given this functional overlap, it's possible that changes to imprinting signatures caused by traumatic insult could contribute to deficits in learning and memory and thus precipitate mental illness. This hypothesis is supported by evidence that certain imprinting disorders, such as Angelman syndrome, are associated with neurocognitive symptoms (Davies et al. 2001; Isles and Wilkinson 2000).

Altogether, there is promising data suggesting that offspring phenotypes are influenced by paternal experience via epigenetic modifications in spermatozoa. All of the the current evidence comes from rodent studies, particularly IVF studies. Increasingly, however, investigators are also interrogating epigenetic differences in human spermatozoa to understand whether gametic biological differences associate with prior stress exposures. These studies will continue to be informed by basic research aimed at revealing which epigenetic marks in gametes are transferred to embryos and the process by which this transfer occurs. Human studies will be further enriched with the inclusion and phenotyping of multiple generations while also including semen sampling in analyses. By rigorously addressing the potential confounds inherent to studying the behavior of multiple generations, some of which are outlined in this section, we believe that the field will continue to develop convincing evidence that multigenerational behaviors are influenced via nongenetic mechanisms carried by gametes.

References

Almli LM, Fani N, Smith AK, Ressler KJ (2014) Genetic approaches to understanding post-traumatic stress disorder. Int J Neuropsychopharmacol 17(2):355–370. https://doi.org/10.1017/S1461145713001090

Bandura A (1962) Social-learning through imitation of adult models. Am J Orthopsychiatry 32(2):290–291

Bandura A (1978) Social learning theory of aggression. J Commun 28(3):12–29

Bohacek J, Mansuy IM (2015) Molecular insights into transgenerational non-genetic inheritance of acquired behaviours. Nat Rev Genet 16(11):641–652. https://doi.org/10.1038/nrg3964

Bohacek J, Farinelli M, Mirante O, Steiner G, Gapp K, Coiret G, Mansuy IM (2015) Pathological brain plasticity and cognition in the offspring of males subjected to postnatal traumatic stress. Mol Psychiatry 20(5):621–631. https://doi.org/10.1038/mp.2014.80

Bowers ME, Yehuda R (2016) Intergenerational transmission of stress in humans. Neuropsychopharmacology 41(1):232–244. https://doi.org/10.1038/npp.2015.247

Brunner AM, Nanni P, Mansuy IM (2014) Epigenetic marking of sperm by post-translational modification of histones and protamines. Epigenetics Chromatin 7(1):2. https://doi.org/10.1186/1756-8935-7-2

Campos EI, Reinberg D (2009) Histones: annotating chromatin. Annu Rev Genet 43:559–599. https://doi.org/10.1146/annurev.genet.032608.103928

Chirat F, Arkhis A, Martinage A, Jaquinod M, Chevaillier P, Sautiere P (1993) Phosphorylation of human sperm protamines HP1 and HP2: identification of phosphorylation sites. Biochim Biophys Acta 1203(1):109–114

Chorbov VM, Todorov AA, Lynskey MT, Cicero TJ (2011) Elevated levels of DNA methylation at the OPRM1 promoter in blood and sperm from male opioid addicts. J Opioid Manag 7(4):258–264

Davies W, Isles AR, Wilkinson LS (2001) Imprinted genes and mental dysfunction. Ann Med 33(6):428–436

Daxinger L, Whitelaw E (2012) Understanding transgenerational epigenetic inheritance via the gametes in mammals. Nat Rev Genet 13(3):153–162. https://doi.org/10.1038/nrg3188

Dias BG, Ressler KJ (2014a) Experimental evidence needed to demonstrate inter- and transgenerational effects of ancestral experiences in mammals. BioEssays 36(10):919–923. https://doi.org/10.1002/bies.201400105

Dias BG, Ressler KJ (2014b) Parental olfactory experience influences behavior and neural structure in subsequent generations. Nat Neurosci 17(1):89–96. https://doi.org/10.1038/nn.3594

Dias BG, Goodman JV, Ahluwalia R, Easton AE, Andero R, Ressler KJ (2014) Amygdala-dependent fear memory consolidation via miR-34a and Notch signaling. Neuron 83(4):906–918. https://doi.org/10.1016/j.neuron.2014.07.019

Dietz DM, Laplant Q, Watts EL, Hodes GE, Russo SJ, Feng J, Nestler EJ (2011) Paternal transmission of stress-induced pathologies. Biol Psychiatry 70(5):408–414. https://doi.org/10.1016/j.biopsych.2011.05.005

Ehrlich DE, Josselyn SA (2016) Plasticity-related genes in brain development and amygdala-dependent learning. Genes Brain Behav 15(1):125–143. https://doi.org/10.1111/gbb.12255

Feil R, Fraga MF (2012) Epigenetics and the environment: emerging patterns and implications. Nat Rev Genet 13(2):97–109. https://doi.org/10.1038/nrg3142

Feng L, Chen X (2015) Epigenetic regulation of germ cells-remember or forget? Curr Opin Genet Dev 31:20–27. https://doi.org/10.1016/j.gde.2015.04.003

Finegersh A, Homanics GE (2014) Paternal alcohol exposure reduces alcohol drinking and increases behavioral sensitivity to alcohol selectively in male offspring. PLoS One 9(6):e99078. https://doi.org/10.1371/journal.pone.0099078

Flory JD, Bierer LM, Yehuda R (2011) Maternal exposure to the holocaust and health complaints in offspring. Dis Markers 30(2-3):133–139. https://doi.org/10.3233/DMA-2011-0748

Franklin TB, Russig H, Weiss IC, Graff J, Linder N, Michalon A, Mansuy IM (2010) Epigenetic transmission of the impact of early stress across generations. Biol Psychiatry 68(5):408–415. https://doi.org/10.1016/j.biopsych.2010.05.036

Franklin TB, Linder N, Russig H, Thony B, Mansuy IM (2011) Influence of early stress on social abilities and serotonergic functions across generations in mice. PLoS One 6(7):e21842. https://doi.org/10.1371/journal.pone.0021842

Gapp K, Jawaid A, Sarkies P, Bohacek J, Pelczar P, Prados J, Mansuy IM (2014a) Implication of sperm RNAs in transgenerational inheritance of the effects of early trauma in mice. Nat Neurosci 17(5):667–669. https://doi.org/10.1038/nn.3695

Gapp K, Soldado-Magraner S, Alvarez-Sanchez M, Bohacek J, Vernaz G, Shu H, Mansuy IM (2014b) Early life stress in fathers improves behavioural flexibility in their offspring. Nat Commun 5:5466. https://doi.org/10.1038/ncomms6466

Gapp K, Bohacek J, Grossmann J, Brunner AM, Manuella F, Nanni P, Mansuy IM (2016) Potential of environmental enrichment to prevent transgenerational effects of paternal trauma. Neuropsychopharmacology 41(11):2749–2758. https://doi.org/10.1038/npp.2016.87

Goldberg RB, Geremia R, Bruce WR (1977) Histone synthesis and replacement during spermatogenesis in the mouse. Differentiation 7(3):167–180

Goldberg J, True WR, Eisen SA, Henderson WG (1990) A twin study of the effects of the Vietnam war on posttraumatic stress disorder. JAMA 263(9):1227–1232

Gutteling BM, de Weerth C, Buitelaar JK (2004) Maternal prenatal stress and 4-6 year old children's salivary cortisol concentrations pre- and post-vaccination. Stress 7(4):257–260. https://doi.org/10.1080/10253890500044521

Gutteling BM, de Weerth C, Buitelaar JK (2005) Prenatal stress and children's cortisol reaction to the first day of school. Psychoneuroendocrinology 30(6):541–549. https://doi.org/10.1016/j.psyneuen.2005.01.002

Hammoud SS, Nix DA, Zhang H, Purwar J, Carrell DT, Cairns BR (2009) Distinctive chromatin in human sperm packages genes for embryo development. Nature 460(7254):473–478. https://doi.org/10.1038/nature08162

Heijmans BT, Tobi EW, Stein AD, Putter H, Blauw GJ, Susser ES, Lumey LH (2008) Persistent epigenetic differences associated with prenatal exposure to famine in humans. Proc Natl Acad Sci U S A 105(44):17046–17049. https://doi.org/10.1073/pnas.0806560105

Heldt SA, Zimmermann K, Parker K, Gaval M, Weinshenker D, Ressler KJ (2014) BDNF deletion or TrkB impairment in amygdala inhibits both appetitive and aversive learning. J Neurosci 34(7):2444–2450. https://doi.org/10.1523/JNEUROSCI.4085-12.2014

Isles AR, Wilkinson LS (2000) Imprinted genes, cognition and behaviour. Trends Cogn Sci 4(8):309–318

Kawano M, Kawaji H, Grandjean V, Kiani J, Rassoulzadegan M (2012) Novel small noncoding RNAs in mouse spermatozoa, zygotes and early embryos. PLoS One 7(9):e44542. https://doi.org/10.1371/journal.pone.0044542

Klengel T, Pape J, Binder EB, Mehta D (2014) The role of DNA methylation in stress-related psychiatric disorders. Neuropharmacology 80:115–132. https://doi.org/10.1016/j.neuropharm.2014.01.013

Klose RJ, Bird AP (2006) Genomic DNA methylation: the mark and its mediators. Trends Biochem Sci 31(2):89–97. https://doi.org/10.1016/j.tibs.2005.12.008

Krawetz SA, Kruger A, Lalancette C, Tagett R, Anton E, Draghici S, Diamond MP (2011) A survey of small RNAs in human sperm. Hum Reprod 26(12):3401–3412. https://doi.org/10.1093/humrep/der329

Lane N, Dean W, Erhardt S, Hajkova P, Surani A, Walter J, Reik W (2003) Resistance of IAPs to methylation reprogramming may provide a mechanism for epigenetic inheritance in the mouse. Genesis 35(2):88–93. https://doi.org/10.1002/gene.10168

Lehrner A, Bierer LM, Passarelli V, Pratchett LC, Flory JD, Bader HN, Yehuda R (2014) Maternal PTSD associates with greater glucocorticoid sensitivity in offspring of Holocaust survivors. Psychoneuroendocrinology 40:213–220. https://doi.org/10.1016/j.psyneuen.2013.11.019

Liu WM, Pang RT, Chiu PC, Wong BP, Lao K, Lee KF, Yeung WS (2012) Sperm-borne microRNA-34c is required for the first cleavage division in mouse. Proc Natl Acad Sci U S A 109(2):490–494. https://doi.org/10.1073/pnas.1110368109

Maguschak KA, Ressler KJ (2011) Wnt signaling in amygdala-dependent learning and memory. J Neurosci 31(37):13057–13067. https://doi.org/10.1523/JNEUROSCI.3248-11.2011

Maguschak KA, Ressler KJ (2012) A role for WNT/beta-catenin signaling in the neural mechanisms of behavior. J Neuroimmune Pharmacol 7(4):763–773. https://doi.org/10.1007/s11481-012-9350-7

Mashoodh R, Franks B, Curley JP, Champagne FA (2012) Paternal social enrichment effects on maternal behavior and offspring growth. Proc Natl Acad Sci U S A 109(Suppl 2):17232–17238. https://doi.org/10.1073/pnas.1121083109

Oh G, Wang SC, Pal M, Chen ZF, Khare T, Tochigi M, Petronis A (2015) DNA modification study of major depressive disorder: beyond locus-by-locus comparisons. Biol Psychiatry 77(3):246–255. https://doi.org/10.1016/j.biopsych.2014.06.016

Ostermeier GC, Miller D, Huntriss JD, Diamond MP, Krawetz SA (2004) Reproductive biology: delivering spermatozoan RNA to the oocyte. Nature 429(6988):154. https://doi.org/10.1038/429154a

Ouko LA, Shantikumar K, Knezovich J, Haycock P, Schnugh DJ, Ramsay M (2009) Effect of alcohol consumption on CpG methylation in the differentially methylated regions of H19 and IG-DMR in male gametes: implications for fetal alcohol spectrum disorders. Alcohol Clin Exp Res 33(9):1615–1627. https://doi.org/10.1111/j.1530-0277.2009.00993.x

Pantano L, Jodar M, Bak M, Ballesca JL, Tommerup N, Oliva R, Vavouri T (2015) The small RNA content of human sperm reveals pseudogene-derived piRNAs complementary to protein-coding genes. RNA 21(6):1085–1095. https://doi.org/10.1261/rna.046482.114

Plasschaert RN, Bartolomei MS (2014) Genomic imprinting in development, growth, behavior and stem cells. Development 141(9):1805–1813. https://doi.org/10.1242/dev.101428

Rodgers AB, Morgan CP, Bronson SL, Revello S, Bale TL (2013) Paternal stress exposure alters sperm microRNA content and reprograms offspring HPA stress axis regulation. J Neurosci 33(21):9003–9012. https://doi.org/10.1523/JNEUROSCI.0914-13.2013

Rodgers AB, Morgan CP, Leu NA, Bale TL (2015) Transgenerational epigenetic programming via sperm microRNA recapitulates effects of paternal stress. Proc Natl Acad Sci U S A 112(44):13699–13704. https://doi.org/10.1073/pnas.1508347112

Roseboom T, de Rooij S, Painter R (2006) The Dutch famine and its long-term consequences for adult health. Early Hum Dev 82(8):485–491. https://doi.org/10.1016/j.earlhumdev.2006.07.001

Schulz LC (2010) The Dutch Hunger Winter and the developmental origins of health and disease. Proc Natl Acad Sci U S A 107(39):16757–16758. https://doi.org/10.1073/pnas.1012911107

Skinner MK (2016) Endocrine disruptors in 2015: epigenetic transgenerational inheritance. Nat Rev Endocrinol 12(2):68–70. https://doi.org/10.1038/nrendo.2015.206

Smoller JW (2016) The genetics of stress-related disorders: PTSD, depression, and anxiety disorders. Neuropsychopharmacology 41(1):297–319. https://doi.org/10.1038/npp.2015.266

Smythies J, Edelstein L, Ramachandran V (2014) Molecular mechanisms for the inheritance of acquired characteristics-exosomes, microRNA shuttling, fear and stress: Lamarck resurrected? Front Genet 5:133. https://doi.org/10.3389/fgene.2014.00133

Solomon Z, Kotler M, Mikulincer M (1988) Combat-related posttraumatic stress disorder among second-generation Holocaust survivors: preliminary findings. Am J Psychiatry 145(7):865–868. https://doi.org/10.1176/ajp.145.7.865

Soubry A, Guo L, Huang Z, Hoyo C, Romanus S, Price T, Murphy SK (2016) Obesity-related DNA methylation at imprinted genes in human sperm: results from the TIEGER study. Clin Epigenetics 8:51. https://doi.org/10.1186/s13148-016-0217-2

Sweatt JD (2013) The emerging field of neuroepigenetics. Neuron 80(3):624–632. https://doi.org/10.1016/j.neuron.2013.10.023

Teperek M, Simeone A, Gaggioli V, Miyamoto K, Allen GE, Erkek S, Jullien J (2016) Sperm is epigenetically programmed to regulate gene transcription in embryos. Genome Res 26(8):1034–1046. https://doi.org/10.1101/gr.201541.115

Vassoler FM, White SL, Schmidt HD, Sadri-Vakili G, Pierce RC (2013) Epigenetic inheritance of a cocaine-resistance phenotype. Nat Neurosci 16(1):42–47. https://doi.org/10.1038/nn.3280

Vecoli C, Montano L, Andreassi MG (2016) Environmental pollutants: genetic damage and epigenetic changes in male germ cells. Environ Sci Pollut Res Int 23(23):23339–23348. https://doi.org/10.1007/s11356-016-7728-4

Weaver IC, Cervoni N, Champagne FA, D'Alessio AC, Sharma S, Seckl JR, Meaney MJ (2004) Epigenetic programming by maternal behavior. Nat Neurosci 7(8):847–854. https://doi.org/10.1038/nn1276

Weiss IC, Franklin TB, Vizi S, Mansuy IM (2011) Inheritable effect of unpredictable maternal separation on behavioral responses in mice. Front Behav Neurosci 5:3. https://doi.org/10.3389/fnbeh.2011.00003

Yan W (2014) Potential roles of noncoding RNAs in environmental epigenetic transgenerational inheritance. Mol Cell Endocrinol 398(1-2):24–30. https://doi.org/10.1016/j.mce.2014.09.008

Yehuda R (2002) Post-traumatic stress disorder. N Engl J Med 346(2):108–114. https://doi.org/10.1056/NEJMra012941

Yehuda R, Bierer LM (2008) Transgenerational transmission of cortisol and PTSD risk. Prog Brain Res 167:121–135. https://doi.org/10.1016/S0079-6123(07)67009-5

Yehuda R, Daskalakis NP, Bierer LM, Bader HN, Klengel T, Holsboer F, Binder EB (2016) Holocaust exposure induced intergenerational effects on FKBP5 methylation. Biol Psychiatry 80(5):372–380. https://doi.org/10.1016/j.biopsych.2015.08.005

Zhang TY, Meaney MJ (2010) Epigenetics and the environmental regulation of the genome and its function. Annu Rev Psychol 61(439-466):C431–C433. https://doi.org/10.1146/annurev.psych.60.110707.163625

Case Study: Developmental Physiology at High Altitude

18

Alexandra Jochmans-Lemoine and Vincent Joseph

Abstract

High altitude is a challenging environment mostly characterized by a low pressure of oxygen but also by cold temperatures, air dryness, reduced protection against exposure to solar radiations, and more limited resources than at lower altitudes. During postnatal development, energy requirements are elevated, and reduction of oxygen supply (hypoxia) during this period has profound physiological consequences. Different models of exposure to hypoxia in newborn mammals have been used over the years and have helped to establish the effects of hypoxia during development on the cardiorespiratory system. Exposure to hypoxia during postnatal development has long-term consequences that manifest throughout the life span. These consequences of neonatal hypoxia might help the adults to better withstand the effects of the reduced oxygen pressure or on the contrary impair the subsequent responses to hypoxia. Most experimental research on development at high altitude focuses on the hypoxic environment, and the cardiorespiratory system, while only few data are available concerning thermoregulatory processes, and the interactions between cold and hypoxia during postnatal development at high altitude. In summary developmental hypoxia determines the ability of adult mammals to withstand life at high altitude, and the available data indicate that this might be an important driving force in short-term acclimatization and long-term adaptation to high altitude. Developmental physiology at high altitude should therefore be considered as a central element for physiology and adaptation to this specific environment.

A. Jochmans-Lemoine · V. Joseph (✉)
Centre de Recherche, Inst. Universitaire de Cardiologie et Pneumologie de Quebec and Université Laval, Quebéc, QC, Canada
e-mail: vincent.joseph@fmed.ulaval.ca

© Springer International Publishing AG, part of Springer Nature 2018

W. Burggren, B. Dubansky (eds.), *Development and Environment*,
https://doi.org/10.1007/978-3-319-75935-7_18

18.1 Introduction

Animals permanently living at high altitude include large and small mammals, birds, reptiles, and amphibian species. These animals are exposed to drastic environmental conditions mainly characterized by reduced availability of Oxygen (O_2), less abundant resources, and cold temperatures. The lower O_2 availability is a highly challenging stressor that must be adequately counteracted to sustain the higher energy requirements necessary to survive in this environment. A very large array of research has been conducted to understand how these animals maintain a net flux of O_2 from the upper airways to the cells, sustaining metabolic activity despite the reduced O_2 availability. This is mainly achieved by physiological responses including enhanced ventilation which helps keeping the high alveolar pressure of O_2 (Dempsey et al. 2014; Joseph and Pequignot 2009), larger lungs with extended gas-exchange surface area to maximize O_2 diffusion across the alveolo-capillary barrier (Frisancho 2013; Jochmans-Lemoine et al. 2015), high O_2 binding affinity of hemoglobin to maintain elevated blood O_2 conductance preserving the arterio-venal PO_2 gradient (Storz et al. 2010), high hemoglobin to maintain high O_2 concentration in arterial blood, dense tissue vascular network for enhanced diffusion of O_2, and efficient utilization of O_2 at the cellular level (Beall 2007; Monge and Leon-Velarde 1991; Cheviron et al. 2014; Lui et al. 2015).

The physiological systems that govern these responses are eminently "plastic" and subjected to anatomical and functional changes in response to environmental clues. Different time lines of plasticity are classically recognized to determine the resulting phenotype: on one hand phenotypic plasticity, or acclimatization, allows a gradual improvement of physiological functions after initial exposure to high altitude to optimize O_2 flux through the cardiorespiratory system, its delivery to tissues, and utilization in cells. Typically, this process is fully reversible upon return to previous conditions, and in the classical view of acclimatization, these changes are not heritable from one generation to the next (however, this is being increasingly challenged; see below). Ventilatory and hematological acclimatization to chronic hypoxia are well-known responses that increase O_2 uptake and transport (Dempsey et al. 2014; Joseph and Pequignot 2009). On the other hand the process of genetic adaptation might be the result of natural selection over generations in animals that have been living "high" for thousands to millions years. An additional challenge is encountered by newborn animals at high altitude: their small size enhances the mass-specific O_2 consumption, a substantial part of energy is required for growth, and environmental changes during development can lead to long-term effects that persist at adulthood. Indeed, in some cases postnatal hypoxia has an overwhelming impact on the phenotype observed in adults at high altitude, and there is mounting evidences that it is necessary to take into account the responses to postnatal hypoxia and their long-term consequences (also referred to as "developmental plasticity") to explain the occurrence of a particular phenotype at high altitude. A more rigorous way to express this is that under a specific environment such as high altitude, the expression of a given phenotype in adults is the result of the interactions between the genetic backgrounds (including limits put on phenotypic plasticity) and the developmental plasticity imposed by the environment (Via et al. 1995; Russell et al. 2008).

While the concepts of phenotypic plasticity, developmental plasticity, and genetic adaptation are typically considered as different fields of research, there is a growing trend to consider that "plasticity" and particularly its developmental aspect is one of the major driving forces of genetic adaptation. Indeed "plastic" responses induced by environmental factors experienced at one generation can lead to permanent morphological changes in the next generation through epigenetic variations affecting DNA methylation in germ-line cells (e.g., nongenetic inheritance) (Dias and Ressler 2014). Whether these processes can affect adaptation to altitude still remains unknown, but there is at least a fertile theoretical framework that helps bringing together developmental physiology and evolution (Danchin and Pocheville 2014) that can pave the way to a better integration of the plethoric data and the apparent large diversity of phenotypic responses and genetic adaptation observed at altitude. While we will not further expand these aspects in the present chapter, it should be kept in mind that a broader interpretation of these data is clearly possible.

18.2 Why Bother About Developmental Physiology at High Altitude?

Without doubt specific responses to high altitude during development are strong enough to have profound consequences on human health, but surprisingly they also have imposed important socioeconomics and even cultural changes during the history of colonization in South America (and possibly in other high-altitude regions on earth). In fact, the first capital of Peru (Jauja) was located in the high-altitude valleys at 3300 m above sea level. But the mortality rate among newborn animals (pigs, chickens, horses) was so high that the Spanish invaders decided to move their capital city to Lima, on the harsh environment of the coastal desert boarding the Pacific Ocean (Monge 1948). On the same line of evidence, the first Spaniard settlers in the city of Potosi (4000 m) did not succeed in having children, or their children died "either at birth or in the fortnight thereafter" (Monge 1948). Pregnant women used to go to cities at lower altitude to give birth and raise their children for the 1st year. The first child that was able to survive was born 53 years after the initial establishment of the Spaniards, but his father was considered a fool by relative and friends for wanting his child to be born in Potosi. Regardless, the foolish father dedicated his son to Saint Nicholas of Tolentino; he named his child after the Saint and raised him in Potosi "curing it miraculously of many sick spells, caused not by the cold but other deadly diseases" (Monge 1948). It is said that a whole generation of children was named Nicholas thereafter, hoping that this would offer miraculous protections against the fatal consequences facing newborn babies at high altitude. During the same time, however, the native settlers had no problems in giving birth and raising their child with "customary Indian fertility" (Monge 1948); nobody asked however if this was not a simple reflection of the fact that St Nicholas felt more inclined to protect the babies born from Indian couples rather than from Spanish. Regardless of these considerations, it is clear that postnatal development is a critical aspect of natural history for animals and humans living at high altitude.

18.3 From Acclimatization to Adaptation Across Generations

While being seemingly anecdotal, the abovementioned historical elements highlight the important fact that throughout history, lowland species have migrated to high altitude with divergent success. On one hand, pigs, horses, or fowls are notably intolerant to life at altitude (at least during the first generations of exposure), but on the other hand other species that were introduced in South America by Europeans are now commonly found in high-altitude areas (these animals will be characterized as being tolerant to life at high altitude), while animals endemic from high altitude are considered as being genetically adapted. Most importantly, different phenotypes might be found in these animals (see below). One important question to ask is how to integrate the different time line of the responses to high altitude in a comprehensive manner?

If adults from a lowland migrant species are able to exhibit adequate acclimatization, they will be able to give birth to a first generation of high-altitude natives. Newborns must adequately respond to the challenge of postnatal hypoxia, which may result in long-term beneficial or detrimental consequences into adulthood. Individuals who survive and adequately develop will give birth to several successive generations, which may ultimately result in genetic adaptation, defined as the result of natural selection of a gene or a group of genes that govern functional change(s) in response to a given environment and that improve survival and ultimately reproduction. These changes are passed from one generation to the next through genetic material.

This line of events can be schematized as proposed in Fig. 18.1 that shows the relationship between individual time scale (from birth to adulthood—horizontal) and the processes of responses, acclimatization, and adaptation (vertical). In this figure, the vertical time line has been modified from Hochachka et al. (1998), which presented, in adults, "formally defined relationships between time and physiological responses to environmental changes." It has previously been proposed that developmental events should be added to this drawing to take into account the "programming" effects of postnatal environment (Huicho 2007), and we will use a similar approach to discuss developmental physiology at altitude.

However, it should also be noted that responses that are counterproductive at high altitude are also possible, and in this case the appropriate adaptive response is to avoid or attenuate these responses. For example, in most mammals, pulmonary arterial pressure rises in response to chronic hypoxia, and in species genetically adapted, or tolerant to hypoxia, the rise of pulmonary pressure is limited or absent (Jochmans-Lemoine et al. 2015; Storz et al. 2010); this is due to a reduced muscularization of small pulmonary arteries (Tucker and Rhodes 2001). Additionally, elevated hematocrit levels increase blood viscosity with detrimental consequences (Storz et al. 2010), and theoretical elements indicate that the optimum level of hematocrit that ensures the most adequate O_2 transport and utilization is very close to the normal sea level hematocrit (Villafuerte et al. 2004).

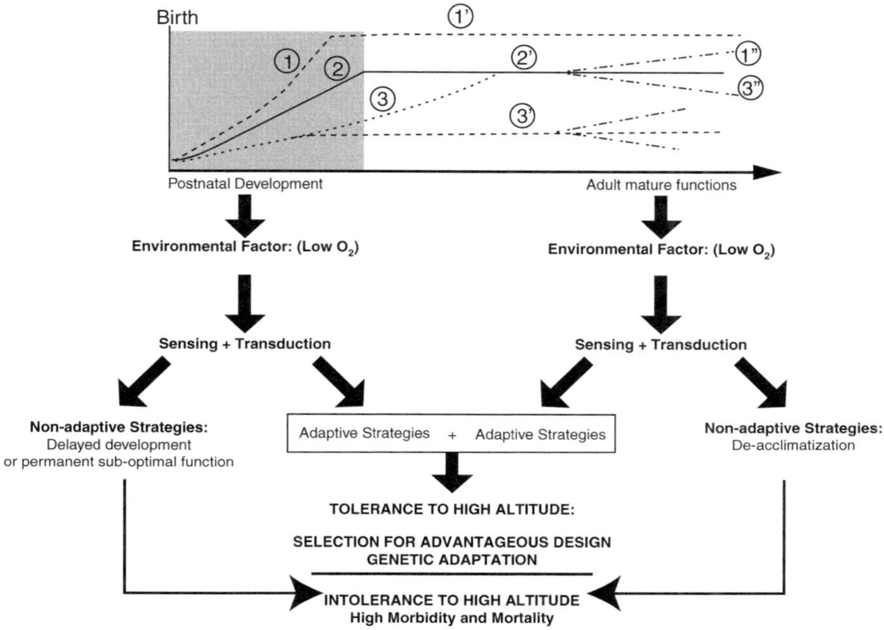

Fig. 18.1 Relationship between individual time scale (from birth to adulthood—horizontal) and the processes of responses, acclimatization, and adaptation to high altitude (vertical arrows). During postnatal life physiological functions such as respiratory control, lung anatomy, or thermoregulation follow a gradual process of development leading to a mature system. Postnatal hypoxia might either accelerate (1), have no effect (2), or delay (3) this development. These different pathways of development lead to a mature system optimized (1′), normal (2′), or dysfunctional (3′) relatively to the requirements of the high-altitude environment. In adults, phenotypic plasticity also contributes to adjust the phenotype (1″), but nonadaptive responses might also occur (3″). At each stage (adult or postnatal development), responses to environmental clues are orchestrated by specific sensors. Depending on these pathways, individuals might either be tolerant or intolerant to life at high altitude. Genetic adaptation is thought to result from selection of the most advantageous designs. Figure adapted from Hochachka et al. (1998) and Huicho (2007)

18.4 Diversity of the Phenotypes Accounting for Adaptation or Acclimatization at High Altitude

18.4.1 Specific Phenotype of Animals Tolerant to High Altitude

High-altitude horses from Columbia have large allelic divergences with a control population at sea level, including in the *epas1* gene (encoding the HIF-2α protein) and in genes involved in central nervous system function (Hendrickson 2013). Peruvian chickens at high altitude have high hemoglobin oxygen affinity (Velarde et al. 1991); house mice have specific physiological responses at high altitude, including large lungs, with an extended gas exchange area and high ventilation (Jochmans-Lemoine et al. 2015); and dogs present at high altitude (4300 m) have been found to have a marked hyperventilation (Banchero et al. 1975) and accelerated lung development induced by postnatal hypoxia (Johnson Jr et al. 1985). Surprisingly, however, this list includes animals that were first described as being intolerant to high altitude (horses, chickens—see above) and makes apparently reference to genetic changes supporting their "tolerance" to altitude. One key difference however between animals listed below as being "adapted" to altitude is the time during which this adaptation took place—4 to 5 centuries vs. 10^5 to 10^6 years, and it might be argued that as a typical example where "physiology meets evolution" (Danchin and Pocheville 2014) it is critical to understand the developmental aspect of "tolerance" to high altitude.

Humans represent a typical example of a lowland migrant species that have reached high-altitude regions at different times and locations over the past 30,000 or 40,000 years, and the different groups found at high altitude have different physiological and genetic adaptations (Beall 2007; Bigham et al. 2010).

18.4.2 Specific Phenotype of Animals Adapted to High Altitude

Among many other examples, llamas (Llanos et al. 2007), guinea pigs (Bartels et al. 1979), deer mice (Snyder 1985; Storz et al. 2009), pika (Pichon et al. 2009; Pichon et al. 2013), Andean foxes (Leon-Velarde et al. 1996), and Himalayan gooses (Ivy and Scott 2014) are mainstream examples of animals endemic at high altitude. These animals have been living "high" for thousands to millions of years and possess clear genetic adaptations that ensure efficient hyperventilation (Ivy and Scott 2014; Storz et al. 2010), high hemoglobin oxygen affinity (Snyder 1985; Storz et al. 2009), small elliptical red cells with high hemoglobin concentration, high muscle myoglobin concentration, and/or more efficient O_2 extraction at the tissue level (Benavides et al. 1989). Additionally, llamas do not present the typical pulmonary hypertension upon exposure to hypoxia (Llanos et al. 2007) and present only a small increase in hemoglobin concentration at high altitude (Banchero et al. 1971). Deer mice from high-altitude regions in North America are considered genetically adapted because of the high O_2 affinity of their hemoglobin (Snyder 1985; Storz et al. 2009), but they also demonstrate hematological acclimatization with elevated hematocrit and hemoglobin levels upon exposure to altitude, which decrease upon return to lower altitude (Tufts et al. 2013). In humans, signs of genetic adaptation to high altitude are also present in populations from the Andes, Tibet, and Ethiopia with important phenotypic divergences between these groups, and while high-altitude residents from the Andes demonstrate clear sign of adaptation (Bigham et al. 2010), other traits of this population such as their low ventilation level and high hematocrit (Storz et al. 2010; Beall 2007) are considered as being maladaptive and in the most extreme cases related to the development of "chronic mountain sickness" (Leon-Velarde et al. 2005).

While being necessarily limited, this extremely brief overview highlights the diversity of specific adaptations to the high-altitude environment and shed light on the complexity of the physiological, molecular, and genetic mechanisms that underlie these responses. While trying to explain this phenotype based on adult physiology and genetics is without doubt extremely helpful, there is a long tradition of research that takes into account the long-lasting influence of postnatal hypoxia to better understand the unique physiology of animals at high altitude, and there is growing evidence that this is indeed a critical aspect to consider. In the following paragraphs, we will present the effects of postnatal hypoxia on each system involved in acclimatization and adaptation to high altitude. When data are available, we will also describe the long-term consequences induced by postnatal hypoxia on these systems.

18.5 Ventilatory Responses to Chronic Hypoxia in Adults and Newborns

In adults, exposure to hypoxia evokes an immediate increase of the ventilation mediated by the peripheral chemoreceptors, mainly localized in the carotid body at the bifurcation of the common carotid artery (Kumar and Prabhakar 2012). Upon reduced cellular O_2 tension, carotid type I cells are depolarized, leading to the activation of voltage-dependent calcium channels and the release of excitatory and modulatory transmitters to activate the postsynaptic nerve endings of the carotid sinus nerve, which in turn sends its efferent projections to the nucleus tractus solitarius, a key element of the brainstem dorsal respiratory group. The depolarization of type I cells is mediated by hypoxic-induced inhibition of potassium channels (Lopez-Barneo et al. 1988), and numerous evidences show that this is mediated by gasotransmitters such as carbon monoxide (CO) and hydrogen sulfide (H_2S) (Peng et al. 2010) and by a subunit of the AMP kinase (Ross et al. 2011), but this issue remains controversial (Buckler 2012), and the relative contribution of these systems to the net hypoxic response in vivo is unclear (Kemp 2006; Peers et al. 2010). This is further complicated by the fact that numerous inhibitory feedback mechanisms are also present in carotid body and activated upon hypoxic exposure, including classical neurotransmitters such as dopamine (Iturriaga et al. 2009; Gonzalez et al. 1994), or nitric oxide (Prabhakar and Peers 2014).

When exposed to hypoxia for a prolonged period of time, minute ventilation continues to rise gradually (Schmitt et al. 1994; Powell et al. 1998; Dempsey et al. 2014). This process of ventilatory acclimatization to chronic hypoxia has been first described in humans as a gradual decline of the arterial pressure of carbon dioxide (CO_2) showing increased ventilation over the course of a few days after an ascent at high altitude (Rahn and Otis 1949). Because this occurs in parallel with a gradual increase of arterial pressure of O_2, it has long been thought that a higher sensitivity of the respiratory system to hypoxia was responsible for this response. Indeed, key experiments in goats have determined that the peripheral chemoreceptors are both necessary and sufficient to induce ventilatory acclimatization (Smith et al. 1986;

Busch et al. 1985; Forster et al. 1981) and that the sensitivity of peripheral chemoreceptors to hypoxia increases during chronic exposure (Vizek et al. 1987). This is accompanied by important structural and biochemical changes (He et al. 2005, 2006; Chen et al. 2002a, b; Joseph and Pequignot 2009), which include neovascularization and proliferation of progenitor stem cells present in the carotid body that will form new chemosensitive type I cells (Platero-Luengo et al. 2014). There is a good agreement that this process depends upon the activation of the hypoxia inducible factor (HIF) system since heterozygous mice KO for HIF-1α (HIF-1α +/−) have a reduced ventilatory acclimatization to hypoxia (Kline et al. 2002).

In newborn the peripheral chemoreceptors are functional in utero, and they become almost silent at birth due to the transition from the low fetal arterial O_2 pressure (PaO_2) to the relatively hyperoxic ex utero environment (Blanco et al. 1984). Within a few hours, or a few days, the peripheral chemoreceptors are reset to their new functional O_2 pressure (PO_2) level and regain their sensitivity to hypoxia (Blanco et al. 1984). During postnatal development, carotid body is essential to maintain the breathing pattern, and acute suppression of carotid body activity by inhalation of a few breaths of hyperoxic gas reduces minute ventilation by about 40–60% in human neonates (Al-Matary et al. 2004). In newborn mammals, carotid body denervation induces a profound hypoventilation, impairs body growth, and leads to high mortality during the next days or weeks (Bureau et al. 1985; Donnelly and Haddad 1990; Hofer 1984). Thus, adequate function of peripheral chemoreceptor is critical for the establishment of normal breathing during postnatal development. Yet, as opposed to the robust and sustained ventilatory response evoked by acute exposure to hypoxia in adults, the typical response in newborn is biphasic with a peak response occurring within a few minutes of exposure to hypoxia, followed by a rapid downregulation of ventilation that might even go below the baseline level (Bissonnette 2000). Then a gradual process of development occurs, during which the ventilatory response to hypoxia reaches its full maturity between the 2nd and 3rd postnatal week (Bissonnette and Knopp 2001; Potvin et al. 2014; Joseph et al. 2000). This is accompanied by an increased sensory response of peripheral chemoreceptors to hypoxia (Blanco et al. 1984; Kholwadwala and Donnelly 1992; Wasicko et al. 1999; Sterni et al. 1995; Niane et al. 2011) and central maturation of the respiratory control system (Simakajornboon and Kuptanon 2005; Bissonnette 2000). Chronic hypoxia delays this pattern of postnatal development (Wasicko et al. 1999; Eden and Hanson 1987; Joseph et al. 2000), and some studies have shown that in rats raised in chronic hypoxia, the sensitivity of the peripheral chemoreceptors to hypoxia is reduced (Hertzberg et al. 1992; Landauer et al. 1995), while other did find a drastic reduction of the hypoxic ventilatory response, without noticeable changes of the carotid sinus nerve response to hypoxia (Eden and Hanson 1987), suggesting that chronic postnatal hypoxia might also alter the integration of peripheral chemoreceptor inputs at the level of the central nervous system. Some studies have shown changes of catecholamine synthesis rate and turnover in the peripheral chemoreceptors of rats raised under chronic hypoxia, which might underlie the delayed maturation (Hertzberg et al. 1992; Joseph et al. 2000), and others have reported that postnatal hypoxia stimulates cellular growth of the carotid body

(Wang and Bisgard 2005), but since the net output of the carotid body in response to hypoxia integrates both excitatory neurotransmission and inhibitory feedback mechanisms, the size of this organ does not predict higher sensitivity. Moreover these alterations persist in adulthood, even when the rats are returned to normoxia for a prolonged period of time (Okubo and Mortola 1990), and interfere with the normal process of ventilatory acclimatization to hypoxia (Lumbroso and Joseph 2009). In humans, studies reported a blunted hypoxic ventilatory response in adult men living at high altitude (in South America) (Severinghaus et al. 1966; Lahiri et al. 1969), and studies performed in children of different ages showed that this altered response, rather than being an innate trait, was acquired during development (Lahiri et al. 1976).

Apparently, these developmental delays and long-term alterations of the respiratory control system would have no adaptive values and might even be detrimental at high altitude. High-altitude natives in South America are well known for their sensitivity to develop chronic mountain sickness, characterized by excessive erythrocytosis and reduced alveolar ventilation, and it has been suggested that the occurrence of this high altitude disease could be linked to perinatal hypoxic events (Moore et al. 2007). In adult rats that have been raised under postnatal hypoxia, the ventilatory acclimatization to chronic hypoxia is impaired, which is associated with excessive erythrocytosis upon exposure to chronic hypoxia (Lumbroso and Joseph 2009). As far as we are aware, very few studies have evaluated if these long-term effects on respiratory control are also present in species adapted to high altitude. Because this is not an adaptive effect, it is possible to hypothesize that this should not be the case. We recently compared the effect of postnatal hypoxia in rats and mice (that have divergent physiological responses when raised at high altitude—(Jochmans-Lemoine et al. 2015)) and showed that the developmental delay of hypoxic ventilatory response in mice is much less important than in rats (Jochmans-Lemoine et al. 2015, unpublished data). Interestingly, a protection against the reduced ventilatory hypoxic sensitivity due to postnatal hypoxia might explain why adult mice living at high altitude have higher ventilation compared to rats (Jochmans-Lemoine et al. 2015), but this remains to be firmly established. In other words, at least for these mice, their tolerance to high altitude could be determined, at least partially, by the reduced sensitivity of their respiratory control system to the detrimental effects of postnatal hypoxia. On the contrary, in rats the delayed development of hypoxic ventilatory response might contribute to their well-known intolerance to high altitude. These data are interesting, but they are not sufficient to conclude that protection against the detrimental effects of postnatal hypoxia is an important feature of genetic adaptation to high altitude. Indeed, while mice are commonly found at high altitude, they do not display genetic markers of adaptation to altitude (Storz et al. 2007) that are found in other rodents, such as the deer mice (Storz et al. 2009).

In humans the genetic background appears to be very important to determine the effect of developmental hypoxia on respiratory control. Despite similar pattern of lifelong exposure to altitude, high-altitude natives from the Himalayan plateau do not have the typical blunted ventilatory response to hypoxia observed in Andean natives (Zhuang et al. 1993; Hackett et al. 1980). In South American subjects that are born and raised at sea level, the proportion of native American ancestry established by genetic markers is inversely proportional to ventilation during exercise, the hypoxic ventilatory response, and the ratio of ventilation to VO_2 (Brutsaert et al. 2005).

In summary, postnatal hypoxia delays the maturation of the peripheral chemoreceptors, drastically reduces the ventilatory response to hypoxia in newborn mammals, and leads to long-lasting reduction of the hypoxic ventilatory response and acclimatization. It is clear that in rats this contributes to their well-known intolerance to hypoxia, and it also affects the respiratory control in human high-altitude natives. Human studies suggest that the genetic background determines the sensitivity and long-term consequences of postnatal hypoxia on the respiratory control system. It remains to be determined if this resistance to the effects of postnatal hypoxia observed in some species or human group is a widespread strategy that differentiates species that are tolerant (or adapted) and species that are intolerant to high altitude. The underlying mechanisms likely involve genes that govern the hypoxic responses in peripheral chemoreceptors and in the central nervous system.

18.6 Effects of Chronic Hypoxia on Lung Morphology and Oxygen Diffusion in Adults and Newborns

During postnatal development, intense structural and functional changes occur in the lungs. In rats, alveolar formation occurs during a critical period starting around postnatal day 4 (Burri et al. 1974). At this age, the lungs are made of a relatively small number of large, thick-walled saccules. This stage is followed by progressive projection of ridges into the airspaces, forming the contours of the future alveoli (septation phase). During the second postnatal week, the alveolar walls become thinner, and alveolar capillary are properly organized allowing the formation of a fully mature lung (alveolarization phase) (Blanco et al. 1991; Bruce et al. 1999; Burri 1984; Burri et al. 1974). This process allows the formation of alveoli, which are more numerous, smaller, and thin-walled compared to the saccules seen in newborn. The development of pulmonary vessels occurs in parallel to these events. In humans, septation and alveolarization begin shortly before birth and develop at least until the end of the 1st or 2nd year of life (Burri 1984).

Human natives from altitude regions have higher diffusing capacity, lung volume, and functional residual capacity than low-altitude subjects (Droma et al. 1991; Frisancho 2013), lower maximum expiratory flow-rate per lung volume, and lower upstream airway conductance (see Massaro and Massaro (2002) for review). These traits suggest that at high altitude there is an excess growth of the gas exchange region compared with conducting airways (Massaro and Massaro 2002). Because the gas exchange regions develop mainly after birth, while the conductive zone develops in-utero, this suggests that exposure to altitude or hypoxia during postnatal development accelerates the maturation of the gas exchange surface in the lungs (Massaro and Massaro 2002). Indeed subjects that are acclimatized to altitude since childhood (in Peru), or that were born at altitude, have a higher forced vital capacity, a VO_2 max roughly 15–20% higher, and arterial O_2 saturation during exercise 5% higher compared to subjects that have migrated to altitude as adults (Brutsaert et al. 2004). Yet, in these subjects the hypoxic ventilatory response is notably lower (Lahiri et al. 1969, 1976; Severinghaus et al. 1966), and ventilation during exercise is reduced by about 16% which indicates that their blunted ventilation is not limiting for O_2 transport, and in fact these subjects extract more O_2 despite their lower pulmonary ventilation (Frisancho et al. 1973; Frisancho 2013). Remarkably, the net effect of developmental exposure to high altitude persists at sea level: adult Peruvian Quechuas from Lima that have been born and raised at high altitude have higher aerobic capacity than subjects born and raised at sea level, after controlling for their level of activity (Kiyamu et al. 2015). Finally, infants between 1 and 24 months of age born at the altitude of 3440 m (La Quiaca, Jujuy, Argentina) have higher lung volume than infants at sea level (Llapur et al. 2013). Therefore, these are permanent anatomical changes imposed by exposure to high altitude during the critical period of postnatal development.

Animal studies consistent with these findings are available. In dogs, exposure at altitude (3100 m) increases the distensibility, diffusion capacity, and volume of the lungs if the exposure starts at 2.5 months of age: early during development, but these effects are not present if dogs are raised at high altitude as adults (Johnson Jr et al. 1985). Rats that have been raised at the high-altitude station of the Jungfraujoch (3450 m in the Swiss Alps) between the 23rd and the 44th postnatal days have larger lungs and higher alveolar and capillary surface areas than control rats raised at low altitude (Burri and Weibel 1971). However, if rats raised at sea level are exposed to 13% O_2 throughout gestation and postnatal development, then the rate of postnatal alveolar formation is reduced between postnatal days 2 and 40 (Blanco et al. 1991), leading to reduced alveolar surface area, while it is effectively increased if exposure to hypoxia is restricted to postnatal days 23–44 as in the previous mentioned study (Burri and Weibel 1971). In guinea pigs born at sea level and raised at high altitude (3800 m, Barcroft Laboratory of the University of California White Mountain Research Station) for up to 6 months starting from the 3rd postnatal week, lung growth is accelerated, resulting in increased lung volume, alveolar surface, capillary surface area, and higher membrane and lung diffusing capacities (Hsia et al. 2005). On the same line of evidence, high-altitude living deer mice have larger lungs than deer mice from sea level (Hammond et al. 2001; Shirkey and Hammond 2014). Therefore, the effects of high altitude residence on lungs appear to be mostly due to a developmental influence that results in increased lung volume, with enhanced diffusion capacity and higher alveolar surfaces.

On the other hand, some data clearly indicate that postnatal hypoxia might delay alveolar maturation and permanently reduce the gas exchange surface. In adult rats living at high altitude, lung volume and the alveolar surface area are apparently suboptimal (Lumbroso et al. 2012) and much lower than in mice living at the same altitude (Jochmans-Lemoine et al. 2015). However, high-altitude rats that have been exposed to 32% O_2 (similar to sea level PO_2) during the first 2 postnatal weeks, and raised thereafter under ambient hypoxic conditions until adulthood, have a reduced hematocrit level, lower right ventricular hypertrophy, and lung structure with smaller airspace to tissue ratio that most probably reflect higher alveolar exchange surface area (Lumbroso et al. 2012). When we exposed sea level rats to the low O_2 pressure found in La Paz during postnatal days 4–14 (period of alveolar development), the alveolar surface area was reduced and arterial oxygen saturation was lower in response to acute hypoxia compared to control rats (Jochmans-Lemoine et al. 2018), much like previously reported when hypoxic exposure is restricted to this early postnatal period (Blanco et al. 1991). Thus, the impaired gas exchange function of the lungs found in rats at high altitude is induced by postnatal hypoxia and appears to play a key role to determine the maladaptive phenotype of rats living at 3600 m above sea level (La Paz, Bolivia) that have an excessive hematocrit, extremely high right ventricular hypertrophy, low ventilation, and low metabolic rate (Lumbroso et al. 2012; Jochmans-Lemoine et al. 2015). This is a striking difference compared with species that are endemic and show genetic adaptation to high altitude (deer mice), or to species that are tolerant to high altitude (house mice—(Jochmans-Lemoine et al. 2015), probably reflecting differences on the cellular and molecular interactions between hypoxia and the processes governing postnatal lung development. Interestingly, it has been suggested that because postnatal hypoxic exposure reduces O_2 consumption rate (see below), and because there is a very tight relationship between O_2 consumption and gas exchange surface in the lungs, the impaired alveolar development in rats raised under chronic hypoxia might simply reflect the reduced O_2 needs (Massaro and Massaro 2002).

18.7 Metabolic Effects of Hypoxia in Adults and Newborns

Apart from increasing lung ventilation and optimizing the lung exchange surface to maintain a high flux of O_2 at high altitude, another response to hypoxic exposure is to reduce the rate of O_2 consumption. This response might be highly efficient and is conserved across evolution, and under extreme conditions of hypoxia, it ensures survival in species that show stunningly long-lasting resistance to anoxia such as aquatic turtles (Hochachka and Lutz 2001; Hochachka et al. 1996). While in mammals this extreme response is not possible, a certain degree of metabolic rate reduction is observed upon exposure to acute hypoxia, particularly in newborn or in small adults, in which metabolic rate is high relatively to their body mass (Mortola 1999, 2004; Singer 2004). The decrease of metabolic rate might be accompanied by a reduction of core body temperature, due to a reduction of the thermoregulatory set point. Current models are consistent with the hypothesis that hypoxia activates cAMP-dependent pathways in the hypothalamic preoptic area, causing an elevation of the thermal sensitivity of preoptic warm-sensitive neurons, in turn leading to an inhibition of thermogenesis and activation of heat loss (Steiner and Branco 2002; Branco et al. 2006; Bicego et al. 2006): in other words, this response is a decrease of the thermoregulatory set point initiated in the preoptic hypothalamic area. Gasotransmitters such as nitric oxide, CO and H_2S appear as critical mediators of this response (Branco et al. 1997, 2014; Steiner and Branco 2002; Paro et al. 2001). Interestingly, blockade of NMDA receptors amplifies the hypoxic-induced fall of body temperature, and this effect is present in developing rats between the 4th and the 20th postnatal day (Baig and Joseph 2008). These data clearly illustrate that this process is tightly regulated by the nervous system rather than a passive response to O_2 limitation (Branco et al. 2014; Barros et al. 2001; Steiner and Branco 2002; Gordon and Fogelson 1991).

However, under prolonged hypoxia, this effect does not persist, and after only 1 day of hypoxic exposure, body temperature in rats reaches pre-hypoxic levels but with a blunted circadian variations (Mortola and Seifert 2000; Bishop et al. 2000). After 4 weeks of hypoxic exposure (in mice), body temperature and its circadian variations returned to a normoxic pattern (Beaudry and McClelland 2010). Furthermore, the evidence indicate that reduction of metabolic rate is not a valuable strategy for animals that should respond to the everyday challenges of hunting or foraging for their food, escaping predators, responding to period of cold exposures, and engaging into social behaviors. In fact, animals that are tolerant or adapted to hypoxia appear to have a normal metabolic rate (Jochmans-Lemoine et al. 2015), and if anything adaptation at high altitude favors the maintenance of high aerobic capacity (Brutsaert et al. 2004; Cheviron et al. 2012, 2014; Hochachka et al. 1998). In rats raised at high altitude (3600 m—La Paz, Bolivia), we have found a low metabolic rate that seems mainly to be a response to impaired gas exchange function, in order to maintain arterial oxygen saturation (Jochmans-Lemoine et al. 2015). The basal metabolic rate in these animals was indeed so low that when they were challenged with exposure to severe hypoxia (12% O_2—or a PO_2 of 60 mmHg, that would be less than 8% O_2 at a sea level atmospheric pressure), they were not able to further reduce O_2 consumption or rectal temperature and exhibited high mortality. On the other hand, under similar conditions, mice had a normal metabolic rate, were able to reduce O_2 consumption and rectal temperature, and tolerated the most sever level of hypoxia without mortality.

As a general rule, body size appears as an important determinant of the ability to decrease metabolic rate in hypoxia, and this is best explained by the higher mass-specific O_2 consumption in small animals compared to large animals (Hill 1959; Frappell et al. 1992; Mortola 2004; Singer 2004); thus, the hypometabolic response is increased in newborn mammals compared to adults of the same species. There are species-specific differences in the magnitude of the hypometabolic response to hypoxia that are inversely related to the hypoxic ventilatory response (Mortola et al. 1989), leading to the concept that the ratio of minute ventilation to O_2 consumption (Ve/VO_2) is the best estimate of the hypoxic ventilatory response. During postnatal development the hypometabolic response to hypoxia becomes gradually less important, and this evolves in parallel with the maturation of the hypoxic ventilatory response (Mortola 1999; Baig and Joseph 2008). It might be hypothesized that because maturation of the hypoxic ventilatory response is delayed at high altitude, the hypometabolic response to hypoxia should be maintained for a longer period of time. It might also be questioned whether there are differences between adapted (or tolerant) species and intolerant species in the maturation of these responses at high altitude. As far as we are aware, however, only a limited number of studies have addressed these questions. Hamsters are an interesting example because they live underground, in burrows that are notable for being hypoxic and hypercapnic environment. Compared with rats raised under postnatal hypoxia from birth to postnatal days 21–22, hamsters do not demonstrate a reduced metabolic rate, and they maintain higher minute ventilation (Frappell and Mortola 1994). Because rats demonstrate a blunted postnatal growth in these conditions, while the hamsters had a normal postnatal body growth, the specific response of the hamster to postnatal hypoxia has clearly an adaptive value (Frappell and Mortola 1994). Interestingly, in 1-day-old human infants that are born in La Paz (Bolivia—3600 m), minute ventilation and metabolic rate are both similar than in sea level infants, implying that the hypometabolic response to hypoxia is not present and that higher O_2 extraction from the lungs compensates for the lower inspired PO_2 at altitude (Mortola et al. 1992). As noted above it has been suggested that O_2 consumption rate and lung growth under chronic postnatal hypoxia are tightly linked (Massaro and Massaro 2002); if this is indeed the case, then a systematic exploration of metabolic rate during postnatal development on animals showing divergent lung growth at high altitude could help establish whether this hypothesis is relevant.

18.8 Interaction Between Multiple Stressors: Responses to Cold Exposure at High Altitude

Exposure to cold temperatures at high altitude poses an additional challenge. In endotherm animals (birds and mammals), the physiological response to cold under normoxic conditions is to enhance heat production, which requires enhanced O_2 consumption, and is necessary to maintain body temperature at its high optimal level. Under intense cold exposure, the regulatory mechanisms ensuring heat production are overridden by heat losses, and internal temperature drops (hypothermia), which might lead to respiratory and cardiac arrest, and ultimately to death (Fong 2010).

Despite their immature thermoregulatory responses and their small size (that increases body surface-volume ratio), newborn mammals are more resistant to cold exposure than adults. Indeed, body temperature decreases more rapidly and more profoundly in newborn compared to adult mammals (Mortola 2005), but in parallel, cardiorespiratory arrest occurs at much lower body temperature in newborn than in adults (Fong 2010; Tattersall and Milsom 2003). In the absence of systematic literature studying thermoregulatory responses during postnatal development at high altitude, it is only possible to have some hints on the interaction between hypoxia and cold exposure during postnatal development from two types of experimental paradigms that can be more easily used in sea level laboratory settings: some studies asked what are the effects of cold exposure on hypoxic responses; other studies asked what are the effects of hypoxia on responses to cold exposure.

18.8.1 Effects of Cold Exposure on Hypoxic Responses

The intricate interactions between cold and hypoxic exposure on metabolic control have been investigated in the late 1950s (Hill 1959), and since these initial studies, all available evidence are in the same direction: compared to thermoneutral conditions, the effects of hypoxic exposure on metabolic rate are much more important under cold conditions. In 6-day-old mice, maintained under thermoneutral conditions at 33 °C, brief exposure to hypoxia (10% O_2—3 min) almost doubles minute ventilation with only a marginal reduction of O_2 consumption (−10 to −15%) (Bollen et al. 2009), which is consistent with the notion that metabolic and respiratory depression require longer exposure time to be apparent. When mice are placed at 26 °C, basal O_2 consumption and minute ventilation are increased (which is the normal thermogenic response), but hypoxic exposure caused only a small increase of minute ventilation, followed by a profound ventilatory and metabolic depression that extended for a few minutes after the return to normoxia, accompanied by a small reduction of body temperature. In this study, hypoxic exposure at 26 °C also induced an important behavioral response characterized by body movements and vocalization of the pups that were not present at 33 °C. This behavioral response has been interpreted as reflecting a more stressful effect of hypoxia under cold exposure, when O_2 consumption is increased (Bollen et al. 2009). In newborn dogs (1–2 weeks) exposed to cold conditions (20 °C) and normoxia, O_2 consumption and minute ventilation increase compared to dogs maintain at 30 °C. Both values decreased to a much greater extent when the dogs are exposed gradually to hypoxia in cold conditions compared to warm conditions, but below 10% O_2, metabolic rate is no longer different no matter what the temperature is (Rohlicek et al. 1998). Remarkably, the arterial pressure of CO_2 and the arterial content of O_2 were maintained at similar levels under cold and warm conditions, for all levels of hypoxia (Rohlicek et al. 1998), while the PaO_2 was only slightly lower in cold hypoxic compared to warm hypoxic conditions. This highlight that there is a highly efficient matching between minute ventilation and metabolic rate to adjust the hyperventilatory response under cold and warm conditions (leading to the similar $PaCO_2$), and, as noted above, it is necessary to take into account both minute ventilation and O_2 consumption (or CO_2 production rate) to have a reliable estimate of the hypoxic ventilatory response under cold exposure.

18.8.2 Effects of Hypoxia on Responses to Cold Exposure

Studies conducted in newborn rats (2-day-old) have shown that in hypoxia the typical increase of metabolic rate observed in response to cold exposure is suppressed (Mortola and Dotta 1992); a similar effect was reported in newborn infants (1-day-old) born at the altitude of 4300 m (Cerro de Pasco—Peru) (Frappell et al. 1998). Therefore, cold-induced thermogenesis is suppressed during exposure to acute hypoxia at sea level and in 1-day-old human at high altitude. However, studies performed in newborn rats tend to indicate that longer exposure to hypoxia might, on the contrary, increase the thermogenic response to hypoxia (Mortola and Naso 1998). Indeed, when 8-day-old rats are raised under chronic hypoxia between postnatal days 2 and 8 (at an inspired PO_2 level of 92 mmHg, equivalent to 12% O_2), the thermogenic response evoked by cold exposure is more elevated than in control rats. In 2-day-old rats whose mother has been exposed to hypoxia throughout gestation, the thermogenic response to cold exposure was not different from control rats. Surprisingly however there were no signs of tissue or cellular responses that could explain the findings in the 8-day-old hypoxic pups; on the contrary, the content of uncoupling protein (UCP—a major determinant of thermogenesis capacity) was decreased in the interscapular brown adipose tissue of rat pups raised in postnatal hypoxia (Mortola and Naso 1998). As far as we are aware, there is no report of the developmental profile of thermogenic capacity and cold responses during postnatal development at high altitude. In adult deer mice that have been captured at high altitude and thereafter maintained at low altitude for 6 weeks, the thermogenic capacity measured as maximum O_2 consumption induced by cold exposure under hypoxia was about 60% higher than in lowland deer mice (Cheviron et al. 2012), but it seems that this effect is determined by genetics rather than by development at high altitude. In high-altitude mole rats (3200 m), the thermogenic response to cold exposure was less important than in low-altitude control mole rats (Broekman et al. 2006), but high-altitude animals had lower thermal conductance (higher isolation or efficient behavioral response to limit heat loss during cold exposure), and their response to stimulation of non-shivering thermogenesis by noradrenaline was more important (higher increased of metabolic rate and body temperature), showing overall greater thermoregulatory capabilities (Broekman et al. 2006).

Conclusion

The data reviewed in this chapter highlight the critical importance of the period of postnatal development for physiological responses at high altitude. It is evident that for newborn animals living under the constant exposure to reduced O_2 availability, there is a delay of maturation of the respiratory control system leading to a blunted hypoxic ventilatory response in adults, but this long-term detrimental effect might be, at least in some cases, compensated by accelerated lung growth, thus leading to enhanced aerobic performances despite the reduced minute ventilation. However, some data tend to indicate that genetic adaptations to altitude might contribute to eliminate the effects of postnatal hypoxia on the respiratory control system and maintain high O_2 consumption rate, therefore ensuring accelerated lung growth. There are also strong evidences showing interactions between cold exposure and hypoxia, both factors (cold or hypoxia) modulating the expected physiological response to the other factor. We've only been able to identify a limited number of studies describing thermoregulation in human newborn at high altitude, or on the effects of postnatal hypoxia on thermoregulation in newborn rats, but the postnatal maturation of thermoregulation at high altitude remains poorly known. Similarly it remains to establish whether there are specific adaptations during postnatal development, which is without doubt an intriguing line of research to pursue. Overall, this field of research is important to better understand the natural history, and the adaptation of animals living at high altitude, and it might also provide relevant information for human health at high altitude.

Acknowledgments The author acknowledges Dr. Jorge Soliz, for a careful revision of the manuscript and fruitful discussions. Alexandra Joachmans-Lemoine performed some experiments cited in this manuscript (published and unpublished materials) and provided critical insights on the manuscript. Drs. Salinas, Soria, and Gonzalez (Bolivian Institute for Altitude Biology) provided key supports for high-altitude studies cited in the manuscript. The Natural Science and Engineering Research Council of Canada for providing research fund (grant # RGPGP-2014-00083).

References

Al-Matary A, Kutbi I, Qurashi M, Khalil M, Alvaro R, Kwiatkowski K, Cates D, Rigatto H (2004) Increased peripheral chemoreceptor activity may be critical in destabilizing breathing in neonates. Semin Perinatol 28(4):264–272

Baig MS, Joseph V (2008) Age specific effect of MK-801 on hypoxic body temperature regulation in rats. Respir Physiol Neurobiol 160(2):181–186. https://doi.org/10.1016/j.resp.2007.09.008

Banchero N, Grover RF, Will JA (1971) Oxygen transport in the llama (*Lama glama*). Respir Physiol 13(1):102–115

Banchero N, Cruz J, Bustinza J (1975) Mechanisms of O2 transport in Andean dogs. Respir Physiol 23(3):361–370

Barros RC, Zimmer ME, Branco LG, Milsom WK (2001) Hypoxic metabolic response of the golden-mantled ground squirrel. J Appl Physiol 91(2):603–612

Bartels H, Bartels R, Rathschlag-Schaefer AM, Robbel H, Ludders S (1979) Acclimatization of newborn rats and guinea pigs to 3000 to 5000 m simulated altitudes. Respir Physiol 36(3):375–389

Beall CM (2007) Two routes to functional adaptation: Tibetan and Andean high-altitude natives. Proc Natl Acad Sci U S A 104(Suppl 1):8655–8660

Beaudry JL, McClelland GB (2010) Thermogenesis in CD-1 mice after combined chronic hypoxia and cold acclimation. Comp Biochem Physiol B Biochem Mol Biol 157(3):301–309. https://doi.org/10.1016/j.cbpb.2010.07.004

Benavides CE, Perez R, Espinoza M, Cabello G, Riquelme R, Parer JT, Llanos AJ (1989) Cardiorespiratory functions in the fetal llama. Respir Physiol 75(3):327–334

Bicego KC, Barros RC, Branco LG (2006) Physiology of temperature regulation: comparative aspects. Comp Biochem Physiol A Mol Integr Physiol 147(3):616–639

Bigham A, Bauchet M, Pinto D, Mao X, Akey JM, Mei R, Scherer SW, Julian CG, Wilson MJ, Lopez Herraez D, Brutsaert T, Parra EJ, Moore LG, Shriver MD (2010) Identifying signatures of natural selection in Tibetan and Andean populations using dense genome scan data. PLoS Genet 6(9):e1001116. https://doi.org/10.1371/journal.pgen.1001116

Bishop B, Silva G, Krasney J, Salloum A, Roberts A, Nakano H, Shucard D, Rifkin D, Farkas G (2000) Circadian rhythms of body temperature and activity levels during 63 h of hypoxia in the rat. Am J Physiol Regul Integr Comp Physiol 279(4):R1378–R1385

Bissonnette JM (2000) Mechanisms regulating hypoxic respiratory depression during fetal and postnatal life. Am J Physiol Regul Integr Comp Physiol 278(6):R1391–R1400

Bissonnette JM, Knopp SJ (2001) Developmental changes in the hypoxic ventilatory response in C57BL/6 mice. Respir Physiol 128(2):179–186

Blanco CE, Dawes GS, Hanson MA, McCooke HB (1984) The response to hypoxia of arterial chemoreceptors in fetal sheep and new-born lambs. J Physiol 351:25–37

Blanco LN, Massaro D, Massaro GD (1991) Alveolar size, number, and surface area: developmentally dependent response to 13% O_2. Am J Phys 261(6 Pt 1):L370–L377
Bollen B, Bouslama M, Matrot B, Rotrou Y, Vardon G, Lofaso F, Van den Bergh O, D'Hooge R, Gallego J (2009) Cold stimulates the behavioral response to hypoxia in newborn mice. Am J Physiol Regul Integr Comp Physiol 296(5):R1503–R1511. https://doi.org/10.1152/ajpregu.90582.2008
Branco LG, Carnio EC, Barros RC (1997) Role of the nitric oxide pathway in hypoxia-induced hypothermia of rats. Am J Phys 273(3 Pt 2):R967–R971
Branco LG, Gargaglioni LH, Barros RC (2006) Anapyrexia during hypoxia. J Therm Biol 31(1-2):82–89
Branco LG, Soriano RN, Steiner AA (2014) Gaseous mediators in temperature regulation. Compr Physiol 4(4):1301–1338. https://doi.org/10.1002/cphy.c130053
Broekman M, Bennett NC, Jackson CR, Scantlebury M (2006) Mole-rats from higher altitudes have greater thermoregulatory capabilities. Physiol Behav 89(5):750–754. https://doi.org/10.1016/j.physbeh.2006.08.023
Bruce MC, Honaker CE, Cross RJ (1999) Lung fibroblasts undergo apoptosis following alveolarization. Am J Respir Cell Mol Biol 20(2):228–236. https://doi.org/10.1165/ajrcmb.20.2.3150
Brutsaert TD, Parra E, Shriver M, Gamboa A, Palacios JA, Rivera M, Rodriguez I, Leon-Velarde F (2004) Effects of birthplace and individual genetic admixture on lung volume and exercise phenotypes of Peruvian Quechua. Am J Phys Anthropol 123(4):390–398. https://doi.org/10.1002/ajpa.10319
Brutsaert TD, Parra EJ, Shriver MD, Gamboa A, Rivera-Ch M, Leon-Velarde F (2005) Ancestry explains the blunted ventilatory response to sustained hypoxia and lower exercise ventilation of Quechua altitude natives. Am J Physiol Regul Integr Comp Physiol 289(1):R225–R234. https://doi.org/10.1152/ajpregu.00105.2005
Buckler KJ (2012) Effects of exogenous hydrogen sulphide on calcium signalling, background (TASK) K channel activity and mitochondrial function in chemoreceptor cells. Pflugers Arch 463(5):743–754. https://doi.org/10.1007/s00424-012-1089-8
Bureau MA, Lamarche J, Foulon P, Dalle D (1985) Postnatal maturation of respiration in intact and carotid body-chemodenervated lambs. J Appl Physiol 59(3):869–874
Burri PH (1984) Fetal and postnatal development of the lung. Annu Rev Physiol 46:617–628. https://doi.org/10.1146/annurev.ph.46.030184.003153
Burri PH, Weibel ER (1971) Morphometric estimation of pulmonary diffusion capacity. II. Effect of Po2 on the growing lung, adaption of the growing rat lung to hypoxia and hyperoxia. Respir Physiol 11(2):247–264
Burri PH, Dbaly J, Weibel ER (1974) The postnatal growth of the rat lung. I. Morphometry. Anat Rec 178(4):711–730
Busch MA, Bisgard GE, Forster HV (1985) Ventilatory acclimatization to hypoxia is not dependent on arterial hypoxemia. J Appl Physiol 58(6):1874–1880
Chen J, He L, Dinger B, Stensaas L, Fidone S (2002a) Role of endothelin and endothelin A-type receptor in adaptation of the carotid body to chronic hypoxia. Am J Physiol Lung Cell Mol Physiol 282(6):L1314–L1323. https://doi.org/10.1152/ajplung.00454.2001
Chen Y, Tipoe GL, Liong E, Leung S, Lam SY, Iwase R, Tjong YW, Fung ML (2002b) Chronic hypoxia enhances endothelin-1-induced intracellular calcium elevation in rat carotid body chemoreceptors and up-regulates ETA receptor expression. Pflugers Arch 443(4):565–573. https://doi.org/10.1007/s00424-001-0728-2
Cheviron ZA, Bachman GC, Connaty AD, McClelland GB, Storz JF (2012) Regulatory changes contribute to the adaptive enhancement of thermogenic capacity in high-altitude deer mice. Proc Natl Acad Sci U S A 109(22):8635–8640. https://doi.org/10.1073/pnas.1120523109
Cheviron ZA, Connaty AD, McClelland GB, Storz JF (2014) Functional genomics of adaptation to hypoxic cold-stress in high-altitude deer mice: transcriptomic plasticity and thermogenic performance. Evolution 68(1):48–62. https://doi.org/10.1111/evo.12257
Danchin E, Pocheville A (2014) Inheritance is where physiology meets evolution. J Physiol 592(Pt 11):2307–2317. https://doi.org/10.1113/jphysiol.2014.272096

Dempsey JA, Powell FL, Bisgard GE, Blain GM, Poulin MJ, Smith CA (2014) Role of chemoreception in cardiorespiratory acclimatization to, and deacclimatization from, hypoxia. J Appl Physiol 116(7):858–866. https://doi.org/10.1152/japplphysiol.01126.2013

Dias BG, Ressler KJ (2014) Parental olfactory experience influences behavior and neural structure in subsequent generations. Nat Neurosci 17(1):89–96. https://doi.org/10.1038/nn.3594

Donnelly DF, Haddad GG (1990) Prolonged apnea and impaired survival in piglets after sinus and aortic nerve section. J Appl Physiol 68(3):1048–1052

Droma T, McCullough RG, McCullough RE, Zhuang JG, Cymerman A, Sun SF, Sutton JR, Moore LG (1991) Increased vital and total lung capacities in Tibetan compared to Han residents of Lhasa (3,658 m). Am J Phys Anthropol 86(3):341–351. https://doi.org/10.1002/ajpa.1330860303

Eden GJ, Hanson MA (1987) Effects of chronic hypoxia from birth on the ventilatory response to acute hypoxia in the newborn rat. J Physiol 392:11–19

Fong AY (2010) Postnatal changes in the cardiorespiratory response and ability to autoresuscitate from hypoxic and hypothermic exposure in mammals. Respir Physiol Neurobiol 174(1-2):146–155. https://doi.org/10.1016/j.resp.2010.08.012

Forster HV, Bisgard GE, Klein JP (1981) Effect of peripheral chemoreceptor denervation on acclimatization of goats during hypoxia. J Appl Physiol Respir Environ Exerc Physiol 50(2):392–398

Frappell PB, Mortola JP (1994) Hamsters vs. rats: metabolic and ventilatory response to development in chronic hypoxia. J Appl Physiol 77(6):2748–2752

Frappell P, Lanthier C, Baudinette RV, Mortola JP (1992) Metabolism and ventilation in acute hypoxia: a comparative analysis in small mammalian species. Am J Phys 262(6 Pt 2):R1040–R1046

Frappell PB, Leon-Velarde F, Aguero L, Mortola JP (1998) Response to cooling temperature in infants born at an altitude of 4,330 meters. Am J Respir Crit Care Med 158(6):1751–1756. https://doi.org/10.1164/ajrccm.158.6.9803071

Frisancho AR (2013) Developmental functional adaptation to high altitude: review. Am J Hum Biol 25(2):151–168. https://doi.org/10.1002/ajhb.22367

Frisancho AR, Martinez C, Velasquez T, Sanchez J, Montoye H (1973) Influence of developmental adaptation on aerobic capacity at high altitude. J Appl Physiol 34(2):176–180

Gonzalez C, Almaraz L, Obeso A, Rigual R (1994) Carotid body chemoreceptors: from natural stimuli to sensory discharges. Physiol Rev 74:829–898

Gordon CJ, Fogelson L (1991) Comparative effects of hypoxia on behavioral thermoregulation in rats, hamsters, and mice. Am J Phys 260(1 Pt 2):R120–R125

Hackett PH, Reeves JT, Reeves CD, Grover RF, Rennie D (1980) Control of breathing in Sherpas at low and high altitude. J Appl Physiol Respir Environ Exerc Physiol 49(3):374–379

Hammond KA, Szewczak J, Krol E (2001) Effects of altitude and temperature on organ phenotypic plasticity along an altitudinal gradient. J Exp Biol 204(Pt 11):1991–2000

He L, Dinger B, Fidone S (2005) Effect of chronic hypoxia on cholinergic chemotransmission in rat carotid body. J Appl Physiol 98(2):614–619. https://doi.org/10.1152/japplphysiol.00714.2004

He L, Chen J, Dinger B, Stensaas L, Fidone S (2006) Effect of chronic hypoxia on purinergic synaptic transmission in rat carotid body. J Appl Physiol 100(1):157–162. https://doi.org/10.1152/japplphysiol.00859.2005

Hendrickson SL (2013) A genome wide study of genetic adaptation to high altitude in feral Andean Horses of the paramo. BMC Evol Biol 13:273. https://doi.org/10.1186/1471-2148-13-273

Hertzberg T, Hellstrom S, Holgert H, Lagercrantz H, Pequignot JM (1992) Ventilatory response to hyperoxia in newborn rats born in hypoxia--possible relationship to carotid body dopamine. J Physiol 456:645–654

Hill JR (1959) The oxygen consumption of new-born and adult mammals. Its dependence on the oxygen tension in the inspired air and on the environmental temperature. J Physiol 149:346–373

Hochachka PW, Lutz PL (2001) Mechanism, origin, and evolution of anoxia tolerance in animals. Comp Biochem Physiol B Biochem Mol Biol 130(4):435–459

Hochachka PW, Buck LT, Doll CJ, Land SC (1996) Unifying theory of hypoxia tolerance: molecular/metabolic defense and rescue mechanisms for surviving oxygen lack. Proc Natl Acad Sci U S A 93(18):9493–9498

Hochachka PW, Gunga HC, Kirsch K (1998) Our ancestral physiological phenotype: an adaptation for hypoxia tolerance and for endurance performance? Proc Natl Acad Sci U S A 95(4):1915–1920
Hofer MA (1984) Lethal respiratory disturbance in neonatal rats after arterial chemoreceptor denervation. Life Sci 34(5):489–496
Hsia CC, Carbayo JJ, Yan X, Bellotto DJ (2005) Enhanced alveolar growth and remodeling in Guinea pigs raised at high altitude. Respir Physiol Neurobiol 147(1):105–115. https://doi.org/10.1016/j.resp.2005.02.001
Huicho L (2007) Postnatal cardiopulmonary adaptations to high altitude. Respir Physiol Neurobiol 158(2-3):190–203. https://doi.org/10.1016/j.resp.2007.05.004
Iturriaga R, Alcayaga J, Gonzalez C (2009) Neurotransmitters in carotid body function: the case of dopamine--invited article. Adv Exp Med Biol 648:137–143. https://doi.org/10.1007/978-90-481-2259-2_16
Ivy CM, Scott GR (2014) Control of breathing and the circulation in high-altitude mammals and birds. Comp Biochem Physiol A Mol Integr Physiol. https://doi.org/10.1016/j.cbpa.2014.10.009
Jochmans-Lemoine A, Villalpando G, Gonzales M, Valverde I, Soria R, Joseph V (2015) Divergent physiological responses in laboratory rats and mice raised at high altitude. J Exp Biol. 218:1035–1043. https://doi.org/10.1242/jeb.112862
Jochmans-Lemoine A, Revollo S, Villalpando G, Valverde I, Gonzales M, Laouafa S, Soliz J, Joseph V (2018) Divergent mitochondrial antioxidant activities and lung alveolar architecture in the lungs of rats and mice at high altitude. Front Physiol 9:311
Johnson RL Jr, Cassidy SS, Grover RF, Schutte JE, Epstein RH (1985) Functional capacities of lungs and thorax in beagles after prolonged residence at 3,100 m. J Appl Physiol 59(6):1773–1782
Joseph V, Pequignot JM (2009) Breathing at high altitude. CMLS 66(22):3565–3573. https://doi.org/10.1007/s00018-009-0143-y
Joseph V, Soliz J, Pequignot J, Sempore B, Cottet-Emard JM, Dalmaz Y, Favier R, Spielvogel H, Pequignot JM (2000) Gender differentiation of the chemoreflex during growth at high altitude: functional and neurochemical studies. Am J Physiol Regul Integr Comp Physiol 278(4):R806–R816
Kemp PJ (2006) Detecting acute changes in oxygen: will the real sensor please stand up? Exp Physiol 91(5):829–834. https://doi.org/10.1113/expphysiol.2006.034587
Kholwadwala D, Donnelly DF (1992) Maturation of carotid chemoreceptor sensitivity to hypoxia: in vitro studies in the newborn rat. J Physiol 453:461–473
Kiyamu M, Rivera-Chira M, Brutsaert TD (2015) Aerobic capacity of Peruvian Quechua: a test of the developmental adaptation hypothesis. Am J Phys Anthropol 156(3):363–373. https://doi.org/10.1002/ajpa.22655
Kline DD, Peng YJ, Manalo DJ, Semenza GL, Prabhakar NR (2002) Defective carotid body function and impaired ventilatory responses to chronic hypoxia in mice partially deficient for hypoxia-inducible factor 1 alpha. Proc Natl Acad Sci U S A 99(2):821–826. https://doi.org/10.1073/pnas.022634199
Kumar P, Prabhakar NR (2012) Peripheral chemoreceptors: function and plasticity of the carotid body. Compr Physiol 2(1):141–219. https://doi.org/10.1002/cphy.c100069
Lahiri S, Kao FF, Velasquez T, Martinez C, Pezzia W (1969) Irreversible blunted respiratory sensitivity to hypoxia in high altitude natives. Respir Physiol 6(3):360–374
Lahiri S, DeLaney RG, Brody JS, Simpser M, Velasquez T, Motoyama EK, Polgar C (1976) Relative role of environmental and genetic factors in respiratory adaptation to high altitude. Nature 261(5556):133–135
Landauer RC, Pepper DR, Kumar P (1995) Effect of chronic hypoxaemia from birth upon chemosensitivity in the adult rat carotid body in vitro. J Physiol 485(Pt 2):543–550
Leon-Velarde F, de Muizon C, Palacios JA, Clark D, Monge C (1996) Hemoglobin affinity and structure in high-altitude and sea-level carnivores from Peru. Comp Biochem Physiol A Physiol 113(4):407–411
Leon-Velarde F, Maggiorini M, Reeves JT, Aldashev A, Asmus I, Bernardi L, Ge RL, Hackett P, Kobayashi T, Moore LG, Penaloza D, Richalet JP, Roach R, Wu T, Vargas E, Zubieta-Castillo

G, Zubieta-Calleja G (2005) Consensus statement on chronic and subacute high altitude diseases. High Alt Med Biol 6(2):147–157. https://doi.org/10.1089/ham.2005.6.147

Llanos AJ, Riquelme RA, Herrera EA, Ebensperger G, Krause B, Reyes RV, Sanhueza EM, Pulgar VM, Behn C, Cabello G, Parer JT, Giussani DA, Blanco CE, Hanson MA (2007) Evolving in thin air--lessons from the llama fetus in the altiplano. Respir Physiol Neurobiol 158(2-3):298–306

Llapur CJ, Martinez MR, Caram MM, Bonilla F, Cabana C, Yu Z, Tepper RS (2013) Increased lung volume in infants and toddlers at high compared to low altitude. Pediatr Pulmonol 48(12):1224–1230. https://doi.org/10.1002/ppul.22764

Lopez-Barneo J, Lopez-Lopez JR, Urena J, Gonzalez C (1988) Chemotransduction in the carotid body: K+ current modulated by PO2 in type I chemoreceptor cells. Science 241(4865):580–582

Lui MA, Mahalingam S, Patel P, Connaty AD, Ivy CM, Cheviron ZA, Storz JF, McClelland GB, Scott GR (2015) High-altitude ancestry and hypoxia acclimation have distinct effects on exercise capacity and muscle phenotype in deer mice. Am J Physiol Regul Integr Comp Physiol 308(9):R779–R791. https://doi.org/10.1152/ajpregu.00362.2014

Lumbroso D, Joseph V (2009) Impaired acclimatization to chronic hypoxia in adult male and female rats following neonatal hypoxia. Am J Physiol Regul Integr Comp Physiol 297(2):R421–R427. https://doi.org/10.1152/ajpregu.00068.2009

Lumbroso D, Lemoine A, Gonzales M, Villalpando G, Seaborn T, Joseph V (2012) Life-long consequences of postnatal normoxia exposure in rats raised at high altitude. J Appl Physiol 112(1):33–41. https://doi.org/10.1152/japplphysiol.01043.2011

Massaro D, Massaro GD (2002) Invited review: pulmonary alveoli: formation, the "call for oxygen," and other regulators. Am J Physiol Lung Cell Mol Physiol 282(3):L345–L358. https://doi.org/10.1152/ajplung.00374.2001

Monge C (1948) Acclimatization in the Andes. Historical confirmation of climate aggression in the development of Andean man. John Hopkins Press, Baltimore

Monge C, Leon-Velarde F (1991) Physiological adaptation to high altitude: oxygen transport in mammals and birds. Physiol Rev 71(4):1135–1172

Moore LG, Niermeyer S, Vargas E (2007) Does chronic mountain sickness (CMS) have perinatal origins? Respir Physiol Neurobiol 158(2-3):180–189

Mortola JP (1999) How newborn mammals cope with hypoxia. Respir Physiol 116(2-3):95–103

Mortola JP (2004) Implications of hypoxic hypometabolism during mammalian ontogenesis. Respir Physiol Neurobiol 141(3):345–356. https://doi.org/10.1016/j.resp.2004.01.011

Mortola JP (2005) Influence of temperature on metabolism and breathing during mammalian ontogenesis. Respir Physiol Neurobiol 149(1-3):155–164. https://doi.org/10.1016/j.resp.2005.01.012

Mortola JP, Dotta A (1992) Effects of hypoxia and ambient temperature on gaseous metabolism of newborn rats. Am J Phys 263(2 Pt 2):R267–R272

Mortola JP, Naso L (1998) Thermogenesis in newborn rats after prenatal or postnatal hypoxia. J Appl Physiol 85(1):84–90

Mortola JP, Seifert EL (2000) Hypoxic depression of circadian rhythms in adult rats. J Appl Physiol 88(2):365–368

Mortola JP, Rezzonico R, Lanthier C (1989) Ventilation and oxygen consumption during acute hypoxia in newborn mammals: a comparative analysis. Respir Physiol 78(1):31–43

Mortola JP, Frappell PB, Frappell DE, Villena-Cabrera N, Villena-Cabrera M, Pena F (1992) Ventilation and gaseous metabolism in infants born at high altitude, and their responses to hyperoxia. Am Rev Respir Dis 146(5 Pt 1):1206–1209. https://doi.org/10.1164/ajrccm/146.5_Pt_1.1206

Niane LM, Donnelly DF, Joseph V, Bairam A (2011) Ventilatory and carotid body chemoreceptor responses to purinergic P2X receptor antagonists in newborn rats. J Appl Physiol 110(1):83–94. https://doi.org/10.1152/japplphysiol.00871.2010

Okubo S, Mortola JP (1990) Control of ventilation in adult rats hypoxic in the neonatal period. Am J Phys 259(4 Pt 2):R836–R841

Paro FM, Steiner AA, Branco LGS (2001) Thermoregulatory response to hypoxia after inhibition of the central carbon monoxide-heme oxygenase pathway. J Thermal Biol 26:339–343

Peers C, Wyatt CN, Evans AM (2010) Mechanisms for acute oxygen sensing in the carotid body. Respir Physiol Neurobiol 174(3):292–298. https://doi.org/10.1016/j.resp.2010.08.010

Peng YJ, Nanduri J, Raghuraman G, Souvannakitti D, Gadalla MM, Kumar GK, Snyder SH, Prabhakar NR (2010) H2S mediates O2 sensing in the carotid body. Proc Natl Acad Sci U S A 107(23):10719–10724

Pichon A, Zhenzhong B, Favret F, Jin G, Shufeng H, Marchant D, Richalet JP, Ge RL (2009) Long-term ventilatory adaptation and ventilatory response to hypoxia in plateau pika (Ochotona curzoniae): role of nNOS and dopamine. Am J Physiol Regul Integr Comp Physiol 297(4):R978–R987. https://doi.org/10.1152/ajpregu.00108.2009

Pichon A, Zhenzhong B, Marchant D, Jin G, Voituron N, Haixia Y, Favret F, Richalet JP, Ge RL (2013) Cardiac adaptation to high altitude in the plateau pika (Ochotona curzoniae). Phys Rep 1(2):e00032. https://doi.org/10.1002/phy2.32

Platero-Luengo A, Gonzalez-Granero S, Duran R, Diaz-Castro B, Piruat JI, Garcia-Verdugo JM, Pardal R, Lopez-Barneo J (2014) An O2-sensitive glomus cell-stem cell synapse induces carotid body growth in chronic hypoxia. Cell 156(1-2):291–303. https://doi.org/10.1016/j.cell.2013.12.013

Potvin C, Rossignol O, Uppari N, Dallongeville A, Bairam A, Joseph V (2014) Reduced hypoxic ventilatory response in newborn mice knocked-out for the progesterone receptor. Exp Physiol 99(11):1523–1537. https://doi.org/10.1113/expphysiol.2014.080986

Powell FL, Milsom WK, Mitchell GS (1998) Time domains of the hypoxic ventilatory response. Respir Physiol 112(2):123–134

Prabhakar NR, Peers C (2014) Gasotransmitter regulation of ion channels: a key step in O2 sensing by the carotid body. Physiology 29(1):49–57. https://doi.org/10.1152/physiol.00034.2013

Rahn H, Otis AB (1949) Man's respiratory response during and after acclimatization to high altitude. Am J Phys 157(3):445–462

Rohlicek CV, Saiki C, Matsuoka T, Mortola JP (1998) Oxygen transport in conscious newborn dogs during hypoxic hypometabolism. J Appl Physiol 84(3):763–768

Ross FA, Rafferty JN, Dallas ML, Ogunbayo O, Ikematsu N, McClafferty H, Tian L, Widmer H, Rowe IC, Wyatt CN, Shipston MJ, Peers C, Hardie DG, Evans AM (2011) Selective expression in carotid body type I cells of a single splice variant of the large conductance calcium- and voltage-activated potassium channel confers regulation by AMP-activated protein kinase. J Biol Chem 286(14):11929–11936. https://doi.org/10.1074/jbc.M110.189779

Russell GA, Rezende EL, Hammond KA (2008) Development partly determines the aerobic performance of adult deer mice, *Peromyscus maniculatus*. J Exp Biol 211(Pt 1):35–41. https://doi.org/10.1242/jeb.012658

Schmitt P, Soulier V, Pequignot JM, Pujol JF, Denavit-Saubie M (1994) Ventilatory acclimatization to chronic hypoxia: relationship to noradrenaline metabolism in the rat solitary complex. J Physiol 477(Pt 2):331–337

Severinghaus JW, Bainton GR, Carcelen A (1966) Respiratory insensitivity to hypoxia in chronically hypoxic man. Respir Physiol 1:308–334

Shirkey NJ, Hammond KA (2014) The relationship between cardiopulmonary size and aerobic performance in adult deer mice at high altitude. J Exp Biol 217(Pt 20):3758–3764. https://doi.org/10.1242/jeb.103713

Simakajornboon N, Kuptanon T (2005) Maturational changes in neuromodulation of central pathways underlying hypoxic ventilatory response. Respir Physiol Neurobiol 149(1-3):273–286. https://doi.org/10.1016/j.resp.2005.05.005

Singer D (2004) Metabolic adaptation to hypoxia: cost and benefit of being small. Respir Physiol Neurobiol 141(3):215–228. https://doi.org/10.1016/j.resp.2004.02.009

Smith C, Bisgard G, Nielsen A, Daristotle L, Kressin N, Forster H, Dempsey J (1986) Carotid bodies are required for ventilatory acclimatization to chronic hypoxia. J Appl Physiol 60:1003–1010

Snyder LR (1985) Low P50 in deer mice native to high altitude. J Appl Physiol 58(1):193–199

Steiner AA, Branco LG (2002) Hypoxia-induced anapyrexia: implications and putative mediators. Annu Rev Physiol 64:263–288

Sterni LM, Bamford OS, Tomares SM, Montrose MH, Carroll JL (1995) Developmental changes in intracellular Ca2+ response of carotid chemoreceptor cells to hypoxia. Am J Phys 268(5 Pt 1):L801–L808

Storz JF, Baze M, Waite JL, Hoffmann FG, Opazo JC, Hayes JP (2007) Complex signatures of selection and gene conversion in the duplicated globin genes of house mice. Genetics 177(1):481–500. https://doi.org/10.1534/genetics.107.078550

Storz JF, Runck AM, Sabatino SJ, Kelly JK, Ferrand N, Moriyama H, Weber RE, Fago A (2009) Evolutionary and functional insights into the mechanism underlying high-altitude adaptation of deer mouse hemoglobin. Proc Natl Acad Sci U S A 106(34):14450–14455. https://doi.org/10.1073/pnas.0905224106

Storz JF, Scott GR, Cheviron ZA (2010) Phenotypic plasticity and genetic adaptation to high-altitude hypoxia in vertebrates. J Exp Biol 213(Pt 24):4125–4136. https://doi.org/10.1242/jeb.048181

Tattersall GJ, Milsom WK (2003) Hypothermia-induced respiratory arrest and recovery in neonatal rats. Respir Physiol Neurobiol 137(1):29–40

Tucker A, Rhodes J (2001) Role of vascular smooth muscle in the development of high altitude pulmonary hypertension: an interspecies evaluation. High Alt Med Biol 2(2):173–189. https://doi.org/10.1089/152702901750265288

Tufts DM, Revsbech IG, Cheviron ZA, Weber RE, Fago A, Storz JF (2013) Phenotypic plasticity in blood-oxygen transport in highland and lowland deer mice. J Exp Biol 216(Pt 7):1167–1173. https://doi.org/10.1242/jeb.079848

Velarde FL, Espinoza D, Monge C, de Muizon C (1991) A genetic response to high altitude hypoxia: high hemoglobin-oxygen affinity in chicken (*Gallus gallus*) from the Peruvian Andes. C R Acad Sci III 313(9):401–406

Via S, Gomulkiewicz R, De Jong G, Scheiner SM, Schlichting CD, Van Tienderen PH (1995) Adaptive phenotypic plasticity: consensus and controversy. Trends Ecol Evol 10(5):212–217

Villafuerte FC, Cardenas R, Monge CC (2004) Optimal hemoglobin concentration and high altitude: a theoretical approach for Andean men at rest. J Appl Physiol 96(5):1581–1588. https://doi.org/10.1152/japplphysiol.00328.2003

Vizek M, Pickett CK, Weil JV (1987) Increased carotid body hypoxic sensitivity during acclimatization to hypobaric hypoxia. J Appl Physiol 63:2403–2410

Wang ZY, Bisgard GE (2005) Postnatal growth of the carotid body. Respir Physiol Neurobiol 149(1-3):181–190. https://doi.org/10.1016/j.resp.2005.03.016

Wasicko MJ, Sterni LM, Bamford OS, Montrose MH, Carroll JL (1999) Resetting and postnatal maturation of oxygen chemosensitivity in rat carotid chemoreceptor cells. J Physiol 514(Pt 2):493–503

Zhuang J, Droma T, Sun S, Janes C, McCullough RE, McCullough RG, Cymerman A, Huang SY, Reeves JT, Moore LG (1993) Hypoxic ventilatory responsiveness in Tibetan compared with Han residents of 3,658 m. J Appl Physiol 74(1):303–311

Index

© Springer International Publishing AG, part of Springer Nature 2018

W. Burggren, B. Dubansky (eds.), *Development and Environment*,
https://doi.org/10.1007/978-3-319-75935-7

MIX
Papier aus verantwortungsvollen Quellen
Paper from responsible sources
FSC® C105338

If you have any concerns about our products,
you can contact us on
ProductSafety@springernature.com

In case Publisher is established outside the EU,
the EU authorized representative is:
Springer Nature Customer Service Center GmbH
Europaplatz 3, 69115 Heidelberg, Germany

Printed by Libri Plureos GmbH
in Hamburg, Germany